ROBERT SIMSON

THE ELEMENTS OF EUCLID VIZ THE FIRST SIX BOOKS TOGETHER WITH THE ELEVENTH AND TWELFTH ALSO THE BOOK OF EUCLID'S DATA

Elibron Classics
www.elibron.com

Elibron Classics series.

ISBN 1-4021-5149-7 (paperback)
ISBN 1-4212-7789-1 (hardcover)

This Elibron Classics Replica Edition is an unabridged facsimile of the edition published in 1830 by C.J.G. Rivington; etc., London.

THE

ELEMENTS OF EUCLID;

VIZ.

THE FIRST SIX BOOKS,

TOGETHER WITH THE

ELEVENTH AND TWELFTH:

ALSO THE BOOK OF

EUCLID'S DATA.

By ROBERT SIMSON, M.D.

EMERITUS PROFESSOR OF MATHEMATICS IN THE UNIVERSITY OF GLASGOW.

TO WHICH ARE ADDED,

THE ELEMENTS

OF

PLANE AND SPHERICAL TRIGONOMETRY;

AND

A TREATISE ON THE CONSTRUCTION OF THE

TRIGONOMETRICAL CANON

ALSO

A CONCISE ACCOUNT OF LOGARITHMS,

By the Rev. A. ROBERTSON, D.D. F.R.S.

SAVILIAN PROFESSOR OF ASTRONOMY IN THE UNIVERSITY OF OXFORD.

THE TWENTY-THIRD EDITION,

CAREFULLY REVISED AND CORRECTED.

LONDON:

C. J. G. AND F. RIVINGTON; J. NUNN; LONGMAN, REES, ORME, BROWN, AND GREEN; T. CADELL; J. RICHARDSON; J. M. RICHARDSON; BALDWIN AND CRADOCK; HURST, CHANCE, AND CO.; R. WILLIAMS; HAMILTON, ADAMS, AND CO.; WHITTAKER, TREACHER, AND ARNOT; SIMPKIN AND MARSHALL; J. COLLINGWOOD; AND J. AND J. J. DEIGHTON, CAMBRIDGE.

1830.

G. WOODFALL, ANGEL COURT, SKINNER STREET, LONDON.

DR. SIMSON'S

PREFACE TO THE ELEMENTS.

THE opinions of the moderns concerning the author of the Elements of Geometry, which go under Euclid's name, are very different and contrary to one another. Peter Ramus ascribes the Propositions, as well as their Demonstrations, to Theon; others think the Propositions to be Euclid's, but that the Demonstrations are Theon's; and others maintain, that all the Propositions and their Demonstrations are Euclid's own. John Buteo and Sir Henry Savile are the authors of greatest note who assert this last; and the greater part of geometers have ever since been of this opinion, as they thought it the most probable. Sir Henry Savile, after the several arguments he brings to prove it, makes this conclusion (page 13, Prælect.) "That excepting a very few interpolations, explications and additions, Theon altered nothing in Euclid." But, by often considering and comparing together the Definitions and Demonstrations as they are in the Greek editions we now have, I found that Theon, or whoever was the editor of the present Greek text, by adding some things, suppressing others, and mixing his own with Euclid's Demonstrations, has changed more things to the worse than is commonly supposed, and those not of small moment, especially in the fifth and eleventh books of the Elements, which this editor has greatly vitiated; for instance, by substituting a shorter, but insufficient Demonstration of the 18th Prop. of the 5th Book, in place of the legitimate one which Euclid had given; and by taking out of this Book, besides other things, the good definition which Eudoxus or Euclid had given of compound ratio, and giving an absurd one in place of it in the 5th Definition of the 6th Book, which neither Euclid, Archimedes, Apollonius, nor any geometer before Theon's time, ever made use of, and of which there is not to be found the least appearance in any of their writings; and, as this Definition did much embarrass beginners, and is quite useless, it is now thrown out of the Elements, and another, which, without doubt, Euclid had given, is put in its proper place among the Definitions of the 5th Book, by which the doctrine of compound ratios is rendered plain and easy. Besides, among the Definitions of the 11th Book, there is this, which is the tenth, viz. "Equal and similar solid figures are those which are contained by similar planes of the same number and magnitude." Now this proposition is a Theorem, not a Definition; because the equality of figures of any kind must be demonstrated, and not assumed; and therefore, though this were a true proposition, it ought to have been demonstrated. But indeed this Proposition, which makes the 10th definition of the 11th Book, is not true universally, except in the case in which each of the solid angles of the figures is contained by no more than three plane angles; for in other cases, two solid figures may be contained by similar planes of the same number and magnitude, and yet be unequal to one another, as shall be made evident in the Notes subjoined to these Elements. In like manner, in the Demonstration of the 26th Prop. of the 11th Book, it is taken for granted that those solid angles are equal to one another which are contained by plane angles of the same number and magnitude, placed in the same order: but neither is this universally true, except in the case in which the solid angles are contained by no more than three plane angles; nor of this case is there any demonstration in the Elements we now have, though it be quite necessary there should be one. Now, upon the 10th Definition of this Book depend the 25th and 28th Propositions of it; and upon the 25th and 26th depend other eight, viz. the 27th, 31st, 32d, 33d, 34th, 36th, 37th, and 40th of the same Book: and the 12th of the 12th Book depends upon the 8th of the same: and this 8th, and the Corollary of Proposition 17th and Proposition 18th of the 12th Book, depend upon the 9th Definition of the 11th Book, which is not a right definition; because there may be solids contained by the same number of similar plane figures, which are not similar to one another, in the true sense of similarity received by geometers; and all these Propositions have, for these reasons, been insufficiently demonstrated since Theon's time hitherto. Besides, there are several other things, which have nothing of Euclid's accuracy, and which plainly shew that his Elements have been much corrupted by unskilful geometers; and though these are not so gross as the others now mentioned, they ought by no means to remain uncorrected.

Upon these accounts it appeared necessary, and I hope will prove acceptable to all lovers of accurate reasoning and of mathematical learning, to remove such blemishes, and restore the principal Books of the Elements to their original accuracy, as far as I was able; especially since these Elements are the foundation of a science, by which the investigation and discovery of useful truths, at least in mathematical learning, is promoted as far as the limited powers of the mind allow: and which likewise is of the greatest use in the arts both of peace and war, to many of which geometry is absolutely necessary. This I have endeavoured to do, by taking away the inaccurate and false reasonings which unskilful editors have put into the place of some of the genuine Demonstrations of Euclid, who has ever been justly celebrated as the most accurate of geometers, and by restoring to him those things which Theon or others have suppressed, and which have these many ages been buried in oblivion.

* In this Edition, Ptolemy's Proposition concerning a property of quadrilateral figures in a circle, is added at the end of the sixth Book. Also the Note on the 29th Proposition, Book 1st, is altered, and made more explicit, and a more general Demonstration is given, instead of that which was in the Note on the 10th Definition of Book 11th; besides, the translation is much amended by the friendly assistance of a learned gentleman.

* This Paragraph (it is believed) was added by Dr. Simson to the Preface for the second edition in 1763.

PREFACE TO THE DATA.

EUCLID'S Data is the first in order of the books written by the ancient geometers to facilitate and promote the method of resolution or analysis. In the general, a thing is said to be given, which is either actually exhibited, or can be found out, that is, which is either known by hypothesis, or that can be demonstrated to be known; and the propositions in the book of Euclid's Data shew what things can be found out or known from those that by hypothesis are already known; so that in the analysis or investigation of a problem, from the things that are laid down to be known or given, by the help of these propositions other things are demonstrated to be given, and from these, other things are again shewn to be given, and so on, until that which was proposed to be found out in the problem is demonstrated to be given; and when this is done the problem is solved, and its composition is made and derived from the compositions of the Data which were made use of in the analysis. And thus the Data of Euclid are of the most general and necessary use in the solution of problems of every kind.

EUCLID is reckoned to be the author of the Book of the Data, both by the ancient and modern geometers: and there seems to be no doubt of his having written a book on this subject, but which, in the course of so many ages has been much vitiated by unskilful editors in several places, both in the order of the propositions, and in the definitions and demonstrations themselves. To correct the errors which are now found in it, and bring it nearer to the accuracy with which it was no doubt at first written by Euclid, is the design of this edition, that so it may be rendered more useful to geometers, at least to beginners who desire to learn the investigatory method of the ancients. And for their sake, the compositions of most of the Data are subjoined to their demonstrations, that the compositions of problems solved by help of the Data may be the more easily made.

MARINUS', the philosopher's preface, which, in the Greek edition, is prefixed to the Data, is here left out, as being of no use to understand them. At the end of it he says, that Euclid has not used the synthetical but the analytical method in delivering them: in which he is quite mistaken; for in the analysis of a theorem, the thing to be demonstrated is assumed in the analysis; but in the demonstrations of the Data, the thing to be demonstrated, which is, that something or other is given, is never once assumed in the demonstration, from which it is manifest, that every one of them is demonstrated synthetically; though indeed, if a proposition of the Data be turned into a problem, (for example, the 84th or 85th in the former editions, which here are the 85th and 86th,) the demonstration of the proposition becomes the analysis of the problem.

WHEREIN this edition differs from the Greek, and the reasons of the alterations from it, will be shewn in the Notes at the end of the Data.

ADVERTISEMENT TO THIS EDITION.

*** THE PROPRIETORS of this Work have spared neither pains nor expense to render the present Edition superior to all former ones. It will be found to be much improved in appearance, being printed in a neater type, with a new set of letters cast expressly for the purpose. Some trifling alterations are made in the punctuation, &c., and great care has been taken that it should issue from the press free from errors.

London, June, 1830.

SHORT ACCOUNTS

OF

THE LIFE AND WRITINGS

OF

EUCLID,

AND OF HIS TRANSLATOR AND COMMENTATOR,

DR. ROBERT SIMSON.

OF EUCLID.

EUCLID, the celebrated mathematician, according to the account of Pappus and Proclus, was born at Alexandria, in Egypt, where he flourished and taught mathematics with great applause, under the reign of Ptolemy Lagos, about 280 years before Christ. Some Arabian historians, however, inform us, that he was born at Tyre, that his father's name was Naucrates, an inhabitant of Damas.—The particular place of his nativity appears, therefore, to be uncertain: but whether or not Alexandria had the honour of giving birth to this celebrated mathematician, all historians agree, that he flourished and taught mathematics there at the time above mentioned; which city, from that period to the conquest of it by the Saracens, seems to have been the residence, if not the birth-place, of all the most eminent mathematicians of that time. Euclid reduced into regularity and order all the fundamental principles of pure mathematics, which had been delivered down by Thales, Pythagoras, Eudoxus, and other mathematicians before him, and added many others of his own: on which account it is said he was the first who reduced arithmetic and geometry into the form of a science. He likewise applied himself to the study of mixed mathematics, particularly to astronomy and optics.

His works, as we learn from Pappus and Proclus, are the Elements, Data, Introduction to Harmony, Phænomena, Optics, Catoptrics, a Treatise on the Division of Superficies, Porisms, Loci ad Superficiem, Fallacies, and Four Books of

Conics. The most celebrated of these, is the first work, the Elements of Geometry; of which there have been numberless editions in all languages; and a fine edition of all his works, now extant, was printed in 1703, by David Gregory, Savilian Professor of Astronomy at Oxford.

The Elements, as commonly published, consist of fifteen books, of which the last two, it is suspected, are not Euclid's, but a comment of Hypsicles of Alexandria, who lived two hundred years after him. They are divided into three parts, viz. the Contemplation of Superficies, Numbers, and Solids: the first four Books treat of planes only; the fifth of the proportions of magnitudes in general; the 6th of the proportion of plane figures; the 7th, 8th, and 9th, give us the fundamental properties of numbers; the 10th contains the theory of commensurable and incommensurable lines and spaces; the 11th, 12th, 13th, 14th and 15th, treat of the doctrine of solids.

There is no doubt but, before Euclid, Elements of Geometry were compiled by Hippocrates of Chios, Eudoxus, Leon, and many others, mentioned by Proclus in the beginning of his second book; for he affirms that Euclid new ordered many things in the Elements of Eudoxus, completed many things in those of Theatetus, and besides, strengthened such propositions as before were too slightly or but superficially established, with the most firm and convincing demonstrations.

History is silent as to the time of Euclid's death or his age. But Pappus represents him as a person of a courteous and agreeable behaviour, and in great esteem with Ptolemy Lagos, king of Egypt; who one day asking him, whether there was not any shorter way of coming at geometry than by his Elements, Euclid is said to have answered, "*that there was no royal road to geometry.*"

OF DR. ROBERT SIMSON.

Dr. Robert Simson, Professor of Mathematics in the University of Glasgow, was the eldest son of Mr. John Simson, of Kirtonhall in Ayrshire, and was born on the 14th of October 1687. Being designed by his father for the church, he was sent to the University of Glasgow about the year 1701, where he was distinguished by his proficiency in classical learning, and in the sciences.

Having procured a copy of Euclid's Elements, with the aid only of a few preliminary explanations from some more advanced students, he entered on the study of that oldest and best introduction to mathematics. In a short time he read and understood the first six, with the 11th and 12th books, and afterwards proceeding still farther in his mathematical pursuits, by his progress in the more difficult branches he laid the foundation of his future eminence. His reputation as a mathematician in a few years became so high, and his general character so much respected, that in 1710, when he was only twenty-two years of age, the members of the college voluntarily made him an offer of the mathematical chair, in which a vacancy in a short time was expected to take place. From his natural modesty however, he felt much reluctance, at so early an age, to advance abruptly from the state of a student to that of a professor in the same college, and therefore solicited permission to spend one year at least in London, where, besides other obvious advantages, he might have opportunities of becoming acquainted with some of the eminent mathematicians of England, who were then the most distinguished in Europe. In this request he was readily indulged; and without delay he proceeded to London, where he remained about a year, diligently employed in the improvement of his mathematical knowledge.

When the vacancy in the professorship of Mathematics at Glasgow did occur, the University, while Mr. Simson was still in London, appointed him to fill it; and the minute of election, which is dated March 11, 1711, concluded with this very proper condition, "That they will admit the said Mr. Robert Simson, providing always that he give satisfactory proof of his skill in mathematics, previous to his admission." He was duly admitted professor of mathematics on the 20th of November of that year.

His manner of teaching was uncommonly clear and successful; and among his scholars, several rose to distinction as mathematicians; among whom may be mentioned the celebrated Dr. Matthew Stewart, professor of Mathematics at Edinburgh: the two rev. Drs. Williamson, one of whom succeeded Dr. Simson at Glasgow; the rev. Dr. Trail, formerly professor of Mathematics at Aberdeen; Dr. James Moor,

Greek professor at Glasgow; and professor Robison of Edinburgh, with many others of distinguished merit. In the year 1758, Dr. S. being then 71 years of age, found it necessary to employ an assistant in teaching; and in 1761, on his recommendation, the rev. Dr. Williamson was appointed his assistant and successor.

His only publication, after resigning his office, was a new and improved edition of Euclid's Data, which in 1762 was annexed to the second edition of the Elements. But from that period, although much solicited to bring forward some of his other works on the ancient geometry, and notwithstanding he was fully apprised of the universal curiosity excited respecting his discovery, of Euclid's Porisms, he resisted every importunity on the subject.

Through Dr. Jurin, then secretary of the Royal Society, Dr. Simson had some intercourse with Dr. Halley, and other distinguished members of that society. And about the same time and afterwards he had frequent correspondence with Mr. Maclaurin, Mr. James Stirling, Dr. James Moor, Dr. Matthew Stewart, Dr. William Trail, Mr. Williamson of Lisbon, and with Mr. John Nourse, his bookseller and publisher in London.

Dr. S. was originally possessed of great intellectual powers, an accurate and distinguishing understanding, an inventive genius, and a retentive memory; and these powers being excited by an ardent curiosity, produced a singular capacity for investigating the truths of mathematical science. By such talents, and with a correct taste, formed by the study of the Greek geometers, he was also peculiarly qualified for communicating his knowledge, both in his lectures and in his writings, with perspicuity and elegance.

He was esteemed a good classical scholar; and though the simplicity of geometrical demonstration does not admit of much variety of style, yet in his works a good taste in that respect may be distinguished. In his Latin prefaces also, in which there is some history and discussion, the purity of language has been generally approved. It is to be regretted indeed, that he had not had an opportunity of employing in early life his Greek and mathematical learning, in giving an edition of Pappus in the original language.

Strict integrity and private worth, with corresponding purity of morals, gave the highest value to a character which, from other qualities and attainments, was much respected and esteemed. On all occasions, even in the gayest hours of social intercourse, the Doctor maintained a constant attention to propriety. He had serious and just impressions of religion; but he was uniformly reserved in expressing particular opinions about it; and from his sentiments of decorum, he never introduced religion as a subject of conversation in mixed society, and all attempts to do so in his clubs were, by him, checked with gravity and decision.

He was seriously indisposed only for a few weeks before his death, having through a very long life enjoyed a uniform state of good health. He died on the 1st of October, 1768, when his 81st year was almost completed.

The writings and publications of Dr. S. were almost exclusively of the pure geometrical kind, after the genuine manner of the ancients. He has only two pieces printed in the volumes of the Philosophical Transactions: viz.

1. Two general Propositions of Pappus, in which many of Euclid's Porisms are included, vol. 32, ann. 1723.—These two propositions were afterwards incorporated into the author's posthumous works, printed in 1776, by Philip, Earl Stanhope.

2. On the Extraction of the Approximate Roots of Numbers by infinite Series; vol. 48, ann. 1753.

The separate publications in his life-time were:

3. Conic Sections, in 1735, 4to.

4. The Loci Plani of Apollonius, restored; in 1749, 4to.

5. Euclid's Elements; in 1756, 4to, and since that time, many editions in 8vo, with the addition of Euclid's Data.

6. After his death, Earl Stanhope was at the expense of printing in 1776, under the title of "Opera Reliqua," several of Dr. S.'s posthumous pieces: which were (1) Apollonius's Determinate Section: (2) A Treatise on Porisms*: (3) A Tract on Logarithms: (4) On the Limits of Quantities and Ratios: (5) Some Select Geometrical Problems.

* A part of this Treatise was translated by the Rev. John Lawson, and is in his Mathematical Tracts.

THE

ELEMENTS OF EUCLID.

BOOK I.

DEFINITIONS.

I. A Point is that which hath no parts, or which hath no magnitude. See Notes.

II. A line is length without breadth.

III. The extremities of a line are points.

IV. A straight line is that which lies evenly between its extreme points.

V. A superficies is that which hath only length and breadth.

VI. The extremities of a superficies are lines.

VII. A plane superficies is that in which any two points being taken, the straight line between them lies wholly in that superficies. See N.

VIII. "A plane angle is the inclination of two lines to one "another in a plane, which meet together, but are not in "the same direction." See N.

IX. A plane rectilineal angle is the inclination of two straight lines to one another, which meet together, but are not in the same straight line.

N.B. ' When several angles are at one point B, any one of ' them is expressed by three letters, of which the letter that ' is at the vertex of the angle, that is, at the point in which ' the straight lines that contain the angle meet one another, ' is put between the other two letters, ' and one of these two is somewhere ' upon one of those straight lines, and ' the other upon the other line: Thus, ' the angle which is contained by the straight lines AB, CB, ' is named the angle ABC, or CBA; that which is contained ' by AB, DB, is named the angle ABD, or DBA; and that ' which is contained by DB, CB, is called the angle DBC, or ' CBD; but, if there be only one angle at a point, it may be ' expressed by a letter placed at that point; as the angle at ' E.'

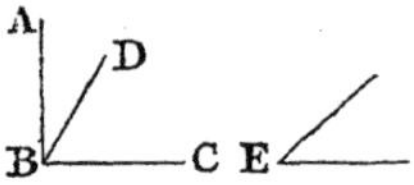

X. When a straight line, standing on another straight line, makes the adjacent angles equal to one another, each of the angles is called a right angle; and the straight line which stands on the other, is called a perpendicular to it.

XI. An obtuse angle is that which is greater than a right angle.

XII. An acute angle is that which is less than a right angle.

XIII. " A term or boundary is the extremity of any thing."

XIV. A figure is that which is enclosed by one or more boundaries.

XV. A circle is a plane figure contained by one line, which is called the circumference, and is such, that all straight lines drawn from a certain point within the figure to the circumference, are equal to one another.

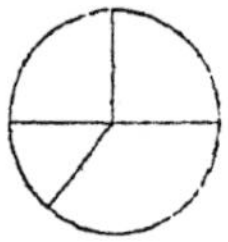

XVI. And this point is called the centre of the circle. See last fig.

XVII. A diameter of a circle is a straight line drawn through the centre, and terminated both ways by the circumference. See N. See last fig.

XVIII. A semicircle is the figure contained by a diameter and the part of the circumference cut off by the diameter. See last fig.

XIX. "A segment of a circle is the figure contained by a "straight line, and the circumference it cuts off."

XX. Rectilineal figures are those which are contained by straight lines.

XXI. Trilateral figures, or triangles, by three straight lines.

XXII. Quadrilateral, by four straight lines.

XXIII. Multilateral figures, or polygons, by more than four straight lines.

XXIV. Of three-sided figures, an equilateral triangle is that which has three equal sides.

XXV. An isosceles triangle is that which has only two sides equal.

XXVI. A scalene triangle is that which has three unequal sides.

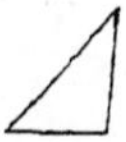

XXVII. A right-angled triangle is that which has a right angle.

XXVIII. An obtuse-angled triangle is that which has an obtuse angle.

XXIX. An acute-angled triangle is that which has three acute angles.

XXX. Of four-sided figures, a square is that which has all its sides equal, and all its angles right angles.

XXXI. An oblong is that which has all its angles right angles, but has not all its sides equal.

XXXII. A rhombus is that which has all its sides equal, but its angles are not right angles.

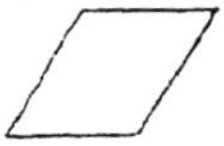

See N. XXXIII. A rhomboid is that which has its opposite sides equal to one another, but all its sides are not equal, nor its angles right angles.

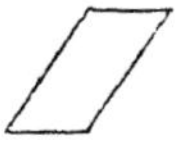

XXXIV. All other four-sided figures besides these, are called Trapeziums.

XXXV. Parallel straight lines are such as are in the same plane, and which, being produced ever so far both ways, do not meet.

POSTULATES.

I. Let it be granted, that a straight line may be drawn from any one point to any other point.

II. That a terminated straight line may be produced to any length in a straight line.

III. And that a circle may be described from any centre, at any distance from that centre.

AXIOMS.

I. Things which are equal to the same thing, are equal to one another.

II. If equals be added to equals, the wholes are equal.

III. If equals be taken from equals, the remainders are equal.

IV. If equals be added to unequals, the wholes are unequal.

V. If equals be taken from unequals, the remainders are unequal.

VI. Things which are double of the same, are equal to one another.

VII. Things which are halves of the same, are equal to one another.

VIII. Magnitudes which coincide with one another, that is, which exactly fill the same space, are equal to one another.

IX. The whole is greater than its part.

X. Two straight lines cannot enclose a space.

XI. All right angles are equal to one another.

XII. " If a straight line meet two straight lines, so as to
" make the two interior angles on the same side of it taken
" together, less than two right angles, these straight lines
" being continually produced, shall at length meet upon
" that side on which are the angles which are less than two
" right angles." See the notes on Prop. XXIX. of Book I.

PROPOSITION I.

PROBLEM.—*To describe an equilateral triangle upon a given finite straight line.*

Let AB be the given straight line; it is required to describe an equilateral triangle upon AB.

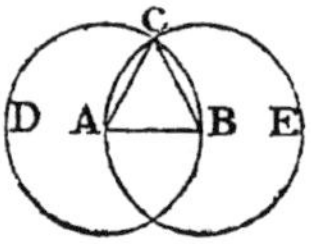

From the centre A, at the distance AB, describe * the circle BCD; and from the centre B, at the distance BA, describe the circle ACE; and from the point C, in which the circles cut one another, draw the straight lines * CA, CB, to the points A, B: ABC shall be an equilateral triangle.

* 3 Postulate.
* 1 Post.

Because the point A is the centre of the circle BCD, AC is equal * to AB; and because the point B is the centre of the circle ACE, BC is equal to BA: but it has been proved that CA is equal to AB; therefore CA, CB are each of them equal to AB: but things which are equal to the same thing are equal * to one another; therefore CA is equal to CB: wherefore CA, AB, BC are equal to one another: and the triangle ABC is therefore equilateral, and it is described upon the given straight line AB. Which was required to be done.

* 15 Definition.
* 1 Axiom.

PROPOSITION II.

PROB.—*From a given point to draw a straight line equal to a given straight line.*

Let A be the given point, and BC the given straight line; it is required to draw from the point A, a straight line equal to BC.

From the point A to B, draw * the straight line AB; and upon it describe * the equilateral triangle DAB, and produce * the straight lines DA, DB, to E and F; from the centre B, at the distance BC, describe * the circle CGH, and from the cen-

* 1 Post.
* 1. 1.
* 2 Post.
* 3 Post.

tre D, at the distance DG, describe the circle GKL: AL shall be equal to BC.

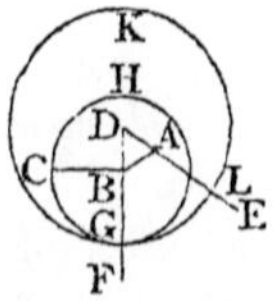

Because the point B is the centre of the circle CGH, BC is equal* to BG; and because D is the centre of the circle GKL, DL is equal to DG; and† DA, DB, parts of them, are equal; therefore the remainder AL is equal to the remainder* BG: but it has been shewn, that BC is equal to BG; wherefore AL and BC are each of them equal to BG: and things that are equal to the same thing are equal† to one another; therefore the straight line AL is equal to BC. Wherefore, from the given point A, a straight line AL has been drawn equal to the given straight line BC. Which was to be done.

* 15 Def.
† Constr.
* 3 Ax.
† 1 Ax.

PROPOSITION III.

Prob.—*From the greater of two given straight lines, to cut off a part equal to the less.*

Let AB and C be the two given straight lines, whereof AB is the greater; it is required to cut off from AB, the greater, a part equal to C, the less.

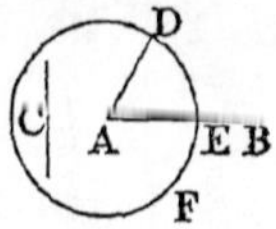

From the point A, draw* the straight line AD equal to C; and from the centre A, and at the distance AD, describe* the circle DEF: AE shall be equal to C.

Because A is the centre of the circle DEF, AE is equal† to AD; but the straight line C is likewise equal to† AD; whence AE and C are each of them equal to AD; wherefore, the straight line AE is equal to* C, and from AB, the greater of two straight lines, a part AE has been cut off, equal to C the less. Which was to be done.

* 2. 1.
* 3 Post.
† 15 Def.
† Constr.
* 1 Ax.

PROPOSITION IV.

Theorem.—*If two triangles have two sides of the one, equal to two sides of the other, each to each, and have likewise the angles contained by those sides equal to one another, they shall likewise have their bases,* or third sides, *equal; and the two triangles shall be equal, and their other angles shall be equal, each to each, viz. those to which the equal sides are opposite.*

Let ABC, DEF be two triangles, which have the two sides AB, AC equal to the two sides DE, DF, each to each, viz. AB to DE, and AC to DF; and the angle BAC equal to the angle EDF: the base BC shall be equal to the base EF; and the triangle ABC to the triangle DEF; and the other angles, to which the equal sides are opposite, shall be equal, each to each, viz. the angle ABC to the angle DEF, and the angle ACB to DFE.

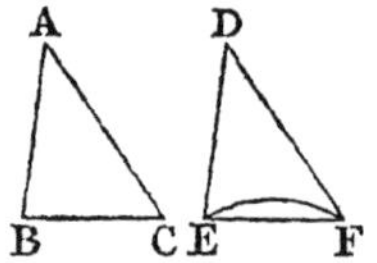

For if the triangle ABC be applied to DEF, so that the point A may be on D, and the straight line AB upon DE, the point
† Hyp. B shall coincide with the point E, because AB is equal† to DE: and AB coinciding with DE, AC shall coincide with DF,
† Hyp. because the angle BAC is equal† to the angle EDF; wherefore also the point C shall coincide with the point F, because
† Hyp. the straight line AC† is equal to DF: but the point B was proved to coincide with the point E; wherefore the base BC shall coincide with the base EF; because, the point B coinciding with E, and C with F, if the base BC does not coincide with the base EF, two straight lines would inclose a space,
* 10 Ax. which is impossible:* therefore the base BC coincides with
† 8 Ax. the base EF, and therefore is equal † to it. Wherefore the whole triangle ABC coincides with the whole triangle DEF, and is equal to it; and the other angles of the one, coincide with the remaining angles of the other, and are equal to them, viz. the angle ABC to the angle DEF, and the angle ACB to DFE. Therefore, if two triangles have two sides of the one, equal to two sides of the other, each to each, and have likewise the angles contained by those sides equal to one another, their bases shall likewise be equal; and the triangles shall be equal, and their other angles, to which the equal sides are opposite, shall be equal, each to each. Which was to be demonstrated.

PROPOSITION V.

THEOR.—*The angles at the base of an isosceles triangle are equal to one another; and if the equal sides be produced, the angles upon the other side of the base shall be equal.*

Let ABC be an isosceles triangle, of which the side AB is equal to AC; and let the straight lines AB, AC be produced to D and E: the angle ABC shall be equal to the angle ACB, and the angle CBD to the angle BCE.

In BD take any point F; and from AE, the greater, cut off
AG equal * to AF, the less, and join FC, GB. * 3. 1.

Because AF is equal to † AG, and AB to ‡ AC, the two † Constr.
sides FA, AC are equal to the two GA, AB, each to each: ‡ Hyp.
and they contain the angle FAG common to the two tri-
angles AFC, AGB; therefore the base FC is equal* to the * 4. 1.
base GB, and the triangle AFC to the triangle AGB; and
the remaining angles of the one, are equal* to the re- * 4. 1.
maining angles of the other, each to each, to which the equal sides are opposite; viz. the angle ACF to the angle ABG, and the angle AFC to the angle AGB: and because the whole AF is equal to the whole AG, of which the parts AB, AC are equal, the remainder BF is

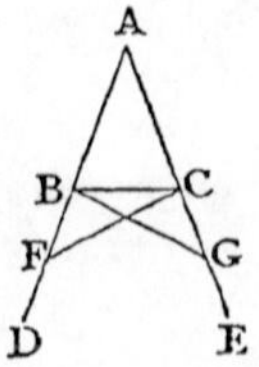

equal * to the remainder CG: and FC was proved * 3 Ax.
to be equal to GB; therefore the two sides BF, FC are equal
to the two CG, GB, each to each: and the angle BFC was
proved to be equal to the angle CGB, and the base BC is com-
mon to the two triangles BFC, CGB; wherefore these triangles
are equal, * and their remaining angles, each to each, to which * 4. 1.
the equal sides are opposite: therefore the angle FBC is equal
to the angle GCB, and the angle BCF to the angle CBG. And,
since it has been demonstrated, that the whole angle ABG is
equal to the whole ACF, the parts of which, the angles CBG,
BCF are also equal; therefore the remaining angle ABC is
equal † to the remaining angle ACB, which are the angles at † 3 Ax.
the base of the triangle ABC: and it has also been proved that
the angle FBC is equal to the angle GCB, which are the angles
upon the other side of the base. Therefore, the angles at the
base, &c. Q. E. D.

COROLLARY.—Hence, every equilateral triangle is also equiangular.

PROPOSITION VI.

THEOR.—*If two angles of a triangle be equal to one another, the sides also which subtend, or* are opposite to, *the equal angles, shall be equal to one another.*

Let ABC be a triangle, having the angle ABC equal to the angle ACB: the side AB shall be equal to the side AC.

For if AB be not equal to AC, one of them is greater than
* 3. 1. the other: let AB be the greater; and from it cut * off DB equal to AC, the less, and join DC: therefore, because in the triangles DBC, ACB, DB is equal to AC, and BC common to both, the two sides DB, BC are equal to the two AC, CB, each to each; and the angle DBC is equal to the
† Hyp. angle † ACB; therefore the base DC is equal to the base AB, and the triangle DBC is equal to
* 4. 1. the triangle * ACB, the less to the greater, which is absurd. Therefore AB is not unequal to AC, that is, it is equal to it. Wherefore, if two angles, &c. Q. E. D.

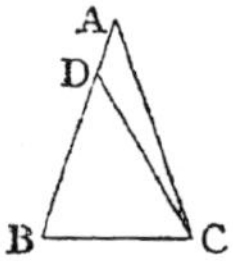

COROLLARY.—Hence, every equiangular triangle is also equilateral.

PROPOSITION VII.

See N. THEOR.—*Upon the same base, and on the same side of it, there cannot be two triangles that have their sides which are terminated in one extremity of the base, equal to one another, and likewise those which are terminated in the other extremity.*

If it be possible, upon the same base AB, and upon the same side of it, let there be two triangles ACB, ADB, which have their sides CA, DA, terminated in the extremity A of the base, equal to one another, and likewise their sides CB, DB, that are terminated in B.

Join CD; then, in the case in which the vertex of each of the triangles is without the other
† Hyp. triangle, because AC is equal † to AD, the angle
* 5. 1. ACD is equal * to the angle ADC: but the angle
† 9 Ax. ACD is greater † than the angle BCD; therefore the angle ADC is greater also than BCD; much more then is the angle BDC greater than the angle BCD.
† Hyp. Again, because CB is equal † to DB, the angle BDC is equal *
* 5. 1. to the angle BCD; but it has been demonstrated to be greater than it; which is impossible.

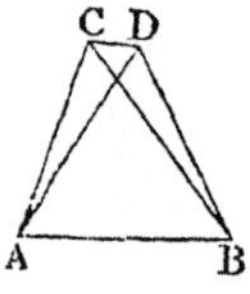

But if one of the vertices, as D, be within the other triangle ACB, produce AC, AD to E, F; therefore, because AC is
† Hyp. equal † to AD in the triangle ACD, the angles ECD, FDC,
* 5. 1. upon the other side of the base CD, are equal * to one another:
† 9 Ax. but the angle ECD is greater † than the angle BCD; wherefore

the angle FDC is likewise greater than BCD; much more then is the angle BDC greater than the angle BCD. Again, because CB is equal † to DB, the angle BDC is equal * to the angle BCD; but BDC has been proved to be greater than the same BCD; which is impossible. The case in which the vertex of one triangle is upon a side of the other, needs no demonstration.

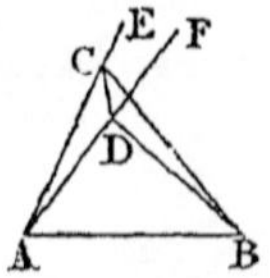

† Hyp.
* 5. 1.

Therefore, upon the same base, and on the same side of it, there cannot be two triangles that have their sides, which are terminated in one extremity of the base, equal to another, and likewise those which are terminated in the other extremity. Q. E. D.

PROPOSITION VIII.

THEOR.—*If two triangles have two sides of the one, equal to two sides of the other, each to each, and have likewise their bases equal; the angle which is contained by the two sides of the one, shall be equal to the angle contained by the two sides equal to them, of the other.*

Let ABC, DEF be two triangles, having the two sides AB, AC equal to the two sides DE, DF, each to each, viz. AB to DE, and AC to DF; and also the base BC equal to the base EF: the angle BAC shall be equal to the angle EDF.

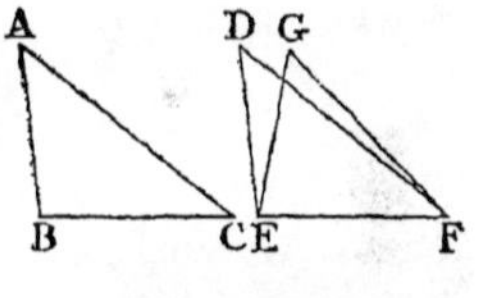

For if the triangle ABC be applied to DEF, so that the point B may be on E, and the straight line BC upon EF, the point C shall also coincide with the point F, because BC is equal† to EF. Therefore BC coinciding with EF, BA and AC shall coincide with ED and DF: for if the base BC coincides with the base EF, but the sides BA, CA do not coincide with the sides ED, FD, but have a different situation, as EG, FG; then upon the same base EF, and upon the same side of it, there can be two triangles that have their sides which are terminated in one extremity of the base, equal to one another, and likewise their sides terminated in the other extremity; but this is * impossible; therefore, if the base BC

† Hyp.
* 7. 1.

coincides with the base EF, the sides BA, AC cannot but coincide with the sides ED, DF; wherefore likewise the angle BAC
* 8 Ax. coincides with the angle EDF, and is equal* to it. Therefore, if two triangles, &c. Q. E. D.

PROPOSITION IX.

Prob.—*To bisect a given rectilineal angle; that is, to divide it into two equal angles.*

Let BAC be the given rectilineal angle; it is required to bisect it.

* 3. 1. Take any point D in AB, and from AC, cut* off AE equal
* 1. 1. to AD; join DE, and upon it describe* an equilateral triangle DEF; then join AF: the straight line AF shall bisect the angle BAC.

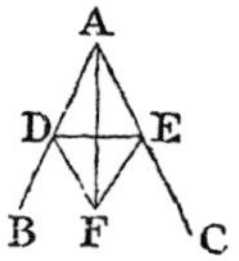

† Constr. Because AD is equal† to AE, and AF is common to the two triangles DAF, EAF, the two sides DA, AF are equal to the two sides EA, AF, each to each; and the base DF is
† Constr. equal† to the base EF; therefore the angle
* 8. 1. DAF is equal* to the angle EAF: wherefore, the given rectilineal angle BAC is bisected by the straight line AF. Which was to be done.

PROPOSITION X.

Prob.—*To bisect a given finite straight line; that is, to divide it into two equal parts.*

Let AB be the given straight line; it is required to divide it into two equal parts.

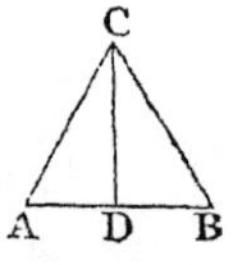

* 1. 1. Describe* upon it an equilateral triangle
* 9. 1. ABC, and bisect* the angle ACB, by the straight line CD: AB shall be cut into two equal parts in the point D.

† Constr. Because AC is equal† to CB, and CD common to the two triangles ACD, BCD, the two sides AC, CD are
† Constr. equal to BC, CD, each to each; and the angle ACD is equal†
* 4. 1. to the angle BCD; therefore the base AD is equal to the base* DB, and the straight line AB is divided into two equal parts in the point D. Which was to be done.

PROPOSITION XI.

PROB.—*To draw a straight line at right angles to a given straight line, from a given point in the same.*

Let AB be a given straight line, and C a point given in it; it is required to draw a straight line from the point C, at right angles to AB. See N.

Take any point D in AC, and make * CE equal to CD; and upon DE, describe * the equilateral triangle DFE, and join FC: the straight line FC, drawn from the given point C, shall be at right angles to the given straight line AB. * 3. 1. * 1. 1.

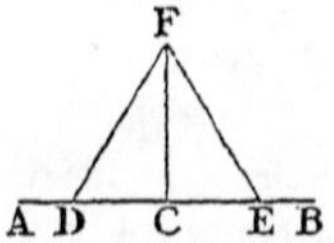

Because DC is equal † to CE, and FC common to the two triangles DCF, ECF, the two sides DC, CF are equal to the two EC, CF, each to each; and the base DF is equal † to the base EF; therefore the angle DCF is equal * to the angle ECF; and they are adjacent angles. But when the adjacent angles which one straight line makes with another straight line, are equal to one another, each of them is called a right * angle; therefore each of the angles DCF, ECF, is a right angle. Wherefore, from the given point C, in the given straight line AB, FC has been drawn at right angles to AB. Which was to be done. † Constr. † Constr. * 8. 1. * 10 Def.

COR.—By help of this problem it may be demonstrated, that two straight lines cannot have a common segment.

If it be possible, let the two straight lines ABC, ABD have the segment AB common to both of them. From the point B, draw † BE at right angles to AB; and because ABC is a straight line, the angle CBE is equal * to the angle EBA; in the same manner, because ABD is a straight line, the angle DBE is equal to the angle EBA; wherefore † the angle DBE is equal to the angle CBE, the less to the greater; which is impossible: therefore, two straight lines cannot have a common segment. † 11. 1. * 10 Def. † 1 Ax.

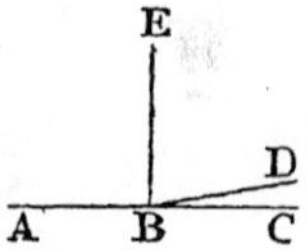

PROPOSITION XII.

PROB.—*To draw a straight line perpendicular to a given straight line of an unlimited length, from a given point without it.*

Let AB be the given straight line, which may be produced to any length both ways, and let C be a point without it; it is required to draw a straight line perpendicular to AB, from the point C.

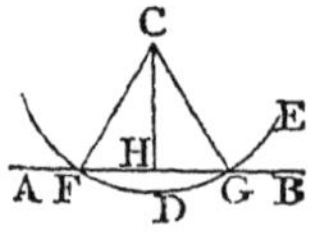

Take any point D upon the other side of AB, and from the centre C, at the distance CD, describe * the circle EGF, meeting AB in F, G; bisect * FG in H, and join CH: the straight line CH, drawn from the given point C, shall be perpendicular to the given straight line AB.

* 3 Post.
* 10. 1.

Join CF, CG: and because FH is equal † to HG, and HC common to the two triangles FHC, GHC, the two sides FH, HC are equal to the two GH, HC, each to each; and the base CF is equal * to the base CG: therefore the angle CHF is equal * to the angle CHG; and they are adjacent angles: but when a straight line, standing on another straight line, makes the adjacent angles equal to one another, each of them is a right angle, and the straight line which stands upon the other, is called a perpendicular † to it: therefore, from the given point C, a perpendicular CH has been drawn to the given straight line AB. Which was to be done.

† Constr.
* 15 Def.
* 8. 1.
† 10 Def.

PROPOSITION XIII.

THEOR.—*The angles which one straight line makes with another upon one side of it, are either two right angles, or are together equal to two right angles.*

Let the straight line AB make with CD, upon one side of it, the angles CBA, ABD: these shall either be two right angles, or shall together be equal to two right angles.

For if the angle CBA be equal to ABD, each of them is a right † angle: but if not, from the point B draw BE at right angles * to CD; therefore the angles CBE, EBD * are two right angles: and because the angle CBE is equal to the two angles CBA, ABE together, add the angle EBD to each of these equals; therefore the angles CBE, EBD are equal * to the three angles CBA, ABE, EBD. Again, because the angle DBA is equal to the two angles DBE, EBA, add to each of these equals the angle ABC; therefore the angles

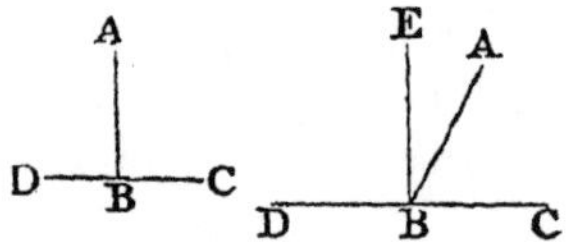

† 10 Def.
* 11. 1.
* 10 Def.
* 2 Ax.

DBA, ABC are equal † to the three angles DBE, EBA, ABC: † 2 Ax.
but the angles CBE, EBD have been demonstrated to be equal to the same three angles; and things that are equal to the same thing, are equal * to one another; therefore the angles * 1 Ax.
CBE, EBD are equal to the angles DBA, ABC: but CBE, EBD are two right angles; therefore, DBA, ABC are together equal † † 1 Ax.
to two right angles. Wherefore, the angles which one straight line, &c. Q. E. D.

PROPOSITION XIV.

THEOR.—*If, at a point in a straight line, two other straight lines, upon the opposite sides of it, make the adjacent angles together equal to two right angles, these two straight lines shall be in one and the same straight line.*

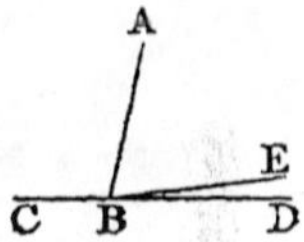

At the point B, in the straight line AB, let the two straight lines BC, BD upon the opposite sides of AB, make the adjacent angles ABC, ABD equal together to two right angles: BD shall be in the same straight line with CB.

For if BD be not in the same straight line with CB, let BE be in the same straight line with it: therefore, because the straight line AB makes with the straight line CBE, upon one side of it, the angles ABC, ABE, these angles are together equal * to two right angles; but the angles ABC, ABD * 13. 1.
are likewise together equal † to two right angles; therefore, the † Hyp.
angles CBA, ABE are equal † to the angles CBA, ABD: take † 1 Ax.
away the common angle ABC, and the remaining angle ABE is equal * to the remaining angle ABD, the less to the greater, * 3 Ax.
which is impossible; therefore BE is not in the same straight line with BC. And in like manner, it may be demonstrated, that no other can be in the same straight line with it but BD, which, therefore, is in the same straight line with CB. Wherefore, if at a point, &c. Q. E. D.

PROPOSITION XV.

THEOR.—*If two straight lines cut one another, the vertical, or* opposite, *angles shall be equal.*

Let the two straight lines AB, CD cut one another in the point E: the angle AEC shall be equal to the angle DEB, and CEB to AED.

Because the straight line AE makes with CD the angles
* 13. 1. CEA, AED, these angles are together equal * to two right angles. Again, because the straight line DE makes with AB the angles AED, DEB, these also are
* 13. 1. together equal * to two right angles; and CEA, AED have been demonstrated to be equal to two right angles; wherefore the
† 1 Ax. angles CEA, AED are equal † to the angles AED, DEB: take away the common angle AED, and the re-
* 3 Ax. maining angle CEA is equal * to the remaining angle DEB. In the same manner it can be demonstrated, that the angles CEB, AED are equal. Therefore, if two straight lines, &c. Q. E. D.

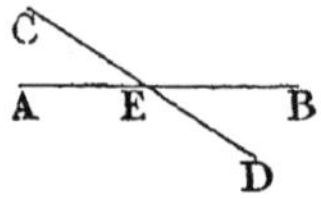

Cor. 1. From this it is manifest, that if two straight lines cut one another, the angles which they make at the point where they cut, are together equal to four right angles.

Cor. 2. And consequently, that all the angles made by any number of lines meeting in one point, are together equal to four right angles.

PROPOSITION XVI.

Theor.—*If one side of a triangle be produced, the exterior angle is greater than either of the interior opposite angles.*

Let ABC be a triangle, and let its side BC be produced to D: the exterior angle ACD shall be greater than either of the interior opposite angles CBA, BAC.

* 10. 1. Bisect* AC in E; join BE and produce it to F, and make EF
† 3. 1. equal† to BE, and join FC.
† Constr. Because AE is equal† to EC, and BE† to EF; AE, EB are equal to CE, EF, each
* 15. 1. to each; and the angle AEB is equal* to the angle CEF, because they are opposite vertical angles; therefore the base AB is
* 4. 1. equal* to the base CF, and the triangle AEB to the triangle CEF, and the remaining angles to the remaining angles, each to each, to which the equal sides are opposite: wherefore the angle BAE is equal to the angle ECF:
† 9 Ax. but the angle ECD is greater† than the angle ECF, therefore the angle ACD is greater than BAE. In the same manner, if the side BC be bisected, and AC be produced to G, it may

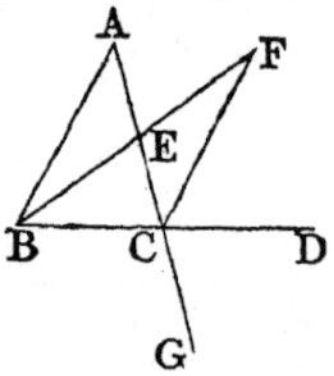

be demonstrated, that the angle BCG, that is, the angle* ACD, is greater than the angle ABC. Therefore, if one side, &c. Q. E. D. * 15. 1.

PROPOSITION XVII.

THEOR.—*Any two angles of a triangle are together less than two right angles.*

Let ABC be any triangle: any two of its angles together shall be less than two right angles.

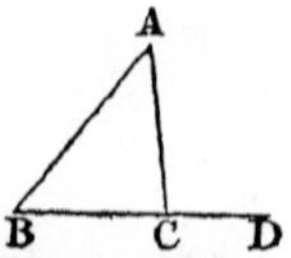

Produce BC to D: and because ACD is the exterior angle of the triangle ABC, ACD is greater* than the interior and opposite angle ABC; to each of these add the angle ACB; therefore the angles ACD, ACB are greater† than the angles ABC, ACB: but ACD, ACB are together equal* to two right angles; therefore the angles ABC, BCA are less than two right angles. In like manner, it may be demonstrated, that BAC, ACB, as also CAB, ABC, are less than two right angles. Therefore, any two angles, &c. Q. E. D.

* 16. 1.
† 4 Ax.
* 13. 1.

PROPOSITION XVIII.

THEOR.—*The greater side of every triangle is opposite to the greater angle.*

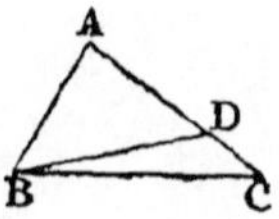

Let ABC be a triangle, of which the side AC is greater than the side AB: the angle ABC shall be greater than the angle BCA.

Because AC is greater than AB, make* AD equal to AB, and join BD: and because ADB is the exterior angle of the triangle BDC, it is greater* than the interior and opposite angle DCB; but ADB is equal* to ABD, because the side AB is equal† to the side AD; therefore the angle ABD is likewise greater than the angle ACB; therefore much more is the angle ABC greater than ACB. Therefore, the greater side, &c. Q. E. D.

* 3. 1.
* 16. 1.
* 5. 1.
† Constr.

PROPOSITION XIX.

THEOR.—*The greater angle of every triangle is subtended by the greater side,* or has the greater side opposite to it.

Let ABC be a triangle, of which the angle ABC is greater

than the angle BCA: the side AC shall be greater than the side AB.

For if it be not greater, AC must either be equal to AB, or less than it: it is not equal, because then the angle ABC
* 5. 1. would be equal * to the angle CAB; but it
† Hyp. is † not; therefore AC is not equal to AB: neither is it less, because then, the angle
* 18. 1. ABC would be less * than the angle ACB; but it is not; therefore the side AC is not less than AB: and it has been shewn that it is not equal to AB: therefore AC is greater than AB. Wherefore, the greater angle, &c. Q. E. D.

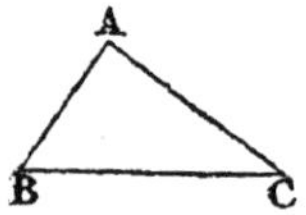

PROPOSITION XX.

See N. THEOR.—*Any two sides of a triangle are together greater than the third side.*

Let ABC be a triangle: any two sides of it together shall be greater than the third side; viz. the sides BA, AC greater than the side BC; and AB, BC greater than AC; and BC, CA greater than AB.

* 3. 1. Produce BA to the point D, and make * AD equal to AC; and join DC.

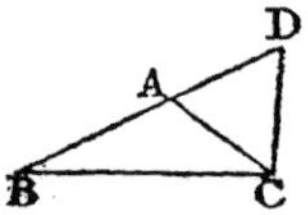

Because DA is equal to AC, the angle
* 5. 1. ADC is equal * to ACD; but the angle BCD
† 9 Ax. is greater † than the angle ACD; therefore the angle BCD is greater than the angle ADC: and because the angle BCD of the triangle DCB, is greater than its angle BDC,
* 19. 1. and that the greater * angle is subtended by the greater side, therefore the side DB is greater than the side BC: but DB is equal to BA ‡ and AC; therefore the sides BA, AC are greater than BC. In the same manner it may be demonstrated, that the sides AB, BC are greater than CA; and BC, CA greater than AB. Therefore, any two sides, &c. Q. E. D.

PROPOSITION XXI.

See N. THEOR.—*If from the ends of the side of a triangle, there be drawn two straight lines to a point within the triangle, these shall be less than the other two sides of the triangle, but shall contain a greater angle.*

Let ABC be a triangle, and from the points B, C, the ends

* Constr. ‡ Because AD is equal * to AC, add BA to each, therefore the whole BD
† 2 Ax. is equal † to the two BA, AC.

of the side BC, let the two straight lines BD, CD be drawn to the point D within the triangle: BD and DC shall be less than the other two sides BA, AC of the triangle, but shall contain an angle BDC greater than the angle BAC.

Produce BD to E: and because two sides of a triangle* are greater than the third side, the two sides BA, AE of the triangle ABE, are greater than BE: to each of these add EC; therefore the sides BA, AC are greater† than BE, EC. Again, because the two sides CE, ED of the triangle CED, are greater† than CD, add DB to each of these; therefore the sides CE, EB are greater† than CD, DB: but it has been shewn that BA, AC are greater than BE, EC; much more then are BA, AC greater than BD, DC.

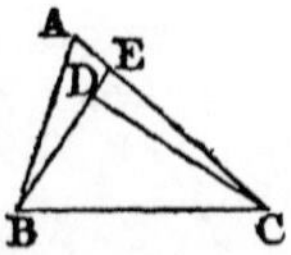

* 20. 1.

† 4 Ax.

† 20. 1.

† 4 Ax.

Again, because the exterior angle of a triangle* is greater than the interior and opposite angle, the exterior angle BDC of the triangle CDE, is greater than CED: for the same reason, the exterior angle CEB of the triangle ABE, is greater than BAC: and it has been demonstrated, that the angle BDC is greater than the angle CEB; much more then is the angle BDC greater than the angle BAC. Therefore, if from the ends of, &c. Q. E. D. * 16. 1.

PROPOSITION XXII.

PROB.—*To make a triangle of which the sides shall be equal to three given straight lines, but any two whatever of these must be greater than the third*.* See N.

* 20. 1.

Let A, B, C be the three given straight lines, of which any two whatever are greater than the third; viz. A and B greater than C; A and C greater than B; and B and C greater than A; it is required to make a triangle, of which the sides shall be equal to A, B, C, each to each.

Take a straight line DE terminated at the point D, but unlimited towards E, and make* DF equal to A, FG equal to B, and GH equal to C: from the centre F, at the distance FD, describe* the circle DKL; and from the centre G, at the distance GH, describe* another circle HLK; and join KF, KG: the triangle KFG shall have its sides equal to the three straight lines A, B, C.

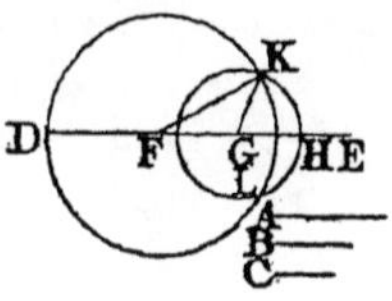

* 3. 1.

* 3 Post.

* 3 Post.

Because the point F is the centre of the circle DKL, FD is
* 15 Def. equal * to FK; but FD is equal † to the straight line A; there-
† Constr. fore FK is equal † to A: again, because G is the centre of the
† 1 Ax.
* 15 Def. circle LKH, GH is equal * to GK; but GH is equal to C;
† Constr. therefore also GK is equal to C: and FG is equal † to B; therefore the three straight lines KF, FG, GK are equal to the three A, B, C: and therefore the triangle KFG has its three sides KF, FG, GK equal to the three given straight lines A, B, C. Which was to be done.

PROPOSITION XXIII.

Prob.—*At a given point in a given straight line, to make a rectilineal angle equal to a given rectilineal angle.*

Let AB be the given straight line, and A the given point in it, and DCE the given rectilineal angle; it is required to make an angle at the given point A, in the given straight line AB, that shall be equal to the given rectilineal angle DCE.

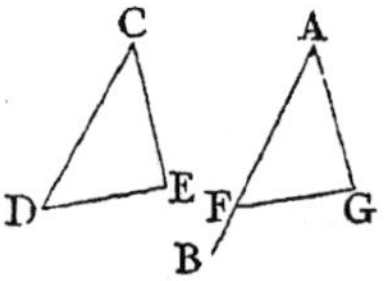

In CD, CE, take any points D, E,
* 22. 1. and join DE; and make * the triangle AFG, the sides of which shall be equal to the three straight lines CD, DE, EC; so that CD be equal to AF, CE to AG, and DE to FG: the angle FAG shall be equal to the angle DCE.

Because DC, CE are equal to FA, AG, each to each, and the
* 8. 1. base DE to the base FG, the angle DCE is equal * to the angle FAG. Therefore, at the given point A, in the given straight line AB, the angle FAG is made equal to the given rectilineal angle DCE. Which was to be done.

PROPOSITION XXIV.

See N. Theor.—*If two triangles have two sides of the one, equal to two sides of the other, each to each, but the angle contained by the two sides of one of them, greater than the angle contained by the two sides equal to them, of the other, the base of that which has the greater angle, shall be greater than the base of the other.*

Let ABC, DEF be two triangles, which have the two sides AB, AC equal to the two DE, DF, each to each; viz. AB equal to DE, and AC to DF, but the angle BAC greater than the angle EDF: the base BC shall be greater than the base EF.

Of the two sides DE, DF, let DE be the side which is not greater than the other; and at the point D, in the straight line DE, make* the angle EDG equal to the angle BAC; and * 23. 1.
make DG equal* to AC or DF, and join EG, GF. * 3. 1.

Because AB is equal † to DE, and AC † to DG, the two sides † Hyp.
BA, AC are equal to the two ED, DG, each to each, and the † Constr.
angle BAC is equal † to the angle EDG; therefore the base BC is † Constr.
equal * to the base EG. And because DG is equal to DF, the angle * 4. 1.
DFG is equal * to the angle DGF; * 5. 1.
but the angle DGF is greater † than † 9 Ax.
the angle EGF; therefore the angle DFG is greater than EGF; therefore, much more is the angle EFG greater than the angle EGF: and because the angle EFG of the triangle EFG, is greater than its angle EGF, and that the greater* angle is * 19. 1.
subtended by the greater side, therefore the side EG is greater than the side EF: but EG was proved to be equal to BC; therefore BC is greater than EF. Therefore, if two triangles, &c. *Q. E. D.*

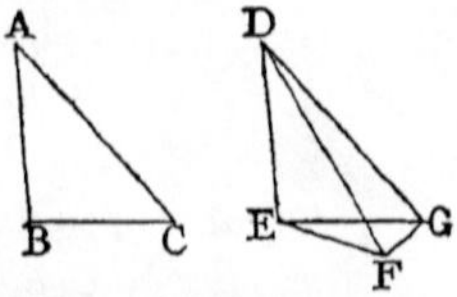

PROPOSITION XXV.

THEOR.—*If two triangles have two sides of the one, equal to two sides of the other, each to each, but the base of the one, greater than the base of the other, the angle contained by the sides of that which has the greater base, shall be greater than the angle contained by the sides equal to them, of the other.* See N.

Let ABC, DEF be two triangles, which have the two sides AB, AC equal to the two sides DE, DF, each to each; viz. AB equal to DE, and AC to DF, but the base BC greater than the base EF: the angle BAC shall be greater than the angle EDF.

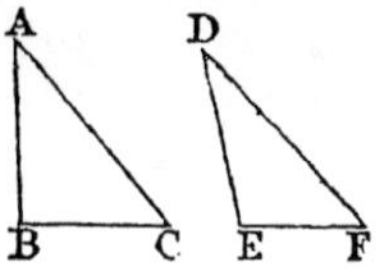

For if it be not greater, it must either be equal to it, or less than it: but the angle BAC is not equal to the angle EDF, because then the base BC would be equal* to EF; but it is† not; * 4. 1.
therefore the angle BAC is not equal to the angle EDF: neither † Hyp.

* 24. 1. is it less, because then, the base BC would be less * than the
† Hyp. base EF; but it is † not; therefore the angle BAC is not less than the angle EDF: and it was shewn, that it is not equal to it; therefore the angle BAC is greater than the angle EDF. Wherefore, if two triangles, &c. Q. E. D.

PROPOSITION XXVI.

THEOR.—*If two triangles have two angles of the one, equal to two angles of the other, each to each, and one side equal to one side, viz. either the sides adjacent to the equal angles, or the sides opposite to equal angles in each, then shall the other sides be equal, each to each, and also the third angle of the one, to the third angle of the other.*

Let ABC, DEF be two triangles, which have the angles ABC, BCA equal to the angles DEF, EFD, each to each, viz. ABC to DEF, and BCA to EFD; also one side equal to one side: and first, let those sides be equal which are adjacent to the angles that are equal in the two triangles, viz. BC to EF: the other sides shall be equal, each to each, viz. AB to DE, and AC to DF, and the third angle BAC to the third angle EDF.

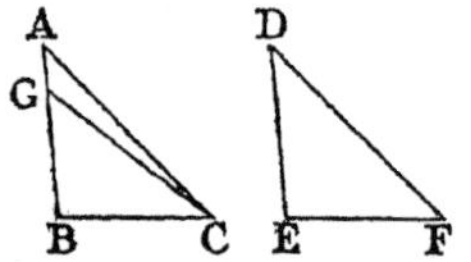

For if AB be not equal to DE, one of them must be greater than the other: let AB be the greater of the two, and make
† 3. 1. BG equal † to DE, and join GC: therefore, because BG is equal
† Hyp. to DE, and BC † to EF, the two sides GB, BC are equal to the
† Hyp. two DE, EF, each to each; and the angle GBC is equal † to
* 4. 1. the angle DEF; therefore the base GC is equal * to the base DF, and the triangle GBC to the triangle DEF, and the other angles to the other angles, each to each, to which the equal sides are opposite; therefore the angle GCB is equal to the angle DFE: but DFE is, by the hypothesis, equal to the angle
† 1 Ax. BCA; wherefore also the angle BCG is equal † to the angle BCA, the less to the greater; which is impossible: therefore AB is not unequal to DE, that is, it is equal to it: and BC is
† Hyp. equal † to EF: therefore the two AB, BC are equal to the two
† Hyp. DE, EF, each to each; and the angle ABC is equal † to the
* 4. 1. angle DEF; therefore the base AC is equal * to the base DF, and the third angle BAC to the third angle EDF.

Next, let the sides which are opposite to equal angles in each triangle be equal to one another, viz. AB to DE: likewise in this case, the other sides shall be equal, AC to DF, and BC to EF; and also the third angle BAC to the third angle EDF.

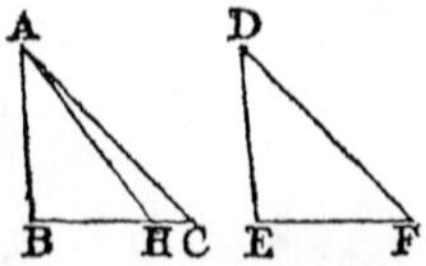

For if BC be not equal to EF, let BC be the greater of them, and make BH equal † to EF, and join AH; and because BH is equal to EF, and AB to † DE, the two AB, BH are equal to the two DE, EF, each to each; and they contain equal † angles; therefore the base AH is equal † to the base DF, and the triangle ABH to the triangle DEF, and the other angles to the other angles, each to each, to which the equal sides are opposite; therefore the angle BHA is equal to the angle EFD: but EFD is equal † to the angle BCA; therefore also the angle BHA is equal † to the angle BCA; that is, the exterior angle BHA of the triangle AHC, is equal to its interior and opposite angle BCA; which is impossible *: wherefore BC is not unequal to EF, that is, it is equal to it: and AB is equal † to DE; therefore the two AB, BC are equal to the two DE, EF, each to each; and they contain † equal angles; wherefore the base AC is equal † to the base DF, and the third angle BAC, to the third angle EDF. Therefore, if two triangles, &c. Q. E. D.

† 3. 1. † Hyp. † Hyp. † 4. 1. † Hyp. † 1 Ax. * 16. 1 † Hyp. † Hyp. † 4. 1.

PROPOSITION XXVII.

THEOR.—*If a straight line, falling upon two other straight lines, make the alternate angles equal to one another, these two straight lines shall be parallel.*

Let the straight line EF, which falls upon the two straight lines AB, CD, make the alternate angles AEF, EFD equal to one another: AB shall be parallel to CD.

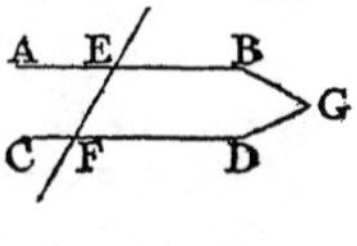

For if it be not parallel, AB and CD, being produced, will meet either towards B, D, or towards A, C: let them be produced and meet towards B, D, in the point G; therefore GEF is a triangle, and its exterior angle AEF is greater * than the interior and opposite angle EFG; but it is also equal † to it; which is impossible; therefore, AB and CD, being produced, do not meet towards B, D. In like manner

* 16. 1. † Hyp.

it may be demonstrated, that they do not meet towards A, C: but those straight lines which meet neither way, though pro-
* 35 Def. duced ever so far, are parallel * to one another: therefore AB is parallel to CD. Wherefore, if a straight line, &c. Q. E. D.

PROPOSITION XXVIII.

THEOR.—*If a straight line, falling upon two other straight lines, make the exterior angle equal to the interior and opposite upon the same side of the line, or make the interior angles upon the same side together equal to two right angles, the two straight lines shall be parallel to one another.*

Let the straight line EF, which falls upon the two straight lines AB, CD, make the exterior angle EGB equal to the interior and opposite angle GHD upon the same side; or make the interior angles on the same side, BGH, GHD together equal to two right angles: AB shall be parallel to CD.

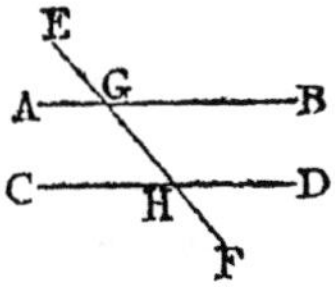

† Hyp. Because the angle EGB is equal † to the
* 15. 1. angle GHD, and the angle EGB is equal *
† 1 Ax. to the angle AGH, therefore the angle AGH is equal † to the angle GHD; and they are the alternate angles; therefore AB
* 27. 1. is parallel * to CD. Again, because the angles BGH, GHD are
* Hyp. equal * to two right angles, and that AGH, BGH are also
* 13. 1. equal * to two right angles, therefore the angles AGH, BGH
† 1 Ax. are equal † to the angles BGH, GHD: take away the common
† 3 Ax. angle BGH; therefore the remaining angle AGH is equal † to the remaining angle GHD; and they are alternate angles;
† 27. 1. therefore AB is parallel † to CD. Wherefore, if a straight line, &c. Q. E. D.

PROPOSITION XXIX.

See the notes on this proposition.

THEOR.—*If a straight line fall upon two parallel straight lines, it makes the alternate angles equal to one another; and the exterior angle equal to the interior and opposite upon the same side; and likewise the two interior angles upon the same side together equal to two right angles.*

Let the straight line EF fall upon the parallel straight lines

AB, CD: the alternate angles AGH, GHD shall be equal to one another; and the exterior angle EGB shall be equal to the interior and opposite, upon the same side, GHD; and the two interior angles BGH, GHD upon the same side, shall be together equal to two right angles.

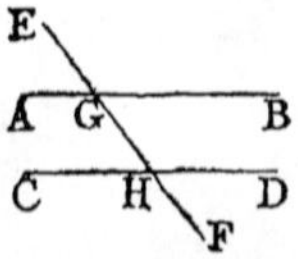

For if AGH be not equal to GHD, one of them must be greater than the other; let AGH be the greater: and because the angle AGH is greater than the angle GHD, add to each of them the angle BGH; therefore, the angles AGH, BGH are greater † than the angles BGH, GHD: but the angles † 4 Ax.
AGH, BGH are equal * to two right angles; therefore the * 13. 1.
angles BGH, GHD are less than two right angles: but those straight lines which, with another straight line falling upon them, make the interior angles on the same side less than two right angles, will meet * together if continually produced; * 12 Ax.
therefore the straight lines AB, CD, if produced far enough, will meet: but they never meet, since they are parallel by the hypothesis; therefore the angle AGH is not unequal to the angle GHD, that is, it is equal to it: but the angle AGH is (See the notes on this proposition.)
equal * to the angle EGB; therefore likewise EGB is equal † * 15. 1.
to GHD: add to each of these the angle BGH; therefore the † 1 Ax.
angles EGB, BGH are equal † to the angles BGH, GHD: but † 2 Ax.
EGB, BGH are equal * to two right angles; therefore also * 13. 1.
BGH, GHD are equal † to two right angles. Wherefore, if a † 1 Ax.
straight line, &c. *Q. E. D.*

PROPOSITION XXX.

THEOR.—*Straight lines which are parallel to the same straight line, are parallel to each other.*

Let AB, CD be each of them parallel to EF: AB shall be parallel to CD.

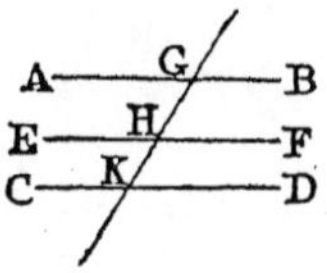

Let the straight line GHK cut AB, EF, CD: and because GHK cuts the parallel straight lines AB, EF, the angle AGH is equal * to the angle GHF. Again, because * 29. 1.
the straight line GK cuts the parallel straight lines EF, CD, the angle GHF is equal * to the angle GKD: * 29. 1.
and it was shewn, that the angle AGK is equal to the angle

† 1 Ax. GHF; therefore also AGK is equal † to GKD: and they
* 27. 1. are alternate angles; therefore AB is parallel * to CD. Wherefore, straight lines, &c. Q. E. D.

PROPOSITION XXXI.

PROB.—*To draw a straight line, through a given point, parallel to a given straight line.*

Let A be the given point, and BC the given straight line; it is required to draw a straight line through the point A, parallel to the straight line BC.

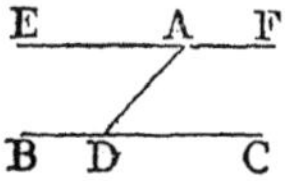

In BC take any point D, and join AD; and at the point A, in the straight line AD,
* 23. 1. make * the angle DAE equal to the angle ADC; and produce the straight line EA to F: EF shall be parallel to BC.

Because the straight line AD, which meets the two straight lines, BC, EF, makes the alternate angles EAD, ADC equal to
* 27. 1. one another, EF is parallel * to BC. Therefore, the straight line EAF is drawn through the given point A, parallel to the given straight line BC. Which was to be done.

PROPOSITION XXXII.

THEOR.—*If a side of any triangle be produced, the exterior angle is equal to the two interior and opposite angles; and the three interior angles of every triangle are equal to two right angles.*

Let ABC be a triangle, and let one of its sides BC be produced to D: the exterior angle ACD shall be equal to the two interior and opposite angles CAB, ABC; and the three interior angles of the triangle, viz. ABC, BCA, CAB shall together be equal to two right angles.

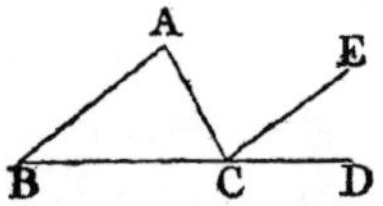

Through the point C, draw CE pa-
* 31. 1. rallel * to the straight line AB: and because AB is parallel to CE, and AC meets them, the alternate angles BAC,
* 29. 1. ACE are equal *. Again, because AB is parallel to CE, and BD falls upon them, the exterior
† 29. 1. angle ECD is equal † to the interior and opposite angle

ABC: but the angle ACE was shewn to be equal to the angle BAC; therefore the whole exterior angle ACD is equal † to the two interior and opposite angles CAB, ABC: to each of these equals add the angle ACB; and the angles ACD, ACB are equal † to the three angles CBA, BAC, ACB; but the angles ACD, ACB are equal * to two right angles; therefore also the angles CBA, BAC, ACB are equal † to two right angles. Wherefore, if a side of any triangle, &c. Q. E. D.

† 2 Ax.
† 2 Ax.
* 13. 1.
† 1 Ax.

COR. 1. All the interior angles of any rectilineal figure, together with four right angles, are equal to twice as many right angles as the figure has sides.

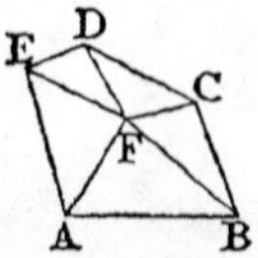

For any rectilineal figure ABCDE, can be divided into as many triangles as the figure has sides, by drawing straight lines from a point F within the figure, to each of its angles. And, by the preceding proposition, all the angles of these triangles are equal to twice as many right angles as there are triangles, that is, as there are sides of the figure: and the same angles are equal to the angles of the figure, together with the angles at the point F, which is the common vertex of the triangles; that is *, together with four right angles. Therefore all the angles of the figure, together with four right angles, are equal to twice as many right angles as the figure has sides.

* 2 Cor. 15. 1.

COR. 2. All the exterior angles of any rectilineal figure are together equal to four right angles.

Because every interior angle ABC, with its adjacent exterior ABD, is equal* to two right angles, therefore all the interior, together with all the exterior angles of the figure, are equal to twice as many right angles as there are sides of the figure; that is, by the foregoing corollary, they are equal to all the interior angles of the figure, together with four right angles; therefore all the exterior angles are equal to four right angles.

* 13. 1.

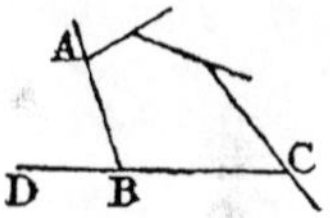

PROPOSITION XXXIII.

THEOR.—*The straight lines which join the extremities of two equal and parallel straight lines towards the same parts, are also themselves equal and parallel.*

Let AB, CD be equal and parallel straight lines, and joined towards the same parts by the straight lines AC, BD: AC, BD shall be equal and parallel.

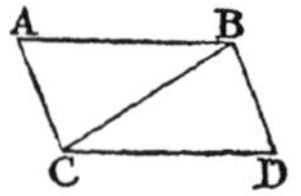

Join BC: and because AB is parallel to CD, and BC meets them, the alternate an-
* 29. 1. gles* ABC, BCD are equal: and because AB is equal to CD, and BC common to the two triangles ABC, DCB, the two sides AB, BC are equal to the two DC, CB, each to each: and the angle ABC was proved to be equal to the angle BCD; therefore the base AC is equal to the base BD, and the triangle ABC to the triangle BCD, and the other angles to the other
* 4. 1. angles*, each to each, to which the equal sides are opposite: therefore the angle ACB is equal to the angle CBD: and because the straight line BC meets the two straight lines AC, BD, and makes the alternate angles ACB, CBD equal to one
* 27. 1. another, AC is parallel* to BD; and it was shewn to be equal to it. Therefore, straight lines, &c. Q. E. D.

PROPOSITION XXXIV.

THEOR.—*The opposite sides and angles of parallelograms are equal to one another, and the diameter bisects them, that is, divides them into two equal parts.*

N.B. A parallelogram is a four-sided figure, of which the opposite sides are parallel: and the diameter is the straight line joining two of its opposite angles.

Let ACDB be a parallelogram, of which BC is a diameter: the opposite sides and angles of the figure shall be equal to one another, and the diameter BC shall bisect it.

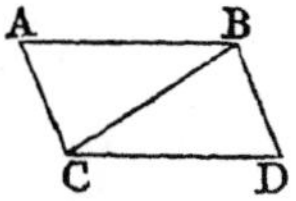

Because AB is parallel to CD, and BC meets them, the alternate angles ABC, BCD
* 29. 1. are equal* to one another: and because AC is parallel to BD, and BC meets them, the
* 29. 1. alternate angles ACB, CBD are equal* to one another: wherefore the two triangles ABC, CBD have two angles ABC, BCA in the one, equal to two angles BCD, CBD in the other, each to each, and one side BC common to the two triangles, which is adjacent to their equal angles; therefore their other sides are equal, each to each, and the third
* 26. 1. angle of the one, to the third angle of the other*, viz. the side AB to the side CD, and AC to BD, and the angle BAC equal to the angle BDC: and because the angle ABC is equal

to the angle BCD, and the angle CBD to the angle ACB, the whole angle ABD is equal† to the whole angle ACD: and the angle BAC has been shewn to be equal to the angle BDC; therefore the opposite sides and angles of parallelograms are equal to one another. Also, their diameter bisects them: for AB being equal to CD, and BC common, the two AB, BC are equal to the two DC, CB, each to each; and the angle ABC has been proved equal to the angle BCD: therefore, the triangle ABC is equal* to the triangle BCD, and the diameter BC divides the parallelogram ACDB into two equal parts. Q. E. D.

† 2 Ax.

* 4. 1.

PROPOSITION XXXV.

THEOR.—*Parallelograms upon the same base, and between the same parallels, are equal to one another.* See N.

Let the parallelograms ABCD, EBCF be upon the same base BC, and between the same parallels AF, BC: the parallelogram ABCD shall be equal to the parallelogram EBCF. See the 2d and 3d figures.

If the sides AD, DF of the parallelograms ABCD, DBCF, opposite to the base BC, be terminated in the same point D, it is plain that each of the parallelograms is double* of the triangle BDC; and they are therefore equal† to one another.

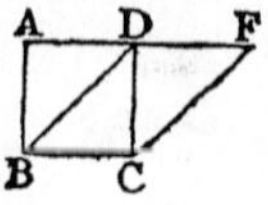

* 34. 1.
† 6 Ax.

But, if the sides AD, EF, opposite to the base BC of the parallelograms ABCD, EBCF, be not terminated in the same point, then, because ABCD is a parallelogram, AD is equal* to BC; for the same reason, EF is equal to BC; wherefore AD is equal* to EF, and DE is common; therefore the whole, or the remainder AE, is equal* to the whole, or the remainder DF: AB also is equal† to DC; therefore the two EA, AB are equal to the two FD, DC, each to each; and the exterior angle FDC is equal* to the interior EAB: therefore the base EB, is equal to the base FC, and the triangle EAB equal* to the triangle FDC. Take the triangle FDC from the trapezium ABCF, and from the same trapezium take the triangle EAB, and the remainders* are equal; that is, the parallelogram ABCD is equal to the parallelogram EBCF. Therefore, parallelograms upon the same base, &c. Q. E. D.

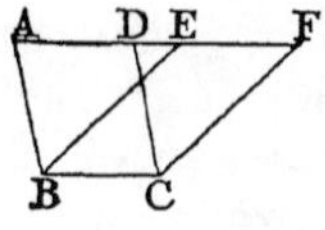

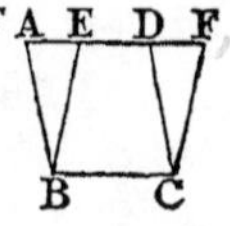

* 34. 1.
* 1 Ax.
* 2 or 3 Ax.
† 34. 1.
* 29. 1.
* 4. 1.
* 3 Ax.

PROPOSITION XXXVI.

THEOR.—*Parallelograms upon equal bases, and between the same parallels, are equal to one another.*

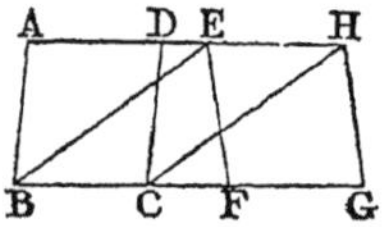

Let ABCD, EFGH be parallelograms upon equal bases BC, FG, and between the same parallels AH, BG: the parallelogram ABCD shall be equal to EFGH.

Join BE, CH: and because BC is
† Hyp. equal † to FG, and FG to * EH, BC is equal to † EH; and
* 34. 1. † 1 Ax. they are † parallels, and joined towards the same parts by the
† Hyp. straight lines BE, CH: but straight lines which join the extremities of equal and parallel straight lines towards the same
* 33. 1. parts, are themselves * equal and parallel; therefore EB, HC
† Def. 34. 1. are both equal and parallel; and therefore EBCH is a † parallel-
* 35. 1. ogram; and it is equal * to ABCD, because they are upon the same base BC, and between the same parallels BC, AH: for the like reason, the parallelogram EFGH is equal to the same
† 1 Ax. EBCH: therefore the parallelogram ABCD is equal † to EFGH. Wherefore, parallelograms, &c. Q. E. D.

PROPOSITION XXXVII.

THEOR.—*Triangles upon the same base, and between the same parallels, are equal to one another.*

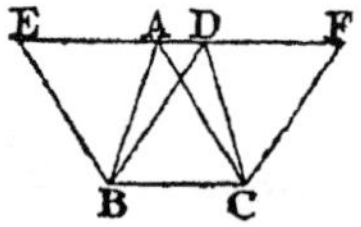

Let the triangles ABC, DBC be upon the same base BC, and between the same parallels AD, BC: the triangle ABC shall be equal to the triangle DBC.

† 2 Post. Produce † AD both ways to the points
* 31. 1. E, F; and through B, draw * BE parallel to CA; and through C, draw CF parallel to BD: therefore each of the figures
† Def. 34. 1. EBCA, DBCF, is a † parallelogram; and EBCA is equal * to
* 35. 1. DBCF, because they are upon the same base BC, and between the same parallels BC, EF; and the triangle ABC is the half
* 34. 1. of the parallelogram EBCA, because the diameter AB bisects * it; and the triangle DBC is the half of the parallelogram DBCF, because the diameter DC bisects it: but the halves of
* 7 Ax. equal things are * equal; therefore the triangle ABC is equal to the triangle DBC. Wherefore, triangles, &c. Q. E. D.

PROPOSITION XXXVIII.

THEOR.—*Triangles upon equal bases, and between the same parallels, are equal to one another.*

Let the triangles ABC, DEF be upon equal bases BC, EF, and between the same parallels BF, AD: the triangle ABC shall be equal to the triangle DEF.

Produce † AD both ways to the points G, H; and through B, draw BG parallel * to CA, and through F, draw FH parallel to ED: then each of the figures GBCA, DEFH, is a † parallelogram; and they are equal * to one another, because they are upon equal bases BC, EF, and between the same parallels BF, GH; and the triangle ABC is the half of the parallelogram GBCA, because the diameter AB bisects * it; and the triangle DEF is the half of the parallelogram DEFH, because the diameter DF bisects it: but the halves of equal things are * equal; therefore the triangle ABC is equal to the triangle DEF. Wherefore, triangles, &c. Q. E. D.

† 2 Post. * 31. 1. † Def. 34. 1. * 36. 1. * 34. 1. * 7 Ax.

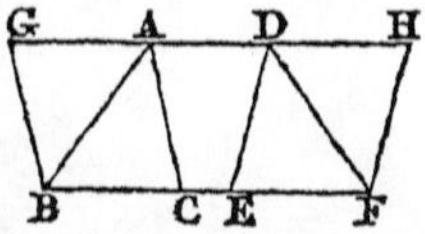

PROPOSITION XXXIX.

THEOR.—*Equal triangles upon the same base, and upon the same side of it, are between the same parallels.*

Let the equal triangles ABC, DBC be upon the same base BC, and upon the same side of it: they shall be between the same parallels.

Join AD: AD shall be parallel to BC. For if it is not, through the point A, draw * AE parallel to BC, and join EC. The triangle ABC is equal * to the triangle EBC, because they are upon the same base BC, and between the same parallels BC, AE: but the triangle ABC is equal † to the triangle DBC; therefore also the triangle DBC is equal † to the triangle EBC, the greater to the less; which is impossible; therefore AE is not parallel to BC. In the same manner it can be demonstrated, that no other line but AD is parallel to BC; AD is therefore parallel to it. Wherefore, equal triangles, &c. Q. E. D.

* 31. 1. * 37. 1. † Hyp. † 1 Ax.

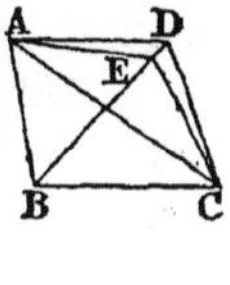

PROPOSITION XL.

THEOR.—*Equal triangles, upon equal bases in the same straight line, and towards the same parts, are between the same parallels.*

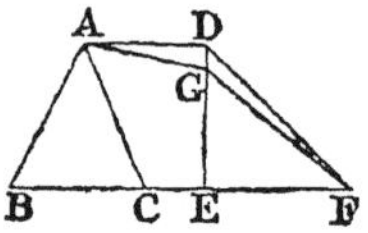

Let the equal triangles ABC, DEF be upon equal bases BC, EF, in the same straight line BF, and towards the same parts: they shall be between the same parallels.

Join AD: AD shall be parallel to BC. For if it is not,
* 31. 1. through A, draw* AG parallel to BF, and join GF. The tri-
* 38. 1. angle ABC is equal* to the triangle GEF, because they are upon equal bases BC, EF, and between the same parallels BF,
† Hyp. AG: but the triangle ABC is equal† to the triangle DEF;
† 1 Ax. therefore also the triangle DEF is equal† to the triangle GEF, the greater to the less; which is impossible: therefore AG is not parallel to BF. And in the same manner it can be demonstrated, that there is no other parallel to it but AD; AD is therefore parallel to BF. Wherefore, equal triangles, &c. Q. E. D.

PROPOSITION XLI.

THEOR.—*If a parallelogram and a triangle be upon the same base, and between the same parallels, the parallelogram shall be double of the triangle.*

Let the parallelogram ABCD and the triangle EBC be upon the same base BC, and between the same parallels BC, AE: the parallelogram ABCD shall be double of the triangle EBC.

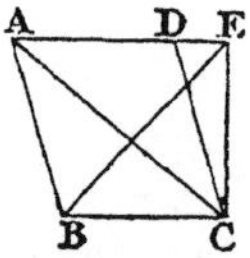

* 37. 1. Join AC: then the triangle ABC is equal* to the triangle EBC, because they are upon the same base BC, and between the same parallels BC, AE: but the parallelogram ABCD is double of the triangle ABC, because the diameter AC divides it into
* 34. 1. two equal* parts; wherefore ABCD is also double of the triangle EBC. Therefore, if a parallelogram, &c. Q. E. D.

PROPOSITION XLII.

PROB.—*To describe a parallelogram that shall be equal to a given triangle, and have one of its angles equal to a given rectilineal angle.*

Let ABC be the given triangle, and D the given rectilineal angle; it is required to describe a parallelogram that shall be equal to the given triangle ABC, and have one of its angles equal to D.

Bisect * BC in E, join AE, and at the point E, in the * 10. 1.
straight line EC, make * the angle CEF equal to D; and * 23. 1.
through A, draw * AFG parallel to EC, and through C, draw * 31. 1.
CG parallel to EF; therefore FECG is a † parallelogram. And † Def. 34. 1.
because BE is equal † to EC, the triangle † Constr.
ABE is equal * to the triangle AEC, since * 38. 1.
they are upon equal bases BE, EC, and between the same parallels BC, AG; therefore the triangle ABC is double of the triangle AEC: but the parallelogram FECG is likewise double * of the triangle AEC, * 41. 1.
because they are upon the same base EC, and between the same parallels EC, AG; therefore the parallelogram FECG is equal † to the triangle ABC; and it has one of its angles † 6 Ax.
CEF equal † to the given angle D: wherefore, a parallelogram † Constr.
FECG has been described equal to the given triangle ABC, having one of its angles CEF equal to the given angle D. Which was to be done.

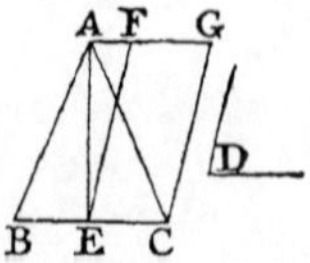

PROPOSITION XLIII.

THEOR.—*The complements of the parallelograms which are about the diameter of any parallelogram, are equal to one another.*

Let ABCD be a parallelogram, of which the diameter is AC; and EH, GF parallelograms about AC, *that is, through which* AC *passes;* and BK, KD the other parallelograms which make up the whole figure ABCD, which are therefore called the complements: the complement BK shall be equal to the complement KD.

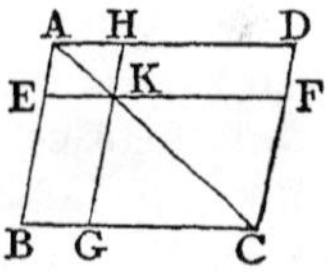

Because ABCD is a parallelogram, and AC its diameter, the triangle ABC is equal * to the triangle ADC. Again, because * 34. 1.
EKHA is a parallelogram, the diameter of which is AK, the

† 34. 1. triangle AEK is equal† to the triangle AHK; and for the same reason, the triangle KGC is equal to the triangle KFC. Therefore, because the triangle AEK is equal to the triangle AHK, and the triangle KGC to KFC, the triangle AEK to-
† 2 Ax. gether with the triangle KGC is equal † to the triangle AHK together with the triangle KFC: but the whole triangle ABC was proved equal to the whole ADC; therefore the remaining
† 3 Ax. complement BK is equal † to the remaining complement KD. Wherefore, the complements, &c. Q. E. D.

PROPOSITION XLIV.

PROB.—*To a given straight line, to apply a parallelogram which shall be equal to a given triangle, and have one of its angles equal to a given rectilineal angle.*

Let AB be the given straight line, and C the given triangle, and D the given rectilineal angle; it is required to apply to the straight line AB, a parallelogram equal to the triangle C, and having an angle equal to D.

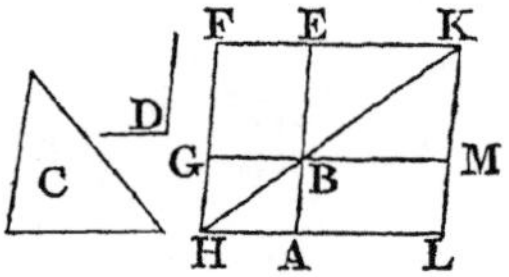

* 42. 1. Make* the parallelogram BEFG equal to the triangle C, and having the angle EBG equal to the angle D, so that BE be in the same straight line with AB; and produce FG to H; and through A,
* 31. 1. draw * AH parallel to BG or EF, and join HB. Then, because the straight line HF falls upon the parallels AH, EF, the
* 29. 1. angles AHF, HFE are together equal * to two right angles; wherefore the angles BHF, HFE are less than two right angles; but straight lines, which, with another straight line, make the interior angles upon the same side less than two right angles,
* 12 Ax. do meet * if produced far enough; therefore HB, FE shall meet if produced; let them meet in K, and through K, draw KL parallel to EA or FH, and produce HA, GB to the points L, M: then HLKF is a parallelogram, of which the diameter is HK; and AG, ME are parallelograms about HK; and LB,
* 43. 1. BF are the complements; therefore LB is equal * to BF: but
† Constr. BF is equal † to the triangle C; wherefore LB is equal † to
† 1 Ax. * 15. 1. the triangle C: and because the angle GBE is equal * to the
† Constr. angle ABM, and likewise † to the angle D, the angle ABM is
† 1 Ax. equal † to the angle D. Therefore, to the straight line AB,

the parallelogram LB is applied, equal to the triangle C, and having the angle ABM equal to the angle D. Which was to be done.

PROPOSITION XLV.

PROB.—*To describe a parallelogram equal to a given rectilineal figure, and having an angle equal to a given rectilineal angle.* See N.

Let ABCD be the given rectilineal figure, and E the given rectilineal angle; it is required to describe a parallelogram equal to ABCD, and having an angle equal to E.

Join DB; and describe * the parallelogram FH equal to the triangle ADB, and having the angle FKH equal to the angle E; and to the straight line GH, apply * the parallelogram GM equal to the triangle DBC, having the angle GHM equal to the angle E: the figure FKML shall be the parallelogram required. * 42. 1. * 44. 1.

Because the angle E is equal † to each of the angles FKH, GHM, the angle FKH is equal † to GHM: add to each of these the angle KHG; therefore the angles FKH, KHG are equal † to the angles KHG, GHM: but FKH, KHG are equal * to two right angles; therefore also KHG, GHM are equal † to two right angles: and because at the point H, in the straight line GH, the two straight lines KH, HM, upon the opposite sides of it, make the adjacent angles equal to two right angles, KH is in the same straight line * with HM: and because the straight line HG meets the parallels KM, FG, the alternate angles MHG, HGF * are equal; add to each of these the angle HGL; therefore the angles MHG, HGL are equal † to the angles HGF, HGL: but the angles MHG, HGL are equal * to two right angles; wherefore also the angles HGF, HGL are equal † to two right angles, and therefore FG is in the same straight line † with GL: and because KF is parallel to HG †, and HG to ML, KF is parallel * to ML: and KM, FL are † parallels; wherefore KFLM is † a parallelogram: and because the triangle ABD is equal to the parallelogram HF †, and the triangle DBC to the parallelogram GM; the whole rectilineal figure ABCD is equal † † Constr. † 1 Ax. † 2 Ax. * 29. 1. † 1 Ax. * 14. 1. * 29. 1. † 2 Ax. * 29. 1. † 1 Ax. † 14. 1. † Constr. * 30. 1. † Constr. † Def. 34. 1. † Constr. † 2 Ax.

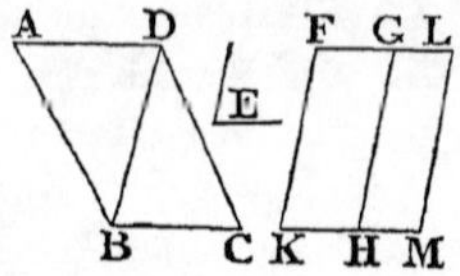

to the whole parallelogram KFLM. Therefore, the parallelogram KFLM has been described equal to the given rectilineal figure ABCD, having the angle FKM equal to the given angle E. Which was to be done.

COR. From this it is manifest, how, to a given straight line, to apply a parallelogram, which shall have an angle equal to a given rectilineal angle, and shall be equal to a
* 44. 1. given rectilineal figure; viz. by applying * to the given straight line a parallelogram equal to the first triangle ABD, and having an angle equal to the given angle.

PROPOSITION XLVI.

PROB.—*To describe a square upon a given straight line.*

Let AB be the given straight line; it is required to describe a square upon AB.

* 11. 1. From the point A, draw * AC at right angles to AB, and
* 3. 1. make * AD equal to AB; through the point D, draw DE pa-
* 31. 1. rallel * to AB; and through B, draw BE parallel to AD;
† Def. 34. 1. therefore ADEB is a † parallelogram: whence AB is equal *
* 34. 1. † Constr. to DE, and AD to BE: but BA is equal † to AD; therefore the four straight lines BA,
† 1 Ax. AD, DE, EB are equal † to one another, and the parallelogram ADEB is equilateral: likewise all its angles are right angles; for, since the straight line AD meets the parallels AB, DE,
* 29. 1. the angles BAD, ADE are equal * to two right
† Constr. angles: but BAD is a † right angle; therefore also ADE is
† 3 Ax. * 34. 1. a † right angle: but the opposite angles of parallelograms * are equal; therefore each of the opposite angles ABE, BED
† 1 Ax. is a † right angle; wherefore the figure ADEB is rectangular: and it has been demonstrated that it is equilateral; it is
† 30 Def. therefore a † square, and it is described upon the given straight line AB. Which was to be done.

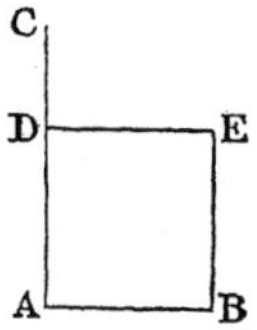

COR. Hence, every parallelogram that has one right angle, has all its angles right angles.

PROPOSITION XLVII.

THEOR.—*In any right-angled triangle, the square which is described upon the side subtending the right angle, is equal to the squares described upon the sides which contain the right angle.*

Let ABC be a right-angled triangle, having the right angle BAC: the square described upon the side BC, shall be equal to the squares described upon BA, AC.

On BC describe * the square BDEC; and on BA, AC, the squares GB, HC; and through A, draw * AL parallel to BD, or CE, and join AD, FC. Then, because the angle BAC is a † right angle, and that the angle BAG is also a * right angle, the two straight lines AC, AG upon the opposite sides of AB, make with it at the point A, the adjacent angles equal to two right angles: therefore CA is in the same straight line * with AG: for the same reason, AB and AH are in the same straight line. And because the angle DBC is equal † to the angle FBA, each of them being a right † angle, add to each the angle ABC; therefore the whole angle DBA is equal * to the whole FBC: and because the two sides AB, BD are equal † to the two FB, BC, each to each, and the angle DBA equal to the angle FBC, therefore the base AD is equal * to the base FC, and the triangle ABD to the triangle FBC: now the parallelogram BL is double * of the triangle ABD, because they are upon the same base BD, and between the same parallels BD, AL; and the square GB is double of the triangle FBC, because these also are upon the same base FB, and between the same parallels FB, GC: but the doubles of equals are equal * to one another; therefore the parallelogram BL is equal to the square GB. In the same manner, by joining AE, BK, it can be demonstrated, that the parallelogram CL is equal to the square HC; therefore the whole square BDEC is equal † to the two squares GB, HC: and the square BDEC is described upon the straight line BC, and the squares GB, HC, upon BA, AC; therefore, the square upon the side BC, is equal to the squares upon the sides BA, AC. Therefore, in any right-angled triangle, &c. Q. E. D.

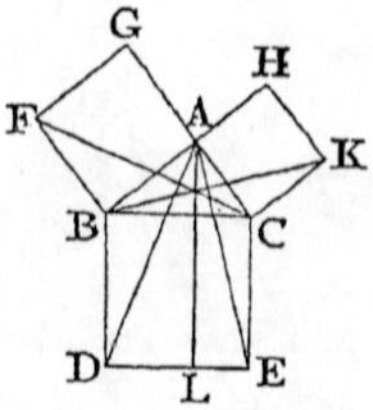

* 46. 1.
* 31. 1.
† Hyp.
* 30 Def.
* 14. 1.
† 11 Ax.
† 30 Def.
* 2 Ax.
† 30 Def.
* 4. 1.
* 41. 1.
* 6 Ax.
† 2 Ax.

PROPOSITION XLVIII.

THEOR.—*If the square described upon one of the sides of a triangle, be equal to the squares described upon the other two sides of it, the angle contained by these two sides is a right angle.*

Let the square described upon BC, one of the sides of the triangle ABC, be equal to the squares upon the other sides BA, AC: the angle BAC shall be a right angle.

* 1. 1. From the point A, draw * AD at right angles to AC, and
† 3. 1. make † AD equal to BA, and join DC. Then, because DA is equal to AB, the square of DA is equal to the square of AB: to each of these add the square of AC; therefore the squares
† 2 Ax. of DA, AC are equal † to the squares of BA,
* 47. 1. AC: but the square of DC is equal * to the
† Constr. squares of DA, AC, because DAC is a † right angle; and the square of BC, by hypothesis, is equal to the squares of BA, AC; therefore the
† 1 Ax. square of DC is equal † to the square of BC; and therefore also the side DC is equal to the side BC. And
† Constr. because the side DA is equal † to AB, and AC common to the two triangles DAC, BAC, the two DA, AC are equal to the two BA, AC, each to each; and the base DC has been proved
* 8. 1. equal to the base BC: therefore the angle DAC is equal * to
† Constr. the angle BAC: but DAC is a † right angle; therefore also
† 1 Ax. BAC is a † right angle. Therefore, if the square, &c. Q. E. D.

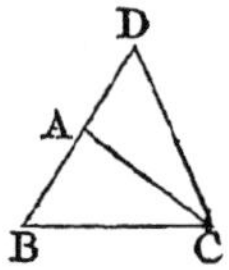

THE

ELEMENTS OF EUCLID.

BOOK II.

DEFINITIONS.

I. *Every right-angled parallelogram, or* rectangle, *is said to be contained by any two of the straight lines which contain one of the right angles.*

II. In every parallelogram, any of the parallelograms about a diameter, together with the two complements, is called a Gnomon. 'Thus 'the parallelogram HG, together with 'the complements AF, FC, is the gno- 'mon, which is more briefly expressed 'by the letters AGK, or EHC, which 'are at the opposite angles of the parallelograms which 'make the gnomon.'

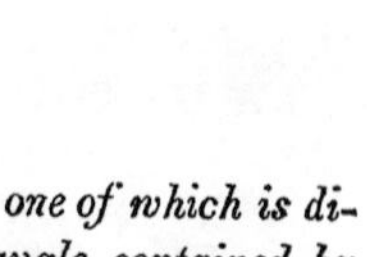

PROPOSITION I.

Theorem.—*If there be two straight lines, one of which is divided into any number of parts, the rectangle contained by the two straight lines, is equal to the rectangles contained by the undivided line, and the several parts of the divided line.*

Let A and BC be two straight lines; and let BC be divided into any parts in the points D, E: the rectangle contained by the straight lines A, BC, shall be equal to the rectangle contained by A, BD, together with that contained by A, DE, and that contained by A, EC.

* 11. 1. From the point B, draw * BF at right
* 3. 1. angles to BC, and make BG equal * to A;
* 31. 1. and through G, draw * GH parallel to
* 31. 1. BC; and through D, E, C, draw * DK, EL, CH parallel to BG: then the rectangle BH is equal to the rectangles BK, DL, EH: but BH is contained by A, BC, for it is contained by

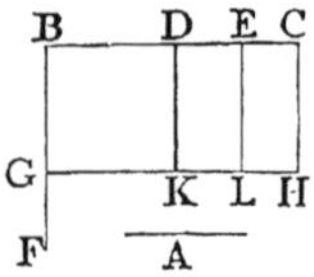

† Constr. GB, BC, and GB is equal † to A; and BK is contained by A, BD, for it is contained by GB, BD, of which GB is equal to
* 34. 1. A; and DL is contained by A, DE, because DK, that is * BG, is equal to A; and in like manner the rectangle EH is contained by A, EC: therefore the rectangle contained by A, BC, is equal to the several rectangles contained by A, BD, and by A, DE, and by A, EC. Wherefore, if there be two straight lines, &c. Q. E. D.

PROPOSITION II.

THEOR.—*If a straight line be divided into any two parts, the rectangles contained by the whole and each of the parts, are together equal to the square of the whole line.*

Let the straight line AB be divided into any two parts in the point C: the rectangle contained by AB, BC, together with the rectangle ‡ AB, AC, shall be equal to the square of AB.

* 46. 1. Upon AB, describe * the square ADEB; and
* 31. 1. through C, draw * CF, parallel to AD or BE: then AE is equal to the rectangles AF, CE: but AE is the square of AB; and AF is the rectangle contained by BA, AC, for it is con-

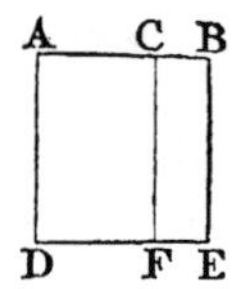

† 30 Def. tained by DA, AC, of which AD is equal † to AB; and CE is contained by AB, BC, for BE is equal to AB: therefore the rectangle contained by AB, AC, together with the rectangle AB, BC, is equal to the square of AB. If, therefore, a straight line, &c. Q. E. D.

‡ N.B.—To avoid repeating the word *contained* too frequently, the rectangle contained by two straight lines AB, AC is sometimes simply called the rectangle AB, AC.

PROPOSITION III.

THEOR.—*If a straight line be divided into any two parts, the rectangle contained by the whole and one of the parts, is equal to the rectangle contained by the two parts, together with the square of the aforesaid part.*

Let the straight line AB be divided into any two parts in the point C: the rectangle AB, BC shall be equal to the rectangle AC, CB, together with the square of BC.

Upon BC, describe * the square CDEB, and produce ED to F; and through A, draw * AF parallel to CD or BE: then the rectangle AE is equal to the rectangles AD, CE: but AE is the rectangle contained by AB, BC, for it is contained by AB, BE, of which BE is equal † to BC; and AD is contained by AC, CB, for CD is equal to CB; and DB is the square of BC: therefore the rectangle AB, BC is equal to the rectangle AC, CB, together with the square of BC. If, therefore, a straight line, &c. Q. E. D.

* 46. 1.
* 31. 1.
† 30 Def.

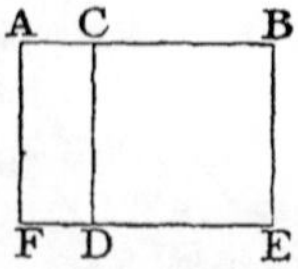

PROPOSITION IV.

THEOR.—*If a straight line be divided into any two parts, the square of the whole line is equal to the squares of the two parts, together with twice the rectangle contained by the parts.*

Let the straight line AB be divided into any two parts in C: the square of AB shall be equal to the squares of AC, CB, and to twice the rectangle contained by AC, CB.

Upon AB, describe * the square ADEB, and join BD; and through C, draw * CGF parallel to AD or BE, and through G, draw HK parallel to AB or DE. And because CF is parallel † to AD, and BD falls upon them, the exterior angle BGC is equal * to the interior and opposite angle ADB; but ADB is equal * to the angle ABD, because BA is equal to AD †, being sides of a square; wherefore the angle CGB is equal † to the angle CBG; and therefore the side BC is equal * to the side CG: but CB is equal * also to GK, and CG to BK; wherefore the figure

* 46. 1.
* 31. 1.
† Constr.
* 29. 1.
* 5. 1.
† 30 Def.
† 1 Ax.
* 6. 1.
* 34. 1.

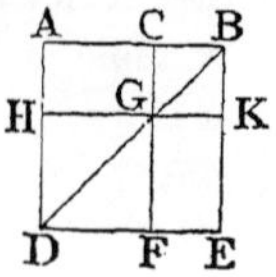

† 1 Ax. CGKB is † equilateral: it is likewise rectangular; for, since CG is parallel to BK, and CB meets them, therefore the angles † 29. 1. KBC, GCB are equal † to two right angles: but KBC is a † † 30 Def. right angle; wherefore GCB is a † right angle: and there- † 3 Ax. fore also the angles * CGK, GKB, opposite to these are right * 34. 1. and 1 Ax. angles; and therefore CGKB is rectangular; but it is also † 30 Def. equilateral, as was demonstrated; wherefore it is a † square, and it is upon the side CB: for the same reason HF also is a † 34. 1. square, and it is upon the side HG, which is equal † to AC: therefore HF, CK are the squares of AC, CB: and because the * 43. 1. complement AG is equal * to the complement GE, and that † 30 Def. AG is the rectangle contained by AC, CB, for GC is equal † † 1 Ax. to CB, therefore GE is also equal † to the rectangle AC, CB; wherefore AG, GE are equal to twice the rectangle AC, CB; and HF, CK are the squares of AC, CB; wherefore the four figures HF, CK, AG, GE are equal to the squares of AC, CB, and to twice the rectangle AC, CB: but HF, CK, AG, GE make up the whole figure ADEB, which is the square of AB: † 1 Ax. therefore the square of AB is equal † to the squares of AC, CB, and twice the rectangle AC, CB. Wherefore, if a straight line, &c. Q. E. D.

COR. From the demonstration it is manifest, that parallelograms about the diameter of a square are likewise squares.

PROPOSITION V.

THEOR.—*If a straight line be divided into two equal parts and also into two unequal parts, the rectangle contained by the unequal parts, together with the square of the line between the points of section, is equal to the square of half the line.*

Let the straight line AB be divided into two equal parts in the point C, and into two unequal parts at the point D: the rectangle AD, DB, together with the square of CD, shall be equal to the square of CB.

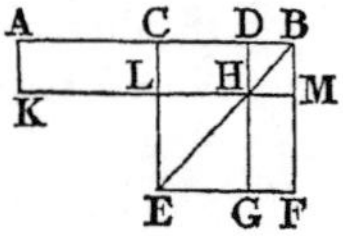

* 46. 1. Upon CB, describe * the square CEFB, * 31. 1. join BE; and through D, draw * DHG parallel to CE or BF; and through H, draw KLM parallel to CB or EF; and also through A, draw AK parallel to * 43. 1. CL or BM. And because the complement CH is equal * to the complement HF, to each of these add DM; therefore the

whole CM is equal † to the whole DF: but CM is equal * to AL, because AC is equal † to CB; therefore also AL is equal † to DF: to each of these add CH, and the whole AH is equal † to DF and CH: but AH is the rectangle contained by AD, DB, for DH is equal * to DB; and DF together with CH is the gnomon CMG; therefore the gnomon CMG is equal † to the rectangle AD, DB: to each of these add LG, which is equal * to the square of CD; therefore the gnomon CMG, together with LG is equal † to the rectangle AD, DB, together with the square of CD: but the gnomon CMG and LG make up the whole figure CEFB, which is the square of CB; therefore the rectangle AD, DB, together with the square of CD, is equal to the square of CB. Wherefore, if a straight line, &c. Q. E. D.

† 2 Ax.
* 36. 1.
† Hyp.
† 1 Ax.
† 2 Ax.
* Cor. 4. 2. & 30 Def.
† 1 Ax.
* Cor. 4. 2. and 34. 1.
† 2 Ax.

From this proposition it is manifest, that the difference of the squares of two unequal lines, AC, CD, is equal to the rectangle contained by their sum and difference.

PROPOSITION VI.

THEOR.—*If a straight line be bisected, and produced to any point, the rectangle contained by the whole line thus produced, and the part of it produced, together with the square of half the line bisected, is equal to the square of the straight line which is made up of the half and the part produced.*

Let the straight line AB be bisected in C, and produced to the point D: the rectangle AD, DB, together with the square of CB, shall be equal to the square of CD.

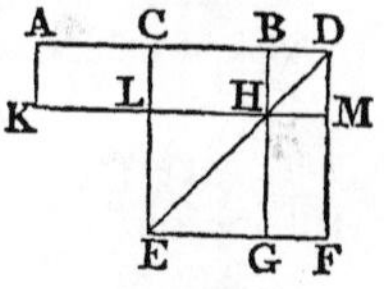

Upon CD, describe * the square CEFD, join DE; and through B, draw * BHG parallel to CE or DF; and through H, draw KLM parallel to AD or EF; and also through A, draw AK parallel to CL or DM. And because AC is equal † to CB, the rectangle AL is equal * to CH; but CH is equal * to HF; therefore also AL is equal † to HF: to each of these add CM; therefore the whole AM is equal † to the gnomon CMG: but AM is the rectangle contained by AD, DB, for DM is equal * to DB: therefore the gnomon CMG is equal † to the rectangle AD, DB: add to each of these LG, which is equal † to the square of CB; therefore the rectangle AD, DB, to-

* 46. 1.
* 31. 1.
† Hyp.
* 36. 1.
* 43. 1.
† 1 Ax.
† 2 Ax.
* Cor. 4. 2. & 30 Def.
† 1 Ax.
† Cor. 4. 2. and 34. 1.

† 2 Ax. gether with the square of CB, is equal † to the gnomon CMG, and the figure LG: but the gnomon CMG and LG make up the whole figure CEFD, which is the square of CD; therefore the rectangle AD, DB, together with the square of CB, is equal to the square of CD. Wherefore, if a straight line, &c. Q. E. D.

PROPOSITION VII.

THEOR.—*If a straight line be divided into any two parts, the squares of the whole line, and of one of the parts, are equal to twice the rectangle contained by the whole and that part, together with the square of the other part.*

Let the straight line AB be divided into any two parts in the point C: the squares of AB, BC shall be equal to twice the rectangle AB, BC, together with the square of AC.

* 46. 1. Upon AB, describe * the square ADEB, and construct the figure as in the preceding propositions. And because AG is
* 43. 1. equal * to GE, add to each of them CK; therefore the whole AK is equal to the whole CE; therefore AK, CE are double of AK: but AK, CE are the gnomon AKF, together with the square CK; therefore the gnomon AKF, together with the square CK,
† 1 Ax. is † double of AK: but twice the rectangle
* Cor. 4. 2. & 30 Def. AB, BC is double of AK, for BK is equal * to BC; therefore the gnomon AKF, together
† 1 Ax. with the square CK, is equal † to twice the rectangle AB, BC:
† Cor. 4. 2. & 34. 1. to each of these equals add HF, which is equal † to the square of AC; therefore the gnomon AKF, together with the squares
† 2 Ax. CK, HF, is equal † to twice the rectangle, AB, BC, and the square of AC: but the gnomon AKF, together with the squares CK, HF, make up the whole figure ADEB and CK, which are the squares of AB and BC; therefore the squares of AB and BC are equal to twice the rectangle AB, BC, together with the square of AC. Wherefore, if a straight line, &c. Q. E. D.

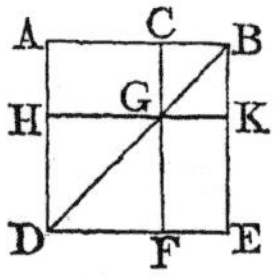

PROPOSITION VIII.

THEOR.—*If a straight line be divided into any two parts, four times the rectangle contained by the whole line and one of the parts, together with the square of the other part, is equal to the square of the straight line which is made up of the whole and that part.*

Let the straight line AB be divided into any two parts in the point C: four times the rectangle AB, BC, together with the square of AC, shall be equal to the square of the straight line made up of AB and BC together.

Produce † AB to D, so that BD be equal † to CB; and upon † 2 Post.
AD, describe † the square AEFD; and construct two figures † 3. 1. † 46. 1.
such as in the preceding. Because CB is equal to BD, and
that CB is equal * to GK, and BD to KN, therefore GK is * 34. 1.
equal † to KN: for the same reason, PR is equal to RO: and † 1 Ax.
because CB is equal to BD, and GK to KN, the rectangle CK
is equal * to BN, and GR to RN: but CK is equal * to RN, * 36. 1. * 43. 1.
because they are the complements of the parallelogram CO;
therefore also BN is equal † to GR: therefore the four rect- † 1 Ax.
angles BN, CK, GR, RN are equal to one another, and so are
quadruple of one of them CK. Again, because CB is equal to
BD, and that BD is equal * to BK, that is †, * Cor. 4. 2. & 30 Def.
to CG, and CB equal † to GK, that is †, to GP, † 34. 1.
therefore CG is equal † to GP: and because † 34. 1.
CG is equal to GP, and PR to RO, the rect- † Cor. 4. 2. & 30 Def.
angle AG is equal † to MP, and PL to RF: † 1 Ax.
but MP is equal * to PL, because they are † 36. 1. * 43. 1.
the complements of the parallelogram ML;

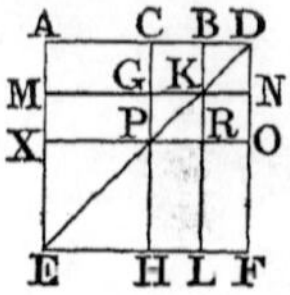

wherefore AG is equal † also to RF: therefore the four rect- † 1 Ax.
angles AG, MP, PL, RF are equal to one another, and so are
quadruple of one of them AG. And it was demonstrated that
the four CK, BN, GR, and RN are quadruple of CK: there-
fore the eight rectangles which contain the gnomon AOH, are
quadruple of AK: and because AK is the rectangle contained
by AB, BC, for BK, ‡ is equal to BC, therefore four times the
rectangle AB, BC, is quadruple of AK: but the gnomon AOH
was demonstrated to be quadruple of AK; therefore four
times the rectangle AB, BC, is equal † to the gnomon AOH: † 1 Ax.
to each of these add XH, which is equal * to the square of * Cor. 4. 2.
AC; therefore four times the rectangle AB, BC, together with and 34. 1.
the square of AC, is equal † to the gnomon AOH and the † 2 Ax.
square XH: but the gnomon AOH and XH make up the
figure AEFD, which is the square of AD; therefore four times
the rectangle AB, BC, together with the square of AC, is equal
to the square of AD, that is, of AB and BC added together in
one straight line. Wherefore, if a straight line, &c. Q. E. D.

‡ Because BN is a † square, therefore BK is equal † to BD: but BD is † Cor. 4. 2. † 30 Def.
equal † to BC; therefore BK is equal † to BC. † Constr. † 1 Ax.

PROPOSITION IX.

THEOR.—*If a straight line be divided into two equal, and also into two unequal parts, the squares of the two unequal parts are together double of the square of half the line, and of the square of the line between the points of section.*

Let the straight line AB be divided into two equal parts at the point C, and into two unequal parts at the point D: the squares of AD, DB, shall together be double of the squares of AC, CD.

* 11. 1. From the point C, draw* CE at right angles to AB, and
† 3. 1. make it equal † to AC or CB, and join EA, EB; through D,
* 31. 1. draw* DF parallel to CE; and through F, draw FG parallel to BA; and join AF. Then, because AC is equal to CE, the
* 5. 1. angle EAC is equal* to the angle AEC: and because the angle ACE is a right angle, the two others AEC, EAC to-
* 32. 1. gether make one* right angle; and they are equal to one another; therefore each of them is half of a right angle. For the same reason, each of the angles CEB, EBC is half a right angle, and therefore the whole AEB is a right angle. And because the angle GEF is half a right angle, and EGF a right angle, for it
* 29. 1. is equal* to the interior and opposite angle ECB, the remaining angle EFG is half a right angle; therefore the angle GEF is equal to
* 6. 1. the angle EFG, and the side EG equal* to the side GF. Again, because the angle at B is half a right angle, and FDB
* 29. 1. a right angle, for it is equal* to the interior and opposite angle ECB, the remaining angle BFD is half a right angle; therefore the angle at B is equal to the angle BFD, and the
* 6. 1. side DF to* the side DB. And because AC is equal to CE, the square of AC is equal to the square of CE; therefore the squares of AC, CE are double of the square of AC: but the
* 47. 1. square of AE is equal* to the squares of AC, CE, because ACE is a right angle; therefore the square of AE is double of the square of AC: again, because EG is equal to GF, the square of EG is equal to the square of GF; therefore the squares of EG, GF are double of the square of GF: but the square of EF
† 47. 1. is equal † to the squares of EG, GF; therefore the square of
* 34. 1. EF is double of the square of GF; and GF is equal* to CD; therefore the square of EF is double of the square of CD: but the square of AE is likewise double of the square of AC;

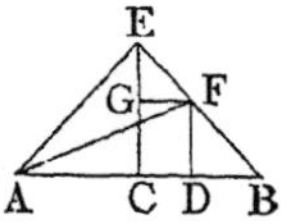

therefore the squares of AE, EF are double of the squares of AC, CD: but the square of AF is equal * to the squares of AE, EF, because AEF is a right angle; therefore the square of AF is double of the squares of AC, CD: but the squares of AD, DF are equal to the square of AF, because the angle ADF is a right angle; therefore the squares of AD, DF are double of the squares of AC, CD: and DF is equal to DB; therefore the squares of AD, DB are double of the squares of AC, CD. If, therefore, a straight line, &c. Q. E. D. * 47. 1.

PROPOSITION X.

THEOR.—*If a straight line be bisected, and produced to any point, the square of the whole line thus produced, and the square of the part of it produced, are together double of the square of half the line bisected, and of the square of the line made up of the half and the part produced.*

Let the straight line AB be bisected in C, and produced to the point D: the squares of AD, DB shall be double of the squares of AC, CD.

From the point C draw * CE at right angles to AB, and make † it equal to AC or CB, and join AE, EB; through E, draw * EF parallel to AD, and through D, draw DF parallel to CE. And because the straight line EF meets the parallels EC, FD, the angles CEF, EFD are equal * to two right angles; and therefore the angles BEF, EFD are less than two right angles: but straight lines, which with another straight line make the interior angles upon the same side less than two right angles, will meet * if produced far enough; therefore EB, FD will meet, if produced, towards B, D: let them meet in G, and join AG. Then, because AC is equal to CE, the angle CEA is equal * to the angle EAC; and the angle ACE is a right angle; therefore each of the angles CEA, EAC is half * a right angle. For the same reason, each of the angles CEB, EBC is half a right angle; therefore AEB is a right angle. And because EBC is half a right angle, DBG is also * half a right angle, for they are vertically opposite: but BDG is a right angle, because it is equal * to the alternate angle DCE; therefore the remaining

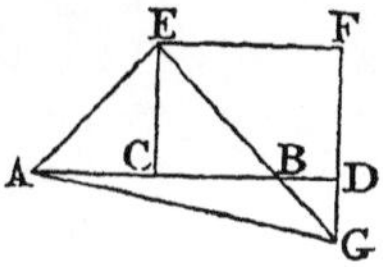

* 11. 1.
† 3. 1.
* 31. 1.
* 29. 1.
* 12 Ax.
* 5. 1.
* 32. 1.
* 15. 1.
* 29. 1.

angle DGB is half a right angle, and is therefore equal to
* 6. 1. the angle DBG; wherefore also the side BD is equal * to the side DG. Again, because EGF is half a right angle, and that the
* 34. 1. angle at F is a right angle, because it is equal * to the opposite angle ECD, the remaining angle FEG is half a right angle, and therefore equal to the angle EGF; wherefore also
* 6. 1. the side GF is equal * to the side FE. And because EC is equal to CA, the square of EC is equal to the square of CA; therefore the squares of EC, CA are double of the square of
* 47. 1. CA; but the square of EA is equal * to the squares of EC, CA; therefore the square of EA is double of the square of AC: again, because GF is equal to FE, the square of GF is equal to the square of FE; and therefore the squares of GF, FE are double of the square of EF: but the square of EG is
* 47. 1. equal * to the squares of GF, FE; therefore the square of EG
† 34. 1. is double of the square of EF; and EF is equal † to CD; wherefore the square of EG is double of the square of CD: but it was demonstrated, that the square of EA is double of the square of AC; therefore the squares of AE, EG are double
* 47. 1. of the squares of AC, CD: but the square of AG is equal * to the squares of AE, EG; therefore the square of AG is double of the squares of AC, CD: but the squares of AD, GD are
* 47. 1. equal * to the square of AG; therefore the squares of AD, DG are double of the squares of AC, CD: but DG is equal to DB; therefore the squares of AD, DB are double of the squares of AC, CD. Wherefore, if a straight line, &c. Q. E. D.

PROPOSITION XI.

Problem.—*To divide a given straight line into two parts, so that the rectangle contained by the whole and one of the parts, shall be equal to the square of the other part.*

Let AB be the given straight line; it is required to divide it into two parts, so that the rectangle contained by the whole and one of the parts, shall be equal to the square of the other part.

* 46. 1. * 10. 1. * 3. 1. * 46. 1. Upon AB, describe * the square ABDC; bisect * AC in E, and join BE; produce CA to F, and make * EF equal to EB; and upon AF describe * the square FGHA: AB shall be divided in H, so that the rectangle AB, BH is equal to the square of AH.

Produce GH to K: and because the straight line AC is bisected in E, and produced to the point F, the rectangle CF, FA, together with the square of AE, is equal * to the square of EF: but EF is equal † to EB; therefore the rectangle CF, FA, together with the square of AE, is equal to the square of EB: but the squares of BA, AE are equal * to the square of EB, because the angle EAB is a right angle; therefore the rectangle CF, FA, together with the square of AE, is equal to the squares of BA, AE: take away the square of AE, which is common to both; therefore the remaining rectangle CF, FA is equal † to the square of AB: but the figure FK is the rectangle contained by CF, FA, for FA is equal † to FG; and AD is the square of AB; therefore FK is equal † to AD: take away the common part AK, and the remainder FH is equal † to the remainder HD: but HD is the rectangle contained by AB, BH, for AB is equal † to BD; and FH is the square of AH; therefore the rectangle AB, BH is equal to the square of AH. Wherefore, the straight line AB is divided in H, so that the rectangle AB, BH is equal to the square of AH. Which was to be done.

* 6. 2.
† Constr.
* 47. 1.
† 3 Ax.
† 30 Def.
† 1 Ax.
† 3 Ax.
† 30 Def.

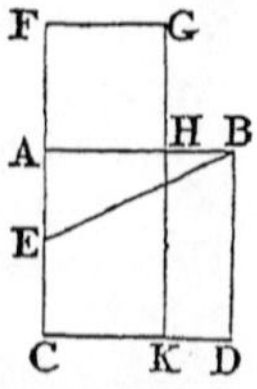

PROPOSITION XII.

THEOR.—*In obtuse-angled triangles, if a perpendicular be drawn from either of the acute angles to the opposite side produced, the square of the side subtending the obtuse angle, is greater than the squares of the sides containing the obtuse angle, by twice the rectangle contained by the side upon which, when produced, the perpendicular falls, and the straight line intercepted without the triangle, between the perpendicular and the obtuse angle.*

Let ABC be an obtuse-angled triangle, having the obtuse angle ACB; and from the point A, let AD be drawn * perpendicular to BC produced: the square of AB shall be greater than the squares of AC, CB, by twice the rectangle BC, CD. * 12. 1.

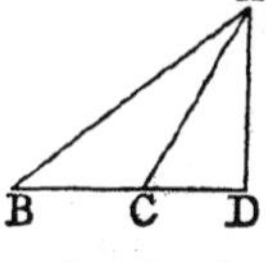

Because the straight line BD is divided into two parts in the point C, the square of BD is equal * to the squares of BC, CD, and twice the rectangle BC, CD: to each of these equals add the * 4. 2.

† 2 Ax. square of DA; therefore the squares of BD, DA are equal † to the squares of BC, CD, DA, and twice the rectangle
* 47. 1. BC, CD: but the square of BA is equal * to the squares of BD, DA, because the angle at D is a right angle; and the
* 47. 1. square of CA is equal * to the squares of CD, DA; therefore the square of BA is equal to the squares of BC, CA, and twice the rectangle BC, CD; that is, the square of BA is greater than the squares of BC, CA, by twice the rectangle BC, CD. Therefore, in obtuse-angled triangles, &c. Q. E. D.

PROPOSITION XIII.

See N. THEOR.—*In every triangle, the square of the side subtending either of the acute angles, is less than the squares of the sides containing that angle, by twice the rectangle contained by either of these sides, and the straight line intercepted between the perpendicular let fall upon it from the opposite angle, and the acute angle.*

Let ABC be any triangle, and the angle at B one of its acute angles; and upon BC, one of the sides containing it,
* 12. 1. let fall the perpendicular * AD from the opposite angle: the square of AC, opposite to the angle B, shall be less than the squares of CB, BA, by twice the rectangle CB, BD.

First, let AD fall within the triangle ABC: and because the straight line CB is divided into two parts in the point D,
* 7. 2. the squares of CB, BD are equal * to twice the rectangle contained by CB, BD, and the square of DC: to each of these equals add the square of AD; therefore the squares of
† 2 Ax. CB, BD, DA are equal † to twice the rectangle CB, BD, and the squares of AD, DC:
* 47. 1. but the square of AB is equal * to the squares of BD, DA, because the angle BDA is a right angle; and the square of AC is equal to the squares of AD, DC; therefore the squares of CB, BA are equal to the square of AC, and twice the rectangle CB, BD; that is, the square of AC alone, is less than the squares of CB, BA, by twice the rectangle CB, BD.

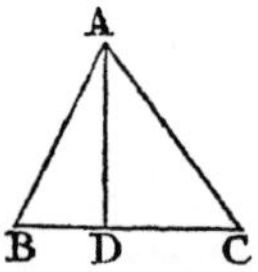

Secondly, let AD fall without the triangle ABC: then, because the angle at D is a right angle, the angle ACB is
* 16. 1. greater * than a right angle; and therefore the square of

AB is equal * to the squares of AC, CB, and twice the rectangle BC, CD: to each of these equals add the square of BC; therefore the squares of AB, BC are equal † to the square of AC, and twice the square of BC, and twice the rectangle BC, CD: but because BD is divided into two parts in C, the rectangle, DB, BC is equal * to the rectangle BC, CD and the square of BC; and the doubles of these are equal; therefore the squares of AB, BC are equal to the square of AC, and twice the rectangle DB, BC: therefore the square of AC alone, is less than the squares of AB, BC, by twice the rectangle DB, BC. * 12. 2. † 2 Ax. * 3. 2.

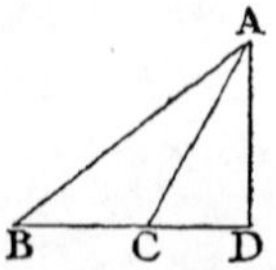

Lastly, let the side AC be perpendicular to BC: then BC is the straight line between the perpendicular and the acute angle at B: and it is manifest, that the squares of AB, BC are equal * to the square of AC, and twice the square of BC. Therefore, in every triangle, &c. Q. E. D. * 47. 1. and 2 Ax.

PROPOSITION XIV.

PROB.—*To describe a square that shall be equal to a given rectilineal figure.* See N.

Let A be the given rectilineal figure; it is required to describe a square that shall be equal to A.

Describe * the rectangular parallelogram BCDE equal to the rectilineal figure A. Then if the sides of it, BE, ED are equal to one another, it is † a square, and what was required is now done: but if they are not equal, produce one of them BE to F, and make † EF equal to ED, and bisect † BF in G; and from the centre G, at the distance GB, or GF, describe the semicircle BHF, and produce DE to H: the square described upon EH shall be equal to the given rectilineal figure A. * 45. 1. † 30 Def. † 3. 1. † 10. 1.

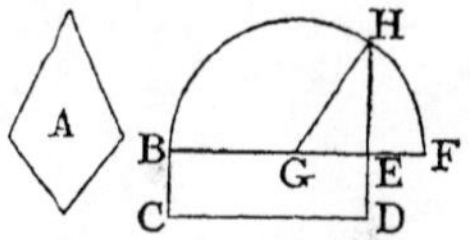

Join GH: and because the straight line BF is divided into two equal parts in the point G, and into two unequal at E, the rectangle BE, EF, together with the square of EG, is equal * to the square of GF: but GF is equal † to GH; therefore the rectangle BE, EF, together with the square of * 5. 2. † 15 Def.

EG, is equal to the square of GH: but the squares of HE,
* 47. 1. EG are equal* to the square of GH; therefore the rectangle
† 1 Ax. BE, EF, together with the square of EG, is equal† to the squares of HE, EG: take away the square of EG, which is common to both; and the remaining rectangle BE, EF is
† 3 Ax. equal† to the square of EH: but the rectangle contained by BE, EF is the parallelogram BD, because EF is equal to ED;
† Constr. therefore BD is equal to the square of EH: but BD is equal† to the rectilineal figure A; therefore the rectilineal figure A
† 1 Ax. is equal† to the square of EH. Wherefore, a square has been made equal to the given rectilineal figure A; viz. the square described upon EH. Which was to be done.

THE

ELEMENTS OF EUCLID.

BOOK III.

DEFINITIONS.

I. Equal circles are those of which the diameters are equal, or from the centres of which the straight lines to the circumferences are equal.

"This is not a definition, but a theorem, the truth of which is evident; for, if the circles be applied to one another, so that their centres coincide, the circles must likewise coincide, since the straight lines from the centres are equal."

II. A straight line is said to touch a circle, when it meets the circle, and being produced does not cut it.

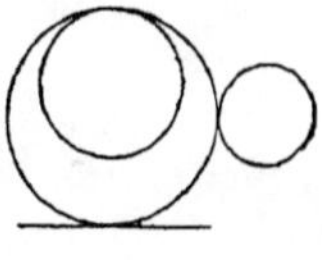

III. Circles are said to touch one another, which meet but do not cut one another.

IV. Straight lines are said to be equally distant from the centre of a circle, when the perpendiculars drawn to them from the centre are equal.

V. And the straight line on which the greater perpendicular falls, is said to be farther from the centre.

VI. A segment of a circle is the figure contained by a straight line and the circumference it cuts off.

VII. "The angle of a segment is that which is contained by the straight line and the circumference."

VIII. An angle in a segment, is the angle contained by two straight lines drawn from any point in the circumference of the segment, to the extremities of the straight line which is the base of the segment.

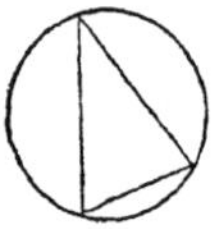

IX. And an angle is said to insist or stand upon the circumference intercepted between the straight lines that contain the angle.

X. A sector of a circle is the figure contained by two straight lines drawn from the centre, and the circumference between them.

XI. Similar segments of circles are those in which the angles are equal, or which contain equal angles.

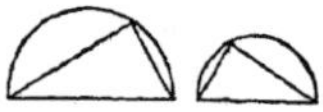

PROPOSITION I.

See N. PROBLEM.—*To find the centre of a given circle.*

Let ABC be the given circle; it is required to find its centre.

* 10. 1. Draw within it any straight line AB, and bisect * it in
* 11. 1. D; from the point D, draw * DC at right angles to AB, and produce it to E, and bisect CE in F: the point F shall be the centre of the circle ABC.

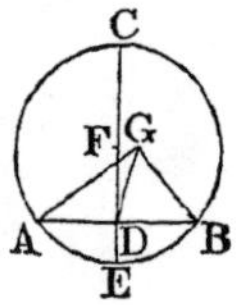

For if it be not, let, if possible, G be the centre, and join GA, GD, GB: then, be-
† Constr. cause DA is equal † to DB, and DG common to the two triangles ADG, BDG, the two sides AD, DG are equal to the two BD, DG, each to each;
† 15 Def. 1. and the base GA is equal † to the base GB, because they are

drawn from the centre G ‡: therefore the angle ADG is equal * to the angle GDB: but when a straight line, standing upon another straight line, makes the adjacent angles equal to one another, each of the angles is a * right angle; therefore the angle GDB is a right angle: but FDB is likewise a † right angle; wherefore the angle FDB is equal † to the angle GDB, the greater to the less; which is impossible: therefore G is not the centre of the circle ABC. In the same manner it can be shewn, that no other point but F is the centre; that is, F is the centre of the circle ABC. Which was to be found.

* 8. 1.
* 10 Def. 1.
† Constr.
† 1 Ax.

Cor. From this it is manifest, that if in a circle a straight line bisect another at right angles, the centre of the circle is in the line which bisects the other.

PROPOSITION II.

Theorem.—*If any two points be taken in the circumference of a circle, the straight line which joins them shall fall within the circle.*

Let ABC be a circle, and A, B any two points in the circumference: the straight line drawn from A to B shall fall within the circle.

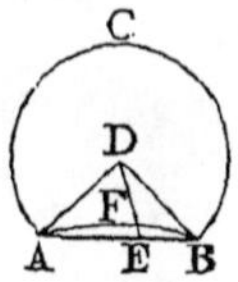

For if it do not, let it fall, if possible, without, as AEB: find * D the centre of the circle ABC, and join DA, DB; in the circumference AB take any point F, join DF, and produce it to E. Then, because DA is equal † to DB, the angle DAB is equal * to the angle DBA: and because AE, a side of the triangle DAE, is produced to B, the exterior angle DEB is greater * than the angle DAE: but DAE was proved equal to the angle DBE; therefore the angle DEB is greater than the angle DBE: but to the greater angle the greater side is * opposite, therefore DB is greater than DE: but DB is equal † to DF; wherefore DF is greater than DE, the less than the greater; which is impossible: therefore the straight line drawn from A to B does not fall without

* 1. 3.
† 15 Def. 1.
* 5. 1.
* 16. 1.
* 19. 1.
† 15 Def. 1.

‡ N. B.—Whenever the expression "straight lines from the centre" or "drawn from the centre" occurs, it is to be understood that they are drawn to the circumference.

the circle. In the same manner it may be demonstrated, that it does not fall upon the circumference; therefore it falls within it. Wherefore, if any two points, &c. Q. E. D.

PROPOSITION III.

THEOR.—*If a straight line drawn through the centre of a circle, bisect a straight line in it which does not pass through the centre, it shall cut it at right angles: and if it cut it at right angles, it shall bisect it.*

Let ABC be a circle; and let CD, a straight line drawn through the centre, bisect any straight line AB, which does not pass through the centre, in the point F: it shall cut AB at right angles.

* 1. 3. Take * E the centre of the circle, and join EA, EB. Then,
† Hyp. because AF is equal † to FB, and FE common to the two triangles AFE, BFE, there are two sides in the one, equal to two sides in the other, each
* 15 Def. 1. to each; and the base EA is equal † to the
* 8. 1. base EB; therefore the angle AFE is equal * to the angle BFE: but when a straight line, standing upon another straight line, makes the adjacent angles equal to one another, each of
* 10 Def. 1. them is a right * angle; therefore each of the angles AFE, BFE is a right angle: wherefore the straight line CD drawn through the centre, bisecting another AB that does not pass through the centre, cuts the same at right angles.

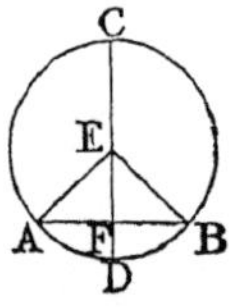

But let CD cut AB at right angles: CD shall also bisect it, that is, AF shall be equal to FB.

The same construction being made, because EA, EB, from
† 15 Def. 1. the centre, are equal † to one another, the angle EAF is
* 5. 1. equal * to the angle EBF; and the right angle AFE is equal †
† 10 Def. 1. to the right angle BFE; therefore, in the two triangles EAF, EBF, there are two angles in the one, equal to two angles in the other, each to each; and the side EF, which is opposite to one of the equal angles in each, is common to both; there-
* 26. 1. fore the other sides are * equal; therefore AF is equal to FB. Wherefore, if a straight line, &c. Q. E. D.

PROPOSITION IV.

THEOR.—*If in a circle two straight lines cut one another, which do not both pass through the centre, they do not bisect each other.*

Let ABCD be a circle, and AC, BD two straight lines in it, which cut one another in the point E, and do not both pass through the centre: AC, BD shall not bisect one another.

For, if it be possible, let AE be equal to EC, and BE to ED. If one of the lines pass through the centre, it is plain that it cannot be bisected by the other which does not pass through the centre: but if neither of them pass through the centre, take * F the centre of the circle, and join EF: and because FE, a straight line through the centre, bisects another † AC which does not pass through the centre, it cuts it at right * angles; wherefore FEA is a right angle: again, because the straight line FE bisects † the straight line BD, which does not pass through the centre, it cuts it at right * angles; wherefore FEB is a right angle: but FEA was shewn to be a right angle: therefore FEA is equal † to the angle FEB, the less to the greater; which is impossible: therefore AC, BD do not bisect one another. Wherefore, if in a circle, &c. Q. E. D.

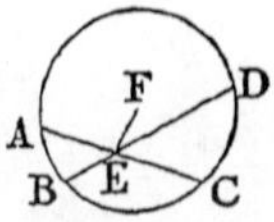

* 1. 3.
† Hyp.
* 3. 3.
† Hyp.
* 3. 3.
† 1 Ax.

PROPOSITION V.

THEOR.—*If two circles cut one another, they shall not have the same centre.*

Let the two circles ABC, CDG cut one another in the points B, C: they shall not have the same centre.

For, if it be possible, let E be their centre; join EC, and draw any straight line EFG meeting them in F and G: and because E is the centre of the circle ABC, EC is equal † to EF: again, because E is the centre of the circle CDG, EC is equal † to EG: but EC was shewn to be equal to EF; therefore EF is equal † to EG, the less to the greater; which is impossible: therefore E is not the centre of the circles ABC, CDG. Wherefore, if two circles, &c. Q. E. D.

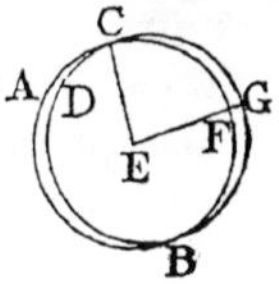

† 15 Def. 1.
† 15 Def. 1.
† 1 Ax.

PROPOSITION VI.

THEOR.—*If two circles touch one another internally, they shall not have the same centre.*

Let the two circles ABC, CDE touch one another internally in the point C: they shall not have the same centre.

For if they have, let it be F: join FC, and draw any straight line FEB, meeting them in E and B: and because F is the

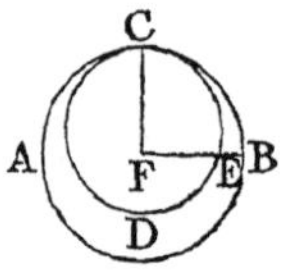

† 15 Def. 1. centre of the circle ABC, FC is equal † to FB: also, because F is the centre of the
† 15 Def. 1. circle CDE, FC is equal † to FE: but FC
† 1 Ax. was shewn to be equal to FB; therefore FE is equal † to FB, the less to the greater; which is impossible: therefore F is not the centre of the circles ABC, CDE. Therefore, if two circles, &c. Q. E. D.

PROPOSITION VII.

THEOR.—*If any point be taken in the diameter of a circle, which is not the centre, of all the straight lines which can be drawn from it to the circumference, the greatest is that in which the centre is, and the other part of that diameter is the least; and, of any others, that which is nearer to the line which passes through the centre, is always greater than one more remote: and, from the same point there can be drawn only two straight lines that are equal to one another, one upon each side of the shortest line.*

Let ABCD be a circle, and AD its diameter, in which let any point F be taken which is not the centre; let the centre be E: of all the straight lines FB, FC, FG, &c. that can be drawn from F to the circumference, FA shall be the greatest, and FD, the other part of the diameter AD, shall be the least: and of the others, FB shall be greater than FC, and FC greater than FG.

Join BE, CE, GE: and because two sides of
* 20. 1. a triangle are greater * than the third, therefore BE, EF are
† 15 Def. 1. greater than BF: but AE is equal † to BE; therefore AE,

EF, that is AF, is greater than BF: again, because BE is equal to CE, and FE common to the triangles BEF, CEF, the two sides BE, EF are equal to the two CE, EF, each to each; but the angle BEF is greater † than the angle CEF; † 9 Ax. 1.
therefore the base BF is greater * than the base CF: for the * 24. 1.
same reason, CF is greater than GF: again, because GF, FE are greater * than EG, and EG is equal to ED, therefore GF, * 20. 1.
FE are greater than ED: take away the common part FE, and the remainder GF is greater † than the remainder FD: † 5 Ax.
therefore FA is the greatest, and FD the least, of all the straight lines from F to the circumference; and BF is greater than CF, and CF than GF.

Also there can be drawn only two equal straight lines from the point F to the circumference, one upon each side of the shortest line FD. At the point E, in the straight line EF, make * the angle FEH equal to the angle FEG, and join FH: * 23. 1.
then, because GE is equal † to EH, and EF common to the † 15 Def. 1.
two triangles GEF, HEF, the two sides GE, EF are equal to the two HE, EF, each to each; and the angle GEF is equal † † Constr.
to the angle HEF; therefore the base FG is equal * to the * 4. 1.
base FH: but, besides FH, no other straight line can be drawn from F to the circumference, equal to FG: for if there can, let it be FK: and because FK is equal to FG, and FG to FH, FK is equal † to FH; that is, a line nearer to that which † 1 Ax.
passes through the centre, is equal to one which is more remote; which has been proved to be impossible. Therefore, if any point be taken, &c. Q. E. D.

PROPOSITION VIII.

THEOR.—*If any point be taken without a circle, and straight lines be drawn from it to the circumference, whereof one passes through the centre; of those which fall upon the concave circumference, the greatest is that which passes through the centre, and of the rest, that which is nearer to the one passing through the centre is always greater than one more remote; but of those which fall upon the convex circumference, the least is that between the point without the circle and the diameter; and of the rest, that which is nearer to the least is always less than one more remote: and only two equal straight lines can be drawn from the same point to the circumference, one upon each side of the least line.*

Let ABC be a circle, and D any point without it, from which let the straight lines DA, DE, DF, DC be drawn to the circumference, whereof DA passes through the centre. Of those which fall upon the concave part of the circumference AEFC, the greatest shall be DA, which passes through the centre; and the nearer to it shall be greater than the more remote, viz. DE greater than DF, and DF greater than DC: but of those which fall upon the convex circumference HLKG, the least shall be DG, between the point D and the diameter AG; and the nearer to it shall be less than the more remote, viz. DK less than DL, and DL less than DH.

* 1. 3. Take * M the centre of the circle ABC, and join ME, MF,
MC, MK, ML, MH. And because AM is equal to ME, add
† 2 Ax. MD to each, therefore AD is equal † to EM, MD: but EM,
* 20. 1. MD are greater * than ED; therefore also AD is greater than
ED. Again, because ME is equal to MF, and MD common
to the triangles EMD, FMD; EM, MD are equal to FM,
MD, each to each: but the angle EMD is
† 9 Ax. greater † than the angle FMD; therefore the
* 24. 1. base ED is greater * than the base FD. In
like manner it may be shewn, that FD is
greater than CD: therefore DA is the greatest, and DE greater than DF, and DF greater
* 20. 1. than DC. And because MK, KD are greater *
† 15 Def. 1. than MD, and MK is equal † to MG, the re-
* 5 Ax. mainder KD is greater * than the remainder
GD; that is, GD is less than KD: and because MLD is a triangle, and from the points M, D, the extremities of its side
MD, the straight lines MK, DK are drawn to the point K
* 21. 1. within the triangle, therefore MK, KD are less * than ML,
† 15 Def. 1. LD: but MK is equal † to ML; therefore the remainder DK
† 5 Ax. is less † than the remainder DL. In like manner it may be
shewn, that DL is less than DH: therefore DG is the least, and DK less than DL, and DL less than DH. Also, there can be drawn only two equal straight lines from the point D to the circumference, one upon each side of the least line. At
† 23. 1. the point M, in the straight line MD, make † the angle DMB
equal to the angle DMK, and join DB: and because MK is equal to MB, and MD common to the triangles KMD, BMD,

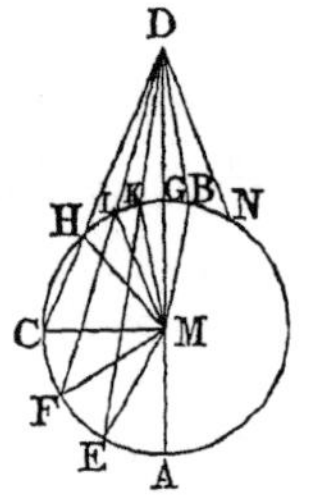

the two sides KM, MD are equal to the two BM, MD, each to each; and the angle KMD is equal † to the angle BMD; † Constr. therefore the base DK is equal * to the base DB: but, besides * 4. 1. DB, there can be no straight line drawn from D to the circumference equal to DK: for if there can, let it be DN: and because DK is equal to DN, and also to DB, therefore DB is equal to DN; that is, a line nearer to the least, is equal to one more remote; which has been proved to be impossible. If, therefore, any point, &c. Q. E. D.

PROPOSITION IX.

THEOR.—*If a point be taken within a circle, from which there fall more than two equal straight lines to the circumference, that point is the centre of the circle.*

Let the point D be taken within the circle ABC, from which to the circumference there fall more than two equal straight lines, viz. DA, DB, DC: the point D shall be the centre of the circle.

For, if not, let E be the centre: join DE, and produce it to the circumference in F, G; then FG is a diameter † of the † 17 Def. 1. circle ABC: and because in FG, the diameter of the circle ABC, there is taken the point D, which is not the centre, DG is the greatest line from it to the circumference, and DC is greater * than DB, and DB greater than DA: * 7. 3. but they are likewise † equal; which is im- † Hyp. possible: therefore E is not the centre of the circle ABC. In like manner it may be demonstrated, that no other point but D is the centre: D therefore is the centre. Wherefore, if a point be taken, &c. Q. E. D.

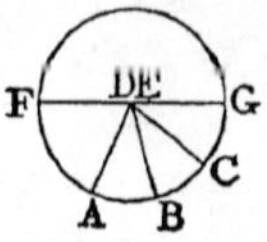

PROPOSITION X.

THEOR.—*One circumference of a circle cannot cut another in more than two points.*

If it be possible, let the circumference FAB cut the circumference DEF in more than two points, viz. in B, G, F: take † the centre K of the circle ABC, and join KB, KG, KF. † 3. 3. Then because K is the centre of the circle ABC, therefore KB,

† 15 Def. 1. KG, KF are all equal † to each other: and because within the circle DEF, there is taken the point K, from which to the circumference DEF fall more than two equal straight lines KB, KG, KF, therefore the point
* 9. 3. K is * the centre of the circle DEF: but K is
† Constr. also the centre † of the circle ABC; therefore the same point is the centre of two circles that cut one another; which is impos-
* 5. 3. sible *. Therefore, one circumference of a circle, &c. Q. E. D.

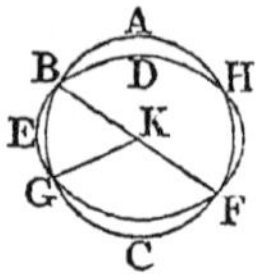

PROPOSITION XI.

THEOR.—*If two circles touch each other internally, the straight line which joins their centres, being produced, shall pass through the point of contact.*

Let the two circles ABC, ADE touch each other internally in the point A; and let F be the centre of the circle ABC, and G the centre of the circle ADE: the straight line which joins the centres F, G, being produced, shall pass through the point A.

For, if not, let it fall otherwise, if possible, as FGDH, and join AF, AG. Then, because
† 20. 1. two sides of a triangle are together greater † than the third side, therefore FG, GA are
† 15 Def. 1. greater than FA: but FA is equal † to FH; therefore FG, GA are greater than FH: take away the common part FG; therefore
† 5 Ax. the remainder AG is greater † than the remainder GH: but
† 15 Def. 1. AG is equal † to GD; therefore GD is greater than GH, the less than the greater; which is impossible: therefore the straight line which joins the points F, G, being produced, cannot fall otherwise than upon the point A, that is, it must pass through it. Therefore, if two circles, &c. Q. E. D.

PROPOSITION XII.

THEOR.—*If two circles touch each other externally, the straight line which joins their centres, shall pass through the point of contact.*

Let the two circles ABC, ADE touch each other externally

in the point A; and let F be the centre of the circle ABC, and G the centre of ADE: the straight line which joins the points F, G, shall pass through the point of contact A.

For, if not, let it pass otherwise, if possible, as FCDG, and join FA, AG. And because F is the centre of the circle ABC, FA is equal to FC: also, because G is the centre of the circle ADE, GA is equal to GD: therefore FA, AG are equal† to FC, DG; wherefore the whole FG is greater than FA, AG: but it is also* less; which is impossible: therefore the straight line which joins the points F, G, cannot pass otherwise than through the point of contact A, that is, it must pass through it. Therefore, if two circles, &c. Q. E. D.

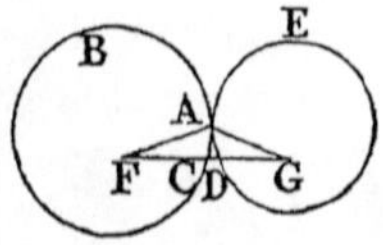

† 2 Ax.

* 20. 1.

PROPOSITION XIII.

THEOR.—*One circle cannot touch another in more points than one, whether it touches it on the inside or outside.* See N.

For, if it be possible, let the circle EBF touch the circle ABC in more points than one; and first on the inside, in the points B, D: join BD, and draw* GH bisecting BD at right angles: therefore, because the points B, D are in the circumferences of each of the circles, the straight line BD falls within each* of them; therefore their centres are* in the straight line GH which bisects BD at right angles; therefore GH passes* through the point of contact: but it does not pass through it, because the points B, D are without the straight line GH; which is absurd: therefore one circle cannot touch another on the inside in more points than one.

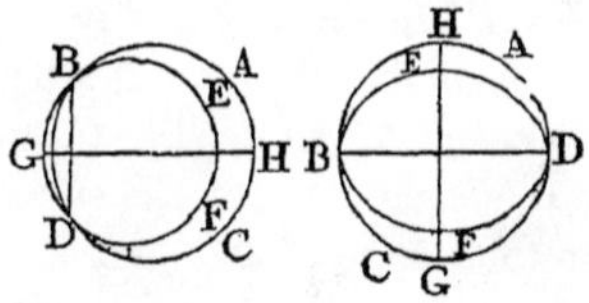

* 10. 11. 1.

* 2. 3.

* Cor. 1. 3.

* 11. 3.

Nor can two circles touch one another on the outside in more than one point. For if it be possible, let the circle ACK touch the circle ABC in the points A, C; join AC: therefore because the two points A, C are in the circumference of the circle ACK, the straight line AC which joins them,

* 2. 3. falls within * the circle ACK: but the circle ACK is without † the circle ABC; therefore the straight
† Hyp.
line AC is without this last circle: but because the points A, C are in the circumference of the circle ABC, the straight line AC must be
* 2. 3. within * the same circle; which is absurd: therefore one circle cannot touch another on the outside in more than one point: and it has been shewn, that they cannot touch on the inside in more points than one. Therefore, one circle, &c. Q. E. D.

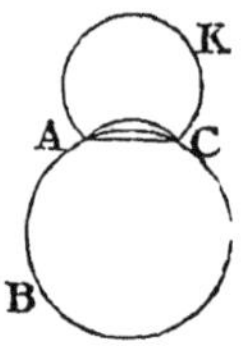

PROPOSITION XIV.

THEOR.—*Equal straight lines in a circle are equally distant from the centre; and those which are equally distant from the centre are equal to one another.*

Let the straight lines AB, CD, in the circle ABDC, be equal to one another: they shall be equally distant from the centre.

† 1. 3. Take † E, the centre of the circle ABDC, and from it †,
† 12. 1. draw EF, EG, perpendiculars to AB, CD, and join EA, EC. Then, because the straight line EF, passing through the centre, cuts the straight line AB, which does not pass through
* 3. 3. the centre, at right angles, it also bisects * it; therefore AF is equal to FB, and AB double of AF; for the same reason,
† Hyp. CD is double of CG: but AB is equal † to CD;
† 7 Ax. therefore AF is equal † to CG. And because
† 15 Def. 1. AE is equal † to EC, the square of AE is equal to the square of EC: but the squares of AF,
* 47. 1. FE are equal * to the square of AE, because the angle AFE is a right angle; and for the like reason, the squ res of EG, GC are equal to the square of
† 1 Ax. EC; therefore the squares of AF, FE are equal † to the squares of CG, GE: but the square of AF is equal to the square of CG, because AF is equal to CG; therefore the re-
† 3 Ax. maining square of EF is equal † to the remaining square of EG, and the straight line EF is therefore equal to EG: but straight lines in a circle are said to be equally distant from the centre, when the perpendiculars drawn to them from the
* 4 Def. 3. centre are * equal; therefore AB, CD are equally distant from the centre.

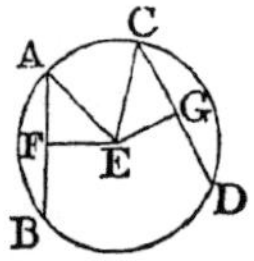

Next, let the straight lines AB, CD be equally distant from the centre, that is† let FE be equal to EG: AB shall be equal to CD. For, the same construction being made, it may, as before, be demonstrated, that AB is double of AF, and CD double of CG, and that the squares of EF, FA are equal to the squares of EG, GC: but the square of FE is equal to the square of EG, because FE is equal† to EG; therefore the remaining square of AF is equal† to the remaining square of CG; and the straight line AF is therefore equal to CG: but AB was shewn to be double of AF, and CD double of CG; wherefore AB is equal† to CD. Therefore, equal straight lines, &c. Q. E. D.

† 4 Def. 3.
† Hyp.
† 3 Ax.
† 6 Ax.

PROPOSITION XV.

THEOR.—*The diameter is the greatest straight line in a circle; and, of all others, that which is nearer to the centre is always greater than one more remote; and the greater is nearer to the centre than the less.* See N.

Let ABCD be a circle, of which the diameter is AD, and the centre E; and let BC be nearer to the centre than FG: AD shall be greater than any straight line BC, which is not a diameter, and BC shall be greater than FG.

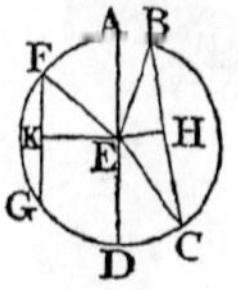

From the centre, draw† EH, EK perpendiculars to BC, FG, and join EB, EC, EF: and because AE is equal† to EB, and ED to EC, therefore AD is equal† to EB, EC: but EB, EC are greater* than BC; wherefore also, AD is greater than BC.

† 12. 1.
† 15 Def. 1.
† 2 Ax.
* 20. 1.

And, because BC is nearer† to the centre than FG, EH is less* than EK: but, as was demonstrated in the preceding, BC is double of BH, and FG double of FK, and the squares of EH, HB are equal to the squares EK, KF: but the square of EH is less than the square of EK, because EH is less than EK; therefore the square of BH is greater than the square of FK, and the straight line BH greater than FK, and therefore BC is greater than FG.

† Hyp.
* 5 Def. 3.

Next, let BC be greater than FG: BC shall be nearer to the centre than FG, that is†, the same construction being made, EH shall be less than EK. Because BC is greater than FG, BH likewise is greater than KF: and the squares of BH,

† 5 Def. 3.

HE are equal to the squares of FK, KE; of which the square of BH is greater than the square of FK, because BH is greater than FK: therefore the square of EH is less than the square of EK, and the straight line EH less than EK; and therefore

† 5 Def. 3. BC is nearer † to the centre than FG. Wherefore, the diameter, &c. Q. E. D.

PROPOSITION XVI.

See N. THEOR.—*The straight line drawn at right angles to the diameter of a circle, from the extremity of it, falls without the circle; and no straight line can be drawn from the extremity, between that straight line and the circumference, so as not to cut the circle; or, which is the same thing, no straight line can make so great an acute angle with the diameter at its extremity, or so small an angle with the straight line which is at right angles to it, as not to cut the circle.*

Let ABC be a circle, the centre of which is D, and the diameter AB: the straight line drawn at right angles to AB from its extremity A, shall fall without the circle.

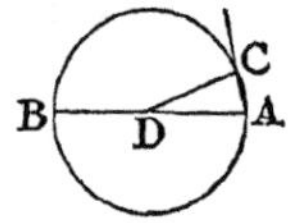

For if it does not, let it fall, if possible, within the circle, as AC; and draw DC to the point C, where it meets the circumference.
† 15 Def. 1. ence. And because DA is equal † to DC,
* 5. 1. the angle DAC is equal * to the angle ACD:
† Hyp. but DAC is a right † angle; therefore ACD is a right angle; and therefore the angles DAC, ACD are
* 17. 1. equal to two right angles; which is * impossible: therefore the straight line drawn from A, at right angles to BA, does not fall within the circle. In the same manner it may be demonstrated, that it does not fall upon the circumference;
* See fig. 2. therefore it must fall without the circle, as AE *.

Also between the straight line AE and the circumference, no straight line can be drawn from the point A, which does not cut the circle. For, if possible, let AF be between them:
* 12. 1. from the point D draw * DG perpendicular to AF, and let it meet the circumference in H. And because AGD is a right
* 17. 1. angle, and DAG less * than a right angle, DA is greater * than
* 19. 1. † 15 Def. 1. DG: but DA is equal † to DH; therefore DH is greater than DG, the less than the greater; which is impossible: there-

fore no straight line can be drawn from the point A, between AE and the circumference, which does not cut the circle; or, which amounts to the same thing, however great an acute angle a straight line makes with the diameter at the point A, or however small an angle it makes with AE, the circumference must pass between that straight line and the perpendicular AE.

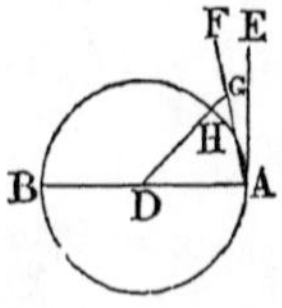

"And this is all that is to be understood, when, in the Greek text, and translations from it, the angle of the semicircle is said to be greater than any acute rectilineal angle, and the remaining angle less than any rectilineal angle." Q. E. D.

COR. From this it is manifest, that the straight line which is drawn at right angles to the diameter of a circle from the extremity of it, touches † the circle; and that it touches it only in one point, because if it did meet the circle in two, it would fall* within it. "Also it is evident, that there can be but one straight line which touches the circle in the same point."

† 2 Def. 3.
* 2. 3.

PROPOSITION XVII.

PROB.—*To draw a straight line from a given point, either without or in the circumference, which shall touch a given circle.*

First, let A be a given point without the given circle BCD; it is required to draw a straight line from A, which shall touch the circle.

Find * the centre E of the circle, and join AE; and from the centre E, at the distance EA, describe the circle AFG; from the point D, draw* DF at right angles to EA, and join EBF, AB: AB shall touch the circle BCD.

* 1. 3.
* 11. 1.

Because E is the centre of the circles BCD, AFG; EA is equal † to EF, and ED to EB; therefore the two sides AE, EB are equal to the two FE, ED, each to each; and they contain the angle at E common to the two triangles AEB, FED; therefore the base DF is equal* to the base AB, and the triangle FED to the triangle AEB, and the other angles to the other angles; therefore the angle EBA is equal to the angle EDF: but EDF is a right † angle, wherefore EBA is a right † angle;

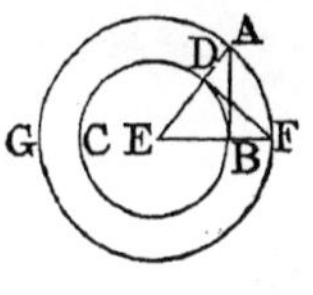

† 15 Def. 1.
* 4. 1.
† Constr.
† 1 Ax.

and EB is drawn from the centre: but a straight line drawn from the extremity of a diameter, at right angles to it,
* Cor. 16. 3. touches * the circle; therefore, AB touches the circle; and it is drawn from the given point A. Which was to be done.

But if the given point be in the circumference of the circle, as the point D, draw DE to the centre E, and DF at
* Cor. 16. 3. right angles to DE: DF touches* the circle.

PROPOSITION XVIII.

THEOR.—*If a straight line touches a circle, the straight line drawn from the centre to the point of contact, shall be perpendicular to the line touching the circle.*

Let the straight line DE touch the circle ABC in the point
† 1. 3. C; take† the centre F, and draw the straight line FC: FC shall be perpendicular to DE.

For if it be not, from the point F, draw FBG perpendi-
* 17. 1. cular to DE: and because FGC is a right angle, GCF is* an
* 19. 1. acute angle; and to the greater angle the greater* side is opposite; therefore FC is greater than FG:
† 15 Def. 1. but FC is equal † to FB; therefore FB is greater than FG, the less than the greater; which is impossible; therefore FG is not perpendicular to DE. In the same manner it may be shewn, that no other is perpendicular to it besides FC; that is, FC is perpendicular to DE. Therefore, if a straight line, &c. Q. E. D.

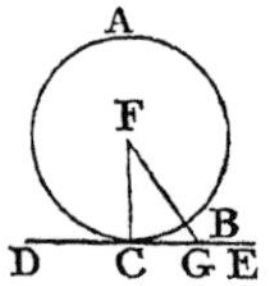

PROPOSITION XIX.

THEOR.—*If a straight line touch a circle, and from the point of contact a straight line be drawn at right angles to the touching line, the centre of the circle shall be in that line.*

Let the straight line DE touch the circle ABC in C; and from C, let CA be drawn at right angles to DE: the centre of the circle shall be in CA.

For, if not, let F be the centre, if possible, and join CF. Because DE touches the circle ABC, and FC is drawn from the centre to
* 18. 3. the point of contact, FC is perpendicular* to DE; therefore FCE is a right angle: but
† Hyp. ACE is also a right † angle; therefore the angle FCE is

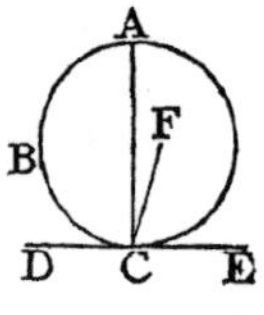

equal † to the angle ACE, the less to the greater; which † 1 Ax.
is impossible; therefore F is not the centre of the circle ABC. In the same manner it may be shewn, that no other point which is not in CA, is the centre; that is, the centre is in CA. Therefore, if a straight line, &c. Q. E. D.

PROPOSITION XX.

THEOR.—*The angle at the centre of a circle is double of the angle at the circumference upon the same base, that is, upon the same part of the circumference.* See N.

Let ABC be a circle, and BEC an angle at the centre, and BAC an angle at the circumference, which have the same circumference BC for their base: the angle BEC shall be double of the angle BAC.

Join AE, and produce it to F. First, let the centre of the circle be within the angle BAC. Because EA is equal to EB,
the angle EAB is equal * to the angle EBA; * 5. 1.
therefore the angles EAB, EBA are double of
the angle EAB: but the angle BEF is equal * to * 32. 1.
the angles EAB, EBA; therefore also the angle BEF is double of the angle EAB: for the same reason, the angle FEC is double of the angle EAC; therefore the whole angle BEC is double of the whole angle BAC.

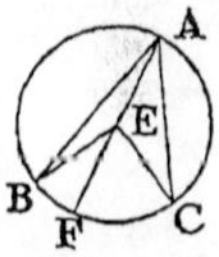

Again, let the centre of the circle be without the angle BAC: it may be demonstrated, as in the first case, that the angle FEC is double of the angle FAC, and that FEB, a part of the first, is double of FAB, a part of the other; therefore the remaining angle BEC is double of the remaining angle BAC. Therefore, the angle at the centre, &c. Q. E. D.

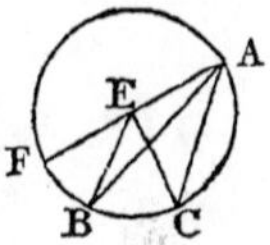

PROPOSITION XXI.

THEOR.—*The angles in the same segment of a circle are equal to one another.* See N.

Let ABCD be a circle, and BAD, BED angles in the same segment BAED: the angles BAD, BED shall be equal to one another.

First, let the segment BAED be greater than a semicircle.
† 1. 3. Take† F, the centre of the circle ABCD, and join BF, FD: and because the angle BFD is at the centre, and the angle BAD at the circumference, and that they have the same part of the circumference, viz. BCD for their base, therefore the
* 20. 3. angle BFD is double * of the angle BAD: for the same reason, the angle BFD is double of the angle BED; therefore the angle BAD is
† 7 Ax. equal † to the angle BED.

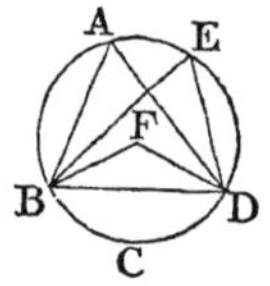

Next, let the segment BAED be not greater than a semicircle. Draw AF to the centre, and produce it to C, and join CE: therefore the segment BADC is greater than a semicircle; and the angles in it, BAC, BEC are equal, by the first case: for the same reason, because CBED is greater than a semicircle, the angles CAD, CED are equal;
† 2 Ax. therefore the whole angle BAD is equal † to the whole angle BED. Wherefore, the angles in the same segment, &c. *Q. E. D.*

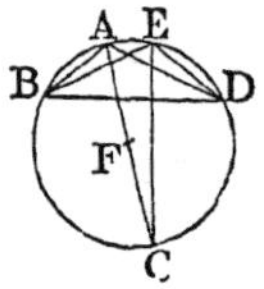

PROPOSITION XXII.

THEOR.—*The opposite angles of any quadrilateral figure inscribed in a circle, are together equal to two right angles.*

Let ABCD be a quadrilateral figure in the circle ABCD: any two of its opposite angles shall together be equal to two right angles.

Join AC, BD: and because the three angles of every tri-
* 32. 1. angle are equal* to two right angles, the three angles of the triangle CAB, viz. the angles CAB, ABC, BCA, are equal to
* 21. 3. two right angles: but the angle CAB is equal * to the angle CDB, because they are in the same segment CDAB; and the angle ACB is equal to the angle ADB, because they are in the same segment ADCB; therefore the two angles CAB,
† 2 Ax. ACB are together equal† to the whole angle ADC: to each of these equals add the angle ABC; therefore
† 2 Ax. the three angles ABC, CAB, BCA are equal† to the two angles ABC, ADC: but ABC, CAB, BCA are equal to two right

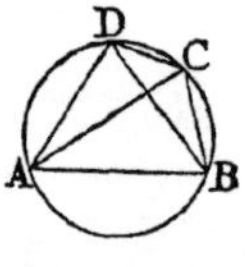

angles; therefore also the angles ABC, ADC are equal† to two right angles. In the same manner the angles BAD, DCB may be shewn to be equal to two right angles. Therefore, the opposite angles, &c. Q. E. D. † 1 Ax.

PROPOSITION XXIII.

THEOR.—*Upon the same straight line, and upon the same side of it, there cannot be two similar segments of circles, not coinciding with one another.* See N.

If it be possible, upon the same straight line AB, and upon the same side of it, let there be two similar segments of circles, ACB, ADB, not coinciding with one another. Then, because the circle ACB cuts the circle ADB in the two points A, B, they cannot cut one another in any other * point: * 10. 3. therefore one of the segments must fall within

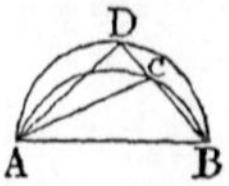

the other: let ACB fall within ADB; draw the straight line BCD, and join CA, DA. And because the segment ACB is similar† to the segment ADB, and that similar segments of † Hyp. circles contain * equal angles, therefore the angle ACB is * 11 Def. 3. equal to the angle ADB ‡, the exterior to the interior; which is impossible *. * 16. 1. Therefore, there cannot be two similar segments of circles upon the same side of the same line, which do not coincide. Q. E. D.

PROPOSITION XXIV.

THEOR.—*Similar segments of circles upon equal straight lines, are equal to one another.* See N.

Let AEB, CFD be similar segments of circles upon the equal straight lines AB, CD: the segment AEB shall be equal to the segment CFD.

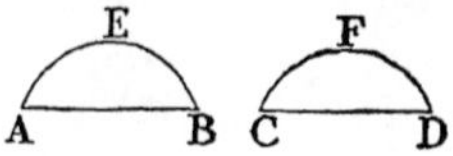

For if the segment AEB be applied to the segment CFD, so that the point A may be on C, and the straight line AB upon CD, the point B shall coincide with the point D, because AB is equal

‡ That is, ACB, the exterior angle of the triangle ACD, is proved to be equal to the interior and opposite angle ADC*; which is impossible. * 16. 1.

to CD: therefore the straight line AB coinciding with CD, the
* 23. 3. segment AEB must * coincide with the segment CFD, and
† 8 Ax. therefore is equal † to it. Wherefore, similar segments, &c.
Q. E. D.

PROPOSITION XXV.

See N. PROB.—*A segment of a circle being given, to describe the circle of which it is the segment.*

Let ABC be the given segment of a circle; it is required to describe the circle of which it is the segment.

* 10. 1. Bisect * AC in D, and from the point D, draw * DB at right
* 11. 1. angles to AC, and join AB. First, let the angles ABD, BAD
† See Fig. 1. be equal † to one another; then the straight line BD is equal *
* 6. 1.

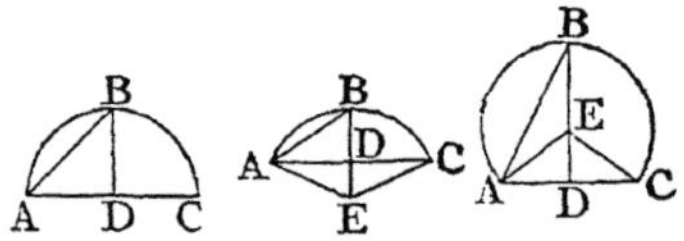

to DA, and therefore ‡ to DC: and because the three straight
* 9. 3. lines DA, DB, DC are all equal, D is the centre of the circle *. From the centre D, at the distance of any of the three DA, DB, DC, describe a circle; this shall pass through the other points, and the circle of which ABC is a segment is described: and because the centre D is in AC, the segment ABC is a semi-
† See Fig. 2. and 3. circle. But if the angles ABD, BAD are not equal † to one
* 23. 1. another, at the point A, in the straight line AB, make * the
† See Fig. 2. angle BAE equal to the angle ABD, and produce † BD, if necessary, to E, and join EC. And because the angle ABE is
* 6. 1. equal to the angle BAE, the straight line BE is equal * to EA: and because AD is equal to DC, and DE common to the triangles ADE, CDE, the two sides AD, DE are equal to the two CD, DE, each to each; and the angle ADE is equal to the
† Constr. angle CDE, for each of them is a right † angle; therefore the
* 4. 1. base AE is equal * to the base EC: but AE was shewn to be
† 1 Ax. equal to EB; wherefore also BE is equal † to EC; and therefore the three straight lines AE, EB, EC are equal to one
* 9. 3. another: wherefore * E is the centre of the circle. From the

‡ For since AC was bisected in D, therefore AD is equal to DC: but
† 1 Ax. BD is proved to be equal to AD, therefore BD is equal † to DC.

centre E, at the distance of any of the three AE, EB, EC, describe a circle; this shall pass through the other points, and the circle, of which ABC is a segment, is described. And it is evident, that if the angle ABD be greater † than the angle † See Fig. 2. BAD, the centre E falls without the segment ABC, which therefore is less than a semicircle: but if the angle ABD be less † than BAD, the centre E falls within the segment ABC, † See Fig. 3. which is therefore greater than a semicircle. Wherefore, a segment of a circle being given, the circle is described of which it is a segment. Which was to be done.

PROPOSITION XXVI.

THEOR.—*In equal circles, equal angles stand upon equal circumferences, whether they be at the centres or circumferences.*

Let ABC, DEF be equal circles, and let the angles BGC, EHF at their centres, and BAC, EDF at their circumferences, be equal to each other: the circumference BKC shall be equal to the circumference ELF.

Join BC, EF: and because the circles ABC, DEF are equal, the straight lines drawn from their centres † are equal: there- † 1 Def. 3. fore the two sides BG, GC are equal to the two EH, HF, each to each: and the angle at G is equal † to the angle at H; † Hyp. therefore the base BC is equal * to the base EF. And because * 4. 1. the angle at A is equal † to the angle at D, the segment BAC † Hyp. is similar * to the segment EDF; and they are upon equal * 11 Def. 3. straight lines BC, EF: but similar segments of circles upon equal straight lines, are equal * to one * 24. 3. another; therefore the segment BAC is equal to the segment EDF: but the whole circle ABC is equal † to the whole DEF; † Hyp.

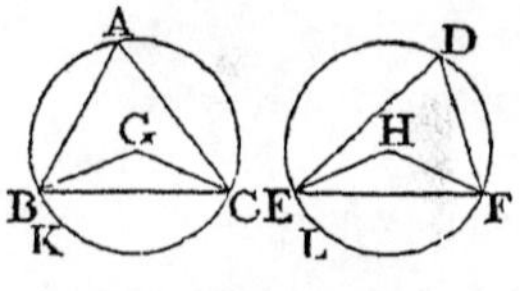

therefore the remaining segment BKC is equal † to the remain- † 3 Ax. ing segment ELF, and the circumference BKC to the circumference ELF. Wherefore, in equal circles, &c. Q. E. D.

PROPOSITION XXVII.

THEOR.—*In equal circles, the angles which stand upon equal circumferences are equal to one another, whether they be at the centres or circumferences.*

Let ABC, DEF be equal circles, and let the angles BGC, EHF at their centres, and BAC, EDF at their circumferences, stand upon the equal circumferences BC, EF: the angle BGC shall be equal to the angle EHF, and the angle BAC to the angle EDF.

If the angle BGC be equal to the angle EHF, it is mani-
* 20. 3. & 7 Ax. 1. fest *, that the angle BAC is also equal to EDF. But, if not, one of them must be greater than the other: let BGC be the greater, and at the point G, in the straight
* 23. 1. line BG, make * the angle BGK equal to the angle EHF. Then, because the angle BGK is equal to the angle EHF, and that equal angles stand upon equal circum-

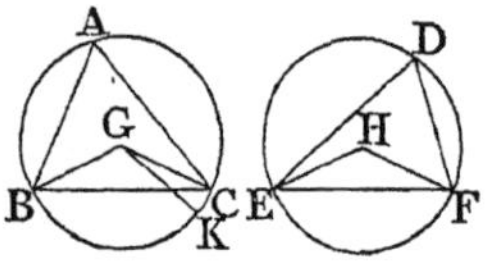

* 26. 3. ferences *, when they are at the centres, therefore the circumference BK is equal to the circumference EF: but EF is
† Hyp. † 1 Ax. equal † to BC; therefore also BK is equal † to BC, the less to the greater; which is impossible; therefore the angle BGC is not unequal to the angle EHF; that is, it is equal to it: and
† 20. 3. the angle at A is half † of the angle BGC, and the angle at D
† 7 Ax. half of the angle EHF; therefore the angle at A is equal † to the angle at D. Wherefore, in equal circles, &c. Q. E. D.

PROPOSITION XXVIII.

THEOR.—*In equal circles, equal straight lines cut off equal circumferences; the greater equal to the greater, and the less to the less.*

Let ABC, DEF be equal circles, and BC, EF equal straight lines in them, which cut off the two greater circumferences BAC, EDF, and the two less BGC, EHF: the greater BAC shall be equal to the greater EDF, and the less BGC to the less EHF.

* 1. 3. Take * K, L, the centres of the circles, and join BK, KC, EL, LF: and because the circles are equal, the straight lines from their centres
† 1 Def. 3. are † equal; therefore BK, KC are equal to EL, LF, each to each; and

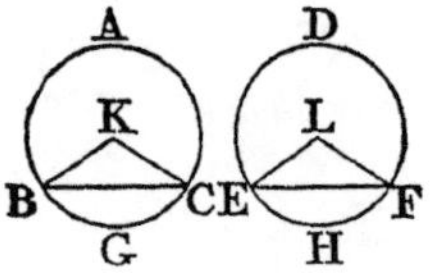

† Hyp. the base BC is equal † to the base EF; therefore the angle
* 8. 1. BKC is equal * to the angle ELF: but equal angles stand
* 26. 3. upon equal * circumferences, when they are at the centres;

therefore the circumference BGC is equal to the circumference EHF: but the whole circle ABC is equal † to the whole EDF; † Hyp.
therefore the remaining part of the circumference, viz. BAC, is equal † to the remaining part EDF. Therefore, in equal † 3 Ax.
circles, &c. Q. E. D.

PROPOSITION XXIX.

THEOR.—*In equal circles, equal circumferences are subtended by equal straight lines.*

Let ABC, DEF be equal circles, and let the circumferences BGC, EHF also be equal; and join BC, EF: the straight line BC shall be equal to the straight line EF.

Take * K, L, the centres of the circles, and join BK, KC, * 1. 3.
EL, LF: and because the circumference BGC is equal to the circumference EHF, the angle BKC is equal * * 27. 3.
to the angle ELF: and because the circles ABC, DEF are equal, the straight lines from their centres are † † 1 Def. 3.
equal; therefore BK, KC are equal to EL, LF, each to each; and they contain equal angles; therefore the base BC is equal * to the base EF. Therefore, in equal * 4. 1.
circles, &c. Q. E. D.

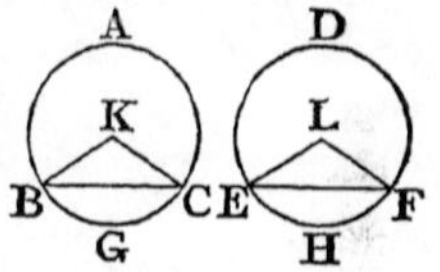

PROPOSITION XXX.

PROB.—*To bisect a given circumference,* that is, to divide it into two equal parts.

Let ADB be the given circumference; it is required to bisect it.

Join AB, and bisect * it in C; from the point C, draw CD * 10. 1.
at right angles † to AB: the circumference ADB shall be † 11. 1.
bisected in the point D.

Join AD, DB: and because AC is equal to CB, and CD common to the triangles ACD, BCD, the two sides AC, CD are equal to the two BC, CD, each to each; and the angle ACD is equal to the angle BCD, because each of them is a right angle; therefore the base AD is equal* to the base BD. But * 4. 1.
equal straight lines cut off equal * circumferences, the greater * 28. 3.

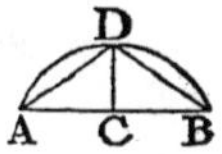

equal to the greater, and the less to the less; and AD, DB
* Cor. 1.3. are each of them less than a semicircle, because DC* passes through the centre, therefore the circumference AD is equal to the circumference DB. Therefore, the given circumference is bisected in D. Which was to be done.

PROPOSITION XXXI.

THEOR.—*In a circle, the angle in a semicircle is a right angle; but the angle in a segment greater than a semicircle, is less than a right angle; and the angle in a segment less than a semicircle, is greater than a right angle.*

Let ABCD be a circle, of which the diameter is BC, and the centre E; and draw CA, dividing the circle into the segments ABC, ADC, and join BA, AD, DC: the angle in the semicircle BAC, shall be a right angle; and the angle in the segment ABC, which is greater than a semicircle, shall be less than a right angle; and the angle in the segment ADC, which is less than a semicircle, shall be greater than a right angle.

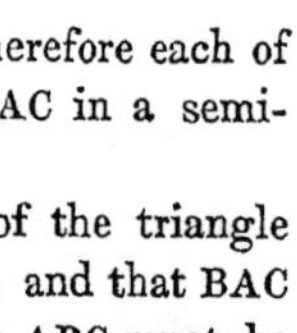

Join AE, and produce BA to F: and be-
† 15 Def. 1. cause BE is equal † to EA, the angle EAB is
* 5. 1. equal* to EBA; also, because AE is equal to EC, the angle EAC is equal to ECA;
† 2 Ax. wherefore the whole angle BAC is equal † to the two angles ABC, ACB: but FAC, the
* 32. 1. exterior angle of the triangle ABC, is equal* to the two angles ABC, ACB; therefore the
† 1 Ax. angle BAC is equal† to the angle FAC; and therefore each of
* 10 Def. 1. them is a right* angle: wherefore the angle BAC in a semicircle, is a right angle.

And because the two angles ABC, BAC of the triangle
* 17. 1. ABC, are together less* than two right angles, and that BAC has been proved to be a right angle, therefore ABC must be less than a right angle; and therefore the angle in a segment ABC greater than a semicircle, is less than a right angle.

And because ABCD is a quadrilateral figure in a circle, any
* 22. 3. two of its opposite angles are equal* to two right angles; therefore the angles ABC, ADC are equal to two right angles: and ABC has been proved to be less than a right angle; wherefore the other ADC is greater than a right angle.

Besides it is manifest, that the circumference of the greater segment ABC falls without the right angle CAB; but the circumference of the less segment ADC falls within the right angle CAF. "And this is all that is meant, when in the Greek text, and the translations from it, the angle of the greater segment is said to be greater, and the angle of the less segment is said to be less, than a right angle." Q. E. D.

COR. From this it is manifest, that if one angle of a triangle be equal to the other two, it is a right angle, because the angle adjacent to it is equal† to the same two; and when † 32. 1.
the adjacent angles are equal, they are† right angles. † 10 Def. 1.

PROPOSITION XXXII.

THEOR.—*If a straight line touch a circle, and from the point of contact a straight line be drawn cutting the circle, the angles which this line makes with the line touching the circle, shall be equal to the angles which are in the alternate segments of the circle.*

Let the straight line EF touch the circle ABCD in B, and from the point B let the straight line BD be drawn, cutting the circle: the angles which BD makes with the touching line EF, shall be equal to the angles in the alternate segments of the circle; that is, the angle DBF shall be equal to the angle which is in the segment DAB, and the angle DBE shall be equal to the angle in the segment DCB.

From the point B, draw* BA at right angles to EF, and * 11. 1.
take any point C, in the circumference DB, and join AD, DC, CB: and because the straight line EF touches the circle ABCD in the point B, and BA is drawn at right angles to the touching line from the point of contact B, the centre of the circle is* in BA; * 19. 3.
therefore the angle ADB in a semicircle is a right* angle; and consequently the other * 31. 3.
two angles BAD, ABD, are equal* to a right * 32. 1.
angle: but ABF is likewise a† right angle; † Constr.
therefore the angle ABF is equal† to the angles BAD, ABD: † 1 Ax.
take from these equals the common angle ABD; therefore the remaining angle DBF is equal† to the angle BAD, which is in † 3 Ax.
the alternate segment of the circle. And because ABCD is a quadrilateral figure in a circle, the opposite angles BAD, BCD are equal* to two right angles: but the angles DBF, DBE are * 22. 3.

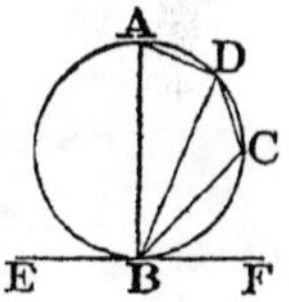

* 13. 1. likewise equal* to two right angles; therefore the angles
† 1 Ax. DBF, DBE are equal† to the angles BAD, BCD: and DBF has been proved equal to BAD; therefore the remaining angle
† 2 Ax. DBE is equal† to the angle BCD in the alternate segment of the circle. Wherefore, if a straight line, &c. Q. E. D.

PROPOSITION XXXIII.

See N. PROB.—*Upon a given straight line to describe a segment of a circle, which shall contain an angle equal to a given rectilineal angle.*

Let AB be the given straight line, and the angle at C the given rectilineal angle; it is required to describe upon the given straight line AB, a segment of a circle, which shall contain an angle equal to the angle C.

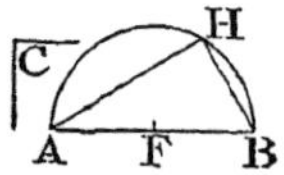

First, let the angle at C be a right angle;
* 10. 1. bisect* AB in F, and from the centre F, at the distance FB, describe the semicircle AHB:
* 31. 3. therefore the angle AHB in a semicircle, is* equal to the right angle at C.

But, if the angle C be not a right angle; at the point A,
* 23. 1. in the straight line AB, make* the angle BAD equal to the
* 11. 1. angle C, and from the point A, draw* AE at right angles to
* 10. 1. AD: bisect* AB in F, and from F, draw* FG at right angles
* 11. 1. to AB, and join GB. And because AF is equal to FB, and FG common to the triangles AFG, BFG, the two sides AF, FG are equal to the two BF, FG, each to
† 10 Def. 1. each; and the angle AFG is equal† to the angle BFG; therefore the base AG is
* 4. 1. equal* to the base GB; and therefore the circle described from the centre G, at the distance GA, shall pass through the point B: let this be the circle AHB: the segment AHB shall contain an angle equal to the given rectilineal angle C.

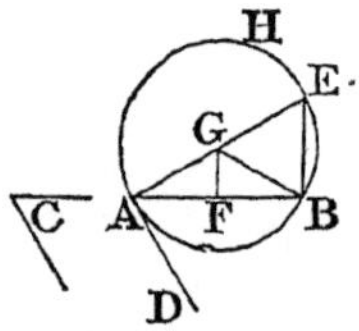

Because from the point A, the extremity of the diameter AE, AD is drawn at right
* Cor. 16. 3. angles to AE, therefore AD* touches the circle: and because AB, drawn from the point of contact A, cuts the circle, the
* 32. 3. angle DAB is equal to the angle in the alternate segment*
† Constr. AHB: but the angle DAB is equal† to the angle C; therefore

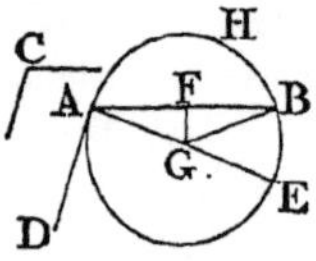

the angle C is equal† to the angle in the segment AHB. † 1 Ax.
Wherefore, upon the given straight line AB, the segment AHB of a circle is described, which contains an angle equal to the given angle at C. Which was to be done.

PROPOSITION XXXIV.

PROB.—*From a given circle to cut off a segment, which shall contain an angle equal to a given rectilineal angle.*

Let ABC be the given circle, and D the given rectilineal angle; it is required to cut off from the circle ABC, a segment that shall contain an angle equal to the given angle D.

Draw* the straight line EF touching the circle ABC in the point B; and at the point B, in the straight line BF, make* the angle FBC equal to the angle D: the segment BAC shall contain an angle equal to the given angle D. * 17. 3. * 23. 1.

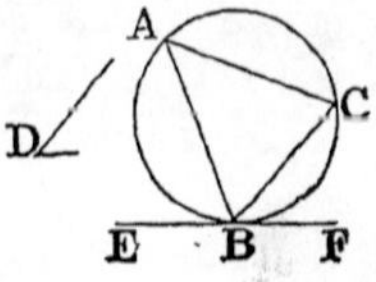

Because the straight line EF touches the circle ABC, and BC is drawn from the point of contact B, the angle FBC is equal* to the angle in the alternate segment BAC of the circle: but the angle FBC is equal † to the angle D; therefore the angle in the segment BAC is equal* to the angle D. Wherefore, from the given circle ABC, the segment BAC is cut off, containing an angle equal to the given angle D. Which was to be done. * 32. 3. † Constr. * 1 Ax.

PROPOSITION XXXV.

THEOR.—*If two straight lines cut one another within a circle, the rectangle contained by the segments of one of them, is equal to the rectangle contained by the segments of the other.* See N.

Let the two straight lines AC, BD cut one another in the point E, within the circle ABCD: the rectangle contained by AE, EC, shall be equal to the rectangle contained by BE, ED.

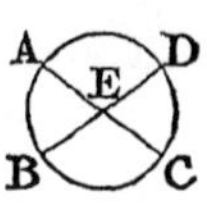

If AC, BD pass each of them through the centre, so that E is the centre, it is evident, that, AE, EC, BE, ED being all† equal, the rectangle AE, EC is likewise equal to the rectangle BE, ED. † 15 Def. 1.

But let one of them BD pass through the centre, and cut the other AC, which does not pass through the centre, at right angles, in the point E: then, if BD be bisected in F, F is the centre of the circle ABCD: join AF: and because BD which passes through the centre, cuts the straight line AC, which does not pass through the centre, at right
* 3. 3. angles in E, AE is equal* to EC: and because the straight line BD is cut into two equal parts in the point F, and into two unequal parts in the point E, the rectangle BE, ED, together
* 5. 2. with the square of EF, is equal* to the square of FB; that is, to the square of FA: but the
* 47. 1. squares of AE, EF are equal* to the square of FA; therefore the rectangle BE, ED, together with the square of EF, is
† 1 Ax. equal† to the squares of AE, EF: take away the common
† 3 Ax. square of EF, and the remaining rectangle BE, ED is equal† to the remaining square of AE; that is, to the‡ rectangle AE, EC.

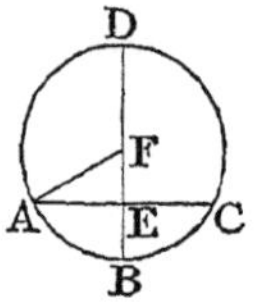

Next, let BD, which passes through the centre, cut the other AC, which does not pass through the centre, in E, but not at right angles: then, as before, if BD be bisected in F,
* 12. 1. F is the centre of the circle. Join AF, and from F draw*
* 3. 3. FG perpendicular to AC; therefore AG is equal* to GC; wherefore the rectangle AE, EC, together with the square
* 5. 2. of EG, is equal* to the square of AG: to each of these equals add the square of GF; therefore the rectangle AE, EC, together with the squares
† 2 Ax. of EG, GF, is equal† to the squares of AG,
* 47. 1. GF: but the squares of EG, GF are equal* to the square of EF; and the squares of AG, GF are equal to the square of AF; therefore the rectangle AE, EC, together with the square of EF, is equal to the square of AF; that is, to the square of FB: but the
* 5. 2. square of FB is equal* to the rectangle BE, ED, together with the square of EF; therefore the rectangle AE, EC,
† 1 Ax. together with the square of EF, is equal† to the rectangle BE, ED, together with the square of EF: take away the common square of EF, and the remaining rectangle AE, EC
† 3 Ax. is therefore equal† to the remaining rectangle BE, ED.

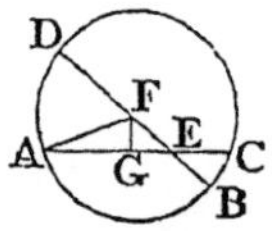

‡ Because AE has been proved to be equal to EC.

Lastly, let neither of the straight lines AC, BD pass through the centre: take† the centre F, and through E, the intersec- † 1. 3.
tion of the straight lines AC, DB, draw the diameter GEFH: and because the rectangle AE, EC is equal, as has been shewn, to the rectangle GE, EH, and, for the same reason, the rectangle BE, ED is equal to the same rectangle GE, EH, therefore the rectangle

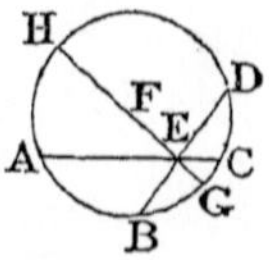

AE, EC is equal† to the rectangle BE, ED. Wherefore, if † 1 Ax.
two straight lines, &c. Q. E. D.

PROPOSITION XXXVI.

THEOR.—*If from any point without a circle two straight lines be drawn, one of which cuts the circle, and the other touches it; the rectangle contained by the whole line which cuts the circle, and the part of it without the circle, shall be equal to the square of the line which touches it.*

Let D be any point without the circle ABC, and let DCA, DB be two straight lines drawn from it, of which DCA cuts the circle, and DB touches the same: the rectangle AD, DC shall be equal to the square of DB.

Either DCA passes through the centre, or it does not: first, let it pass through the centre E, and join EB; therefore the angle EBD is a right* angle: and because the * 18. 3.
straight line AC is bisected in E, and produced to the point D, the rectangle AD, DC, together with the square of EC, is equal* to the square * 6. 2.
of ED: but CE is equal to EB; therefore the rectangle AD, DC, together with the square of EB, is equal to the square of ED: but the square of ED is equal* to the squares of EB, BD, be- * 47. 1.

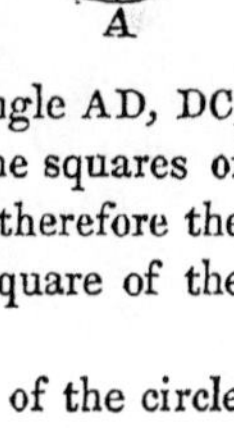

cause EBD is a right angle; therefore the rectangle AD, DC, together with the square of EB, is equal † to the squares of † 1 Ax.
EB, BD: take away the common square of EB; therefore the remaining rectangle AD, DC is equal† to the square of the † 3 Ax.
tangent DB.

But if DCA does not pass through the centre of the circle ABC, take* the centre E, and draw EF perpendicular* to * 1. 3.
AC, and join EB, EC, ED. And because the straight line EF * 12. 1.
which passes through the centre, cuts the straight line AC

which does not pass through the centre, at right angles, it
* 3. 3. also bisects* it; therefore AF is equal to FC: and because the straight line AC is bisected in F, and produced to D, the rectangle AD, DC, together
* 6. 2. with the square of FC, is equal* to the square of FD: to each of these equals add the square of FE; therefore the rectangle AD, DC, to-
† 2 Ax. gether with the squares of CF, FE, is equal† to the squares of DF, FE: but the square of ED
* 47. 1. is equal* to the squares of DF, FE, because EFD is a right angle; and for the same reason, the square of EC is equal to the squares of CF, FE; therefore the rectangle AD, DC,
† 1 Ax. together with the square of EC, is equal† to the square of ED: but CE is equal to EB; therefore the rectangle AD, DC, together with the square of EB, is equal to the square of ED:
* 47. 1. but the squares of EB, BD are equal to the square* of ED, because EBD is a right angle; therefore the rectangle AD, DC, together with the square of EB, is equal to the squares of EB, BD: take away the common square of EB; therefore the re-
† 3 Ax. maining rectangle AD, DC is equal† to the square of DB. Wherefore, if from any point, &c. Q. E. D.

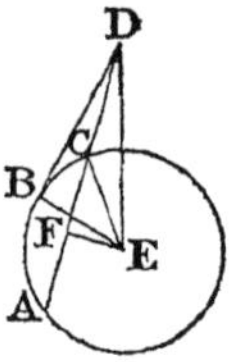

COR. If from any point without a circle, there be drawn two straight lines cutting it, as AB, AC, the rectangles contained by the whole lines and the parts of them without the circle, are equal to one another; viz. the rectangle BA, AE, to the rectangle CA, AF; for each of them is equal to the square of the straight line AD, which touches the circle.

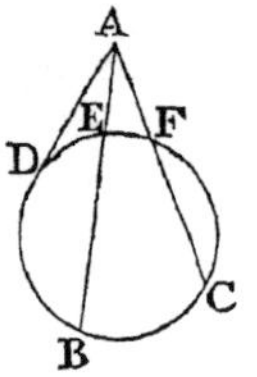

PROPOSITION XXXVII.

See N. THEOR.—*If from a point without a circle there be drawn two straight lines, one of which cuts the circle, and the other meets it; if the rectangle contained by the whole line which cuts the circle, and the part of it without the circle, be equal to the square of the line which meets it, the line which meets shall touch the circle.*

Let any point D be taken without the circle ABC, and from it let two straight lines DCA and DB be drawn, of which DCA cuts the circle, and DB meets it: if the rectangle AD, DC be equal to the square of DB, DB shall touch the circle.

Draw * the straight line DE, touching the circle ABC; find † its centre F, and join FE, FB, FD; then FED is a right * angle: and because DE touches the circle ABC, and DCA cuts it, the rectangle AD, DC is equal * to the square of DE: but the rectangle AD, DC is by hypothesis equal to the square of DB: therefore the square of DE is equal † to the square of DB; and the straight line DE equal to the straight line DB: and FE is equal † to FB; wherefore DE, EF are equal to DB, BF, each to each; and the base FD is common to the two triangles DEF, DBF; therefore the angle DEF is equal * to the angle DBF: but DEF was shewn to be a right angle; therefore also DBF is † a right angle: and BF, if produced, is a diameter; and the straight line which is drawn at right angles to a diameter, from the extremity of it, touches * the circle: therefore DB touches the circle ABC. Wherefore, if from a point, &c. Q. E. D.

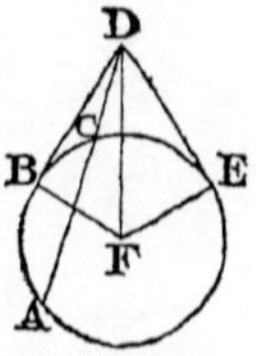

* 17. 3.
† 1. 3.
* 18. 3.
* 36. 3.
† 1 Ax.
† 15 Def. 1.
* 8. 1.
† 1 Ax.
* Cor. 16. 3.

THE

ELEMENTS OF EUCLID.

BOOK IV.

DEFINITIONS.

See N. I. A RECTILINEAL figure is said to be inscribed in another rectilineal figure, when all the angles of the inscribed figure are upon the sides of the figure in which it is inscribed, each upon each.

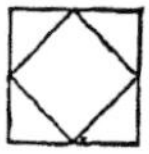

II. In like manner, a figure is said to be described about another figure, when all the sides of the circumscribed figure pass through the angular points of the figure about which it is described, each through each.

III. A rectilineal figure is said to be inscribed in a circle, when all the angles of the inscribed figure are upon the circumference of the circle.

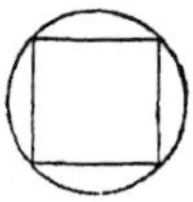

IV. A rectilineal figure is said to be described about a circle, when each side of the circumscribed figure touches the circumference of the circle.

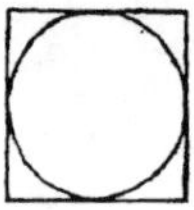

V. In like manner, a circle is said to be inscribed in a rectilineal figure, when the circumference of the circle touches each side of the figure.

VI. A circle is said to be described about a rectilineal figure, when the circumference of the circle passes through all the angular points of the figure about which it is described.

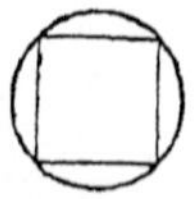

VII. A straight line is said to be placed in a circle, when the extremities of it are in the circumference of the circle.

PROPOSITION I.

PROBLEM.—*In a given circle, to place a straight line, equal to a given straight line, which is not greater than the diameter of the circle.*

Let ABC be the given circle, and D the given straight line, not greater than the diameter of the circle; it is required to place in the circle ABC, a straight line equal to D.

Draw BC ‡ the diameter of the circle ABC: then, if BC is equal to D, the thing required is done; for in the circle ABC, a straight line BC is placed equal to D: but, if it is not, BC is greater † than D: make CE equal* to D; and from the centre C, at the distance CE, describe the circle AEF, and join CA: CA shall be equal to D.

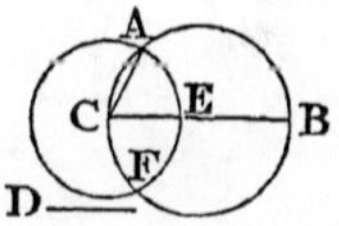

† Hyp.
* 3. 1.

Because C is the centre of the circle AEF, CA is equal† to CE: but D is equal† to CE; therefore D is equal† to CA. Wherefore, in the circle ABC, a straight line CA is placed equal to the given straight line D, which is not greater than the diameter of the circle. Which was to be done.

† 15 Def. 1.
† Constr.
† 1 Ax.

PROPOSITION II.

PROB.—*In a given circle, to inscribe a triangle equiangular to a given triangle.*

Let ABC be the given circle, and DEF the given triangle; it is required to inscribe in the circle ABC, a triangle equiangular to the triangle DEF.

‡ Find the centre, and through it draw any straight line BC, terminated both ways by the circumference; this line is a diameter.

* 17. 3. Draw* the straight line GAH touching the circle in the point A; and at the point A, in the straight line AH,
* 23. 1. make* the angle HAC equal to the angle DEF; and at the point A, in the straight line AG, make the angle GAB equal to the angle DFE; and join BC: ABC shall be the triangle required.

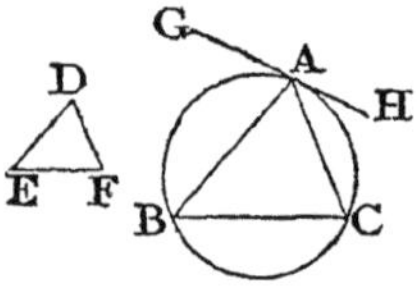

Because HAG touches the circle ABC, and AC is drawn
* 32. 3. from the point of contact, the angle HAC is equal* to the angle ABC in the alternate segment of the circle: but HAC
† Constr. is equal † to the angle DEF; therefore also the angle ABC is
† 1 Ax. equal † to DEF: for the same reason, the angle ACB is equal to the angle DFE; therefore the remaining angle BAC is
* 32. 1. & 1 Ax. equal * to the remaining angle EDF: wherefore the triangle ABC is equiangular to the triangle DEF; and it is inscribed in the circle ABC. Which was to be done.

PROPOSITION III.

PROB.—*About a given circle, to describe a triangle equiangular to a given triangle.*

Let ABC be the given circle, and DEF the given triangle; it is required to describe a triangle about the circle ABC, equiangular to the triangle DEF.

Produce EF both ways to the points
† 1. 3. G, H; find† the centre K of the circle ABC, and from it draw any straight line KB; at the point K,
* 23. 1. in the straight line KB, make* the angle BKA equal to the angle DEG, and the angle BKC equal to the angle DFH; and through the points A, B, C, draw the
* 17. 3. straight lines LAM, MBN, NCL, touching* the circle ABC: LMN shall be the triangle required.

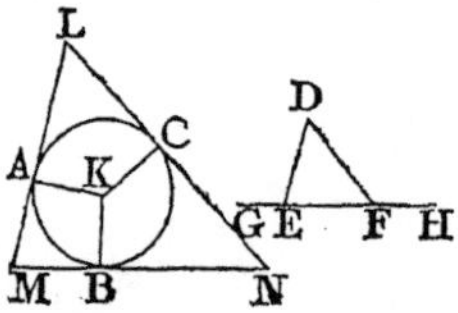

Because LM, MN, NL touch the circle ABC in the points A, B, C, to which from the centre are drawn KA, KB, KC, the
* 18. 3. angles at the points A, B, C are right* angles: and because the four angles of the quadrilateral figure AMBK are equal to

four right angles ‡, for it can be divided into two triangles, and that two of them KAM, KBM are right angles, therefore the other two AKB, AMB are equal † to two right angles: † 3 Ax.
but the angles DEG, DEF are likewise equal * to two right * 13. 1.
angles; therefore the angles AKB, AMB are equal † to the † 1 Ax.
angles DEG, DEF; of which AKB is equal † to DEG; where- † Constr.
fore the remaining angle AMB is equal † to the remaining † 3 Ax.
angle DEF. In like manner, the angle LNM may be demonstrated to be equal to DFE; and therefore the remaining angle MLN is equal * to the remaining angle EDF: there- * 32. 1. & 3 Ax.
fore the triangle LMN is equiangular to the triangle DEF; and it is described about the circle ABC. Which was to be done.

PROPOSITION IV.

B.—*To inscribe a circle in a given triangle.* See N.

Let the given triangle be ABC; it is required to inscribe a circle in ABC.

Bisect * the angles ABC, BCA by the straight lines BD, * 9. 1.
CD meeting one another in the point D, from which draw * * 12. 1.
DE, DF, DG perpendiculars to AB, BC, CA.

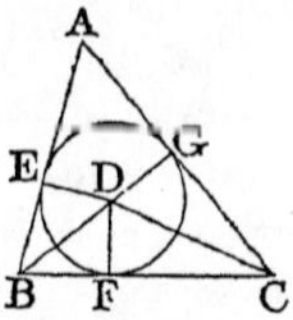

And because the angle EBD is equal to the angle FBD, for the angle ABC is bisected by BD, and that the right angle BED is equal † to the right angle BFD; therefore † 11 Ax.
the two triangles EBD, FBD have two angles of the one, equal to two angles of the other, each to each; and the side BD, which is opposite to one of the equal angles in each, is common to both; therefore their other sides are equal *; wherefore DE is equal to * 26. 1.
DF: for the same reason, DG is equal to DF; therefore DE is equal † to DG: therefore the three straight lines DE, DF, † 1 Ax.
DG are equal to one another, and the circle described from the centre D, at the distance of any of them, will pass through the extremities of the other two, and touch the straight lines AB, BC, CA, because the angles at the points E, F, G, are

‡ Suppose the line MK drawn: then, because the three angles of every triangle are together equal * to two right angles, therefore all the angles of the * 32. 1.
two triangles AMK, BMK are together equal to four right angles; but all the angles of these triangles are together equal to the angles of the figure ABMK; therefore all the angles of the figure AMBK are together equal to four right angles.

right angles, and the straight line which is drawn from the
* 16. 3. extremity of a diameter at right angles to it, touches * the circle; therefore the straight lines AB, BC, CA do each of them touch the circle, and therefore the circle EFG is inscribed in the triangle ABC. Which was to be done.

PROPOSITION V.

See N. PROB.—*To describe a circle about a given triangle.*

Let the given triangle be ABC; it is required to describe a circle about ABC.

* 10. 1. Bisect * AB, AC, in the points D, E, and from these points
* 11. 1. draw DF, EF at right angles * to AB, AC DF, EF produced, meet one another: for, if they do not meet, they are parallel, wherefore AB, AC, which are at right angles to them, are parallel; which is absurd: let them meet in F, and join

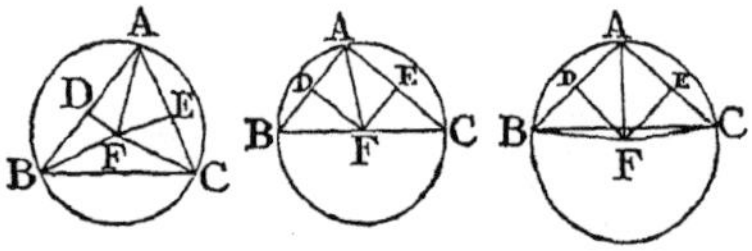

* See fig. 3. FA; also if the point F be not in BC, join BF, CF *. Then, because AD is equal to DB, and DF common, and at right
* 4. 1. angles to AB, the base AF is equal * to the base FB. In like manner it may be shewn, that CF is equal to FA; and there-
† 1 Ax. fore BF is equal † to FC; and FA, FB, FC, are equal to one another: wherefore the circle described from the centre F, at the distance of one of them, will pass through the extremities of the other two, and be described about the triangle ABC. Which was to be done.

COROLLARY. And it is manifest, that when the centre of the circle falls within the triangle, each of its angles is less
† 31. 3. than a right angle †, each of them being in a segment greater than a semicircle; but when the centre is in one of the sides of the triangle, the angle opposite to this side, being in
† 31. 3. a semicircle †, is a right angle; and, if the centre falls without the triangle, the angle opposite to the side beyond which
† 31. 3. it is, being in a segment less than a semicircle †, is greater than a right angle: therefore, conversely, if the given triangle be acute-angled, the centre of the circle falls within it; if it be a right-angled triangle, the centre is in the side opposite to the right angle; and if it be an obtuse-angled triangle,

the centre falls without the triangle, beyond the side opposite to the obtuse angle.

PROPOSITION VI.

PROB.—*To inscribe a square in a given circle.*

Let ABCD be the given circle; it is required to inscribe a square in ABCD.

Draw the diameters AC, BD† at right angles to one another, † 1. 3. & 11. 1.
and join AB, BC, CD, DA: the figure ABCD shall be the square required.

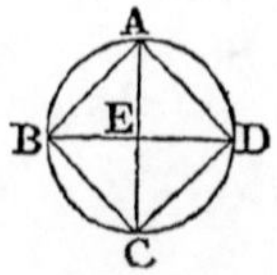

Because BE is equal to ED, for E is the centre, and that EA is common, and at right angles to BD, the base BA is equal * to the base AD: * 4. 1. and for the same reason, BC, CD are each of them equal to BA, or AD; therefore the quadrilateral figure ABCD is equilateral. It is also rectangular; for the straight line BD being the diameter of the circle ABCD, BAD is a semicircle; wherefore the angle BAD is a right * angle: * 31. 3. for the same reason, each of the angles ABC, BCD, CDA is a right angle; therefore the quadrilateral figure ABCD is rectangular: and it has been shewn to be equilateral; therefore it is a† square; † 30 Def. 1. and it is inscribed in the circle ABCD. Which was to be done.

PROPOSITION VII.

PROB.—*To describe a square about a given circle.*

Let ABCD be the given circle; it is required to describe a square about it.

Draw two diameters AC, BD of the circle ABCD, at right angles to one another; and through the points A, B, C, D, draw * FG, GH, HK, KF touching the circle: * 17. 3. the figure GHKF shall be the square required.

Because FG touches the circle ABCD, and EA is drawn from the centre E, to the point of contact A, the angles at A are right * angles; * 18. 3. for the same reason, the angles at the points B, C, D are right angles: and because the angle AEB is a right angle, as likewise is EBG, GH is parallel * to AC; * 28. 1. for the same reason AC is parallel to FK: and in like manner GF, HK may each of them

be demonstrated to be parallel to BED; therefore the figures GK, GC, AK, FB, BK are parallelograms; and therefore GF is
* 34. 1. equal* to HK, and GH to FK: and because AC is equal to BD, and that AC is equal to each of the two GH, FK; and BD to each of the two GF, HK; GH, FK are each of them equal to GF, or HK; therefore the quadrilateral figure FGHK is equilateral. It is also rectangular; for GBEA being a pa-
* 34. 1. rallelogram, and AEB a right angle, AGB * is likewise a right angle: and in the same manner it may be shewn, that the angles at H, K, F, are right angles; therefore the quadrilateral figure FGHK is rectangular: and it was demonstrated
† 30 Def. 1. to be equilateral; therefore it is a † square; and it is described about the circle ABCD. Which was to be done.

PROPOSITION VIII.

PROB.—*To inscribe a circle in a given square.*

Let ABCD be the given square; it is required to inscribe a circle in ABCD.

* 10. 1. Bisect* each of the sides AB, AD in the points F, E; and
† 31. 1. through E, draw † EH parallel to AB or DC; and through F, draw FK parallel to AD or BC: therefore each of the figures AK, KB, AH, HD, AG, GC, BG, GD is a parallelogram; and
* 34. 1. their opposite sides are* equal: and because AD is equal† to
† 30 Def. 1. AB, and that AE is the half of AD, and AF the half of AB,
† 7 Ax. AE is equal† to AF; wherefore the sides opposite to these are equal, viz. FG to GE: in the same manner it may be demonstrated, that GH, GK are each of them equal to FG or GE; therefore the four straight lines GE, GF, GH, GK are equal to one another; and the circle described from the centre G, at the distance of one of them, will pass through the extremities of the other three, and touch the straight lines AB, BC, CD,

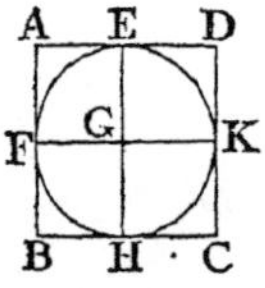

* 29. 1. DA; because the angles at the points ‡ E, F, H, K, are right* angles, and that the straight line which is drawn from the

‡ Because AB is parallel to EH, the two angles BAE, AEH are together
† 29. 1. equal † to two right angles: but BAE † is a right angle; therefore AEH is
† 30 Def. 1. is a right angle: and in the same manner the angles at the points F, H, K, may be proved to be right angles.

extremity of a diameter, at right angles to it, touches* the circle: therefore each of the straight lines AB, BC, CD, DA touches the circle, which therefore is inscribed in the square ABCD. Which was to be done. * Cor. 16. 3.

PROPOSITION IX.

PROB.—*To describe a circle about a given square.*

Let ABCD be the given square; it is required to describe a circle about ABCD.

Join AC, BD, cutting one another in E: and because DA is equal† to AB, and AC common to the triangles DAC, BAC, the two sides DA, AC are equal to the two BA, AC, each to each; and the base DC is equal to the base BC; wherefore the angle DAC is equal* to the angle BAC; and the angle DAB is bisected by the straight line AC: in the same manner it may be demonstrated, that the angles ABC, BCD, CDA are severally bisected by the straight lines BD, AC: therefore, because the angle DAB is equal† to the angle ABC, and that the angle EAB is the half of DAB, and EBA the half of ABC, the angle EAB is equal† to the angle EBA; wherefore the side EA is equal* to the side EB: in the same manner it may be demonstrated, that the straight lines EC, ED are each of them equal to EA, or EB; therefore the four straight lines EA, EB, EC, ED, are equal to one another; and the circle described from the centre E, at the distance of one of them, will pass through the extremities of the other three, and be described about the square ABCD. Which was to be done.

† 30 Def. 1.
* 8. 1.
† 30 Def. 1.
† 7 Ax.
* 6. 1.

PROPOSITION X.

PROB.—*To describe an isosceles triangle, having each of the angles at the base, double of the third angle.*

Take any straight line AB, and divide* it in the point C, so that the rectangle AB, BC may be equal to the square of CA; and from the centre A, at the distance AB, describe the circle BDE, in which, place* the straight line BD equal to AC, which is not greater than the diameter of the circle BDE; and join DA: the triangle ABD shall be such as is required;

* 11. 2.
* 1. 4.

that is, each of the angles ABD, ADB shall be double of the angle BAD.

* 5. 4. Join DC, and about the triangle ADC describe* the circle
ACD: and because the rectangle AB, BC is equal to the
† Constr. square of AC†, and that AC is equal to BD, the rectangle
† 1 Ax. AB, BC is equal† to the square of BD: and
because from the point B, without the circle ACD, two straight lines BCA, BD are drawn to the circumference, one of which cuts, and the other meets the circle, and that the rectangle AB, BC, contained by the whole of the cutting line, and the part of it without the circle, is equal to the square

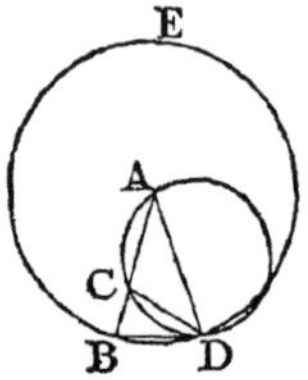

* 37. 3. of BD which meets it, the straight line BD touches* the
circle ACD: and because BD touches the circle, and DC is
* 32. 3. drawn from the point of contact D, the angle BDC is equal*
to the angle DAC in the alternate segment of the circle: to
each of these add the angle CDA; therefore the whole angle
† 2 Ax. BDA is equal† to the two angles CDA, DAC: but the exte-
* 32. 1. rior angle BCD is equal* to the angles CDA, DAC; therefore
† 1 Ax. also BDA is equal† to BCD: but BDA is equal* to the angle
* 5. 1. CBD, because the side AD is equal to the side AB; therefore
† 1 Ax. CBD, or DBA, is equal† to BCD; and consequently the three
angles BDA, DBA, BCD are equal to one another; and be-
cause the angle DBC is equal to the angle BCD, the side BD is
* 6. 1. equal* to the side DC: but BD was made equal to CA; there-
† 1 Ax. fore also CA is equal† to CD, and the angle CDA equal* to
* 5. 1. the angle DAC; therefore the angles CDA, DAC together,
† 32. 1. are double of the angle DAC: but BCD is equal† to the
angles CDA, DAC; therefore also BCD is double of DAC: and BCD was proved to be equal to each of the angles BDA, DBA; therefore each of the angles BDA, DBA is double of the angle DAB. Wherefore, an isosceles triangle ABD is described, having each of the angles at the base, double of the third angle. Which was to be done.

PROPOSITION XI.

PROB.—*To inscribe an equilateral and equiangular pentagon in a given circle.*

Let ABCDE be the given circle; it is required to inscribe an equilateral and equiangular pentagon in the circle ABCDE.

Describe * an isosceles triangle FGH, having each of the angles at G, H, double of the angle at F; and in the circle ABCDE, inscribe * the triangle ACD, equiangular to the triangle FGH, so that the angle CAD may be equal to the angle at F, and each of the angles ACD, CDA equal to the angle at G or H: wherefore each of the angles ACD, CDA is double of the angle CAD. Bisect * the angles ACD, CDA by the straight lines CE, DB; and join AB, BC, DE, EA: ABCDE shall be the pentagon required. * 10. 4. * 2. 4. * 9. 1.

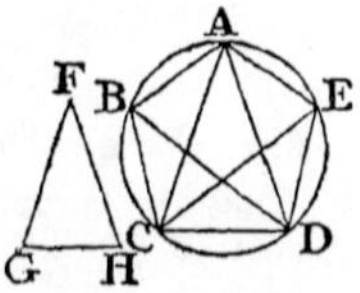

Because each of the angles ACD, CDA is double of CAD, and that they are bisected by the straight lines CE, DB, therefore the five angles DAC, ACE, ECD, CDB, BDA are equal to one another: but equal angles stand upon equal * circumferences; therefore the five circumferences AB, BC, CD, DE, EA are equal to one another: and equal circumferences are subtended by equal * straight lines; therefore the five straight lines AB, BC, CD, DE, EA are equal to one another: wherefore the pentagon ABCDE is equilateral. It is also equiangular; for, because the circumference AB is equal to the circumference DE, if to each be added BCD, the whole ABCD is equal † to the whole EDCB: but the angle AED stands on the circumference ABCD, and the angle BAE on the circumference EDCB; therefore the angle BAE is equal * to the angle AED: for the same reason, each of the angles ABC, BCD, CDE is equal to the angle BAE, or AED; therefore the pentagon ABCDE is equiangular: and it has been shewn, that it is equilateral. Wherefore, in the given circle an equilateral and equiangular pentagon has been inscribed. Which was to be done. * 26. 3. * 29. 3. † 2 Ax. * 27. 3.

PROPOSITION XII.

Prob.—*To describe an equilateral and equiangular pentagon about a given circle.*

Let ABCDE be the given circle; it is required to describe an equilateral and equiangular pentagon about the circle ABCDE.

Let the angles of a pentagon, inscribed in the circle, by the last proposition, be in the points A, B, C, D, E, so that
* 11. 4. the circumferences AB, BC, CD, DE, EA are* equal; and through the points A, B, C, D, E, draw GH, HK, KL, LM,
* 17. 3. MG, touching * the circle; the figure GHKLM shall be the pentagon required.

Take the centre F, and join FB, FK, FC, FL, FD: and because the straight line KL touches the circle ABCDE in the point C, to which FG is drawn from the
* 18. 3. centre F, FC is perpendicular * to KL, therefore each of the angles at C is a right angle; for the same reason, the angles at the points B, D are right angles; and because FCK is a right angle, the square of
* 47. 1. FK is equal * to the squares of FC, CK; for the same reason, the square of FK, is

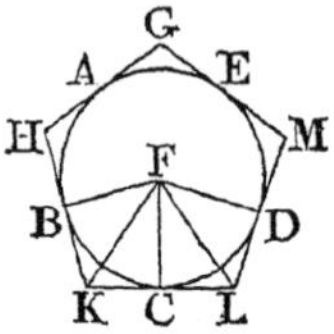

equal to the squares of FB, BK; therefore the squares of FC,
† 1 Ax. CK are equal † to the squares of FB, BK; of which the square of FC is equal to the square of FB; therefore the re-
† 3 Ax. maining square of CK is equal † to the remaining square of BK, and the straight line CK equal to BK: and because FB is equal to FC, and FK common to the triangles BFK, CFK, the two BF, FK are equal to the two CF, FK, each to each; and the base BK was proved equal to the base KC; there-
* 8. 1. fore the angle BFK is equal * to the angle KFC, and the
† 4. 1. angle BKF † to FKC: wherefore the angle BFC is double of the angle KFC, and BKC double of FKC: for the same reason, the angle CFD is double of the angle CFL, and CLD double of CLF: and because the circumference BC
* 27. 3. is equal to the circumference CD, the angle BFC is equal * to the angle CFD; and BFC is double of the angle KFC,
† 7 Ax. and CFD double of CFL; therefore the angle KFC is equal † to the angle CFL: and the right angle FCK is equal to the right angle FCL; therefore in the two triangles FKC, FLC, there are two angles of the one, equal to two angles of the other, each to each; and the side FC, which is adjacent to the equal angles in each, is common to both; therefore the
* 26. 1. other sides are equal * to the other sides, and the third angle to the third angle; therefore the straight line KC is equal to CL, and the angle FKC to the angle FLC: and because KC is equal to CL, KL is double of KC. In the same manner it may be shewn, that HK is double of BK: and because BK is

equal to KC, as was demonstrated, and that KL is double of KC, and HK double of BK, therefore HK is equal † to KL: in like † 6 Ax. manner it may be shewn, that GH, GM, ML are each of them equal to HK, or KL; therefore the pentagon GHKLM is equilateral. It is also equiangular; for, since the angle FKC is equal to the angle FLC, and that the angle HKL is double of the angle FKC, and KLM double of FLC, as was before demonstrated, therefore the angle HKL is equal † to † 6 Ax. KLM: and in like manner it may be shewn, that each of the angles KHG, HGM, GML is equal to the angle HKL or KLM; therefore the five angles GHK, HKL, KLM, LMG, MGH, being equal to one another, the pentagon GHKLM is equiangular: and it is equilateral, as was demonstrated; and it is described about the circle ABCDE. Which was to be done.

PROPOSITION XIII.

PROB.—*To inscribe a circle in a given equilateral and equiangular pentagon.*

Let ABCDE be the given equilateral and equiangular pentagon; it is required to inscribe a circle in the pentagon ABCDE.

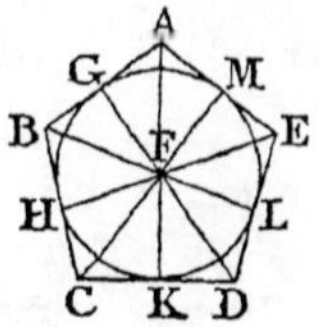

Bisect * the angles BCD, CDE by the straight lines CF, DF, and from the point F, in which they meet, draw the straight lines FB, FA, FE: therefore since BC is equal † to CD, and CF common to the triangles BCF, DCF, the two sides BC, CF are equal to the two DC, CF, each to each; and the angle BCF is equal † to the angle DCF; therefore the base BF is equal * to the base FD, and the other angles to the other angles, to which the equal sides are opposite; therefore the angle CBF is equal to the angle CDF: and because the angle CDE is double of CDF, and that CDE is equal to CBA, and CDF to CBF: CBA is also double of the angle CBF; therefore the angle ABF is equal to the angle CBF; wherefore the angle ABC is bisected by the straight line BF: in the same manner it may be demonstrated, that the angles BAE, AED are bisected by the straight lines AF, FE.

* 9. 1. † Hyp. † Constr. * 4. 1.

From the point F, draw * FG, FH, FK, FL, FM perpendiculars to the straight lines AB, BC, CD, DE, EA: and because * 12. 1.

the angle HCF is equal to KCF, and the right angle FHC equal to the right angle FKC, therefore in the triangles FHC, FKC there are two angles of the one, equal to two angles of the other, each to each; and the side FC, which is opposite to one of the equal angles in each, is common to both; there-
* 26. 1. fore the other sides are equal *, each to each; wherefore the perpendicular FH is equal to the perpendicular FK: in the same manner it may be demonstrated, that FL, FM, FG are each of them equal to FH, or FK; therefore the five straight lines FG, FH, FK, FL, FM are equal to one another; wherefore the circle described from the centre F, at the distance of one of these five, will pass through the extremities of the other four, and touch the straight lines AB, BC, CD, DE, EA, because the angles at the points G, H, K, L, M are right angles, and that a straight line drawn from the extremity of
* 16. 3. the diameter of a circle at right angles to it, touches * the circle: therefore each of the straight lines AB, BC, CD, DE, EA touches the circle: wherefore it is inscribed in the pentagon ABCDE. Which was to be done.

PROPOSITION XIV.

PROB.—*To describe a circle about a given equilateral and equiangular pentagon.*

Let ABCDE be the given equilateral and equiangular pentagon; it is required to describe a circle about it.

* 9. 1. Bisect * the angles BCD, CDE by the straight lines CF, FD, and from the point F, in which they meet, draw the straight lines FB, FA, FE, to the points B, A, E. It may be demonstrated, in the same manner as in the preceding proposition, that the angles CBA, BAE, AED are bisected by the straight lines FB, FA, FE: and because the angle BCD is equal to the angle CDE, and that FCD is the half of the angle BCD, and

† 7 Ax. CDF the half of CDE, therefore the angle FCD is equal † to
* 6. 1. FDC; wherefore the side CF is equal * to the side FD: in like manner it may be demonstrated, that FB, FA, FE, are each of them equal to FC or FD; therefore the five straight lines FA, FB, FC, FD, FE are equal to one another; and the circle described from the centre F, at the distance of one of

them, will pass through the extremities of the other four, and be described about the equilateral and equiangular pentagon ABCDE. Which was to be done.

PROPOSITION XV.

PROB.—*To inscribe an equilateral and equiangular hexagon in a given circle.* See N.

Let ABCDEF be the given circle; it is required to inscribe an equilateral and equiangular hexagon in it.

Find † the centre G of the circle ABCDEF, and draw the diameter AGD; and from D as a centre, at the distance DG, describe the circle EGCH, join EG, CG, and produce them to the points B, F; and join AB, BC, CD, DE, EF, FA: the hexagon ABCDEF shall be equilateral and equiangular. † 1. 3.

Because G is the centre of the circle ABCDEF, GE is equal to GD; and because D is the centre of the circle EGCH, DE is equal to DG; wherefore GE is equal † to ED, and the triangle EGD is equilateral; and therefore its three angles EGD, GDE, DEG, are equal † to one another: but the three angles of a triangle are equal * to two right angles, therefore the angle EGD is the third part of two right angles; in the same manner it may be demonstrated, that the angle DGC is also the third part of two right angles: and because the straight line GC makes with EB the adjacent angles EGC, CGB equal * to two right angles, the remaining angle CGB is the third part of two right angles; therefore the angles EGD, DGC, CGB are equal to one another; and to these are equal * the vertical opposite angles BGA, AGF, FGE; therefore the six angles EGD, DGC, CGB, BGA, AGF, FGE, are equal to one another: but equal angles stand upon equal * circumferences, therefore the six circumferences AB, BC, CD, DE, EF, FA are equal to one another: and equal circumferences are subtended by equal * straight lines, therefore the six straight lines are equal to one another, and the hexagon ABCDEF is equilateral. It is also equiangular; for, since the circumference AF is equal to ED, to each of these add the circumference ABCD; therefore the whole circumference FABCD is equal to the whole EDCBA: and the angle FED

† 1 Ax.

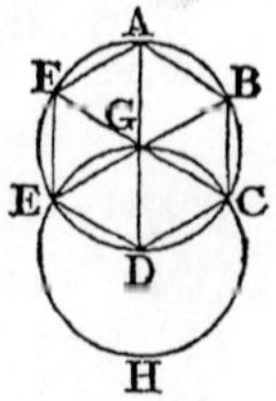

† Cor. 5. 1.
* 32. 1.
* 13. 1.
* 15. 1.
* 26. 3.
* 29. 3.

stands upon the circumference FABCD, and the angle AFE
† 27. 3. upon EDCBA; therefore the angle AFE is equal † to FED: in the same manner it may be demonstrated, that the other angles of the hexagon ABCDEF are each of them equal to the angle AFE or FED; therefore the hexagon is equiangular; and it is equilateral, as was shewn; and it is inscribed in the given circle ABCDEF. Which was to be done.

COR.—From this it is manifest, that the side of the hexagon is equal to the straight line from the centre, that is, to the semidiameter of the circle.

And if through the points A, B, C, D, E, F there be drawn straight lines touching the circle, an equilateral and equiangular hexagon will be described about it, which may be demonstrated from what has been said of the pentagon: and likewise a circle may be inscribed in a given equilateral and equiangular hexagon, and circumscribed about it, by a method like to that used for the pentagon.

PROPOSITION XVI.

See N. PROB.—*To inscribe an equilateral and equiangular quindecagon in a given circle.*

Let ABCD be the given circle; it is required to inscribe an equilateral and equiangular quindecagon in the circle ABCD.

* 2. 4. Let AC be the side of an equilateral triangle inscribed* in the circle, and AB the side of an equilateral and equiangular
* 11. 4. pentagon inscribed* in the same: therefore, of such equal parts, as the whole circumference ABCDF contains fifteen, the circumference ABC, being the third part of the whole, contains five; and the circumference AB, which is the fifth part of the whole, contains three; therefore BC, their difference, con-

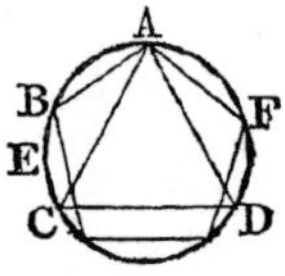

* 30. 3. tains two of the same parts: bisect* BC in E: therefore BE, EC are each of them the fifteenth part of the whole circumference ABCD: therefore if the straight lines BE, EC be drawn,
* 1. 4. and straight lines equal to them be placed round* in the whole circle, an equilateral and equiangular quindecagon will be inscribed in it. Which was to be done.

And in the same manner as was done in the pentagon, if through the points of division made by inscribing the quindecagon, straight lines be drawn touching the circle, an equilateral and equiangular quindecagon will be described about it; and likewise, as in the pentagon, a circle may be inscribed in a given equilateral and equiangular quindecagon, and circumscribed about it.

THE

ELEMENTS OF EUCLID.

BOOK V.

DEFINITIONS.

I. A LESS magnitude is said to be a part of a greater magnitude, when the less measures the greater; that is, 'when 'the less is contained a certain number of times exactly in 'the greater.'

II. A greater magnitude is said to be a multiple of a less, when the greater is measured by the less; that is, 'when 'the greater contains the less a certain number of times 'exactly.'

See N. III. "Ratio is a mutual relation of two magnitudes of the "same kind to one another, in respect of quantity."

IV. Magnitudes are said to have a ratio to one another, when the less can be multiplied so as to exceed the other.

V. The first of four magnitudes is said to have the same ratio to the second, which the third has to the fourth, when any equimultiples whatsoever of the first and third being taken, and any equimultiples whatsoever of the second and fourth, if the multiple of the first be less than that of the second, the multiple of the third is also less than that of the fourth: or, if the multiple of the first be equal to that of the second, the multiple of the third is also equal to that of the fourth: or, if the multiple of the first be greater than that of the second, the multiple of the third is also greater than that of the fourth.

VI. Magnitudes which have the same ratio are called proportionals. 'N. B. When four magnitudes are proportionals, 'it is usually expressed by saying, the first is to the second, 'as the third to the fourth.'

VII. When of the equimultiples of four magnitudes, (taken as in the fifth definition,) the multiple of the first is greater than that of the second, but the multiple of the third is not greater than the multiple of the fourth, then the first is said to have to the second a greater ratio than the third magnitude has to the fourth: and, on the contrary, the third is said to have to the fourth a less ratio than the first has to the second.

VIII. "Analogy or proportion is the similitude of ratios."

IX. Proportion consists in three terms at least.

X. When three magnitudes are proportionals, the first is said to have to the third, the duplicate ratio of that which it has to the second.

XI. When four magnitudes are continual proportionals, the first is said to have to the fourth, the triplicate ratio of that which it has to the second, and so on, quadruplicate, &c. increasing the denomination still by unity, in any number of proportionals. See N.

Definition A, to wit of compound ratio.

When there are any number of magnitudes of the same kind, the first is said to have to the last of them, the ratio compounded of the ratio which the first has to the second, and of the ratio which the second has to the third, and of the ratio which the third has to the fourth, and so on unto the last magnitude.

For example, if A, B, C, D be four magnitudes of the same kind, the first A is said to have to the last D, the ratio compounded of the ratio of A to B, and of the ratio of B to C, and of the ratio of C to D; or, the ratio of A to D is said to be compounded of the ratios of A to B, B to C, and C to D.

And if A has to B the same ratio which E has to F, and B to C the same ratio that G has to H, and C to D the same that K has to L, then, by this definition, A is said to have to D the ratio compounded of ratios which are the same with the ratios of E to F, G to H, and K to L. And the same thing is to be understood when it is more briefly expressed by saying, A has to D the ratio compounded of the ratios of E to F, G to H, and K to L.

In like manner, the same things being supposed, if M has to N the same ratio which A has to D, then, for shortness' sake, M is said to have to N the ratio compounded of the ratios of E to F, G to H, and K to L.

XII. In proportionals, the antecedent terms are called homologous to one another, as also the consequents to one another.

'Geometers make use of the following technical words, to 'signify certain ways of changing either the order or mag-'nitude of proportionals, so that they continue still to be 'proportionals.'

See N. XIII. *Permutando*, or *alternando*, by permutation or alternately. This word is used when there are four proportionals, and it is inferred that the first has the same ratio to the third, which the second has to the fourth; or, that the first is to the third, as the second to the fourth; as is shewn in the 16th Prop. of this fifth Book.

XIV. *Invertendo*, by inversion; when there are four proportionals, and it is inferred, that the second is to the first as the fourth to the third. Prop. B. Book 5.

XV. *Componendo*, by composition; when there are four proportionals, and it is inferred, that the first together with the second, is to the second, as the third together with the fourth, is to the fourth. 18th Prop. Book 5.

XVI. *Dividendo*, by division; when there are four proportionals, and it is inferred, that the excess of the first above the second, is to the second, as the excess of the third above the fourth, is to the fourth. 17th Prop. Book 5.

XVII. *Convertendo,* by conversion; when there are four proportionals, and it is inferred, that the first is to its excess above the second, as the third to its excess above the fourth. Prop. E. Book 5.

XVIII. *Ex æquali,* (sc. distantiâ,) or *ex æquo,* from equality of distance; when there is any number of magnitudes more than two, and as many others, such that they are proportionals when taken two and two of each rank, and it is inferred, that the first is to the last of the first rank of magnitudes, as the first is to the last of the others. 'Of 'this there are the two following kinds, which arise from 'the different order in which the magnitudes are taken, 'two and two.'

XIX. *Ex æquali,* from equality. This term is used simply by itself, when the first magnitude is to the second of the first rank, as the first to the second of the other rank; and as the second is to the third of the first rank, so is the second to the third of the other; and so on in order: and the inference is as mentioned in the preceding definition; whence this is called ordinate proportion. It is demonstrated in the 22d Prop. Book 5.

XX. *Ex æquali in proportione perturbatâ seu inordinatâ,* from equality in perturbate or disorderly proportion *. This term is used when the first magnitude is to the second of the first rank, as the last but one is to the last of the second rank; and as the second is to the third of the first rank, so is the last but two to the last but one of the second rank; and as the third is to the fourth of the first rank, so is the third from the last to the last but two of the second rank; and so on in a cross order: and the inference is as in the 18th Definition. It is demonstrated in the 23d Prop. of Book 5.

* 4 Prop. lib. 2. Archimedis de Sphærâ et Cylindro.

AXIOMS.

I. EQUIMULTIPLES of the same, or of equal magnitudes, are equal to one another.

II. Those magnitudes, of which the same or equal magnitudes are equimultiples, are equal to one another.

III. A multiple of a greater magnitude, is greater than the same multiple of a less.

IV. That magnitude, of which a multiple is greater than the same multiple of another, is greater than that other magnitude.

PROPOSITION I.

THEOREM.—*If any number of magnitudes be equimultiples of as many, each of each; what multiple soever any one of them is of its part, the same multiple shall all the first magnitudes be of all the other.*

Let any number of magnitudes AB, CD be equimultiples of as many others E, F, each of each: whatsoever multiple AB is of E, the same multiple shall AB and CD together, be of E and F together.

Because AB is the same multiple of E that CD is of F, as many magnitudes as there are in AB equal to E, so many are there in CD equal to F. Divide AB into magnitudes equal to E, viz. AG, GB; and CD into CH, HD, equal each of them to F: therefore the number of the magnitudes CH, HD, shall be equal to the number of the others AG, GB: and because AG is equal to E, and CH to F, therefore AG and CH together, are equal

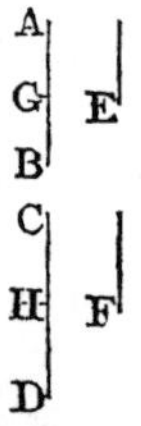

* 2 Ax. 1. to * E and F together: for the same reason, because GB is equal to E, and HD to F, GB and HD together, are equal to E and F together: wherefore, as many magnitudes as there are in AB equal to E, so many are there in AB, CD together, equal to E and F together: therefore, whatsoever multiple AB is of E, the same multiple is AB and CD together, of E and F together.

Therefore, if any magnitudes, how many soever, be equimultiples of as many, each of each; whatsoever multiple any one of them is of its part, the same multiple shall all the first magnitudes be of all the others: 'For the same demonstration holds in any number of magnitudes, which was here 'applied to two.' Q. E. D.

PROPOSITION II.

THEOR.—*If the first magnitude be the same multiple of the second that the third is of the fourth, and the fifth the same multiple of the second that the sixth is of the fourth; then shall the first together with the fifth be the same multiple of the second, that the third together with the sixth is of the fourth.*

Let AB the first be the same multiple of C the second, that DE the third is of F the fourth; and BG the fifth the same multiple of C the second, that EH the sixth is of F the fourth: then shall AG, the first together with the fifth, be the same multiple of C the second, that DH, the third together with the sixth, is of F the fourth.

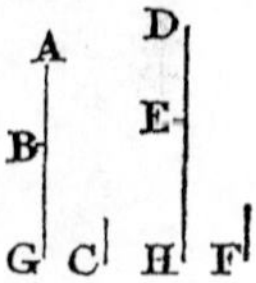

Because AB is the same multiple of C that DE is of F, there are as many magnitudes in AB equal to C, as there are in DE equal to F; in like manner, as many as there are in BG equal to C, so many are there in EH equal to F; therefore as many as there are in the whole AG equal to C, so many are there in the whole DH equal to F: therefore AG is the same multiple of C that DH is of F; that is AG, the first and fifth together, is the same multiple of the second C, that DH, the third and sixth together, is of the fourth F. If, therefore, the first magnitude be the same multiple, &c. Q. E. D.

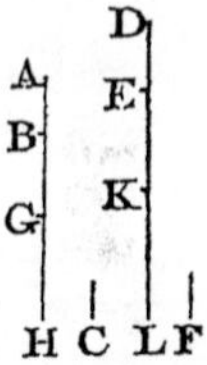

COR. "From this it is plain, that if any number of magnitudes AB, BG, GH, be multiples of another C; and as many DE, EK, KL, be the same multiples of F, each of each; then the whole of the first, viz. AH, is the same multiple of C, that the whole of the last, viz. DL, is of F."

PROPOSITION III.

THEOR.—*If the first be the same multiple of the second, which the third is of the fourth, and if of the first and third there be taken equimultiples, these shall be equimultiples, the one of the second, and the other of the fourth.*

Let A the first be the same multiple of B the second, that C the third is of D the fourth; and of A, C, let equimultiples EF, GH be taken: then EF shall be the same multiple of B, that GH is of D.

Because EF is the same multiple of A, that GH is of C, there are as many magnitudes in EF equal to A, as there are in GH equal to C: let EF be divided into the magnitudes EK, KF, each equal to A; and GH into GL, LH, each equal to C; therefore the number of the magnitudes EK, KF, shall be equal to the number of the others, GL, LH: and because A is the same multiple of B, that C is of D, and that EK is equal to A, and GL equal to C; therefore EK is the same multiple of B, that GL is of D: for the same reason, KF is the same multiple of B, that LH is of D: and so, if there be more parts in EF, GH, equal to A, C: therefore, because the first EK is the same multiple of the second B, which the third GL is of the fourth D, and that the fifth KF is the same multiple of the second B, which the sixth LH is of the fourth D; EF, the first together with the fifth,

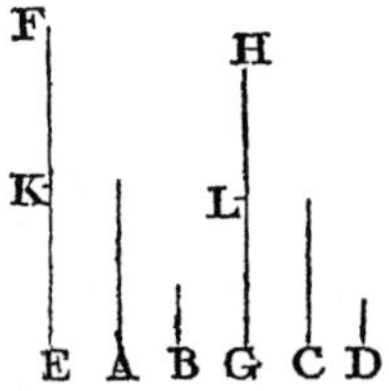

* 2. 5. is the same multiple * of the second B, which GH, the third together with the sixth, is of the fourth D. If, therefore, the first, &c. *Q. E. D.*

PROPOSITION IV.

See N. THEOR.—*If the first of four magnitudes has the same ratio to the second which the third has to the fourth, then any equimultiples whatever of the first and third, shall have the same ratio to any equimultiples of the second and fourth; viz. "the equimultiple of the first shall have the same ratio to that of the second, which the equimultiple of the third has to that of the fourth."*

Let A the first have to B the second, the same ratio which the third C has to the fourth D; and of A and C, let there be taken any equimultiples whatever E, F; and of B and D, any equimultiples whatever G, H: then E shall have the same ratio to G, which F has to H.

Take of E and F, any equimultiples whatever K, L, and of G, H, any equimultiples whatever M, N: then, because E is the same multiple of A, that F is of C, and of E and F have been taken equimultiples K, L, therefore K is the same multiple of A *, that L is of C: for the same reason, M is the same multiple of B, that N is of D. And because *, as A is to B, so is C to D, and of A and C have been taken certain equimultiples K, L, and of B and D have been taken certain equimultiples M, N, therefore if K be greater than M, L is greater than N; and if equal, equal; if less, less *: but K, L are any equimultiples † whatever of E, F, and M, N, any whatever of G, H, therefore as E is to G, so is * F to H. Therefore, if the first, &c. Q. E. D.

* 3. 5.

* Hyp.

* 5 Def. 5.

† Constr.

* 5 Def. 5.

COR. Likewise, if the first has the same ratio to the second, which the third has to the fourth, then also any equimultiples whatever of the first and third, shall have the same ratio to the second and fourth: and in like manner, the first and the third shall have the same ratio to any equimultiples whatever of the second and fourth.

Let A the first have to B the second, the same ratio which the third C has to the fourth D, and of A and C, let E and F be any equimultiples whatever; then E shall be to B, as F to D.

Take of E, F, any equimultiples whatever K, L, and of B, D, any equimultiples whatever G, H: then it may be demonstrated, as before, that K is the same multiple of A, that L is of C: and because † A is to B as C is to D, and of A and C, certain equimultiples have been taken, viz. K and L; and of B and D, certain equimultiples G, H, therefore if K be greater than G, L is greater than H; and if equal, equal; † Hyp.

* 5 Def. 5. if less, less*: but K, L are any† equimultiples whatever of
† Constr. E, F, and G, H any whatever of B, D; therefore, as E is to
† 5 Def. 5. B†, so is F to D. And in the same way the other case is demonstrated.

PROPOSITION V.

See N. THEOR.—*If one magnitude be the same multiple of another, which a magnitude taken from the first is of a magnitude taken from the other, the remainder shall be the same multiple of the remainder, that the whole is of the whole.*

Let the magnitude AB be the same multiple of CD, that AE taken from the first is of CF taken from the other: the remainder EB shall be the same multiple of the remainder FD, that the whole AB is of the whole CD.

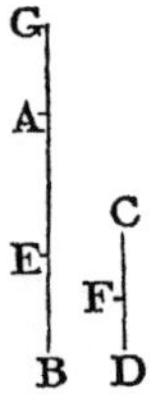

Take AG the same multiple of FD, that AE is
* 1. 5. of CF; therefore AE is* the same multiple of CF, that EG is of CD: but AE, by the hypothesis, is the same multiple of CF, that AB is of CD; therefore EG is the same multiple of CD that AB is of
* 1 Ax. 5. CD; wherefore EG is equal* to AB: take from each of them the common magnitude AE, and the remainder AG is equal to the remainder EB. Wherefore, since AE is the same mul-
† Constr. tiple of CF†, that AG is of FD, and that AG is equal to EB, therefore AE is the same multiple of CF, that EB is of FD:
† Hyp. but AE is the same multiple of CF† that AB is of CD; therefore EB is the same multiple of FD, that AB is of CD. Therefore, if one magnitude, &c. Q. E. D.

PROPOSITION VI.

See N. THEOR.—*If two magnitudes be equimultiples of two others, and if equimultiples of these be taken from the first two, the remainders are either equal to these others, or equimultiples of them.*

Let the two magnitudes AB, CD be equimultiples of the two E, F, and let AG, CH, taken from the first two, be equimultiples of the same E, F: the remainders GB, HD shall be either equal to E, F, or equimultiples of them.

First, let GB be equal to E: HD shall be equal to F. Make CK equal to F: and because AG is the same multiple of E†, that CH is of F, and that GB is equal † Hyp. to E, and CK to F, therefore AB is the same multiple of E, that KH is of F: but AB, by the hypothesis, is the same multiple of E, that CD is of F; therefore KH is the same multiple of F, that CD is of F; wherefore KH is equal* * 1 Ax. 5. to CD: take away the common magnitude CH, then the remainder KC is equal to the remainder HD: but KC is equal† † Constr. to F; therefore HD is equal to F.

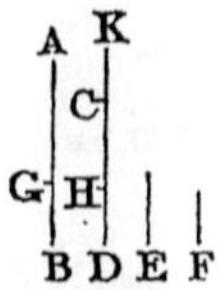

Next, let GB be a multiple of E: HD shall be the same multiple of F. Make CK the same multiple of F, that GB is of E: and because AG is the same multiple of E†, that CH is of F, and GB the † Hyp. same multiple of E, that CK is of F, therefore AB is the same multiple of E*, that KH is of F: * 2. 5. but AB is the same multiple of E†, that CD is † Hyp. of F; therefore KH is the same multiple of F, that CD is of F; wherefore KH is equal* to * 1 Ax. 5. CD: take away CH from both; therefore the remainder KC is equal to the remainder HD: and because GB is the same multiple of E†, that KC is of F, and that KC is equal to HD; † Constr. therefore HD is the same multiple of F, that GB is of E. If, therefore, two magnitudes, &c. Q. E. D.

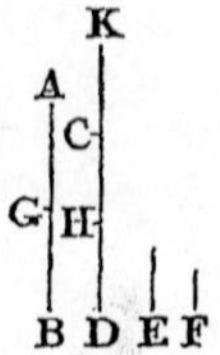

PROPOSITION A.

THEOR.—*If the first of four magnitudes has the same ratio to the second which the third has to the fourth, then, if the first be greater than the second, the third is also greater than the fourth; and if equal, equal; if less, less.* See N.

Take any equimultiples of each of them, as the doubles of each: then, by Def. 5th of this Book, if the double of the first be greater than the double of the second, the double of the third is greater than the double of the fourth: but if the first be greater than the second, the double of the first is greater than the double of the second; wherefore also the double of the third is greater than the double of the fourth; therefore the third is greater than the fourth: in like manner, if the first be equal to the second, or less than it, the third can be proved to be equal to the fourth, or less than it. Therefore, if the first, &c. Q. E. D.

PROPOSITION B.

See N. THEOR.—*If four magnitudes are proportionals, they are proportionals also when taken inversely.*

Let A be to B, as C is to D: then also inversely B shall be to A, as D to C.

Take of B and D, any equimultiples whatever E and F; and of A and C, any equimultiples whatever G and H. First, let E be greater than G, then G is less than E: and
† Hyp. because† A is to B as C is to D, and of A and C the first and third, G and H are equimultiples; and of B and D the second and fourth, E and F are equimultiples; and that G is less than E, there-
* 5 Def. 5. fore H is* less than F; that is, F is greater than H; if, therefore, E be greater than G, F is greater than H: in like manner, if E be equal to G, F may be shewn to be equal to H; and if less, less; but E,
† Constr. F are any equimultiples† whatever of B and D, and G, H any
† 5 Def. 5. whatever of A and C; therefore† as B is to A, so is D to C. Therefore, if four magnitudes, &c. Q. E. D.

PROPOSITION C.

See N. THEOR.—*If the first be the same multiple of the second, or the same part of it, that the third is of the fourth, the first is to the second, as the third is to the fourth.*

Let the first A be the same multiple of the second B, that the third C is of the fourth D: A shall be to B as C is to D.

Take of A and C, any equimultiples whatever E and F; and of B and D, any equimultiples
† Hyp. whatever G and H: then, because A is the same† multiple of B that C is of D, and that E is the
† Constr. same† multiple of A that F is of C, therefore E is
* 3. 5. the same multiple of B* that F is of D; that is, E and F are equimultiples of B and D: but G and
† Constr. H are equimultiples† of B and D; therefore, if E be a greater multiple of B, than G is of B, F is a greater multiple of D, than H is of D; that is, if E be greater than G,

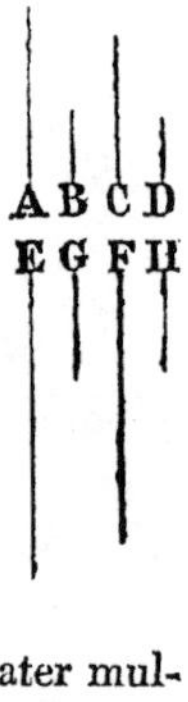

F is greater than H: in like manner, if E be equal to G, or less than it, F may be shewn to be equal to H, or less than it: but E, F are equimultiples†, any whatever, of A, C; and G, † Constr. H any equimultiples whatever of B, D; therefore† A is to B, † 5 Def. 5. as C is to D.

Next, let the first A be the same part of the second B, that the third C is of the fourth D: A shall be to B, as C is to D.

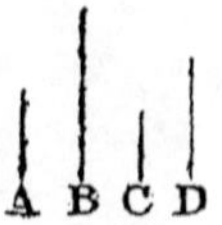

For since A is the same part of B that C is of D, therefore B is the same multiple of A that D is of C: wherefore, by the preceding case, B is to A, as D is to C; and therefore inversely* A is to * B. 5. B, as C is to D. Therefore, if the first be the same multiple, &c. Q. E. D.

PROPOSITION D.

THEOR.—*If the first be to the second as the third to the fourth, and if the first be a multiple, or a part of the second; the third is the same multiple, or the same part of the fourth.* See N.

Let A be to B as C is to D; and first, let A be a multiple of B: C shall be the same multiple of D.

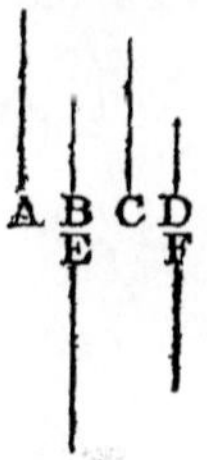

Take E equal to A; and whatever multiple A or E is of B, make F the same multiple of D: then, because† A is to B, as C is to D; and of B † Hyp. the second, and D the fourth, equimultiples have been taken, E and F*; therefore A is to E, as C * Cor. 4. 5. to F: but A is equal† to E, therefore C is equal* † Constr. * A. 5. to F: and F is the same† multiple of D, that A is of B: therefore C is the same multiple of D † Constr. See the last figure. that A is of B.

Next let A be a part of B: C shall be the same part of D.

Because† A is to B, as C is to D; then inversely, B is* to † Hyp. * B. 5. A, as D to C: but A is a part† of B, that is, B is a multiple † Hyp. of A; therefore, by the preceding case, D is the same multiple of C; that is, C is the same part of D, that A is of B. Therefore, if the first, &c. Q. E. D.

PROPOSITION VII.

THEOR.—*Equal magnitudes have the same ratio to the same magnitude: and the same has the same ratio to equal magnitudes.*

Let A and B be equal magnitudes, and C any other: A and B shall each of them have the same ratio to C: and C shall have the same ratio to each of the magnitudes A and B.

Take of A and B, any equimultiples whatever D and E; and of C, any multiple whatever F: then, because D is the

† Constr. same† multiple of A, that E is of B, and that A is equal† to B, therefore D is* equal to E: therefore, if D be greater than F, E is greater than F; and if equal, equal; if less, less; but D, E are any equimultiples of A, B†, and F is any multiple of C; therefore*, as A is to C, so is B to C.

† Hyp.
* 1 Ax. 5.
† Constr.
* 5 Def. 5.

Likewise C shall have the same ratio to A, that it has to B. For, having made the same construction, D may, in like manner, be shewn to be equal to E: therefore, if F be greater than D, it is likewise greater than E; and if equal, equal; if less, less: but F is any multiple whatever of C, and D, E are any equimultiples whatever of A, B; therefore*, C is to A as C is to B. Therefore, equal magnitudes, &c. Q. E. D.

* 5 Def. 5.

PROPOSITION VIII.

See N.

THEOR.—*Of two unequal magnitudes, the greater has a greater ratio to any other magnitude than the less has: and the same magnitude has a greater ratio to the less of two other magnitudes, than it has to the greater.*

Let AB, BC be two unequal magnitudes, of which AB is the greater; and let D be any other magnitude: AB shall have a greater ratio to D, than BC has to D: and D shall have a greater ratio to BC, than it has to AB.

Fig. 1.

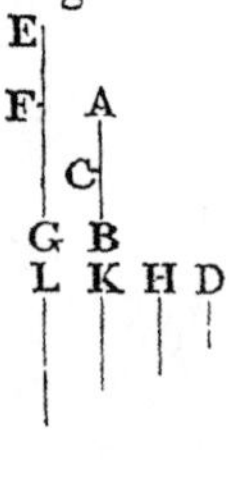

If the magnitude which is not the greater of the two AC, CB, be not less than D, take EF, FG, the doubles of AC, CB, as in Fig. 1. But if that which is not the greater of the two AC, CB, be less than D (as in Fig. 2. and 3.), this magnitude can be multiplied, so as to become greater than D, whether it be AC or CB. Let it be multiplied until it become greater than D, and let the other be multiplied as often; and let EF be the multiple thus taken of AC, and FG the same multiple of CB; therefore EF and FG are each of them greater than D: and in every one of the cases, take H the double of D, K its triple, and so on, till the multiple of D be that which first becomes greater than FG: let L be that multiple of D which is first greater than FG, and K the multiple of D, which is next less than L.

Then, because L is the multiple of D which is the first that becomes greater than FG, the next preceding multiple K is not greater than FG; that is, FG is not less than K: and since EF is the same multiple of AC †, that FG is of CB, † Constr. therefore FG is the same multiple of CB*, that EG is of AB; that is, EG and * L. 5. FG are equimultiples of AB and CB: and since it was shewn, that FG is not less than K, and, by the construction, EF is greater than D, therefore the whole EG is greater than K and D together: but K together with D is equal† to L; therefore EG is greater † Constr. than L: but FG is not greater† than † Constr. L: and EG, FG were proved to be equimultiples of AB, BC; and L is a † multiple of D; therefore* AB has to D a greater † Constr. ratio, than BC has to D. * 7 Def. 5.

Fig. 2. Fig. 3.

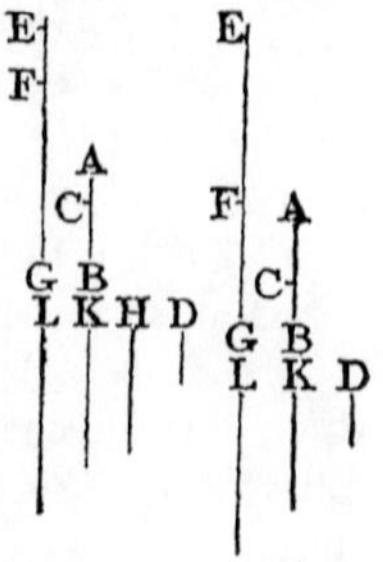

Also, D shall have to BC a greater ratio than it has to AB. For having made the same construction, it may be shewn, in like manner, that L is greater than FG, but that it is not greater than EG: and L is a † multiple of D; and FG, EG † Constr. were proved to be equimultiples of CB, AB; therefore D has to CB a greater ratio * than it has to AB. Wherefore, of two * 7 Def. 5. unequal magnitudes, &c. Q. E. D.

PROPOSITION IX.

THEOR.— *Magnitudes which have the same ratio to the same magnitude, are equal to one another: and those to which the same magnitude has the same ratio, are equal to one another.* See N.

Let A, B have each of them the same ratio to C: A shall be equal to B.

For, if they are not equal, one of them must be greater than the other; let A be the greater: then, by what was shewn in the preceding proposition, there are some equimultiples of A and B, and some multiple of C, such, that the multiple of A is greater than the multiple of C, but the multiple of B is not greater than that of C. Let these multiples be taken; and let D, E be the equimultiples of A, B, and F the multiple of C, such, that D may be greater than F, but E not greater than F: († Hyp.) then, because A is to C as † B is to C, and of A, B, are taken equimultiples D, E, and of C, is taken a multiple F, and that D is greater than F, therefore E is also greater * than F: (* 5 Def. 5.) but E is not † greater than F; († Constr.) which is impossible: therefore A and B are not unequal; that is, they are equal.

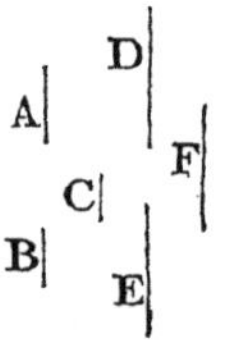

Next, let C have the same ratio to each of the magnitudes A and B: A shall be equal to B.

For, if they are not equal, one of them must be greater than the other; let A be the greater: therefore, as was shown in Prop. 8th, there is some multiple F of C, and some equimultiples E and D, of B and A, such, that F is greater than E, but not greater than D: and because C is to B †, as C is to A, († Hyp.) and that F the multiple of the first is greater than E the multiple of the second *, (* 5 Def. 5.) therefore F the multiple of the third is greater than D the multiple of the fourth: but F is not † († Constr.) greater than D; which is impossible. Therefore A is equal to B. Wherefore, magnitudes which, &c. Q. E. D.

PROPOSITION X.

(See N.) THEOR.—*That magnitude which has a greater ratio than another has unto the same magnitude, is the greater of the two: and that magnitude to which the same has a greater ratio than it has unto another magnitude, is the lesser of the two.*

Let A have to C a greater ratio than B has to C: A shall be greater than B.

For, because A has a greater ratio to C, than B has to C, there are * (* 7 Def. 5.) some equimultiples of A and B, and some multiple of C, such, that the multiple of A is greater than the multiple of C, but the multiple of B is not greater than it: let them be taken; and let D, E be the equimultiples of A, B, and F the multiple of C, such, that D is greater than F, but E is not greater than F; therefore D

is greater than E: and, because D and E are equimultiples of A and B, and that D is greater than E, therefore A is * greater than B. * 4 Ax. 5.

Next, let C have a greater ratio to B than it has to A: B shall be less than A.

For * there is some multiple F of C, and some equimultiples E and D of B and A, such, that F is greater than E, but not greater than D; therefore E is less than D: and because E and D are equimultiples of B and A, and that E is less than D, therefore B is * less than A. Therefore, that magnitude, &c. Q. E. D. * 7 Def. 5. * 4 Ax. 5.

PROPOSITION XI.

THEOR.—*Ratios that are the same to the same ratio, are the same to one another.*

Let A be to B, as C is to D; and as C to D, so let E be to F: A shall be to B, as E to F.

Take of A, C, E, any equimultiples whatever G, H, K; and of B, D, F, any equimultiples whatever L, M, N. Therefore, since A is to B as C to D, and G, H are taken equimultiples of A, C, and L, M, of B, D; if G be greater than L, H is greater than M; and if equal, equal;

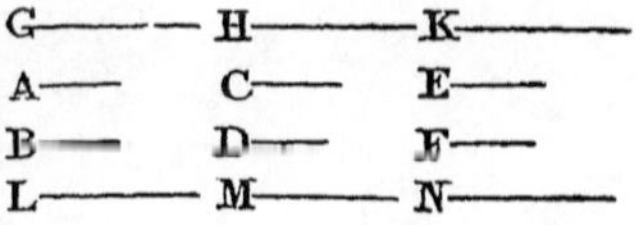

and if less, less *. Again, because C is to D, as E is to F, and H, K are taken equimultiples of C, E, and M, N, of D, F; if H be greater than M, K is greater than N; and if equal, equal; and if less, less: but if G be greater than L, it has been shewn that H is greater than M; and if equal, equal; and if less, less; therefore if G be greater than L, K is greater than N; and if equal, equal; and if less, less: and G, K are any equimultiples whatever of A, E; and L, N any whatever of B, F: therefore * as A is to B, so is E to F. Wherefore, ratios that, &c. Q. E. D. * 5 Def. 5. * 5 Def. 5.

PROPOSITION XII.

THEOR.—*If any number of magnitudes be proportionals, as one of the antecedents is to its consequent, so shall all the antecedents taken together, be to all the consequents.*

Let any number of magnitudes A, B, C, D, E, F be proportionals; that is, as A is to B, so C to D, and E to F: as A is to B, so shall A, C, E together, be to B, D, F together.

Take of A, C, E, any equimultiples whatever G, H, K; and of B, D, F, any equimultiples whatever L, M, N: then because A is to B, as C is to D, and as E to F, and that G, H, K are equimultiples of A, C, E, and L, M, N equimultiples of B, D, F; if G be greater than L, H is greater than M, and K greater than N; and if equal, equal;

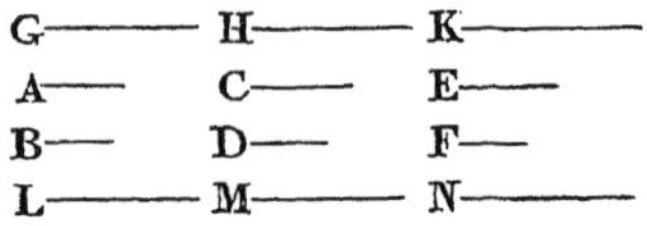

* 5 Def. 5. and if less, less *: wherefore if G be greater than L, then G, H, K together, are greater than L, M, N together; and if equal, equal; and if less, less: but G, and G, H, K together, are any equimultiples of A, and A, C; E together; because if there be any number of magnitudes equimultiples of as many, each of each, whatever multiple one of them is of its

* 1. 5. part, the same multiple is the whole of the whole *: for the same reason, L, and L, M, N are any equimultiples of B, and

† 5 Def. 5. B, D, F: therefore as A is to B †, so are A, C, E together, to B, D, F together. Wherefore, if any number, &c. Q. E. D.

PROPOSITION XIII.

See N. THEOR.—*If the first has to the second the same ratio which the third has to the fourth, but the third to the fourth a greater ratio than the fifth has to the sixth, the first shall also have to the second a greater ratio than the fifth has to the sixth.*

Let A the first have the same ratio to B the second, which C the third has to D the fourth, but C the third a greater ratio to D the fourth, than E the fifth has to F the sixth: also the first A shall have to the second B, a greater ratio than the fifth E has to the sixth F.

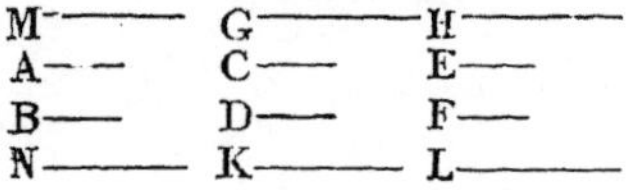

Because C has a greater ratio to D, than E to F, there are some equimultiples of C and E, and some of D and F, such, that the multiple of C is greater than the multiple of D, but the multiple of E is not greater than the multiple of F *: let these be taken, and let G, H be equimultiples of C, E, and K, L equimultiples of D, F, such, that G may be greater than K, but H not greater than L: and whatever multiple G is of C, take M the same multiple of A; and whatever multiple K is of D, take N the same multiple of B: then, because A is to B †, as C to D, and of A and C, M and G are equimultiples; and of B and D, N and K are equimultiples; if M be greater than N, G is greater than K; and if equal, equal; and if less, less *: but G is greater † than K, therefore M is greater than N: but H is not † greater than L: and M, H are equimultiples of A, E; and N, L equimultiples of B, F; therefore A has a greater ratio to B, than E has to F *. Wherefore, if the first, &c. Q. E. D.

* 7 Def. 5.
† Hyp.
* 5 Def. 5.
† Constr.
† Constr.
* 7 Def. 5.

COR. And if the first have a greater ratio to the second, than the third has to the fourth, but the third the same ratio to the fourth, which the fifth has to the sixth, it may be demonstrated, in like manner, that the first has a greater ratio to the second, than the fifth has to the sixth.

PROPOSITION XIV.

THEOR.—*If the first have the same ratio to the second which the third has to the fourth, then, if the first be greater than the third, the second shall be greater than the fourth; and if equal, equal; and if less, less.* See N.

Let the first A have the same ratio to the second B, which the third C has to the fourth D: if A be greater than C, B shall be greater than D.

Because A is greater than C, and B is any other magnitude, A has to B a greater ratio than C has to B *: but, as A is to B †, so is C to D; therefore also C has to D a greater ratio than C has to B *: but of two magnitudes, that to which the same has the greater ratio is the lesser *: therefore D is less than B; that is, B is greater than D.

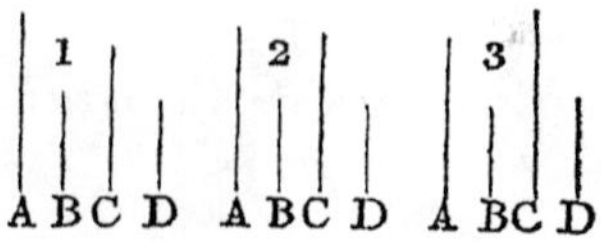

* 8. 5.
† Hyp.
* 13. 5.
* 10. 5.

Secondly, if A be equal to C, B shall be equal to D. For A is to B, as C, that is, A to D; therefore B is equal to D *. * 9. 5.

Thirdly, if A be less than C, B shall be less than D. For C is greater than A; and because C is to D, as A is to B, therefore D is greater than B, by the first case; that is, B is less than D. Therefore, if the first, &c. Q. E. D.

PROPOSITION XV.

THEOR.—*Magnitudes have the same ratio to one another which their equimultiples have.*

Let AB be the same multiple of C, that DE is of F: C shall be to F, as AB to DE.

Because AB is the same multiple of C, that DE is of F, there are as many magnitudes in AB equal to C, as there are in DE equal to F: let AB be divided into magnitudes, each equal to C, viz. AG, GH, HB; and DE into magnitudes, each equal to F, viz. DK, KL, LE: then the number of the first, AG, GH, HB, is equal to the number of the last, DK, KL, LE; and because AG, GH, HB are all equal, and that DK, KL, LE are also equal to one another, there-

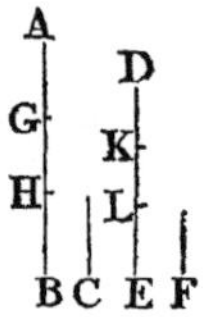

* 7. 5. fore * AG is to DK as GH to KL, and as HB to LE: but as
* 12. 5. one of the antecedents to its consequent *, so are all the antecedents together to all the consequents together; wherefore, as AG is to DK, so is AB to DE: but AG is equal to C, and DK to F: therefore, as C is to F, so is AB to DE. Therefore, magnitudes, &c. Q. E. D.

PROPOSITION XVI.

THEOR.—*If four magnitudes* of the same kind *be proportionals, they shall also be proportionals when taken alternately.*

Let A, B, C, D be four magnitudes of the same kind, which are proportionals; viz. as A to B, so C to D: they shall also be proportionals when taken alternately; that is, A shall be to C, as B to D.

Take of A and B, any equimultiples whatever E and F; and of C and D, take any equimultiples whatever G and H: and because E is the same multiple of A, that F is of B, and that magnitudes have the same ratio to one

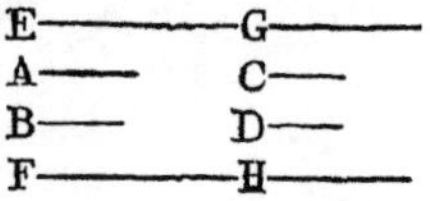

* 15. 5. another * which their equimultiples have, therefore A is to B, as E is to F:
† Hyp. but as A is to B so † is C to D; wherefore as C is to D, so * is
* 11. 5. E to F: again, because G, H are equimultiples of C, D, there-

fore as C is to D, so * is G to H: but it was proved that as C is to D, so is E to F, therefore, as E is to F, so * is G to H. But when four magnitudes are proportionals *, if the first be greater than the third, the second is greater than the fourth; and if equal, equal; if less, less; therefore if E be greater than G, F likewise is greater than H; and if equal, equal; if less, less: and E, F are any † equimultiples whatever of A, B; and G, H any whatever of C, D: therefore *, A is to C as B to D. If, then, four magnitudes, &c. Q. E. D.

* 15. 5.
* 11. 5.
* 14. 5.
† Constr.
* 5 Def. 5.

PROPOSITION XVII.

THEOR.—*If magnitudes, taken jointly, be proportionals, they shall also be proportionals when taken separately; that is, if two magnitudes together, have to one of them, the same ratio which two others have to one of these, the remaining one of the first two shall have to the other, the same ratio which the remaining one of the last two has to the other of these.* See N.

Let AB, BE, CD, DF be the magnitudes, taken jointly, which are proportionals; that is, as AB to BE, so let CD be to DF: they shall also be proportionals taken separately; viz. as AE to EB, so shall CF be to FD.

Take of AE, EB, CF, FD, any equimultiples whatever GH, HK, LM, MN; and again, of EB, FD, take any equimultiples whatever KX, NP: and because GH is the same multiple of AE, that HK is of EB, therefore GH is the same multiple * of AE, that GK is of AB: but GH is the same multiple of AE, that LM is of CF; therefore GK is the same multiple of AB, that LM is of CF. Again, because LM is the same multiple of CF, that MN is of FD, therefore LM is the same multiple * of CF, that LN is of CD: but LM was shewn to be the same multiple of CF, that GK is of AB; therefore GK is the same multiple of AB, that LN is of CD; that is, GK, LN are equimultiples of AB, CD. Next, because HK is the same multiple of EB, that MN is of FD, and that KX is also the same multiple of EB, that NP is of FD, therefore HX is the same multiple * of EB, that MP is of FD. And because AB is to BE †, as CD is to DF, and that of AB and CD, GK and

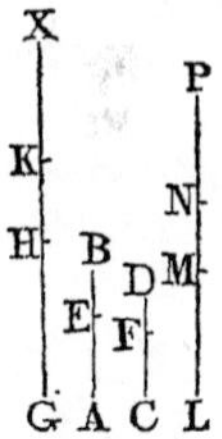

* 1. 5.
* 1. 5.
* 2. 5.
† Hyp.

LN are equimultiples, and of EB and FD, HX and MP are
* 5 Def. 5. equimultiples *, therefore, if GK be greater than HX, then LN is greater than MP; and if equal, equal; and if less, less: but if GH be greater than KX, then, by adding the common
† 4 Ax. 1. part HK to both, GK is greater† than HX: wherefore also LN is greater than MP; and by taking away MN from both, LM
† 5 Ax. 1. is greater† than NP: therefore, if GH be greater than KX, LM is greater than NP. In like manner it may be demonstrated, that if GH be equal to KX, LM is equal to NP; and if less, less:
† Constr. but GH, LM are any equimultiples whatever of AE, CF†,
* 5 Def. 5. and KX, NP are any whatever of EB, FD: therefore *, as AE is to EB, so is CF to FD. If, then, magnitudes, &c. Q. E. D.

PROPOSITION XVIII.

See N. THEOR.—*If magnitudes, taken separately, be proportionals, they shall also be proportionals when taken jointly; that is, if the first be to the second, as the third to the fourth, the first and second together, shall be to the second, as the third and fourth together, to the fourth.*

Let AE, EB, CF, FD be proportionals; that is, as AE to EB, so let CF be to FD: they shall also be proportionals when taken jointly; that is, as AB to BE, so shall CD be to DF.

Take of AB, BE, CD, DF, any equimultiples whatever GH, HK, LM, MN; and again, of BE, DF, take any equimultiples whatever KO, NP: and because KO, NP are equimultiples of BE, DF, and that KH, NM are likewise equimultiples of BE, DF, therefore if KO, the multiple of BE, be greater than KH, which is a multiple of the same BE, then NP, the multiple of DF, is also greater than NM, the multiple of the same DF; and if KO be equal to KH, NP is equal to NM; and if less, less.

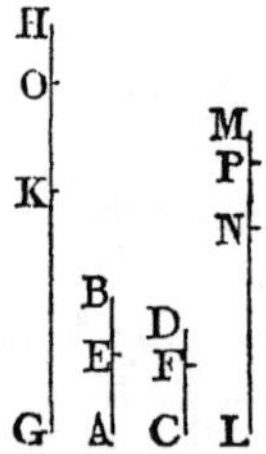

First, let KO be not greater than KH; therefore NP is not greater than NM: and because GH, HK are equimultiples of AB, BE, and that AB is greater than BE, therefore GH is
* 8 Ax. 5. greater* than HK; but KO is not greater than KH; therefore GH is greater than KO. In like manner it may be shewn, that LM is greater than NP: therefore, if KO be not greater than KH, then GH, the multiple of AB, is always greater than

KO, the multiple of BE; and likewise LM, the multiple of CD, is greater than NP, the multiple of DF.

Next, let KO be greater than KH, therefore, as has been shewn, NP is greater than NM: and because the whole GH is the same multiple of the whole AB, that HK is of BE, therefore the remainder GK is the same multiple of the remainder
AE* that GH is of AB; which is the same that * 5. 5.
LM is of CD. In like manner, because LM is
the same multiple of CD, that MN is of DF,
therefore the remainder LN is the same mul-
tiple of the remainder CF*, that the whole * 5. 5.
LM is of the whole CD: but it was shewn,
that LM is the same multiple of CD, that GK
is of AE; therefore GK is the same multiple
of AE, that LN is of CF; that is, GK, LN are equimultiples of AE, CF. And because KO, NP are equimultiples of BE, DF, therefore if from KO, NP there be taken KH, NM, which are likewise equimultiples of BE, DF, the remainders HO, MP
are either equal to BE, DF, or equimultiples of them *. First, * 6. 5.
let HO, MP be equal to BE, DF: then because † AE is to EB, † Hyp.
as CF to FD, and that GK, LN are equimultiples of AE, CF,
therefore GK is to EB*, as LN to FD: but HO is equal to EB, * Cor. 4. 5.
and MP to FD; wherefore GK is to HO, as LN to MP: there-
fore if GK be greater than HO, LN is greater than* MP; and * A. 5.
if equal, equal; and if less, less.

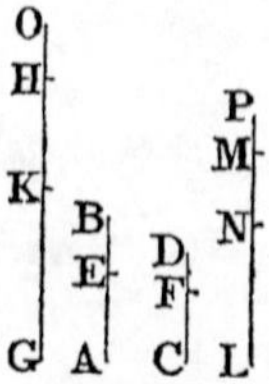

But let HO, MP be equimultiples of EB, FD: then†, be- † Hyp.
cause AE is to EB, as CF to FD, and that of AE, CF are
taken equimultiples GK, LN; and of EB, FD, the equimul-
tiples HO, MP; if GK be greater than HO, LN is greater
than MP; and if equal, equal: and if less, less*; which was * 5 Def. 5.
likewise shewn in the preceding case. But
if GH be greater than KO, taking KH from
both, GK is greater † than HO; wherefore † 5 Ax. 1.
also LN is greater than MP; and conse-
quently, adding NM to both, LM is greater† † 4 Ax. 1.
than NP: therefore, if GH be greater than
KO, LM is greater than NP. In like man-
ner it may be shewn, that if GH be equal to
KO, LM is equal to NP; and if less, less.
And in the case in which KO is not greater than KH, it has been shewn that GH is always greater than KO, and likewise LM greater than NP: but GH, LM are any equimultiples

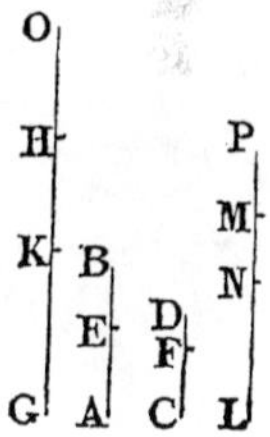

† Constr. whatever of AB, CD†, and KO, NP are any whatever of BE,
* 5 Def. 5. DF; therefore*, as AB is to BE, so is CD to DF. If, then, magnitudes, &c. Q. E. D.

PROPOSITION XIX.

See N. THEOR.—*If a whole magnitude be to a whole, as a magnitude taken from the first is to a magnitude taken from the other, the remainder shall be to the remainder, as the whole to the whole.*

Let the whole AB be to the whole CD, as AE a magnitude taken from AB, is to CF a magnitude taken from CD: the remainder EB shall be to the remainder FD, as the whole AB to the whole CD.

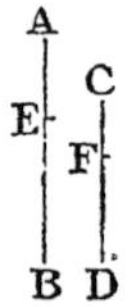

Because AB is to CD, as AE to CF, therefore, al-
* 16. 5. ternately*, BA is to AE, as DC to CF: and because, if magnitudes taken jointly be proportionals, they are
* 17. 5. also proportionals * when taken separately, therefore, as BE is to EA, so is DF to FC; and alternately, as BE is to DF, so is EA to FC: but as AE to CF, so, by the hypothesis, is AB to CD; therefore also,
† 11. 5. BF the remainder is to the remainder DF †, as the whole AB to the whole CD. Wherefore, if the whole, &c. Q. E. D.

COR. If the whole be to the whole, as a magnitude taken from the first, is to a magnitude taken from the other, the remainder shall likewise be to the remainder, as the magnitude taken from the first to that taken from the other. The demonstration is contained in the preceding.

PROPOSITION E.

THEOR.—*If four magnitudes be proportionals, they are also proportionals by conversion; that is, the first is to its excess above the second, as the third to its excess above the fourth.*

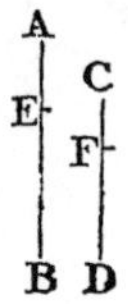

Let AB be to BE, as CD to DF: then BA shall be to AE, as DC to CF.

Because AB is to BE, as CD to DF, therefore by
* 17. 5. division *, AE is to EB, as CF to FD; and by inver-
* B. 5. sion *, BE is to EA, as DF to FC; wherefore, by
* 18. 5. composition *, BA is to AE, as DC is to CF. If, therefore, four, &c. Q. E. D.

PROPOSITION XX.

THEOR.—*If there be three magnitudes, and other three, which, taken two and two, have the same ratio, then if the first be greater than the third, the fourth shall be greater than the sixth; and if equal, equal; and if less, less.* See N.

Let A, B, C be three magnitudes, and D, E, F, other three, which taken two and two have the same ratio; viz. as A is to B, so is D to E; and as B to C, so is E to F: if A be greater than C, D shall be greater than F; and if equal, equal; and if less, less.

Because A is greater than C, and B is any other magnitude, and that the greater has to the same magnitude a greater* ratio than the less has to it, therefore A has to B a greater ratio than C has to B: but as D is to E †, so is A to B; therefore * D has to E a greater ratio than C to B: and because B is to C, as E to F, by inversion †, C is to B, as F is to E: and D was shewn to have to E, a greater ratio than C to B; therefore D has to E a greater * ratio than F to E: but the magnitude which has a greater ratio than another to the same magnitude, is the greater* of the two; therefore D is greater than F.

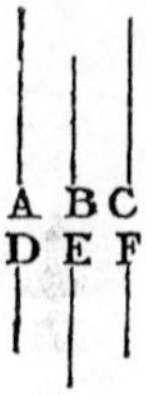

* 8. 5. † Hyp. * 13. 5. † B. 5. * Cor. 13. 5. * 10. 5.

Secondly, let A be equal to C: D shall be equal to F. Because A and C are equal to one another, A is to B, as C is to B*: but† A is to B, as D to E; and† C is to B as F to E; wherefore D is to E, as F to E*; and therefore D is equal to F*.

* 7. 5. † Hyp. † Hyp. & B. 5. * 11. 5. * 9. 5.

Next, let A be less than C: D shall be less than F. For C is greater than A; and, as was shewn in the first case, C is to B, as F to E; and in like manner, B is to A, as E to D; therefore F is greater than D, by the first case; that is, D is less than F. Therefore, if there be three, &c. Q. E. D.

PROPOSITION XXI.

THEOR.—*If there be three magnitudes, and other three, which have the same ratio taken two and two, but in a cross order, then if the first magnitude be greater than the third, the fourth shall be greater than the sixth; and if equal, equal; and if less, less.* See N.

Let A, B, C be three magnitudes, and D, E, F other three, which have the same ratio, taken two and two, but in a cross order; viz. as A is to B, so is to E to F, and as B is to C, so is D to E: if A be greater than C, D shall be greater than F; and if equal, equal; and if less, less.

Because A is greater than C, and B is any other
* 8. 5. magnitude, A has to B a greater ratio* than C has
† Hyp. to B: but† as E to F, so is A to B; therefore*
* 13. 5. E has to F a greater ratio than C to B: and be-
† Hyp. cause† B is to C, as D to E, by inversion, C is to B, as E to D: and E was shewn to have to F a greater ratio than C has to B; therefore E has to F
* Cor. 13. 5. a greater ratio than E has to* D: but the magnitude to which the same has a greater ratio than it has to an-
* 10. 5. other, is the lesser* of the two: therefore F is less than D; that is, D is greater than F.

Secondly, let A be equal to C: D shall be equal to F. Be-
* 7. 5. cause A and C are equal, A is* to B, as C is to B: but A is
† Hyp. to B†, as E to F; and C is to B, as E to D;
* 11. 5. wherefore E is to F*, as E to D; and
* 9. 5. therefore D is equal* to F.

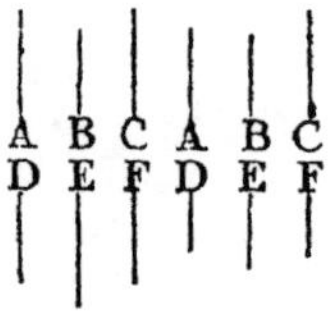

Next, let A be less than C: D shall be less than F. For C is greater than A; and, as was shewn, C is to B, as E to D, and in like manner B is to A, as F to E; therefore F is greater than D, by case first; that is, D is less than F. Therefore, if there be three, &c. Q. E. D.

PROPOSITION XXII.

See N. THEOR.—*If there be any number of magnitudes, and as many others, which, taken two and two in order, have the same ratio, the first shall have to the last of the first magnitudes, the same ratio which the first has to the last of the others. N. B.* This is usually cited by the words "ex æquali," or "ex æquo."

First, let there be three magnitudes, A, B, C, and as many others D, E, F, which, taken two and two, have the same ratio; that is, such, that A is to B as D to E; and as B is to C, so is E to F: A shall be to C, as D to F.

Take of A and D, any equimultiples whatever G and H; and of B and E, any equimultiples whatever K and L; and of C and F, any whatever M and N: then, because A is to B, as D to E, and that G, H are equimultiples of A, D, and K, L equimultiples of B, E, therefore as G is to K, so is * H to L: for the same reason, K is to M as L to N: and because there are three magnitudes G, K, M, and other three H, L, N, which, two and two, have the same ratio *, therefore if G be greater than M, H is greater than N; and if equal, equal; and if less, less: but G, H are any equimultiples whatever of A, D †, and M, N are any equimultiples whatever of C, F; therefore *, as A is to C, so is D to F.

* 4. 5.

* 20. 5.

† Constr.

* 5 Def. 5.

Next, let there be four magnitudes, A, B, C, D, and other four E, F, G, H, which, two and two, have the same ratio; viz. as A is to B, so is E to F: and as B to C, so F to G; and as C to D, so G to H: A shall be to D, as E to H.

Because A, B, C are three magnitudes, and E, F, G, other three, which, taken two and two, have the same ratio, therefore, by the foregoing case, A is to C, as E to G: but C is to D, as G is to H; wherefore again, by the first case, A is to D, as E to H: and so on, whatever be the number of magnitudes. Therefore, if there be any number, &c. Q. E. D.

PROPOSITION XXIII.

THEOR.—*If there be any number of magnitudes, and as many others, which, taken two and two in a cross order, have the same ratio, the first shall have to the last of the first magnitudes, the same ratio which the first has to the last of the others.* *N.B.* This is usually cited by the words "ex æquali in proportione perturbatâ;" or "ex æquo perturbato." See N.

First, let there be three magnitudes A, B, C, and other three, D, E, F, which, taken two and two in a cross order, have the same ratio; that is, such, that A is to B, as E to F; and as B is to C, so is D to E: A shall be to C, as D to F.

Take of A, B, D, any equimultiples whatever G, H, K; and of C, E, F, any equimultiples whatever L, M, N: and because G, H are equimultiples of A, B, and that
* 15. 5. magnitudes have the same ratio* which their equimultiples have, therefore as A is to B, so is G to H: and for the same reason, as E
† Hyp. is to F, so is M to N: but† as A is to B, so
* 11. 5. is E to F; therefore as G is to H*, so is M
† Hyp. to N: and because† as B is to C, so is D to E, and that H, K, are equimultiples of B, D, and L, M of C, E, therefore as H is to
* 4. 5. L, so is* K to M: and it has been shewn that G is to H, as M to N: therefore because there are three magnitudes G, H, L, and other three K, M, N, which have the same ratio taken two and two in a cross order; if G be greater than L, K is greater than N; and if equal, equal; and if less,
* 21. 5. less*: but G, K are any equimultiples† whatever of A, D;
† Constr.
† 5 Def. 5. and L, N any whatever of C, F; therefore as A is to C†, so is D to F.

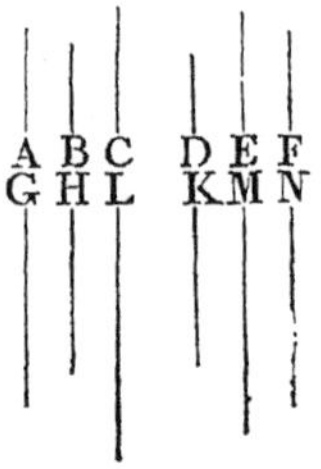

Next, let there be four magnitudes A, B, C, D, and other four E, F, G, H, which, taken two and two in a cross order, have the same ratio; viz. A to B, as G to H; B to C, as F to G; and C to D, as E to F: A shall be to D, as E to H.

A.B.C.D
E.F.G.H

Because A, B, C are three magnitudes, and F, G, H other three, which, taken two and two in a cross order, have the same ratio; by the first case, A is to C, as F to H: but C is to D, as E is to F; wherefore again, by the first case, A is to D, as E to H: and so on, whatever be the number of magnitudes. Therefore, if there be any number, &c. Q. E. D.

PROPOSITION XXIV.

See N. **Theor.**—*If the first has to the second, the same ratio which the third has to the fourth, and the fifth to the second, the same ratio which the sixth has to the fourth, the first and fifth together shall have to the second, the same ratio which the third and sixth together have to the fourth.*

Let AB the first have to C the second, the same ratio which DE the third has to F the fourth; and let BG the fifth have to C the second, the same ratio which EH the sixth has to F the fourth: AG, the first and fifth together, shall have to C the second, the same ratio which DH, the third and sixth together, has to F the fourth.

Because BG is to C, as EH to F, by inversion†, C is to BG, as F to EH: and because, as AB is to C†, so is DE to F; and as C to BG, so F to EH; ex æquali*, AB is to BG, as DE to EH: and because these magnitudes are proportionals, they are likewise proportionals when taken* jointly; therefore as AG is to GB, so is DH to HE: but† as GB to C, so is HE to F; therefore ex æquali* as AG is to C, so is DH to F. Wherefore, if the first, &c. Q. E. D.

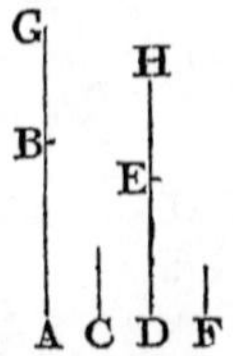

† B. 5.
† Hyp.
* 22. 5.
* 18. 5.
† Hyp.
* 22. 5.

Cor. 1. If the same hypothesis be made as in the proposition, the excess of the first and fifth shall be to the second, as the excess of the third and sixth to the fourth. The demonstration of this is the same with that of the proposition, if division be used instead of composition.

Cor. 2. The proposition holds true of two ranks of magnitudes, whatever be their number, of which each of the first rank has to the second magnitude, the same ratio that the corresponding one of the second rank has to a fourth magnitude; as is manifest.

PROPOSITION XXV.

Theor.—*If four magnitudes of the same kind are proportionals, the greatest and least of them together, are greater than the other two together.* See N.

Let the four magnitudes AB, CD, E, F be proportionals; viz. AB to CD, as E to F; and let AB be the greatest of them, and consequently F the least*: AB together with F, shall be greater than CD together with E. * A. & 14. 5.

Take AG equal to E, and CH equal to F: then because as AB is to CD, so is E to F, and that AG is equal to E, and CH equal to F, therefore AB is to CD†, as AG to CH: and because AB the whole, is to the whole CD, as AG is to CH, likewise the remainder GB is to the remainder HD, as the whole AB is to the whole* CD; but AB is greater† than CD; therefore* GB is greater than HD: and because AG is equal to E, and CH to F, AG and F together, are equal† to CH and E together: therefore if to the unequal magnitudes GB, HD, of which GB is the greater, there be added

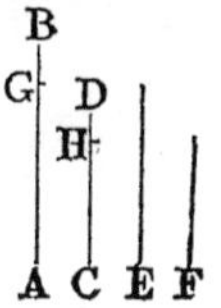

† 7. & 11. 5.
* 19. 5.
† Hyp.
* A. 5.
† 2 Ax. 1.

equal magnitudes, viz. to GB the two AG and F, and CH and
† 4 Ax. 1. E to HD, AB and F together, are greater† than CD and E. Therefore, if four magnitudes, &c. Q. E. D.

PROPOSITION F.

See N. THEOR.—*Ratios which are compounded of the same ratios, are the same to one another.*

Let A be to B, as D to E; and B to C, as E to F: the ratio which is compounded of the ratios of A to B, and B to C, which, by the definition of compound ratio, is the ratio of A to C, shall be the same with the ratio of D to F, which, by the same definition, is compounded of the ratios of D to E, and E to F.

A.	B.	C.
D.	E.	F.

Because there are three magnitudes A, B, C, and three others D, E, F, which, taken two and two in order, have the same
* 22. 5. ratio, ex æquali A is to C, as D to F*.

Next, let A be to B, as E to F, and B to C, as D to E:
* 23. 5. therefore *ex æquali in proportione perturbatâ**, A is to C, as D to F; that is, the ratio of A to C, which is compounded of the ratios of A to B, and B to C, is the same with the ratio of D to F, which is compounded of the ratios of D to E, and E to F.

A.	B.	C.
D.	E.	F.

And in like manner the proposition may be demonstrated, whatever be the number of ratios in either case.

PROPOSITION G.

See N. THEOR.—*If several ratios be the same to several ratios, each to each, the ratio which is compounded of ratios which are the same to the first ratios, each to each, shall be the same to the ratio compounded of ratios which are the same to the other ratios, each to each.*

Let A be to B, as E to F; and C to D, as G to H: and let A be to B, as K to L; and C to D, as L to M: then the ratio of K to M by the definition of compound ratio, is compounded of the ratios of K to L, and L to M, which are the same with the ratios of A to B, and C to D. Again, as E to F, so let N be to O; and as G to H, so let O be to P: then the ratio of N to P, is compounded of the ratios of N to O, and

A.	B.	C.	D.	K.	L.	M.
E.	F.	G.	H.	N.	O.	P.

O to P, which are the same with the ratios of E to F, and G to H: and it is to be shewn that the ratio of K to M, is the same with the ratio of N to P; or that K is to M, as N to P.

Because K is to L, as (A to B, that is, as E to F, that is as) N to O; and as L to M, so is (C to D, and so is G to H, and so is) O to P; ex æquali* K is to M, as N to P. Therefore, if several ratios, &c. Q. E. D. * 22. 5.

PROPOSITION H.

THEOR.—*If a ratio which is compounded of several ratios, be the same to a ratio which is compounded of several other ratios, and if one of the first ratios, or the ratio which is compounded of several of them, be the same to one of the last ratios, or to the ratio which is compounded of several of them, then the remaining ratio of the first, or, if there be more than one, the ratio compounded of the remaining ratios, shall be the same to the remaining ratio of the last, or, if there be more than one, to the ratio compounded of these remaining ratios.* See N.

Let the first ratios be those of A to B, B to C, C to D, D to E, and E to F; and let the other ratios be those of G to H, H to K, K to L, and L to M: also, let the ratio of A to F, which is compounded of* the first ratios, be the same with the ratio of G to M, which is compounded of the other ratios; and besides, let the ratio of A to D, which is compounded of the ratios of A to B, B to C, C to D, be the same with the ratio of G to K, which is compounded of the ratios of G to H, and H to K: then the ratio compounded of the remaining first ratios, to wit, of the ratios of D to E, and E to F, which compounded ratio is the ratio of D to F, shall be the same with the ratio of K to M, which is compounded of the remaining ratios of K to L, and L to M of the other ratios.

* Definition of compound ratio.

A. B. C. D. E. F.
G. H. K. L. M.

Because, by the hypothesis, A is to D, as G to K, by inversion*, D is to A, as K to G; and † as A is to F, so is G to M; therefore * ex æquali, D is to F, as K to M. If, therefore, a ratio which is, &c. Q. E. D.

* B. 5.
† Hyp.
* 22. 5.

PROPOSITION K.

See N.

THEOR.—*If there be any number of ratios, and any number of other ratios, such, that the ratio which is compounded of ratios which are the same to the first ratios, each to each, is the same to the ratio which is compounded of ratios which are the same, each to each, to the last ratios, and if one of the first ratios, or the ratio which is compounded of ratios which are the same to several of the first ratios each to each, be the same to one of the last ratios, or to the ratio which is compounded of ratios which are the same, each to each, to several of the last ratios; then the remaining ratio of the first, or, if there be more than one, the ratio which is compounded of ratios which are the same, each to each, to the remaining ratios of the first, shall be the same to the remaining ratio of the last, or, if there be more than one, to the ratio which is compounded of ratios which are the same each to each to these remaining ratios.*

Let the ratios of A to B, C to D, E to F, be the first ratios; and the ratios of G to H, K to L, M to N, O to P, Q to R, be the other ratios: and let A be to B, as S to T; and C to D, as T to V; and E to F, as V to X: therefore, by the definition of compound ratio, the ratio of S to X is compounded of the

h, k, l.
A, B; C, D; E, F. S, T, V, X.
G, H; K, L; M, N; O, P; Q, R. Y, Z, *a, b, c, d.*
e, f, g. *m, n, o, p.*

ratios of S to T, T to V, and V to X, which are the same to the ratios of A to B, C to D, E to F, each to each. Also, as G to H, so let Y be to Z; and K to L, as Z to *a*; M to N, as *a* to *b*; O to P, as *b* to *c*; and Q to R, as *c* to *d*: therefore, by the same definition, the ratio of Y to *d* is compounded of the ratios of Y to Z, Z to *a*, *a* to *b*, *b* to *c*, and *c* to *d*, which are the same, each to each, to the ratios of G to H, K to L, M to N, O to P, and Q to R: therefore, by the hypothesis, S is to X, as Y to *d*. Also, let the ratio of A to B, that is, the ratio of S to T, which is one of the first ratios, be the same to the ratio of *e* to *g*, which is compounded of the ratios of *e* to *f*, and *f* to *g*, which by the hypothesis, are the same to the ratios of G to H, and K to L, two of the other ratios; and let the ratio of h to l be that which is compounded of the ratios of h to k, and k to l, which are the same to the remaining first ratios, viz. of C to D, and E to F; also, let the ratio of m to p, be that which is compounded of the ratios of m to n, n to o, and o to p, which are the same, each to each, to the remaining other ratios, viz. of M to N, O to P, and Q to R: then the ratio of h to l shall be the same to the ratio of m to p; or h shall be to l, as m to p.

Because e is to f, as (G to H, that is, as) Y to Z; and f is to g, as (K to L, that is, as) Z to a; therefore † ex æquali, † 22. 5.
e is to g, as Y to a: and by the hypothesis, A is to B, that is,
S to T, as e to g: wherefore †, S is to T, as Y to a; and, by † 11. 5.
inversion †, T is to S, as a to Y: but S is to X †, as Y to d: † B. 5.
therefore ex æquali, T is to X, as a to d; also †, because h is † Hyp.
to k as (C to D, that is, as) T to V; and k is to l as (E to F, † Hyp.
that is, as) V to X; therefore ex æquali, h is to l, as T to X: in like manner, it may be demonstrated, that m is to p, as a to d; and it has been shewn, that T is to X, as a to d; therefore * h is to l, as m to p. Q. E. D. * 11. 5.

The propositions G and K are usually, for the sake of brevity, expressed in the same terms with propositions F and H, and therefore it was proper to shew the true meaning of them when they are so expressed; especially, since they are very frequently made use of by geometers.

THE

ELEMENTS OF EUCLID.

BOOK VI.

DEFINITIONS.

I. SIMILAR rectilineal figures are those which have their several angles equal, each to each, and the sides about the equal angles proportionals.

See N. II. "Reciprocal figures, viz. triangles and parallelograms, are "such as have their sides about two of their angles pro- "portionals in such a manner, that a side of the first figure "is to a side of the other, as the remaining side of this "other is to the remaining side of the first."

III. A straight line is said to be cut in extreme and mean ratio, when the whole is to the greater segment, as the greater segment is to the less.

IV. The altitude of any figure is the straight line drawn from its vertex perpendicular to the base.

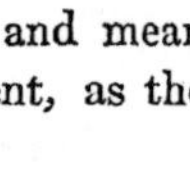

PROPOSITION I.

THEOREM.—*Triangles and parallelograms of the same altitude, are one to another as their bases.* See N.

Let the triangles ABC, ACD, and the parallelograms EC, CF, have the same altitude, viz. the perpendicular drawn from the point A to BD: as the base BC, is to the base CD, so shall the triangle ABC be to the triangle ACD, and the parallelogram EC to the parallelogram CF.

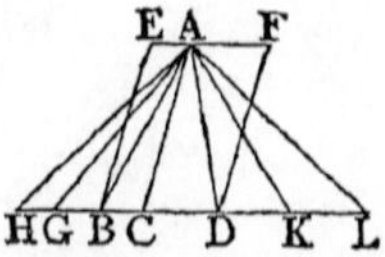

Produce BD both ways to the points H, L, and † take any number of straight lines BG, GH, each equal to the base BC; and DK, KL, any number of them, each equal to the base CD; and join AG, AH, AK, AL: then, because CB, BG, GH are all equal, the triangles AHG, AGB, ABC are all * equal: therefore, whatever multiple the base HC is of the base BC, the same multiple is the triangle AHC of the triangle ABC: for the same reason, whatever multiple the base LC is of the base CD, the same multiple is the triangle ALC of the triangle ADC: and if the base HC be equal to the base CL, the triangle AHC is also equal * to the triangle ALC: and if the base HC be greater than the base CL, likewise the triangle AHC is greater than the triangle ALC; and if less, less: therefore, since there are four magnitudes, viz. the two bases BC, CD, and the two triangles ABC, ACD; and of the base BC, and the triangle ABC, the first and third, any equimultiples whatever have been taken, viz. the base HC and the triangle AHC; and of the base CD and the triangle ACD, the second and fourth, have been taken any equimultiples whatever, viz. the base CL, and the triangle ALC; and since it has been shewn, that if the base HC be greater than the base CL, the triangle AHC is greater than the triangle ALC; and if equal, equal; and if less, less: therefore *, as the base BC is to the base CD, so is the triangle ABC to the triangle ACD.

† 3. 1.
* 38. 1.
* 38. 1.
* 5 Def. 5.

And because the parallelogram CE is double of the triangle ABC *, and the parallelogram CF double of the triangle ACD, and that magnitudes have the same ratio * which their equimultiples have; as the triangle ABC is to the triangle ACD, so is the parallelogram EC to the parallelogram CF: and because it has been shewn, that, as the base BC is to the base

* 41. 1.
* 15. 5.

CD, so is the triangle ABC to the triangle ACD; and as the triangle ABC is to the triangle ACD, so is the parallelogram EC to the parallelogram CF; therefore, as the base BC is to
* 11. 5. the base CD, so is * the parallelogram EC to the parallelogram CF. Wherefore, triangles, &c. Q. E. D.

COROLLARY. From this it is plain, that triangles and parallelograms which have equal altitudes, are one to another as their bases.

Let the figures be placed so as to have their bases in the same straight line; and having drawn perpendiculars from the vertices of the triangles to the bases, the straight line which joins the vertices is parallel to that in which their bases
* 33. 1. are *, because the perpendiculars are both equal and paral-
† 28. 1. lel † to one another. Then, if the same construction be made as in the proposition, the demonstration will be the same.

PROPOSITION II.

See N. THEOR.—*If a straight line be drawn parallel to one of the sides of a triangle, it shall cut the other sides, or these produced, proportionally: and if the sides, or the sides produced, be cut proportionally, the straight line which joins the points of section, shall be parallel to the remaining side of the triangle.*

Let DE be drawn parallel to BC, one of the sides of the triangle ABC: BD shall be to DA, as CE to EA.

* 37. 1. Join BE, CD; then the triangle BDE is equal * to the triangle CDE, because they are on the same base DE, and between the same parallels DE, BC: ADE is another triangle; and equal magni-
* 7. 5. tudes have to the same * the same ratio; therefore, as the triangle BDE is to the triangle ADE, so is the triangle CDE to the triangle ADE: but as the triangle BDE

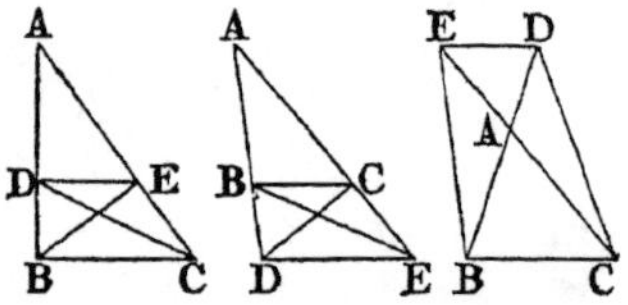

* 1. 6. to the triangle ADE, so is * BD to DA, because, having the same altitude, viz. the perpendicular drawn from the point E to AB, they are to one another as their bases; and for the same reason, as the triangle CDE to the triangle ADE, so is
* 11. 5. CE to EA: therefore as BD to DA, so is CE to EA *.

Next, let the sides AB, AC, of the triangle ABC, or these sides produced, be cut proportionally in the points D, E, that is, so that BD may be to DA as CE to EA; and join DE: DE shall be parallel to BC.

The same construction being made; because as BD to DA, so is CE to EA; and as BD to DA, so is the triangle BDE to the triangle ADE *: and as CE to EA, so is the triangle CDE to the triangle ADE; therefore † the triangle BDE is to the triangle ADE, as the triangle CDE to the triangle ADE; that is, the triangles BDE, CDE have the same ratio to the triangle ADE: therefore * the triangle BDE is equal to the triangle CDE: and they are on the same base DE: but equal triangles on the same base are between the same parallels *, therefore DE is parallel to BC. Wherefore, if a straight line, &c. *Q. E. D.*

* 1. 6.
† 11. 5.
* 9. 5.
* 39. 1.

PROPOSITION III.

THEOR.—*If the angle of a triangle be divided into two equal angles, by a straight line which also cuts the base, the segments of the base shall have the same ratio which the other sides of the triangle have to one another: and if the segments of the base have the same ratio which the other sides of the triangle have to one another, the straight line drawn from the vertex to the point of section, divides the vertical angle into two equal angles.* See N.

Let ABC be a triangle, and let the angle BAC be divided into two equal angles by the straight line AD: BD shall be to DC, as BA to AC.

Through the point C, draw CE parallel* to DA, and let BA produced, meet CE in E. Because the straight line AC meets the parallels AD, EC, the angle ACE is equal* to the alternate angle CAD: but CAD, by the hypothesis, is equal to the angle BAD: wherefore BAD is equal† to the angle ACE. Again, because the straight line BAE meets the parallels AD, EC, the outward angle BAD is equal† to the inward and opposite angle AEC: but the angle ACE has been proved equal to the angle BAD; therefore also ACE is equal† to the angle AEC, and consequently the side AE is equal* to the side AC: and because AD is drawn parallel to one of the sides of the

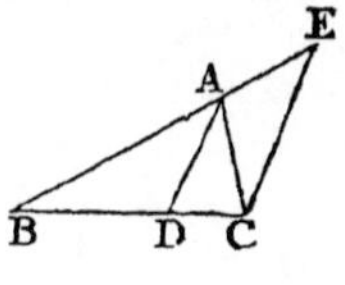

* 31. 1.
* 29. 1.
† 1 Ax.
† 29. 1.
† 1 Ax.
* 6. 1.

* 2. 6. triangle BCE, viz. to EC, therefore * BD is to DC, as BA to
* 7. 5. AE: but AE is equal to AC; therefore *, as BD to DC, so is BA to AC.

Next, let BD be to DC, as BA to AC, and join AD; the angle BAC shall be divided into two equal angles by the straight line AD.

The same construction being made; because, as BD to DC,
* 2. 6. so is BA to AC; and as BD to DC, so is BA to AE *, because
* 11. 5. AD is parallel to EC; therefore * BA is to AC, as BA to AE:
* 9. 5. consequently AC is equal* to AE, and therefore the angle AEC
* 5. 1. is equal* to the angle ACE; but the angle AEC is equal to the outward and opposite angle BAD; and the angle ACE is
* 29. 1. equal to the alternate angle CAD*; wherefore also the angle
† 1 Ax. BAD is equal† to the angle CAD; that is, the angle BAC is cut into two equal angles by the straight line AD. Therefore, if the angle, &c. Q. E. D.

PROPOSITION A.

THEOR.—*If the outward angle of a triangle made by producing one of its sides, be divided into two equal angles by a straight line which also cuts the base produced, the segments between the dividing line and the extremities of the base, have the same ratio which the other sides of the triangle have to one another: and if the segments of the base produced have the same ratio which the other sides of the triangle have, the straight line drawn from the vertex to the point of section, divides the outward angle of the triangle into two equal angles.*

Let ABC be a triangle, and let one of its sides BA be produced to E; and let the outward angle CAE be divided into two equal angles by the straight line AD which meets the base produced in D: BD shall be to DC, as BA to AC.

* 31. 1. Through C, draw* CF parallel to AD: and because the straight line AC meets the parallels AD, FC, the angle ACF
* 29. 1. is equal* to the alternate angle CAD: but
* Hyp. CAD is equal* to the angle DAE; there-
† 1 Ax. fore also DAE is equal† to the angle ACF. Again, because the straight line FAE meets the parallels AD, FC, the outward angle

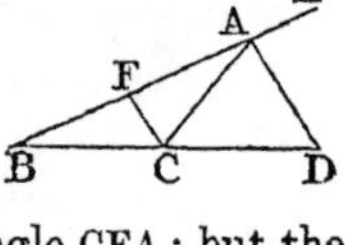

† 29. 1. DAE is equal† to the inward and opposite angle CFA: but the angle ACF has been proved equal to the angle DAE; therefore

also the angle ACF is equal† to the angle CFA; and consequently the side AF is equal* to the side AC: and because AD is parallel to FC, a side of the triangle BCF, therefore* BD is to DC, as BA to AF: but AF is equal to AC; therefore as BD is to DC†, so is BA to AC. † 1 Ax. * 6. 1. * 2. 6. † 7. 5.

Next, let BD be to DC, as BA to AC, and join AD: the angle CAD shall be equal to the angle DAE.

The same construction being made; because BD is to DC, as BA to AC; and that BD is also to DC*, as BA to AF; therefore* BA is to AC, as BA to AF; wherefore AC is equal* to AF, and the angle AFC equal* to the angle ACF: but the angle AFC is equal to the outward angle EAD†, and the angle ACF to the alternate angle CAD; therefore also EAD is equal† to the angle CAD. Wherefore, if the outward, &c. Q. E. D. * 2. 6. * 11. 5. * 9. 5. * 5. 1. † 29. 1. † 1 Ax.

PROPOSITION IV.

THEOR.—*The sides about the equal angles of equiangular triangles are proportionals; and those which are opposite to the equal angles are homologous sides, that is, are the antecedents or consequents of the ratios.*

Let ABC, DCE be equiangular triangles, having the angle ABC equal to the angle DCE, and the angle ACB to the angle DEC, and consequently*, the angle BAC equal to the angle CDE: the sides about the equal angles of the triangles ABC, DCE, shall be proportionals; and those shall be the homologous sides, which are opposite to the equal angles. * 32. 1. and 3 Ax.

Let the triangle DCE be placed* so that its side CE may be contiguous to BC, and in the same straight line with it: then, because the angle BCA is equal† to the angle CED, add to each the angle ABC; therefore the two angles ABC, BCA are equal† to the two angles ABC, CED: but the angles ABC, BCA are together less* than two right angles; therefore the angles ABC, CED are also less than two right angles: wherefore BA, ED if produced will meet*: let them be produced and meet in the point F: then because the angle ABC is equal to the angle DCE, BF is parallel* to CD; and because the angle ACB is equal to the angle DEC, AC is parallel to FE*: therefore FACD is a parallelogram; and consequently*, AF is equal to * 22. 1. † Hyp. † 2 Ax. * 17. 1. * 12 Ax. 1. * 28. 1. * 28. 1. * 34. 1.

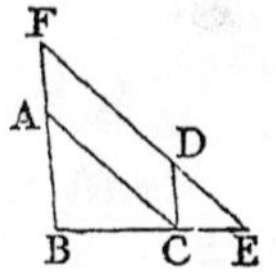

CD, and AC to FD: and because AC is parallel to FE, one of
* 2. 6. the sides of the triangle FBE, BA is to AF*, as BC to CE:
* 7. 5. but AF is equal to CD; therefore * as BA to CD, so is BC to
† 16. 5. CE; and alternately †, as AB to BC, so is DC to CE: again,
* 2. 6. because CD is parallel to BF, as BC to CE *, so is FD to DE:
† 7. 5. but FD is equal to AC; therefore †, as BC to CE, so is AC to
† 16. 5. DE; and alternately †, as BC to CA, so CE to ED: therefore, because it has been proved, that AB is to BC, as DC to CE, and
* 22. 5. as BC to CA, so CE to ED; ex æquali *, BA is to AC, as CD to DE. Therefore, the sides, &c. Q. E. D.

PROPOSITION V.

THEOR.—*If the sides of two triangles, about each of their angles, be proportionals, the triangles shall be equiangular; and the equal angles shall be those which are opposite to the homologous sides.*

Let the triangles ABC, DEF have their sides proportionals, so that AB is to BC, as DE to EF; and BC to CA, as EF to FD; and consequently, ex æquali, BA to AC, as ED to DF: the triangle ABC shall be equiangular to the triangle DEF, and the angles which are opposite to the homologous sides shall be equal, viz. the angle ABC equal to the angle DEF, and BCA to EFD, and also BAC to EDF.

* 23. 1. At the points E, F, in the straight line EF, make * the angle FEG equal to the angle ABC, and the angle EFG, equal to BCA; wherefore the remaining angle
* 32. 1. and 3 Ax. BAC is equal * to the remaining angle EGF, and the triangle ABC is therefore equiangular to the triangle GEF: conse-
* 4. 6. quently * they have their sides opposite to the equal angles proportionals: wherefore,

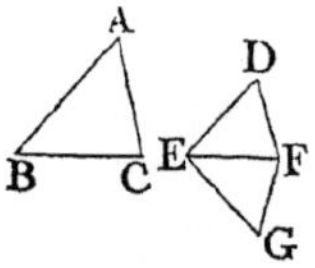

† Hyp. as AB to BC, so is GE to EF: but as AB to BC †, so is DE to
* 11. 5. EF; therefore, as DE to EF, so * GE to EF; that is, DE and
* 9. 5. GE have the same ratio to EF, and consequently * are equal: for the same reason, DF is equal to FG: and because in the triangles DEF, GEF, DE is equal to EG, and EF common, the two sides DE, EF are equal to the two GE, EF, each to each; and the base DF is equal to the base GF; therefore the angle
* 8. 1. DEF is equal * to the angle GEF, and the other angles to the
* 4. 1. other angles which are subtended by the equal sides *: there-

fore the angle DFE is equal to the angle GFE, and EDF to EGF: and because the angle DEF is equal to the angle GEF, and GEF equal† to the angle ABC, therefore the angle ABC is equal† to the angle DEF: for the same reason, the angle ACB is equal to the angle DFE, and the angle at A equal to the angle at D: therefore the triangle ABC is equiangular to the triangle DEF. Wherefore, if the sides, &c. Q. E. D. † Constr. † 1 Ax.

PROPOSITION VI.

THEOR.—*If two triangles have one angle of the one, equal to one angle of the other, and the sides about the equal angles proportionals, the triangles shall be equiangular, and shall have those angles equal which are opposite to the homologous sides.*

Let the triangles ABC, DEF have the angle BAC in the one, equal to the angle EDF in the other, and the sides about those angles proportionals; that is, BA to AC, as ED to DF: the triangles ABC, DEF shall be equiangular, and shall have the angle ABC equal to the angle DEF, and ACB to DFE.

At the points D, F, in the straight line DF, make* the angle FDG equal to either of the angles BAC, EDF, and the angle DFG equal to the angle ACB: wherefore the remaining angle at B is equal* to the remaining angle at G: and consequently the triangle ABC is equiangular to the triangle DGF: therefore as BA to AC, so is* GD to DF: but by the hypothesis, as BA to AC, so is ED to DF; therefore as ED to DF, so is* GD to DF; wherefore ED is equal* to DG: and DF is common to the two triangles EDF, GDF: therefore the two sides ED, DF are equal to the two sides GD, DF, each to each; and the angle EDF is equal† to the angle GDF; wherefore the base EF is equal to the base FG*, and the triangle EDF to the triangle GDF, and the remaining angles to the remaining angles, each to each, which are subtended by the equal sides: therefore the angle DFG is equal to the angle DFE, and the angle at G to the angle at E: but the angle DFG is equal† to the angle ACB; therefore the angle ACB is equal† to the angle DFE: and the angle BAC

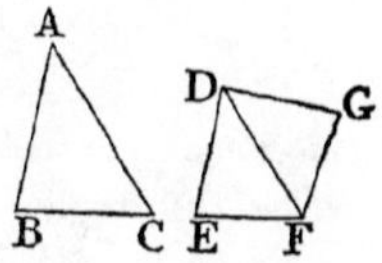

* 23. 1.
* 32. 1. & 3 Ax.
* 4. 6.
* 11. 5.
* 9. 5.
† Constr.
* 4. 1.
† Constr.
† 1 Ax.

* Hyp. is equal* to the angle EDF: wherefore also the remaining
† 32. 1. & 3 Ax. angle at B is equal† to the remaining angle at E: therefore the triangle ABC is equiangular to the triangle DEF. Wherefore, if two triangles, &c. Q. E. D.

PROPOSITION VII.

See N. THEOR.—*If two triangles have one angle of the one, equal to one angle of the other, and the sides about two other angles proportionals; then, if each of the remaining angles be either less, or not less, than a right angle, or if one of them be a right angle, the triangles shall be equiangular, and shall have those angles equal about which the sides are proportionals.*

Let the two triangles ABC, DEF have one angle in the one, equal to one angle in the other, viz. the angle BAC to the angle EDF, and the sides about two other angles ABC, DEF proportionals, so that AB is to BC, as DE to EF; and, in the first case, let each of the remaining angles at C, F, be less than a right angle: the triangle ABC shall be equiangular to the triangle DEF, viz. the angle ABC shall be equal to the angle DEF, and the remaining angle at C equal to the remaining angle at F.

For if the angles ABC, DEF be not equal, one of them must be greater than the other: let ABC be the greater, and
* 23. 1. at the point B, in the straight line AB*, make the angle ABG equal to the angle DEF; and be-
† Hyp. cause the angle at A is equal† to the angle at D, and the angle ABG to the angle DEF, the remaining angle AGB
* 32. 1. & 3 Ax. is equal* to the remaining angle DFE:

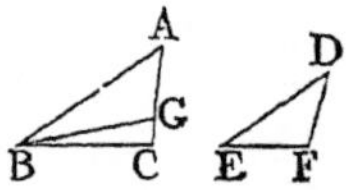

therefore the triangle ABG is equiangular to the triangle
* 4. 6. DEF: wherefore* as AB is to BG, so is DE to EF: but as
* 11. 5. DE to EF so by hypothesis, is AB to BC: therefore* as AB to BC, so is AB to BG: and because AB has the same ratio
* 9. 5. to each of the lines BC, BG; BC is equal* to BG; and there-
* 5. 1. fore the angle BGC is equal* to the angle BCG: but the angle BCG is, by hypothesis, less than a right angle; therefore also the angle BGC is less than a right angle; and there-
* 13. 1. fore the adjacent angle AGB must be greater* than a right angle: but it was proved that the angle AGB is equal to the

angle at F; therefore the angle at F is greater than a right angle: but, by the hypothesis, it is less than a right angle; which is absurd. Therefore the angles ABC, DEF are not unequal, that is, they are equal: and the angle at A is equal† to the angle at D; wherefore the remaining angle at C is equal† to the remaining angle at F: therefore the triangle ABC is equiangular to the triangle DEF. † Hyp. † 32. 1. & 3 Ax.

Next, let each of the angles at C, F be not less than a right angle: the triangle ABC shall also in this case be equiangular to the triangle DEF.

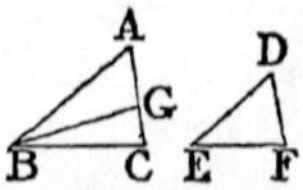

The same construction being made, it may be proved in like manner, that BC is equal to BG, and therefore the angle at C equal to the angle BGC: but the angle at C is not less† than a right angle; therefore the angle BGC is not less than a right angle: wherefore two angles of the triangle BGC are together, not less than two right angles; which is impossible*; and therefore the triangle ABC may be proved to be equiangular to the triangle DEF, as in the first case. † Hyp. * 17. 1.

Lastly, let one of the angles at C, F, viz. the angle at C, be a right angle: in this case likewise, the triangle ABC shall be equiangular to the triangle DEF.

For if they be not equiangular, at the point B, in the straight line AB, make the angle ABG equal to the angle DEF: then it may be proved, as in the first case, that BG is equal to BC; and therefore* the angle BCG equal to the angle BGC: but the angle BCG is a right† angle, therefore† the angle BGC is also a right angle; whence two of the angles of the triangle BGC are together, not less than two right angles; which is impossible*: therefore the triangle ABC is equiangular to the triangle DEF. Wherefore, if two triangles, &c. Q. E. D. * 5. 1. † Hyp. † 1 Ax. * 17. 1.

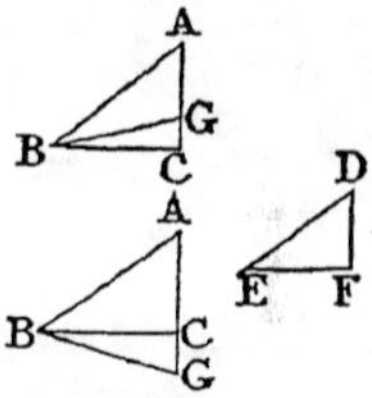

PROPOSITION VIII.

THEOR.—*In a right-angled triangle, if a perpendicular be drawn from the right angle to the base, the triangles on each side of it are similar to the whole triangle, and to one another.* See N.

Let ABC be a right-angled triangle, having the right angle BAC; and from the point A, let AD be drawn perpendicular to the base BC: the triangles ABD, ADC shall be similar to the whole triangle ABC, and to one another.

† 11 Ax. Because the angle BAC is equal to the angle ADB†, each of them being a right angle, and that the angle at B is common to the two triangles ABC, ABD, the remaining angle ACB is equal to the re-

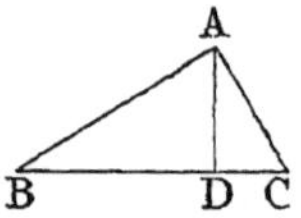

* 32. 1. & 3 Ax. maining angle BAD*: therefore the triangle ABC is equiangular to the triangle ABD, and the sides about their equal

* 4. 6. * 1 Def. 6. angles are proportionals*; wherefore the triangles are* similar: in the like manner it may be demonstrated, that the triangle ADC is equiangular and similar to the triangle ABC‡. And the triangles ABD, ACD, being both equiangular and similar to ABC, are equiangular and similar to each other. Therefore, in a right-angled, &c. Q. E. D.

COR. From this it is manifest, that the perpendicular drawn from the right angle of a right-angled triangle to the base, is a mean proportional between the segments of the base, and also that each of the sides is a mean proportional between the base, and the segment of it adjacent to that

* 4. 6. * 4. 6. * 4. 6. side: because in the triangles BDA, ADC, BD is to DA*, as DA to DC; and in the triangles ABC, DBA, BC is to BA*, as BA to BD; and in the triangles ABC, ACD, BC is to CA*, as CA to CD.

PROPOSITION IX.

See N. PROBLEM.—*From a given straight line to cut off any part required.*

Let AB be the given straight line; it is required to cut off any part from it.

From the point A, draw a straight line AC, making any angle with AB; and in AC take any point D, and take AC the same multiple of AD, that AB is of the part which is to be cut off from it; join BC, and draw DE parallel to it: then AE shall be the part required to be cut off.

† 1 Ax. † 4. 6. † 1 Def. 6. ‡ The triangles ADB, ADC, being both equiangular to ABC, are equiangular† to each other; therefore they have the sides about their equal† angles proportionals; therefore they are† similar to each other.

Because ED is parallel to one of the sides of the triangle ABC, viz. to BC, as CD is to DA so is* BE to EA; and by composition*, CA is to AD, as BA to AE: but CA is a multiple† of AD; therefore* BA is the same multiple of AE: whatever part, therefore, AD is of AC, AE is the same part of AB. Wherefore, from the straight line AB, the part required is cut off. Which was to be done.

* 2. 6.
* 18. 5.
† Constr.
* D. 5.

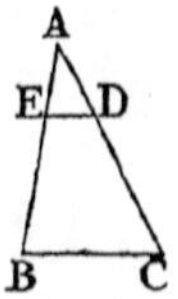

PROPOSITION X.

PROB.—*To divide a given straight line similarly to a given divided straight line, that is, into parts that shall have the same ratios to one another, which the parts of the divided given straight line have.*

Let AB be the straight line given to be divided, and AC the divided line; it is required to divide AB similarly to AC.

Let AC be divided in the points D, E; and let AB, AC be placed so as to contain any angle, and join BC; and through the points D, E, draw* DF, EG parallels to it: AB shall be divided in the points F, G, similarly to AC. * 31. 1.

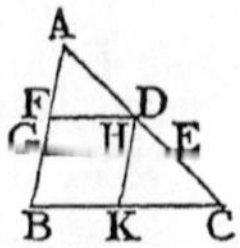

Through D, draw DHK parallel to AB: therefore each of the figures FH, HB is a parallelogram: wherefore DH is equal* to FG, and HK to GB: and because HE is parallel to KC, one of the sides of the triangle DKC, as CE to ED, so is* KH to HD: but KH is equal BG, and HD to GF; therefore, as CE to ED†, so is BG to GF: again, because FD is parallel to GE, one of the sides of the triangle AGE, as ED to DA, so is† GF to FA: but it has been proved that CE is to ED, as BG to GF: therefore as CE is to ED, so is BG to GF, and as ED to DA, so GF to FA. Therefore, the given straight line AB is divided similarly to AC. Which was to be done.

* 34. 1.
* 2. 6.
† 7. 5.
† 2. 6.

PROPOSITION XI.

PROB.—*To find a third proportional to two given straight lines.*

Let AB, AC be the two given straight lines; it is required to find a third proportional to AB, AC.

Let AB, AC be placed so as to contain any angle: produce AB, AC, to the points D, E; and make BD equal to AC; join BC, and through D, draw DE parallel to
* 31. 1. it*: CE shall be a third proportional to AB and AC.

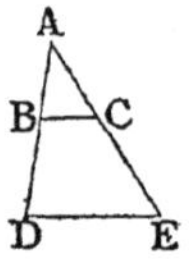

Because BC is parallel to DE, a side of the
* 2. 6. triangle ADE, AB is* to BD, as AC to CE:
† 7. 5. but BD is equal to AC; therefore as AB is to† AC, so is AC to CE. Wherefore, to the two given straight lines AB, AC, a third proportional CE is found. Which was to be done.

PROPOSITION XII.

PROB.—*To find a fourth proportional to three given straight lines.*

Let A, B, C be the three given straight lines; it is required to find a fourth proportional to A, B, C.

Take two straight lines, DE, DF, containing any angle
† 3. 1. EDF; and upon these† make DG equal to A, GE equal to B, and DH equal to C; join GH, and through
* 31. 1. E, draw EF parallel* to it; HF shall be a fourth proportional to A, B, C.

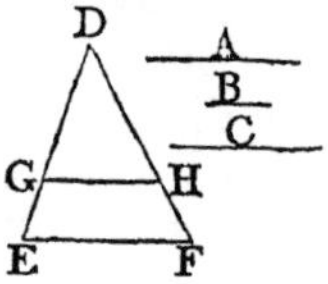

Because GH is parallel to EF, one of the sides of the triangle DEF, DG is to
* 2. 6. GE*, as DH to HF; but DG is equal to A, GE to B, and DH to C; therefore, as
† 7. 5. A is to B†, so is C to HF. Wherefore, to the three given straight lines A, B, C, a fourth proportional HF is found. Which was to be done.

PROPOSITION XIII.

PROB.—*To find a mean proportional between two given straight lines.*

Let AB, BC be the two given straight lines; it is required to find a mean proportional between them.

Place AB, BC in a straight line, and upon AC describe the semicircle ADC; and from
* 11. 1. the point B, draw* BD at right angles to AC: BD shall be a mean proportional between AB and BC.

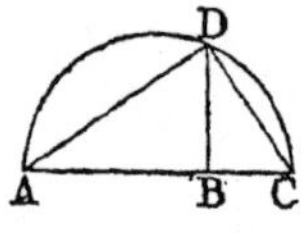

Join AD, DC: and because the angle ADC in a semicircle is a right angle *, and because in the right-angled triangle ADC, BD is drawn from the right angle perpendicular to the base, DB is a mean proportional between AB, BC the segments of the base *. Therefore, between the two given straight lines AB, BC, a mean proportional DB is found. Which was to be done. * 31. 3. * Cor. 8. 6.

PROPOSITION XIV.

THEOR.—*Equal parallelograms which have one angle of the one, equal to one angle of the other, have their sides about the equal angles reciprocally proportional; and parallelograms that have one angle of the one, equal to one angle of the other, and their sides about the equal angles reciprocally proportional, are equal to one another.*

Let AB, BC be equal parallelograms which have the angles at B equal: the sides of the parallelograms AB, BC about the equal angles, shall be reciprocally proportional; that is, DB shall be to BE, as GB to BF.

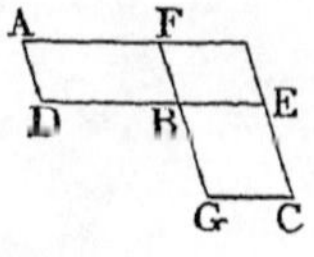

Let the sides DB, BE be placed in the same straight line; wherefore also ‡ FB, BG are in one straight line *; complete the parallelogram FE: and because the parallelogram AB is equal to BC, and that FE is another parallelogram, AB is to FE, as BC to FE *: but as AB to FE, so is the base DB to BE *, and as BC to FE, so is the base GB to BF; therefore, as DB to BE, so is GB to BF *. Wherefore, the sides of the parallelograms AB, BC about their equal angles are reciprocally proportional. * 14. 1. * 7. 5. * 1. 6. * 11. 5.

Next, let the sides about the equal angles be reciprocally proportional, viz. as DB to BE, so GB to BF: the parallelogram AB shall be equal to the parallelogram BC.

Because, as DB to BE, so is GB to BF; and as DB to BE †, so is the parallelogram AB to the parallelogram FE; and as GB to BF, so is the parallelogram BC to the parallelogram † 1. 6.

‡ Because the angle DBF is equal † to the angle GBF, add to each the angle FBE; therefore the two angles DBF, FBE, are together equal † to the two angles FBE, EBG: but DBF, FBE are together equal † to two right angles; therefore FBE, EBG are together equal † to two right angles: therefore FB, BG are in the same * straight line. † Hyp. † 2 Ax. † 13. 1. † 1 Ax. * 14. 1.

* 11. 5. FE; therefore * as AB to FE, so BC to FE: therefore the
* 9. 5. parallelogram AB is equal * to the parallelogram BC. Therefore, equal parallelograms, &c. Q. E. D.

PROPOSITION XV.

THEOR.—*Equal triangles which have one angle of the one, equal to one angle of the other, have their sides about the equal angles reciprocally proportional; and triangles which have one angle in the one, equal to one angle in the other, and their sides about the equal angles reciprocally proportional, are equal to one another.*

Let ABC, ADE be equal triangles, which have the angle BAC equal to the angle DAE: the sides about the equal angles of the triangles shall be reciprocally proportional; that is, CA shall be to AD, as EA to AB.

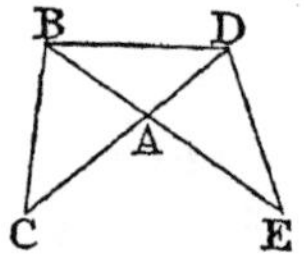

Let the triangles be placed so that their sides CA, AD be in one straight line; where-
† See N. last Prop. fore † also EA and AB are in one straight
* 14. 1. line *; and join BD. Because the triangle ABC is equal to the triangle ADE, and that ABD is another triangle, therefore as the triangle CAB, is to the
* 7. 5. triangle BAD *, so is the triangle AED to the triangle DAB: but as the triangle CAB to the triangle BAD, so is the base
* 1. 6. CA to AD *, and as the triangle EAD to the triangle DAB,
* 1. 6. so is the base EA to AB *; therefore as CA to AD *, so is EA
* 11. 5. to AB: wherefore the sides of the triangles ABC, ADE, about the equal angles, are reciprocally proportional.

Next, let the sides of the triangles ABC, ADE, about the equal angles, be reciprocally proportional, viz. CA to AD, as EA to AB: the triangle ABC shall be equal to the triangle ADE.

Join BD as before: then because, as CA to AD, so is EA to AB; and as CA to AD, so is the triangle ABC to the triangle
* 1. 6. BAD *; and as EA to AB, so is the triangle EAD to the tri-
* 1. 6. angle BAD *; therefore * as the triangle BAC to the triangle
* 11. 5. BAD, so is the triangle EAD to the triangle BAD; that is, the triangles BAC, EAD have the same ratio to the triangle
* 9. 5. BAD: wherefore the triangle ABC is equal * to the triangle ADE. Therefore, equal triangles, &c. Q. E. D.

PROPOSITION XVI.

THEOR.—*If four straight lines be proportionals, the rectangle contained by the extremes is equal to the rectangle contained by the means; and if the rectangle contained by the extremes be equal to the rectangle contained by the means, the four straight lines are proportionals.*

Let the four straight lines AB, CD, E, F, be proportionals, viz. as AB to CD, so E to F: the rectangle contained by AB, F, shall be equal to the rectangle contained by CD, E.

From the points A, C, draw* AG, CH at right angles to AB, * 11. 1.
CD, and make† AG equal to F, and CH equal to E; and com- † 3. 1.
plete† the parallelograms BG, DH. Be- † 31. 1.
cause, as AB to CD, so is E to F; and that
E is equal to CH, and F to AG, AB is* to * 7. 5.
CD as CH to AG: therefore the sides of the parallelograms BG, DH, about the equal angles, are reciprocally proportional; but parallelograms which have their sides about equal angles reciprocally proportional, are equal to one ano-
ther*; therefore the parallelogram BG is equal to the parallelo- * 14. 6.
gram DH: but the parallelogram BG is contained by the straight lines AB, F, because AG is equal to F; and the parallelogram DH is contained by CD and E, because CH is equal to E; therefore the rectangle contained by the straight lines AB, F, is equal to that which is contained by CD and E.

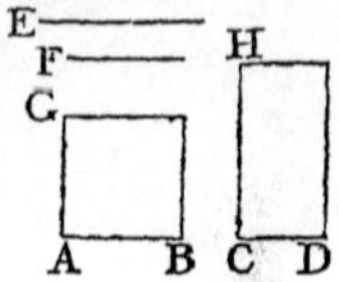

And if the rectangle contained by the straight lines AB, F, be equal to that which is contained by CD, E, these four lines shall be proportional, viz. AB shall be to CD, as E to F.

The same construction being made; because the rectangle contained by the straight lines AB, F, is equal to that which is contained by CD, E, and that the rectangle BG is contained by AB, F, because AG is equal to F; and the rectangle DH by CD, E, because CH is equal to E; therefore the parallelo-
gram BG is equal† to the parallelogram DH; and they are † 1 Ax.
equiangular: but the sides about the equal angles of equal
parallelograms are reciprocally proportional*: wherefore, as * 14. 6.
AB to CD, so is CH to AG: but CH is equal to E, and AG to
F; therefore as AB is to CD†, so is E to F. Wherefore, if † 7. 5.
four, &c. Q. E. D.

PROPOSITION XVII.

THEOR.—*If three straight lines be proportionals, the rectangle contained by the extremes is equal to the square of the mean; and if the rectangle contained by the extremes be equal to the square of the mean, the three straight lines are proportionals.*

Let the three straight lines A, B, C be proportionals, viz. as A to B, so B to C: the rectangle contained by A, C, shall be equal to the square of B.

Take D equal to B: and because as A to B, so B to C; and
* 7. 5. that B is equal to D, A is* to B, as D to C: but if four straight lines be proportionals, the rectangle contained by the extremes is equal to that which is contained by
* 16. 6. the means*; therefore the rectangle contained by A, C, is equal to that contained by B, D: but the rectangle contained by B, D, is the square of B, because B is equal to D; therefore the rectangle contained by A, C, is equal to the square of B.

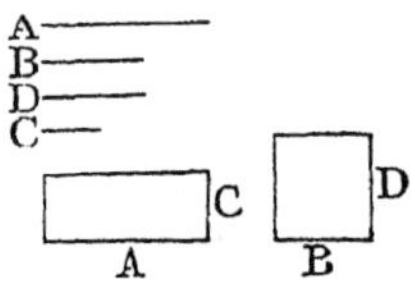

And if the rectangle contained by A, C, be equal to the square of B, A shall be to B, as B to C.

The same construction being made; because the rectangle contained by A, C, is equal to the square of B, and the square of B is equal to the rectangle contained by B, D, because B is equal to D, therefore the rectangle contained by A, C, is equal to that contained by B, D: but if the rectangle contained by the extremes be equal to that contained by the means, the
* 16. 6. four straight lines are proportionals*: therefore A is to B, as D to C: but B is equal to D; wherefore, as A to B, so B to C. Therefore, if three straight lines, &c. Q. E. D.

PROPOSITION XVIII.

See N. PROB.—*Upon a given straight line to describe a rectilineal figure, similar, and similarly situated, to a given rectilineal figure.*

Let AB be the given straight line, and CDEF the given rectilineal figure of four sides; it is required, upon the given straight line AB, to describe a rectilineal figure, similar, and similarly situated, to CDEF.

Join DF, and at the points A, B, in the straight line AB, make* the angle BAG equal to the angle at C, and the angle ABG equal to the angle CDF; therefore the remaining angle CFD is equal to the remaining angle AGB*: therefore the triangle FCD is equiangular to the triangle GAB: again, at the points G, B, in the straight line GB, make* the angle BGH equal to the angle DFE, and the angle GBH equal to FDE; therefore the remaining angle FED is equal to the remaining angle GHB, and the triangle FDE equiangular to the triangle GBH: then, because the angle AGB is equal to the angle CFD, and BGH to DFE, the whole angle AGH is equal† to the whole CFE; for the same reason, the angle ABH is equal to the angle CDE: also the angle at A is equal† to the angle at C, and the angle GHB to FED; therefore the rectilineal figure ABHG is equiangular to CDEF: likewise these figures have their sides about the equal angles proportionals; because the triangles GAB, FCD being equiangular, BA is* to AG, as DC to CF; and because AG is to GB, as CF to FD; and as GB to GH, so, by reason of the equiangular triangles BGH, DFE, is FD to FE; therefore, ex æquali*, AG is to GH, as CF to FE: in the same manner it may be proved that AB is to BH, as CD to DE: and GH is to HB*, as FE to ED. Wherefore, because the rectilineal figures ABHG, CDEF are equiangular, and have their sides about the equal angles proportionals, they are similar* to one another.

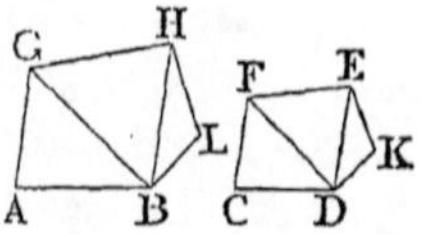

* 23. 1.
* 32. 1. & 3 Ax.
* 23. 1.
† 2 Ax.
† Constr.
* 4. 6.
* 22. 5.
* 4. 6.
* 1 Def. 6.

Next, let it be required to describe upon a given straight line AB, a rectilineal figure, similar, and similarly situated, to the rectilineal figure CDKEF of five sides.

Join DE, and upon the given straight line AB, describe the rectilineal figure ABHG, similar and similarly situated, to the quadrilateral figure CDEF, by the former case: and at the points B, H, in the straight line BH, make the angle HBL equal to the angle EDK, and the angle BHL equal to the angle DEK; therefore the remaining angle at K is equal† to the remaining angle at L: and because the figures ABHG, CDEF are similar, the angle GHB is equal† to the angle FED: and BHL is equal to DEK; wherefore the whole angle GHL is equal to the whole angle FEK: for the same reason, the angle

† 32. 1. & 3 Ax.
† 1 Def. 6.

ABL is equal to the angle CDK: therefore the five-sided figures AGHLB, CFEKD are equiangular: and because the
† 1 Def. 6. figures AGHB, CFED are similar, GH is to HB †, as FE to
* 4. 6. ED; but as HB to HL, so is ED to EK*; therefore, ex æquali*,
* 22. 5. GH is to HL, as FE to EK: for the same reason, AB is to BL
* 4. 6. as CD to DK: and BL is to LH, as* DK to KE, because the triangles BLH, DKE are equiangular: therefore, because the five-sided figures AGHLB, CFEKD are equiangular, and have their sides about the equal angles proportionals, they are similar to one another. In the same manner a rectilineal figure of six sides may be described upon a given straight line similar to one given, and so on. Which was to be done.

PROPOSITION XIX.

THEOR.—*Similar triangles are to one another in the duplicate ratio of their homologous sides.*

Let ABC, DEF be similar triangles, having the angle B equal to the angle E, and let AB be to BC, as DE to EF, so
* 12 Def. 5. that the side BC may be* homologous to EF: the triangle ABC shall have to the triangle DEF the duplicate ratio of that which BC has to EF.

* 11. 6. Take* BG a third proportional to BC, EF, so that BC may be to EF, as EF to BG, and join GA: then, because, as AB to
* 16. 5. BC, so DE to EF; alternately*, AB is
† Constr. to DE, as BC to EF: but as BC to EF†,
* 11. 5. so is EF to BG; therefore*, as AB to DE, so is EF to BG: therefore the sides of the triangles ABG, DEF, which are about the equal angles, are reciprocally proportional: but triangles, which have the sides about

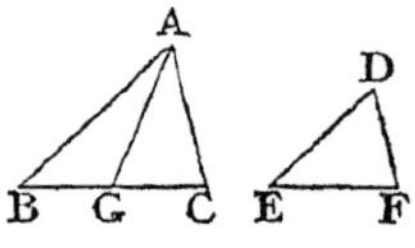

* 15. 6. two equal angles reciprocally proportional, are equal* to one another; therefore the triangle ABG is equal to the triangle DEF: and because as BC is to EF, so EF to BG, and that if
* 10 Def. 5. three straight lines be proportional, the first is said* to have to the third the duplicate ratio of that which it has to the second, therefore BC has to BG the duplicate ratio of that
* 1. 6. which BC has to EF: but as BC to BG, so is* the triangle ABC to the triangle ABG; therefore the triangle ABC has to the triangle ABG the duplicate ratio of that which BC has to EF: but the triangle ABG is equal to the triangle DEF; there-

fore also the triangle ABC has to the triangle DEF the duplicate ratio of that which BC has to EF. Therefore, similar triangles, &c. Q. E. D.

COR. From this it is manifest, that if three straight lines be proportionals, as the first is to the third, so is any triangle upon the first, to a similar and similarly described triangle upon the second.

PROPOSITION XX.

THEOR.—*Similar polygons may be divided into the same number of similar triangles, having the same ratio to one another that the polygons have; and the polygons have to one another, the duplicate ratio of that which their homologous sides have.*

Let ABCDE, FGHKL be similar polygons, and let AB be the homologous side to FG: the polygons ABCDE, FGHKL may be divided into the same number of similar triangles, whereof each shall have to each the same ratio which the polygons have; and the polygon ABCDE shall have to the polygon FGHKL, the duplicate ratio of that which the side AB has to the side FG.

Join BE, EC, GL, LH: and because the polygon ABCDE is similar to the polygon FGHKL, the angle BAE is equal* to the angle GFL, and BA is to AE*, as GF to FL: therefore, because the triangles ABE, FGL have an angle in one, equal to an angle in the other, and their sides about these equal angles proportionals, the triangle ABE is equiangular* to the triangle FGL, and therefore* similar to it; wherefore the angle ABE is equal to the angle FGL: and, because the polygons are similar, the whole angle ABC is equal* to the whole angle FGH, therefore the remaining angle EBC is equal† to the remaining angle LGH: and because the triangles ABE, FGL are similar, EB is to BA*, as LG to GF; and also, because the polygons are similar, AB is to BC*, as FG to GH; therefore, ex æquali*, EB is to BC, as LG to GH; that is, the sides about the equal angles EBC, LGH, are proportionals; therefore the triangle EBC is equiangular* to the triangle LGH, and similar* to it; for the same reason, the triangle ECD likewise is similar to the triangle LHK: therefore, the similar polygons

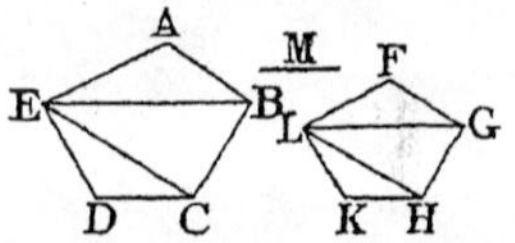

* 1 Def. 6.
* 1 Def. 6.
* 6. 6.
* 4. 6.
* 1 Def. 6.
† 3 Ax.
* 4. 6.
* 1 Def. 6.
* 22. 5.
* 6. 6.
* 4. 6.

ABCDE, FGHKL are divided into the same number of similar triangles.

Also these triangles shall have, each to each, the same ratio which the polygons have to one another, the antecedents being ABE, EBC, ECD, and the consequents FGL, LGH, LHK: and the polygon ABCDE shall have to the polygon FGHKL, the duplicate ratio of that which the side AB has to the homologous side FG.

Because the triangle ABE is similar to the triangle FGL,
* 19. 6. ABE has to FGL, the duplicate ratio* of that which the side BE has to the side GL: for the same reason, the triangle BEC has to GLH the duplicate ratio of that which BE has to GL:
* 11. 5. therefore, as the triangle ABE is to the triangle FGL, so* is the triangle BEC to the triangle GLH. Again, because the triangle EBC is similar to the triangle LGH, EBC has to LGH, the duplicate ratio of that which the side EC has to the side LH: for the same reason, the triangle ECD has to the triangle LHK, the duplicate ratio of that which EC has to LH: there-
* 11. 5. fore, as the triangle EBC to the triangle LGH, so is* the triangle ECD to the triangle LHK: but it has been proved, that the triangle EBC is likewise to the triangle LGH, as the triangle ABE to the triangle FGL; therefore, as the triangle ABE to the triangle FGL, so is triangle EBC to triangle LGH, and triangle ECD to triangle LHK: and therefore, as

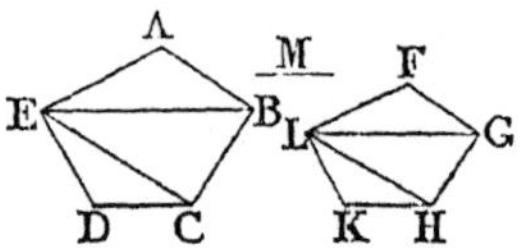

* 12. 5. one of the antecedents to one of the consequents*, so are all the antecedents to all the consequents; that is, as the triangle ABE to the triangle FGL, so is the polygon ABCDE to the polygon FGHKL: but the triangle ABE has to the triangle
† 19. 6. FGL, the duplicate ratio† of that which the side AB has to the homologous side FG; therefore also, the polygon ABCDE has to the polygon FGHKL the duplicate ratio of that which AB has to the homologous side FG. Wherefore, similar polygons, &c. Q. E. D.

Cor. 1. In like manner it may be proved, that similar four-sided figures, or of any number of sides, are one to another in the duplicate ratio of their homologous sides: and it
† 19. 6. has already been proved† in triangles; therefore, universally, similar rectilineal figures are to one another in the duplicate ratio of their homologous sides.

Cor. 2. And if to AB, FG, two of the homologous sides, a

third † proportional M be taken, AB * has to M the duplicate ratio of that which AB has to FG: but the four-sided figure or polygon upon AB, has to the four-sided figure or polygon upon FG, likewise the duplicate ratio † of that which AB has to FG; therefore †, as AB is to M, so is the figure upon AB to the figure upon FG; which was also proved * in triangles; therefore, universally, it is manifest, that if three straight lines be proportionals, as the first is to the third, so is any rectilineal figure upon the first, to a similar and similarly described rectilineal figure upon the second.

† 11. 6.
* 10 Def. 5.
† Cor. 1.
† 11. 5.
* Cor. 19. 6.

PROPOSITION XXI.

THEOR.—*Rectilineal figures which are similar to the same rectilineal figure, are also similar to one another.*

Let each of the rectilineal figures A, B be similar to the rectilineal figure C: the figure A shall be similar to the figure B.

Because A is similar to C, they are equiangular, and also have their sides about the equal angles proportional *: again, because B is similar to C, they are equiangular, and have their sides about the equal angles proportionals *: therefore the figures A, B are each of them equiangular to C, and have the sides about the equal angles of each of them, and of C, proportionals. Wherefore, the rectilineal figures A and C are * equiangular, and have their sides about the equal angles * proportionals: therefore A is similar * to B. Therefore, rectilineal figures, &c. Q. E. D.

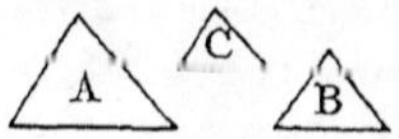

* 1 Def. 6.
* 1 Def. 6.
* 1 Ax. 1.
* 11. 5.
* 1 Def. 6.

PROPOSITION XXII.

THEOR.—*If four straight lines be proportionals, the similar rectilineal figures similarly described upon them shall also be proportionals: and if the similar rectilineal figures similarly described upon four straight lines be proportionals, those straight lines shall be proportionals.*

Let the four straight lines AB, CD, EF, GH be proportionals, viz. AB to CD, as EF to GH; and upon AB, CD, let the similar rectilineal figures KAB, LCD be similarly described; and upon EF, GH, the similar rectilineal figures MF, NH, in like manner: the rectilineal figure KAB shall be to LCD, as MF to NH.

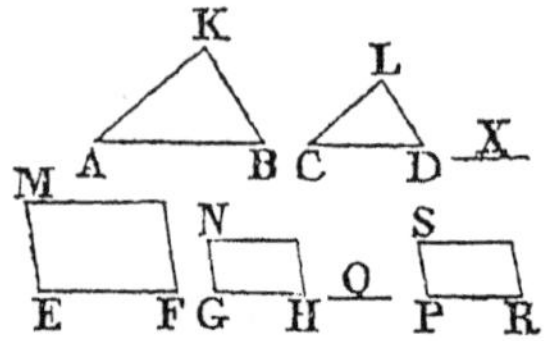

To AB, CD, take a third pro-
* 11. 6. portional* X; and to EF, GH, a third proportional O: and because AB is to CD as EF to GH,
* 11. 5. therefore CD is * to X as GH to
* 22. 5. O; wherefore, ex æquali*, as AB to X, so EF to O: but as AB to X, so is the rectilineal figure KAB to the rectilineal
* 2 Cor. 20. 6. figure LCD, and as EF to O, so is * the rectilineal figure MF
* 11. 5. to the rectilineal figure NH: therefore, as KAB to LCD, so* is MF to NH.

And if the rectilineal figure KAB be to LCD as MF to NH, the straight line AB shall be to CD as EF to GH.

* 12. 6. Make * as AB to CD, so EF to PR; and upon PR describe *
* 18. 6. the rectilineal figure SR, similar and similarly situated to either of the figures MF, NH: then, because as AB to CD, so is EF to PR, and that upon AB, CD, are described the similar and similarly situated rectilineals KAB, LCD, and upon EF, PR, in like manner, the similar rectilineals MF, SR, therefore KAB is to LCD ‡, as MF to SR: but by the hypothesis KAB is to LCD, as MF to NH; and therefore, the rectilineal MF having the same ratio to each of the two NH, SR, these are
* 9. 5. equal* to one another: they are also similar, and similarly situated; therefore GH is equal to PR: and because as AB to
† 7. 5. CD, so is EF to PR, and that PR is equal to GH, AB is to CD †, as EF to GH. If, therefore, four straight lines, &c. Q. E. D.

PROPOSITION XXIII.

See N. THEOR.—*Equiangular parallelograms have to one another, the ratio which is compounded of the ratios of their sides.*

Let AC, CF be equiangular parallelograms, having the angle BCD equal to the angle ECG: the ratio of the parallelogram AC to the parallelogram CF, shall be the same with the ratio which is compounded of the ratios of their sides.

‡ By the preceding case of this Proposition.

Let BC, CG be placed in a straight line; therefore DC‡ and CE are also in a straight line*; and complete the parallelogram DG; and taking any straight line K, make* as BC to CG, so K to L; and as DC to CE, so make* L to M: therefore the ratios of K to L, and L to M, are the same with the ratios of the sides, viz. of BC to CG, and DC to CE: but the ratio of K to M is that which is said to be compounded* of the ratios of K to L, and L to M; therefore K has to M the ratio compounded of the ratios of the sides: and because as BC to CG, so is the parallelogram AC to the parallelogram CH*; but as BC to CG, so is K to L; therefore K is* to L, as the parallelogram AC to the parallelogram CH: again, because as DC to CE, so is the parallelogram CH to the parallelogram CF; but as DC to CE, so is L to M; wherefore L is* to M, as the parallelogram CH to the parallelogram CF: therefore since it has been proved, that as K to L, so is the parallelogram AC to the parallelogram CH; and as L to M, so the parallelogram CH to the parallelogram CF; ex æquali*, K is to M, as the parallelogram AC to the parallelogram CF: but K has to M the ratio which is compounded of the ratios of the sides; therefore also, the parallelogram AC has to the parallelogram CF the ratio which is compounded of the ratios of the sides. Wherefore, equiangular parallelograms, &c. Q. E. D.

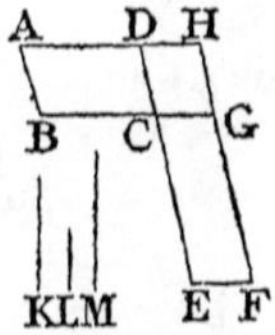

‡ See note at foot of page 145.
* 14. 1.
* 12. 6.
* 12. 6.
* Def. A. 5.
* 1. 6.
* 11. 5.
* 11. 5.
* 22. 5.

PROPOSITION XXIV.

THEOR.—*Parallelograms about the diameter of any parallelogram, are similar to the whole, and to one another.* See N.

Let ABCD be a parallelogram, of which the diameter is AC; and EG, HK parallelograms about the diameter: the parallelograms EG, HK shall be similar both to the whole parallelogram ABCD, and to one another.

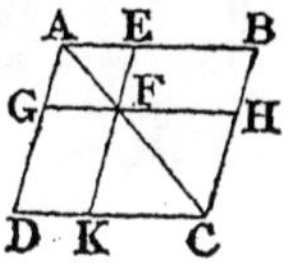

Because DC, GF are parallels, the angle ADC is equal* to the angle AGF: for the same reason, because BC, EF are parallels, the angle ABC is equal to the angle AEF: and each of the angles BCD, EFG is equal to the opposite angle DAB*, and therefore they are equal to one another; wherefore the

* 29. 1.
* 34. 1.

parallelograms ABCD, AEFG are equiangular: and because the angle ABC is equal to the angle AEF, and the angle BAC common to the two triangles BAC, EAF, they are equiangular
* 4. 6. to one another; therefore* as AB to BC, so is AE to EF: and because the opposite sides of parallelograms are equal to
* 34. 1. one another*, AB* is to AD, as AE to AG; and DC to CB,
* 7. 5. as GF to FE; and also CD to DA, as FG to GA: therefore the sides of the parallelograms ABCD, AEFG about the equal
* 1 Def. 6. angles are proportionals; and they are therefore similar* to one another: for the same reason, the parallelogram ABCD is similar to the parallelogram FHCK: wherefore each of the parallelograms GE, KH is similar to DB: but rectilineal figures which are similar to the same rectilineal figure, are
* 21. 6. also similar* to one another; therefore the parallelogram GE is similar to KH. Wherefore, parallelograms, &c. Q. E. D.

PROPOSITION XXV.

See N. PROB.—*To describe a rectilineal figure which shall be similar to one, and equal to another, given rectilineal figure.*

Let ABC be the given rectilineal figure to which the figure to be described is required to be similar, and D that to which it must be equal; it is required to describe a rectilineal figure similar to ABC, and equal to D.

* Cor. 45. 1. Upon the straight line BC, describe* the parallelogram BE
* Cor. 45. 1. equal to the figure ABC; also upon CE, describe* the parallelogram CM equal to D, and having the angle FCE equal to
* 29. 1. and 14. 1. the angle CBL: therefore‡ BC and CF are in a straight line*, as also LE and EM: between BC and CF find* a mean pro-
* 13. 6. portional GH; and upon GH describe* the rectilineal figure
* 18. 6. KGH, similar and similarly situated to the figure ABC.

† Constr. ‡ Because the angle FCE is equal† to the angle CBL, add to each the an-
† 2 Ax. gle ECB; therefore the angles FCE, ECB are equal† to the angles ECB,
* 29. 1. CBL: but the angles ECB, CBL are equal* to two right angles; therefore
† 1 Ax. the angles FCE, ECB are equal† to two right angles: therefore* BC and
* 14. 1. CF are in the same straight line. Again, because the angle LBC is equal to
* 34. 1. the angle FCE, and that the angle LBC is equal to the opposite angle*
† 1 Ax. LEC, therefore the angle LEC is equal† to the angle ECF: add to each the angle CEM; therefore the angles LEC, CEM are equal to the angles
* 29. 1. FCE, CEM: but the angles FCE, CEM are equal* to two right angles;
† 1 Ax. therefore the angles LEC, CEM are equal† to two right angles: therefore†
† 14. 1. LE and EM are in the same straight line.

Because BC is to GH as GH to CF, and that if three straight lines be proportionals, as the first is to the third, so is* the figure upon the first to the similar and similarly described figure upon the second, therefore as BC to CF, so is the rectilineal figure ABC to KGH: but as BC to CF, so is* the parallelogram BE to the parallelogram EF; therefore* as the rectilineal figure ABC is to KGH, so is the parallelogram BE to the parallelogram EF: and the rectilineal figure ABC is equal † to the parallelogram BE; therefore the rectilineal figure KGH is equal* to the parallelogram EF: but EF is equal† to the figure D; wherefore also, KGH is equal to D; and it is similar to ABC. Therefore, the rectilineal figure KGH has been described similar to the figure ABC, and equal to D. Which was to be done.

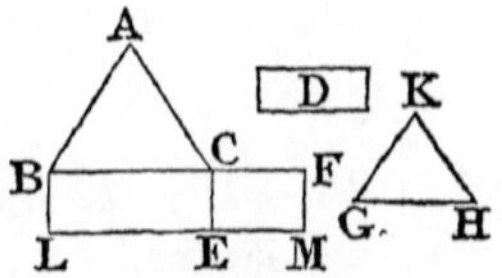

* 2 Cor. 20. 6.
* 1. 6.
* 11. 5.
† Constr.
* 14. 5.
† Constr.

PROPOSITION XXVI.

THEOR.—*If two similar parallelograms have a common angle, and be similarly situated, they are about the same diameter.*

Let the parallelograms ABCD, AEFG be similar and similarly situated, and have the angle DAB common: ABCD and AEFG shall be about the same diameter.

For, if not, let, if possible, the parallelogram BD have its diameter AHC in a different straight line from AF, the diameter of the parallelogram EG, and let GF meet AHC in H; and through H, draw HK parallel to AD or BC: therefore the parallelograms ABCD, AKHG, being about the same diameter, are similar* to one another: wherefore as DA to AB, so is* GA to AK: but because ABCD and AEFG are similar† parallelograms, as DA is to AB, so is GA to AE; therefore* as GA to AE, so GA to AK; that is GA has the same ratio to each of the straight lines AE, AK; and consequently AK is equal* to AE, the less to the greater; which is impossible: therefore ABCD and AKHG are not about the same diameter: wherefore, ABCD and AEFG must be about the same diameter. Therefore, if two similar, &c. Q. E. D.

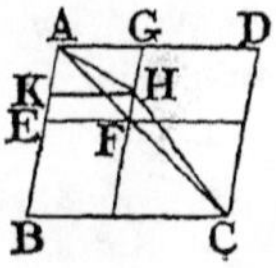

* 24. 6.
* 1 Def. 6.
† Hyp.
* 11. 5.
* 9. 5.

'To understand the three following propositions more 'easily, it is to be observed,

1. 'That a parallelogram is said to be applied to a straight 'line, when it is described upon it as one of its sides. Ex. 'gr. the parallelogram AC is said to be applied to the straight 'line AB.

2. 'But a parallelogram AE is said to be applied to a 'straight line AB, deficient by a parallelogram, when AD the 'base of AE is less than AB, and therefore 'AE is less than the parallelogram AC de'scribed upon AB in the same angle, and 'between the same parallels, by the paral'lelogram DC; and DC is therefore called 'the defect of AE.

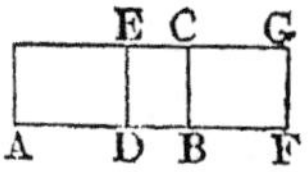

3. 'And a parallelogram AG is said to be applied to a 'straight line AB, exceeding by a parallelogram, when AF 'the base of AG is greater than AB, and therefore AG exceeds 'AC the parallelogram described upon AB in the same angle, 'and between the same parallels, by the parallelogram BG.'

PROPOSITION XXVII.

See N. THEOR.—*Of all parallelograms applied to the same straight line, and deficient by parallelograms similar and similarly situated to that which is described upon the half of the line, that which is applied to the half, and is similar to its defect, is the greatest.*

Let AB be a straight line divided into two equal parts in C; and let the parallelogram AD be applied to the half AC, which is therefore deficient from the parallelogram upon the whole line AB, by the parallelogram CE upon the other half CB: of all the parallelograms applied to any other parts of AB, and deficient by parallelograms that are similar and similarly situated to CE, AD shall be the greatest.

Let AF be any parallelogram applied to AK, any other part of AB than the half, so as to be deficient from the parallelogram upon the whole line AB, by the parallelogram KH similar and similarly situated to CE: AD shall be greater than AF.

First, let AK the base of AF, be greater than AC the half
† Hyp. of AB: and because CE is similar † to the parallelogram KH,

they are about the same* diameter: draw their diameter DB, and complete the scheme: then, because the parallelogram CF is equal * to FE, add KH to both; therefore the whole CH is equal to the whole KE: but CH is equal* to CG, because the base AC is equal to the base CB; therefore CG is equal † to KE: to each of these add CF; then the whole AF is equal † to the gnomon CHL: therefore CE or the parallelogram AD, is greater than the parallelogram AF. * 26. 6. * 43. 1. * 36. 1. † 1 Ax. † 2 Ax.

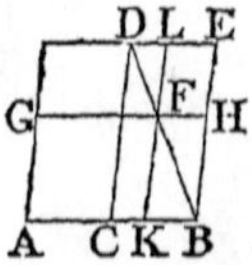

Next, let AK the base of AF, be less than AC: then, the same construction being made, because BC is equal to CA, therefore HM is * equal to MG; therefore the parallelogram DH is equal * to the parallelogram DG; wherefore DH is greater than LG: but DH is equal * to DK; therefore DK is greater than LG: to each of these add AL; then the whole AD is greater than the whole AF. Therefore, of all parallelograms applied, &c. Q. E. D. * 34. 1. * 36. 1. * 43. 1.

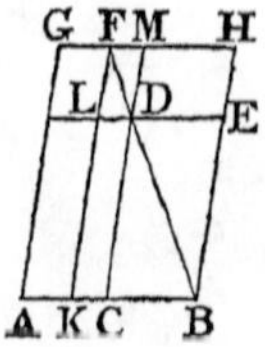

PROPOSITION XXVIII.

PROB.—*To a given straight line to apply a parallelogram equal to a given rectilineal figure, and deficient by a parallelogram similar to a given parallelogram: but the given rectilineal figure to which the parallelogram to be applied is to be equal, must not be greater † than the parallelogram applied to half of the given line, having its defect similar to the defect of that which is to be applied, that is, to the given parallelogram.* See N. † 27. 6.

Let AB be the given straight line, and C the given rectilineal figure to which the parallelogram to be applied is required to be equal, which figure must not be greater than the parallelogram applied to the half of the line, having its defect from that upon the whole line similar to the defect of that which is to be applied; and let D be the parallelogram to which this defect is required to be similar; it is required to apply a parallelogram to the straight line AB, which shall be equal to the figure C, and be deficient from the parallelogram upon the whole line by a parallelogram similar to D.

Divide AB into two equal parts * in the point E, and upon * 10. 1.

* 18. 6. EB describe the parallelogram EBFG similar * and similarly situated to D, and complete the parallelogram AG, which must either be equal to C, or greater than it, by the determination. If AG be equal to C, then what was required is already done: for, upon the straight line AB, the parallelogram AG is applied equal to the figure C, and deficient by the parallelogram EF similar to D. But, if AG be not equal to C, it is greater
† 36. 1. than it: and EF is equal † to AG; therefore EF also is greater than C.
* 25. 6. Make * the parallelogram KLMN equal to the excess of EF above C, and similar and similarly
† Constr. situated to D: then, since D is similar † to EF, therefore *
* 21. 6. also KM is similar to EF: let KL be the homologous side to EG, and LM to GF: and because EF is equal to C and KM together, EF is greater than KM; therefore the straight line
† 3. 1. EG is greater than KL, and GF than LM: make † GX equal
† 31. 1. to LK, and GO equal to LM, and complete † the parallelogram XGOP; therefore XO is equal and similar to KM: but KM is similar to EF; wherefore also XO is similar to EF;
* 26. 6. and therefore * XO and EF are about the same diameter: let GPB be their diameter, and complete the scheme. Then, because EF is equal to C and KM together, and XO a part of the one, is equal to KM a part of the other, the remainder,
† 3 Ax. viz. the gnomon ERO, is equal † to the remainder C: and
* 43. 1. because OR is equal * to XS, by adding SR to each, the whole
* 36. 1. OB is equal to the whole XB: but XB is equal * to TE, because the base AE is equal to the base EB; wherefore also
† 1 Ax. TE is equal † to OB: add XS to each, then the whole TS is equal to the whole, viz. to the gnomon ERO: but it has been proved that the gnomon ERO is equal to C; and therefore also TS is equal to C. Wherefore the parallelogram TS, equal to the given rectilineal figure C, is applied to the given straight line AB, deficient by the parallelogram SR, si-
* 24. 6. milar to the given one D, because SR is similar * to EF. Which was to be done.

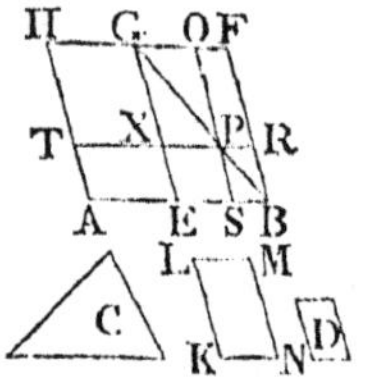

PROPOSITION XXIX.

See N. **Prob.**—*To a given straight line to apply a parallelogram* *equal to a given rectilineal figure, exceeding by a parallelogram similar to another given.*

Let AB be the given straight line, and C the given rectilineal figure to which the parallelogram to be applied is required to be equal, and D the parallelogram to which the excess of the one to be applied above that upon the given line is required to be similar; it is required to apply a parallelogram to the given straight line AB which shall be equal to the figure C, exceeding by a parallelogram similar to D.

Divide AB into two equal parts † in the point E; and † 10. 1.
upon EB, describe * the parallelogram EL similar and simi- * 18. 6.
larly situated to D; and make * the * 25. 6.
parallelogram GH equal to EL and C together, and similar and similarly situated to D: wherefore GH
is similar * to EL: let KH be the * 21. 6.
side homologous to FL, and KG to FE: and because the parallelogram GH is greater than EL, therefore the side KH is greater than FL, and KG than FE: produce FL and FE, and make FLM equal to KH, and FEN to KG, and complete the parallelogram MN; MN is therefore equal and similar to GH: but GH is similar to EL; wherefore MN is similar to EL; and consequently EL and
MN are about the same diameter *: draw their diameter FX, * 26. 6.
and complete the scheme. Therefore, since GH is equal to EL and C together, and that GH is equal to MN, MN is equal to EL and C: take away the common part EL; then the remainder, viz. the gnomon NOL, is equal to C. And because
AE is equal to EB, the parallelogram AN is equal * to the * 36. 1.
parallelogram NB, that is, to BM *: add NO to each; there- * 43. 1.
fore the whole, viz. the parallelogram AX, is equal to the gnomon NOL: but the gnomon NOL is equal to C; therefore also AX is equal to C. Wherefore, to the straight line AB, there is applied the parallelogram AX equal to the given rectilineal figure C, exceeding by the parallelogram PO, which
is similar to D, because PO is similar * to EL. Which was * 24. 6.
to be done.

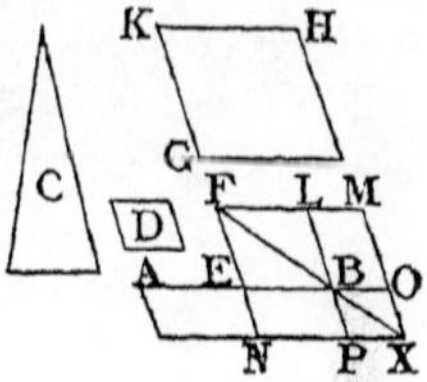

PROPOSITION XXX.

PROB.—*To cut a given straight line in extreme and mean ratio.*

Let AB be the given straight line; it is required to cut it in extreme and mean ratio.

* 46. 1. * 29. 6. Upon AB describe * the square BC, and to AC * apply the parallelogram CD, equal to BC, exceeding by the figure AD similar to BC: then, since BC is a square, therefore also AD is a square: and because BC is equal to CD, by taking the common part CE from each, the remainder BF is equal to the remainder AD: and these figures are equiangular, therefore their sides about the equal angles are

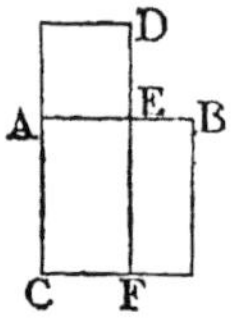

* 14. 6. reciprocally * proportional: therefore, as FE to
* 34. 1. ED, so AE to EB: but FE is equal * to AC,
† 30 Def. that is, to † AB; and ED is equal to AE; therefore as BA to AE, so is AE to EB: but AB is greater than AE; where-
* 14. 5. fore AE is greater than EB *: therefore the straight line
* 3 Def. 6. AB is cut in extreme and mean ratio in E *. Which was to be done.

Otherwise:

Let AB be the given straight line; it is required to cut it in extreme and mean ratio.

Divide AB in the point C, so that the rectangle contained
* 11. 2. by AB, BC, may be equal * to the square of AC: then, because the rectangle AB, BC is equal to

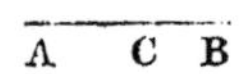

* 17. 6. the square of AC, as BA to AC, so is AC to CB *:
* 3 Def. 6. therefore AB is cut in extreme and mean ratio in C *. Which was to be done.

PROPOSITION XXXI.

See N. THEOR.—*In right-angled triangles, the rectilineal figure described upon the side opposite to the right angle, is equal to the similar and similarly described figures upon the sides containing the right angle.*

Let ABC be a right-angled triangle, having the right angle BAC: the rectilineal figure described upon BC, shall be equal to the similar and similarly described figures upon BA, AC.

Draw the perpendicular† AD: therefore, because in the right-angled triangle ABC, AD is drawn from the right angle at A, perpendicular to the base BC, the triangles ABD, ADC are similar* to the whole triangle ABC, and to one another: and because the triangle ABC is similar to ADB, as CB to BA, so is* BA to BD: and because these three straight lines are proportionals, as the first is to the third, so is the figure upon the first to the similar and similarly described figure* upon the second: therefore as CB to BD, so is the figure upon CB to the similar and similarly described figure upon BA: and inversely*, as DB to BC, so is the figure upon BA to that upon BC: for the same reason, as DC to CB, so is the figure upon CA to that upon CB: therefore as BD and DC together to BC*, so are the figures upon BA, AC, to that upon BC: but BD and DC together are equal to BC; therefore the figure described on BC is equal* to the similar and similarly described figures on BA, AC. Wherefore, in right-angled triangles, &c. Q. E. D.

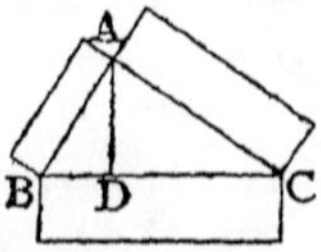

† 12. 1.
* 8. 6.
* 4. 6.
* 2 Cor. 20. 6.
* B. 5.
* 24. 5.
* A. 5.

PROPOSITION XXXII.

THEOR.—*If two triangles, which have two sides of the one, proportional to two sides of the other, be joined at one angle so as to have their homologous sides parallel to one another, the remaining sides shall be in a straight line.* See N.

Let ABC, DCE be two triangles, which have the two sides BA, AC proportional to the two CD, DE, viz. BA to AC, as CD to DE; and let AB be parallel to DC, and AC to DE: BC and CE shall be in a straight line.

Because AB is parallel to DC, and the straight line AC meets them, the alternate angles* BAC, ACD are equal; for the same reason, the angle CDE is equal to the angle ACD; wherefore also BAC is equal† to CDE: and because the triangles ABC, DCE have one angle at A equal to one at D, and the sides about these angles proportionals, viz. BA to AC, as CD to DE, the triangle ABC is equiangular* to DCE; therefore the angle ABC is equal to the angle DCE: and the angle BAC was proved to be equal to ACD; therefore the whole

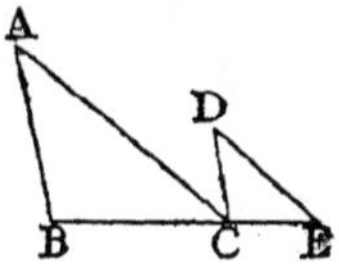

* 29. 1.
† 1 Ax.
* 6. 6.

† 2 Ax. angle ACE is equal† to the two angles ABC, BAC: add the common angle ACB, then the angles ACE, ACB are equal to the
* 32. 1. angles ABC, BAC, ACB: but ABC, BAC, ACB are equal* to two right angles; therefore also the angles ACE, ACB are equal to two right angles: and since at the point C, in the straight line AC, the two straight lines BC, CE, which are on the opposite sides of it, make the adjacent angles ACE, ACB
* 14. 1. equal to two right angles, therefore* BC and CE are in a straight line. Wherefore, if two triangles, &c. Q. E. D.

PROPOSITION XXXIII.

See N. THEOR.—*In equal circles, angles, whether at the centres or circumferences, have the same ratio which the circumferences on which they stand have to one another; so also have the sectors.*

Let ABC, DEF be equal circles; and at their centres the angles BGC, EHF, and the angles BAC, EDF at their circumferences: as the circumference BC to the circumference EF, so shall the angle BGC be to the angle EHF, and the angle BAC to the angle EDF; and also the sector BGC to the sector EHF.

Take any number of circumferences CK, KL, each equal to BC, and any number whatever FM, MN, each equal to EF; and join GK, GL, HM, HN. Because the circumferences BC, CK, KL are all equal, the angles BGC, CGK, KGL are also
* 27. 3. all* equal; therefore what multiple soever the circumference BL is of the circumference BC, the same multiple is the angle BGL of the angle BGC: for the same reason, whatever multiple the circumference EN is of the circumference EF, the same multiple is the angle EHN of the angle EHF: and if the circumference BL be equal to the

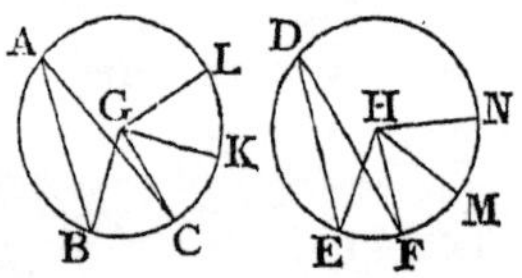

* 27. 3. circumference EN, the angle BGL is also equal* to the angle EHN; and if the circumference BL be greater than EN, likewise the angle BGL is greater than EHN; and if less, less: therefore since there are four magnitudes, the two circumferences BC, EF, and the two angles BGC, EHF; and that of the circumference BC, and of the angle BGC, have been taken

any equimultiples whatever, viz. the circumference BL, and the angle BGL; and of the circumference EF, and of the angle EHF, any equimultiples whatever, viz. the circumference EN, and the angle EHN; and since it has been proved, that if the circumference BL be greater than EN, the angle BGL is greater than EHN; and if equal, equal; and if less, less: therefore as the circumference BC to the circumference EF, so* is the angle BGC to the angle EHF: but as the angle BGC is to the angle EHF, so is* the angle BAC to the angle EDF, for each is double* of each; therefore, as the circumference BC is to EF, so is the angle BGC to the angle EHF, and the angle BAC to the angle EDF.

* 5 Def. 5.
* 15. 5.
* 20. 3.

Also, as the circumference BC to EF, so shall the sector BGC be to the sector EHF. Join BC, CK, and in the circumferences BC, CK take any points X, O, and join BX, XC, CO, OK: then, because in the triangles GBC, GCK, the two sides BG, GC are equal to the two CG, GK, each to each, and that they contain equal angles; the base BC is equal* to the base CK, and the triangle GBC to the triangle GCK: and because the circumference BC is equal to the circumference CK, the remaining part of the whole circumference of the circle ABC, is equal † to the remaining part of the whole circumference of the same circle: therefore the angle BXC is equal* to the angle COK; and the segment BXC is therefore similar to the segment COK*; and they are upon equal straight lines, BC, CK: but similar segments of circles upon equal straight lines, are equal* to one another; therefore the segment BXC is equal to the segment COK: and the triangle BGC was proved to be equal to the triangle CGK; therefore the whole, the sector BGC, is equal to the whole, the sector CGK: for the same reason, the sector KGL is equal to each of the sectors, BGC, CGK: in the same manner, the sectors EHF, FHM, MHN may be proved equal to one another: therefore, what multiple soever the circumference BL is of the circumference BC, the same multiple is the sector BGL of the sector BGC; and for the same reason, whatever multiple the circumference EN is of EF, the same multiple is the sector EHN of the sector EHF: and if the circumference BL be equal to EN, the sector BGL is equal to the sector EHN; and if the

* 4. 1.
† 3 Ax.
* 27. 3.
* 11 Def. 3.
* 24. 3.

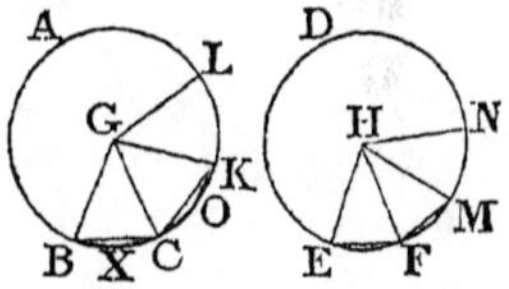

circumference BL be greater than EN, the sector BGL is greater than the sector EHN; and if less, less: since then, there are four magnitudes, the two circumferences BC, EF, and the two sectors BGC, EHF; and that of the circumference BC, and sector BGC, the circumference BL and sector BGL are any equimultiples whatever; and of the circumference EF, and sector EHF, the circumference EN, and sector EHN are any equimultiples whatever; and since it has been proved, that if the circumference BL be greater than EN, the sector BGL is greater than the sector EHN; and if equal, equal; and
* 5 Def. 5. if less, less: therefore*, as the circumference BC is to the circumference EF, so is the sector BGC to the sector EHF. Wherefore, in equal circles, &c. Q. E. D.

PROPOSITION B.

See N. THEOR.—*If an angle of a triangle be bisected by a straight line which likewise cuts the base, the rectangle contained by the sides of the triangle is equal to the rectangle contained by the segments of the base, together with the square of the straight line which bisects the angle.*

Let ABC be a triangle, and let the angle BAC be bisected by the straight line AD: the rectangle BA, AC shall be equal to the rectangle BD, DC, together with the square of AD.

* 5. 4. Describe the circle* ACB about the triangle, and produce AD to the circumference in E, and join EC: then because the
† Hyp. angle BAD is equal† to the angle CAE, and
* 21. 3. the angle ABD to the angle* AEC, for they are in the same segment, the triangles ABD,
† 32. 1. AEC are equiangular † to one another:
* 4. 6. therefore as BA to AD, so is* EA to AC; and consequently the rectangle BA, AC is
* 16. 6. equal* to the rectangle EA, AD, that is*
* 3. 2. to the rectangle ED, DA, together with the square of AD: but
* 35. 3. the rectangle ED, DA is equal to the rectangle* BD, DC; therefore the rectangle BA, AC is equal to the rectangle BD, DC, together with the square of AD. Wherefore, if an angle, &c. Q. E. D.

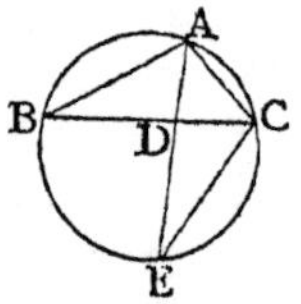

PROPOSITION C.

THEOR.—*If from any angle of a triangle a straight line be drawn perpendicular to the base, the rectangle contained by the sides of the triangle, is equal to the rectangle contained by the perpendicular and the diameter of the circle described about the triangle.* See N.

Let ABC be a triangle, and AD the perpendicular from the angle A to the base BC: the rectangle BA, AC shall be equal to the rectangle contained by AD and the diameter of the circle described about the triangle.

Describe* the circle ACB about the triangle, and draw its diameter AE, and join EC: because the right angle BDA is equal* to the angle ECA in a semicircle, and the angle ABD equal* to the angle AEC in the same segment; the triangles ABD, AEC are equiangular: therefore as* BA to AD, so is EA to AC; and consequently the rectangle BA, AC is equal* to the rectangle EA, AD. If, therefore, from any angle, &c. Q. E. D. * 5. 4. * 31. 3. * 21. 3. * 4. 6. * 16. 6.

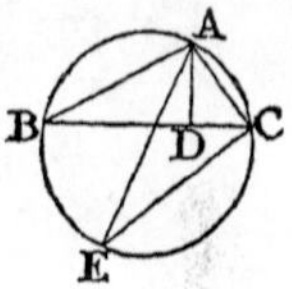

PROPOSITION D.

THEOR.—*The rectangle contained by the diagonals of a quadrilateral figure inscribed in a circle, is equal to both the rectangles contained by its opposite sides.* See N.

Let ABCD be any quadrilateral figure inscribed in a circle, and join AC, BD: the rectangle contained by AC, BD shall be equal to the two rectangles contained by AB, CD, and by AD, BC‡.

Make the angle ABE equal† to the angle DBC; add to each of these the common angle EBD, then the angle ABD is equal to the angle EBC: and the angle BDA is equal* to the angle BCE, because they are in the same segment; therefore the triangle ABD is equiangular to the triangle BCE: where- † 23. 1. * 21. 3.

‡ This is a Lemma of Cl. Ptolomæus, in page 9 of his Μεγάλη Σύνταξις.

* 4. 6. fore*, as BC is to CE, so is BD to DA; and consequently the
* 16. 6. rectangle BC, AD is equal* to the rectangle BD, CE: again, because the angle ABE is equal to the angle DBC, and the
* 21. 3. angle* BAE to the angle BDC, the triangle ABE is equiangular to the triangle BCD; therefore as BA to AE, so is BD to DC; wherefore the rectangle BA, DC is equal to the rectangle BD, AE: but the rectangle BC, AD has been shewn equal to the rectangle BD, CE;
* 1. 2. therefore ‡ the whole rectangle AC, BD* is equal to the rectangle AB, DC, together with the rectangle AD, BC. Therefore, the rectangle, &c. Q. E. D.

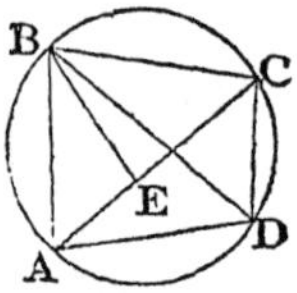

† 2 Ax.
† 1. 2.
‡ Therefore the rectangles BC, AD, and BA, DC are together equal† to the rectangles BD, CE, and BD, AE; that is† to the whole rectangle BD, AC.

THE

ELEMENTS OF EUCLID.

BOOK XI.

DEFINITIONS.

I. A SOLID is that which hath length, breadth, and thickness.

II. That which bounds a solid is a superficies.

III. A straight line is perpendicular, or at right angles, to a plane, when it makes right angles with every straight line in that plane which meets it.

IV. A plane is perpendicular to a plane, when the straight lines drawn in one of the planes perpendicular to the common section of the two planes, are perpendicular to the other plane.

V. The inclination of a straight line to a plane, is the acute angle contained by that straight line, and another drawn from the point in which the first line meets the plane, to the point in which a perpendicular to the plane drawn from any point of the first line above the plane, meets the same plane.

VI. The inclination of a plane to a plane, is the acute angle contained by two straight lines drawn from any the same point of their common section at right angles to it, one upon one plane, and the other upon the other plane.

VII. Two planes are said to have the same or a like inclination to one another which two other planes have, when the said angles of inclination are equal to one another.

VIII. Parallel planes are such as do not meet one another though produced.

See N. IX. A solid angle is that which is made by the meeting of more than two plane angles, which are not in the same plane, in one point.

See N. X. 'The tenth definition is omitted for reasons given in the notes.'

See N. XI. Similar solid figures are such as have all their solid angles equal, each to each, and are contained by the same number of similar planes.

XII. A pyramid is a solid figure contained by planes that are constituted betwixt one plane and one point above it in which they meet.

XIII. A prism is a solid figure contained by plane figures, of which two that are opposite are equal, similar, and parallel to one another; and the others parallelograms.

XIV. A sphere is a solid figure described by the revolution of a semicircle about its diameter, which remains unmoved.

XV. The axis of a sphere is the fixed straight line about which the semicircle revolves.

XVI. The centre of a sphere is the same with that of the semicircle.

XVII. The diameter of a sphere is any straight line which

passes through the centre, and is terminated both ways by the superficies of the sphere.

XVIII. A cone is a solid figure described by the revolution of a right-angled triangle about one of the sides containing the right angle, which side remains fixed.

If the fixed side be equal to the other side containing the right angle, the cone is called a right-angled cone; if it be less than the other side, an obtuse-angled; and if greater, an acute-angled cone.

XIX. The axis of a cone is the fixed straight line about which the triangle revolves.

XX. The base of a cone is the circle described by that side containing the right angle which revolves.

XXI. A cylinder is a solid figure described by the revolution of a right-angled parallelogram about one of its sides which remains fixed.

XXII. The axis of a cylinder is the fixed straight line about which the parallelogram revolves.

XXIII. The bases of a cylinder are the circles described by the two revolving opposite sides of the parallelogram.

XXIV. Similar cones and cylinders are those which have their axes and the diameters of their bases proportionals.

XXV. A cube is a solid figure contained by six equal squares.

XXVI. A tetrahedron is a solid figure contained by four equal and equilateral triangles.

XXVII. An octahedron is a solid figure contained by eight equal and equilateral triangles.

XXVIII. A dodecahedron is a solid figure contained by twelve equal pentagons which are equilateral and equiangular.

XXIX. An icosahedron is a solid figure contained by twenty equal and equilateral triangles.

DEF. A. A parallelopiped is a solid figure contained by six quadrilateral figures, whereof every opposite two are parallel.

PROPOSITION I.

See N. THEOREM.—*One part of a straight line cannot be in a plane, and another part above it.*

If it be possible, let AB, part of the straight line ABC, be in the plane, and the part BC above it: and since the straight line AB is in the plane, it can be produced in that plane: let it be produced to D; and let any plane pass through the straight line AD, and be turned about it until it pass through the point C; and

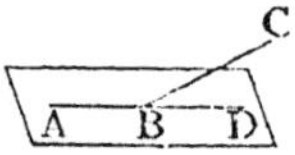

* 7 Def. 1. because the points B, C are in this plane, the straight line* BC is in it: therefore there are two straight lines ABC, ABD in the same plane that have a common segment AB; which

* Cor. 11. 1. is* impossible. Therefore, one part, &c. Q. E. D.

PROPOSITION II.

THEOR.—*Two straight lines which cut one another are in one plane; and three straight lines which meet one another are in one plane.*

Let two straight lines AB, CD cut one another in E: AB, CD shall be in one plane; and three straight lines EC, CB, BE, which meet one another, shall be in one plane.

Let any plane pass through the straight line EB, and let the plane be turned about EB, produced if necessary, until it pass through the point C: then, because the

* 7 Def. 1. points E, C are in this plane, the straight line* EC is in it; for the same reason, the straight line BC is in the same; and by the hypothesis, EB is in it; therefore the three straight lines EC, CB, BE are in one plane: but in the plane

* 1. 11. in which EC, EB are, in the same are* CD, AB: therefore AB, CD are in one plane. Wherefore, two straight lines, &c. Q. E. D.

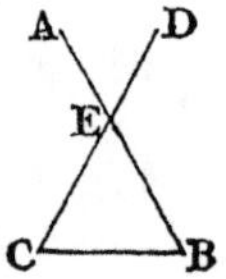

PROPOSITION III.

THEOR.—*If two planes cut one another, their common section is a straight line.* See N.

Let two planes AB, BC cut one another, and let the line DB be their common section: DB shall be a straight line.

If it be not, from the point D to B †, draw, in the plane AB, the straight line DEB, and in the plane BC, the straight line DFB: then two straight lines DEB, DFB have the same extremities, and therefore include a space betwixt them; which is * impossible: therefore BD, the common section of the planes AB, BC, cannot but be a straight line. Wherefore, if two planes, &c. Q. E. D.

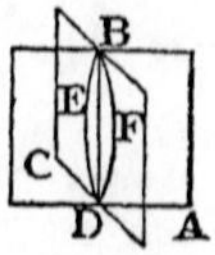

† 1 Post.

* 10 Ax. 1.

PROPOSITION IV.

THEOR.—*If a straight line stand at right angles to each of two straight lines in the point of their intersection, it shall also be at right angles to the plane which passes through them, that is, to the plane in which they are.* See N.

Let the straight line EF stand at right angles to each of the straight lines AB, CD, in E, the point of their intersection: EF shall also be at right angles to the plane passing through AB, CD.

Take the straight lines AE, EB, CE, ED, all equal to one another; and through E, draw, in the plane in which are AB, CD, any straight line GEH, and join AD, CB; then from any point F, in EF, draw FA, FG, FD, FC, FH, FB: and because the two straight lines AE, ED are equal to the two BE, EC, each to each, and that they contain equal angles * AED, BEC, the base AD is equal * to the base BC, and the angle DAE to the angle EBC: and the angle AEG is equal * to the angle BEH; therefore the triangles AEG, BEH have two angles of the one, equal to two angles of the other, each to each, and the sides AE, EB, adjacent to the equal angles, equal to one another; wherefore they have their other sides equal *; therefore GE is

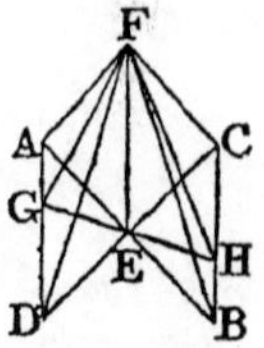

* 15. 1.
* 4. 1.
* 15. 1.
* 26. 1.

equal to EH, and AG to BH: and because AE is equal to EB, and FE common and at right angles to them, the base AF is
* 4. 1. equal * to the base FB; for the same reason, CF is equal to FD: and because AD is equal to BC, and AF to FB, the two sides FA, AD are equal to the two FB, BC, each to each; and the base DF was proved equal to the base FC; therefore the
* 8. 1. angle FAD is equal * to the angle FBC: again, it was proved that GA is equal to BH, and also AF to FB; therefore FA and AG, are equal to FB and BH, each to each; and the angle FAG has been proved equal to the angle FBH; therefore the
* 4. 1. base GF is equal * to the base FH: again, because it was proved that GE is equal to EH, and EF is common, therefore GE, EF are equal to HE, EF, each to each; and the base GF
* 8. 1. is equal to the base FH; therefore the angle GEF is equal * to the angle HEF; and consequently each of these angles is a
* 10 Def. 1. right * angle: therefore FE makes right angles with GH, that is, with any straight line drawn through E, in the plane passing through AB, CD. In like manner it may be proved, that FE makes right angles with every straight line which meets it in that plane. But a straight line is at right angles to a plane when it makes right angles with every straight line
* 3 Def. 11. which meets it in that plane*: therefore EF is at right angles to the plane in which are AB, CD. Wherefore, if a straight line, &c. Q. E. D.

PROPOSITION V.

See N. THEOR.—*If three straight lines meet all in one point, and a straight line stand at right angles to each of them in that point, these three straight lines are in one and the same plane.*

Let the straight line AB stand at right angles to each of the straight lines BC, BD, BE, in B the point where they meet: BC, BD, BE shall be in one and the same plane.

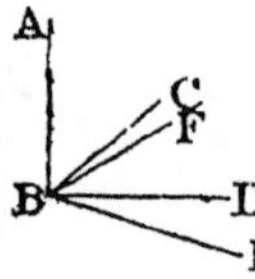

If not, let, if it be possible, BD and BE be in one plane, and BC be above it; and let a plane pass through AB, BC, the common section of which, with the plane in
* 3. 11. which BD and BE are, is a straight * line; let this be BF: therefore the three straight lines AB, BC, BF are all in one plane, viz. that which passes through AB, BC: and

because AB stands at right angles to each of the straight lines BD, BE, it is also at right angles * to the plane passing through them, and therefore makes right angles * with every straight line in that plane which meets it: but BF, which is in that plane, meets it; therefore the angle ABF is a right angle: but the angle ABC, by the hypothesis, is also a right angle; therefore the angle ABF is equal to the angle ABC, and they are both in the same plane; which is † impossible: therefore the straight line BC is not above the plane in which are BD and BE: wherefore the three straight lines BC, BD, BE are in one and the same plane. Therefore, if three straight lines, &c. Q. E. D.

* 4. 11.
* 3 Def. 11.
† 9 Ax.

PROPOSITION VI.

THEOR.—*If two straight lines be at right angles to the same plane, they shall be parallel to one another.*

Let the straight lines AB, CD be at right angles to the same plane: AB shall be parallel to CD.

Let them meet the plane in the points B, D, and draw the straight line BD, to which draw† DE at right angles, in the same plane; and make † DE equal to AB, and join BE, AE, AD. Then, because AB is perpendicular to the plane, it shall make right * angles with every straight line which meets it, and is in that plane: but BD, BE, which are in that plane, do each of them meet AB; therefore each of the angles ABD, ABE is a right angle; for the same reason, each of the angles CDB, CDE is a right angle: and because AB is equal to DE, and BD common, the two sides AB, BD are equal to the two ED, DB, each to each; and they contain right angles; therefore the base AD is equal * to the base BE: again, because AB is equal to DE, and BE to AD; AB, BE are equal to ED, DA, each to each; and, in the triangles ABE, EDA, the base AE is common; therefore the angle ABE is equal * to the angle EDA: but ABE is a right angle; therefore EDA is also a right angle, and ED perpendicular to DA: but it is also perpendicular to each of the two BD, DC; wherefore ED is at right angles to each of the three straight lines BD, DA, DC, in the point in which they meet; therefore * these three straight lines are all in the

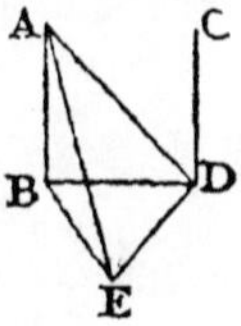

† 11. 1.
† 3. 1.
* 3 Def. 11.
* 4. 1.
* 8. 1.
* 5. 11.

* 2. 11. same plane: but AB is in the plane in which are BD, DA*, because any three straight lines which meet one another, are in one plane; therefore AB, BD, DC are in one plane: and each of the angles ABD, BDC is a right angle; therefore AB
* 28. 1. is parallel* to CD. Wherefore, if two straight lines, &c.
Q. E. D.

PROPOSITION VII.

See N. THEOR.—*If two straight lines be parallel, the straight line drawn from any point in the one to any point in the other, is in the same plane with the parallels.*

Let AB, CD be parallel straight lines, and take any point E in the one, and the point F in the other: the straight line which joins E and F, shall be in the same plane with the parallels.

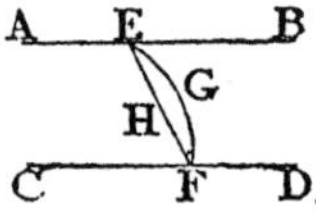

If not, let it be, if possible, above the plane, as EGF; and in the plane ABCD, in which the parallels are, draw the straight line EHF from E to F: and since EGF also is a straight line, the two straight lines
* 10 Ax. 1. EHF, EGF include a space between them; which is* impossible: therefore the straight line joining the points E, F is not above the plane in which the parallels AB, CD are, and is therefore in that plane. Wherefore, if two straight lines, &c.
Q. E. D.

PROPOSITION VIII.

See N. THEOR.—*If two straight lines be parallel, and one of them is at right angles to a plane, the other also shall be at right angles to the same plane.*

Let AB, CD be two parallel straight lines, and let one of them AB be at right angles to a plane: the other CD shall be at right angles to the same plane.

Let AB, CD meet the plane in the points B, D, and join
* 7. 11. BD: therefore* AB, CD, BD are in one plane. In the plane
† 11. 1. to which AB is at right angles, draw † DE at right angles to
† 3. 1. BD, and make † DE equal to AB, and join BE, AE, AD. And because AB is perpendicular to the plane, it is perpendicular
* 3 Def. 11. to every straight line which meets it and is in that plane*;

therefore each of the angles ABD, ABE is a right angle: and because the straight line BD meets the parallel straight lines AB, CD, the angles ABD, CDB are together equal* to two right angles: and ABD is a right angle; therefore also CDB is a right angle, and CD perpendicular to BD: and because AB is equal to DE, and BD common, the two AB, BD are equal to the two ED, DB, each to each; and the angle ABD is equal to the angle EDB, because each of them is a right angle; therefore the base AD is equal* to the base BE: again, because AB is equal to DE, and BE to AD, the two AB, BE are equal to the two ED, DA, each to each; and the base AE is common to the triangles ABE, EDA; wherefore the angle ABE is equal* to the angle EDA: but ABE is a right angle; and therefore EDA is a right angle, and ED perpendicular to DA: but it is also perpendicular to† BD; therefore ED is perpendicular* to the plane which passes through BD, DA; and therefore* makes right angles with every straight line meeting it in that plane: but DC is in the plane passing through BD, DA, because all three are in the plane in which are the parallels AB, CD; wherefore ED is at right angles to DC; and therefore CD is at right angles to DE: but CD is also at right angles to DB; therefore CD is at right angles to the two straight lines DE, DB, in the point of their intersection D; and therefore is at right angles* to the plane passing through DE, DB, which is the same plane to which AB is at right angles. Therefore, if two straight lines, &c. Q. E. D.

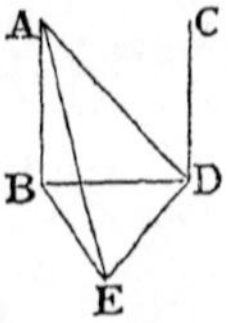

* 29. 1.

* 4. 1.

* 8. 1.

† Constr.

* 4. 11.

* 3 Def. 11.

* 4. 11.

PROPOSITION IX.

THEOR.—*Two straight lines which are each of them parallel to the same straight line, and not in the same plane with it, are parallel to one another.*

Let AB, CD be each of them parallel to EF, and not in the same plane with it: AB shall be parallel to CD.

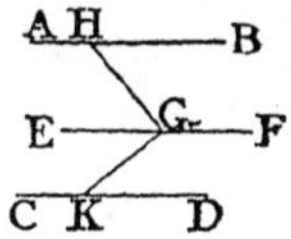

In EF, take any point G, from which† draw, in the plane passing through EF, AB, the straight line GH at right angles to EF; and in the plane passing through EF, CD, draw GK at right angles to the same EF. And because EF

† 11. 1.

* 4. 11. is perpendicular both to GH and GK, EF is perpendicular* to the plane HGK passing through them: and EF is parallel to
* 8. 11. AB; therefore AB is at right angles* to the plane HGK: for the same reason, CD is likewise at right angles to the plane HGK; therefore AB, CD are each of them at right angles to the plane HGK. But if two straight lines are at right
* 6. 11. angles to the same plane, they are parallel* to one another; therefore AB is parallel to CD. Wherefore, two straight lines, &c. Q. E. D.

PROPOSITION X.

THEOR.—*If two straight lines meeting one another be parallel to two others that meet one another, and are not in the same plane with the first two, the first two and the other two shall contain equal angles.*

Let the two straight lines AB, BC, which meet one another, be parallel to the two straight lines DE, EF, that meet one another, and are not in the same plane with AB, BC: the angle ABC shall be equal to the angle DEF.

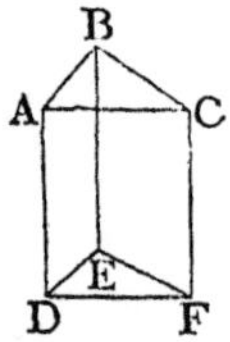

Take BA, BC, ED, EF all equal to one another; and join AD, CF, BE, AC, DF: then because BA is equal and parallel to ED, therefore
* 33. 1. AD is* both equal and parallel to BE: for the same reason, CF is equal and parallel to BE; therefore AD and CF are each of them equal and parallel to BE. But straight lines that are parallel to the same straight line, and not in the same plane
* 9. 11. with it, are parallel* to one another; therefore AD is parallel
* 1 Ax. 1. to CF; and it is equal* to it; and AC, DF join them towards
* 33. 1. the same parts; and therefore* AC is equal and parallel to DF. And because AB, BC are equal to DE, EF, each to each,
* 8. 1. and the base AC to the base DF, the angle ABC is equal* to the angle DEF. Therefore, if two straight lines, &c. Q. E. D.

PROPOSITION XI.

PROBLEM.—*To draw a straight line perpendicular to a plane, from a given point above it.*

Let A be the given point above the plane BH; it is required to draw from the point A, a straight line perpendicular to the plane BH.

In the plane, draw any straight line BC, and from the point
A, draw * AD perpendicular to BC: if then AD be also per- * 12. 1.
pendicular to the plane BH, the thing re-
quired is already done; but if it be not,
from the point D, draw *, in the plane BH, * 11. 1.
the straight line DE at right angles to BC;
and from the point A, draw AF perpen-
dicular to DE: AF shall be perpendicular
to the plane BH.

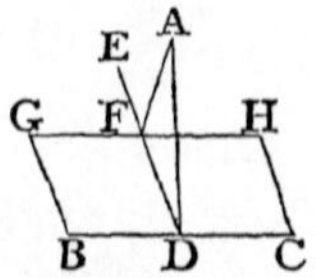

Through F, draw * GH parallel to BC: and because BC is * 31. 1.
at right angles to ED and DA, BC is at right angles * to the * 4. 11.
plane passing through ED, DA; and GH is parallel to BC:
but, if two straight lines be parallel, one of which is at right
angles to a plane, the other is at right * angles to the same * 8. 11.
plane; wherefore GH is at right angles to the plane through
ED, DA; and is perpendicular * to every straight line meet- * 3 Def. 11.
ing it in that plane: but AF, which is in the plane through
ED, DA, meets it; therefore GH is perpendicular to AF; and
consequently AF is perpendicular to GH: and AF is perpen-
dicular to DE; therefore AF is perpendicular to each of the
straight lines GH, DE. But if a straight line stand at right
angles to each of two straight lines in the point of their inter-
section, it is also at right angles † to the plane passing through † 4. 11.
them: but the plane passing through ED, GH, is the plane
BH; therefore AF is perpendicular to the plane BH: therefore,
from the given point A, above the plane BH, the straight line
AF is drawn perpendicular to that plane. Which was to be
done.

PROPOSITION XII.

Prob.—*To erect a straight line at right angles to a given plane, from a point given in the plane.*

Let A be the point given in the plane; it is required to erect a straight line from the point A, at right angles to the plane.

From any point B above the plane, draw * * 11. 11.
BC perpendicular to it; and from A, draw * * 31. 1.
AD parallel to BC. Because, therefore, AD,
CB are two parallel straight lines, and one of
them BC is at right angles to the given plane,
the other AD is also * at right angles to it: therefore, a * 8. 11.

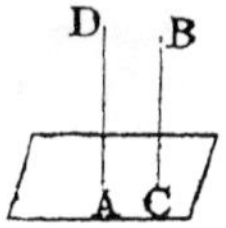

straight line has been erected at right angles to a given plane, from a point given in it. Which was to be done.

PROPOSITION XIII.

THEOR.—*From the same point in a given plane, there cannot be two straight lines at right angles to the plane, upon the same side of it: and there can be but one perpendicular to a plane from a point above the plane.*

For, if it be possible, let the two straight lines AB, AC be at right angles to a given plane, from the same point A in the plane, and upon the same side of it. Let a plane pass through BA, AC; the common section of this with the given plane is
* 3. 11. a straight * line passing through A: let DAE be their common section: therefore the straight lines AB, AC, DAE are in one plane: and because CA is at right angles to the
† 3 Def. 11. given plane, it makes right angles † with every straight line meeting it in that plane: but DAE, which is in that plane, meets CA; therefore CAE is a right angle: for the same reason BAE is a right angle; wherefore the angle
† 11 Ax. CAE is equal † to the angle BAE; and they are in one plane, which is impossible. Also, from a point above a plane, there can be but one perpendicular to that plane; for if there could
* 6. 11. be two, they would be parallel* to one another, which is absurd. Therefore, from the same point, &c. Q. E. D.

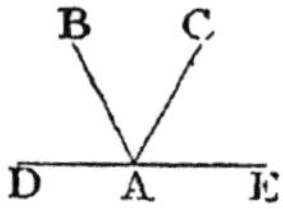

PROPOSITION XIV.

THEOR.—*Planes to which the same straight line is perpendicular, are parallel to one another.*

Let the straight line AB be perpendicular to each of the planes CD, EF: these planes shall be parallel to one another.

If not, they shall meet one another when produced: let them meet; their common section is a straight line GH, in which take any point K, and join AK, BK. Then, because AB is perpendicular to the plane EF, it is perpen-
* 3 Def. 11. dicular * to the straight line BK which is in that plane; therefore ABK is a right angle: for the same

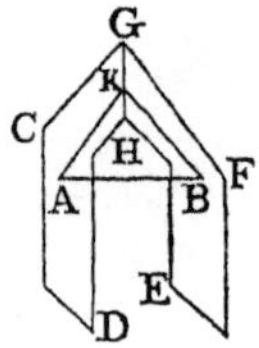

reason BAK is a right angle; wherefore the two angles ABK, BAK of the triangle ABK, are equal to two right angles; which is * impossible: therefore the planes CD, EF, though produced, do not meet one another; that is *, they are parallel. Therefore, planes, &c. Q. E. D.

* 17. 1.
* 8 Def. 11.

PROPOSITION XV.

THEOR.—*If two straight lines meeting one another, be parallel to two other straight lines which meet one another but are not in the same plane with the first two, the plane which passes through these is parallel to the plane passing through the others.* See N.

Let AB, BC, two straight lines meeting one another, be parallel to DE, EF that meet one another, but are not in the same plane with AB, BC: the planes through AB, BC, and DE, EF shall not meet, though produced.

From the point B, draw BG perpendicular * to the plane which passes through DE, EF, and let it meet that plane in G; and through G, draw GH parallel * to ED, and GK parallel to EF. And because BG is perpendicular to the plane through DE, EF, it makes * right angles with every straight line meeting it in that plane: but the straight lines GH, GK in that plane meet it; therefore each of the angles BGH, BGK is a right angle: and because BA is parallel * to GH, (for each of them is parallel to DE, and they are not both in the same plane with it,) the angles GBA, BGH are together equal * to two right angles: and BGH is a right angle; therefore also GBA is a right angle, and GB perpendicular to BA: for the same reason GB is perpendicular to BC; since therefore the straight line GB stands at right angles to the two straight lines BA, BC that cut one another in B, GB is perpendicular * to the plane through BA, BC: and it is perpendicular † to the plane through DE, EF; therefore BG is perpendicular to each of the planes through AB, BC, and DE, EF: but planes to which the same straight line is perpendicular, are parallel * to one another; therefore the plane through AB, BC is parallel to the plane through DE, EF. Wherefore, if two straight lines, &c. Q. E. D.

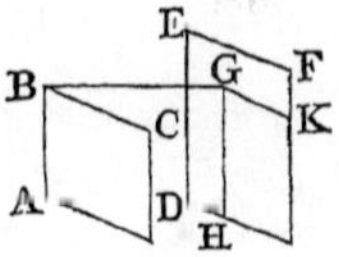

* 11. 11.
* 31. 1.
* 3 Def. 11.
* 9. 11.
* 29. 1.
* 4. 11.
† Constr.
* 14. 11.

PROPOSITION XVI.

See N. THEOR.—*If two parallel planes be cut by another plane, their common sections with it are parallels.*

Let the parallel planes AB, CD be cut by the plane EFHG, and let their common sections with it be EF, GH: EF shall be parallel to GH.

For if it is not, EF, GH shall meet if produced either on the side of FH, or EG. First, let them be produced on the side of FH, and meet in the point K: therefore, since EFK is in the plane AB, every
† I. 11. point† in EFK is in that plane: and K is a point in EFK; therefore K is in the plane AB: for the same reason, K is also in the plane CD; wherefore the planes AB, CD, produced meet one another: but they do not meet, since they are parallel by the hypothesis; therefore the straight lines EF, GH do not meet when produced on the side of FH: in the same manner it may be proved, that EF, GH do not meet when produced on the side of EG. But straight lines which are in the same plane, and do not meet, though produced either way, are parallel; therefore EF is parallel to GH. Wherefore, if two parallel planes, &c. Q. E. D.

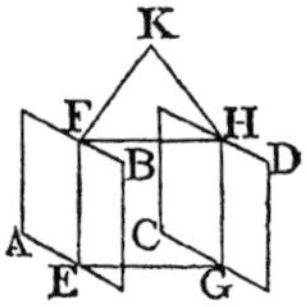

PROPOSITION XVII.

THEOR.—*If two straight lines be cut by parallel planes, they shall be cut in the same ratio.*

Let the straight lines AB, CD be cut by the parallel planes GH, KL, MN, in the points A, E, B; C, F, D: as AE is to EB, so shall CF be to FD.

Join AC, BD, AD, and let AD meet the plane KL in the point X; and join EX, XF. Because the two parallel planes KL, MN are cut by the plane EBDX, the com-
* 16. 11. mon sections EX, BD are* parallel: for the same reason, because the two parallel planes GH, KL are cut by the plane AXFC, the common sections AC, XF are parallel: and because EX is parallel to BD,

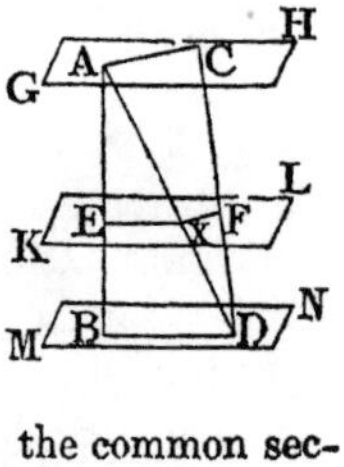

a side of the triangle ABD; as AE to EB, so is* AX to XD: again, because XF is parallel to AC, a side of the triangle ADC; as AX to XD, so is CF to FD: and it was proved, that AX is to XD, as AE to EB; therefore*, as AE to EB, so is CF to FD. Wherefore, if two straight lines, &c. Q. E. D.

* 2. 6.
* 11. 5.

PROPOSITION XVIII.

THEOR.—*If a straight line be at right angles to a plane, every plane which passes through it shall be at right angles to that plane.*

Let the straight line AB be at right angles to the plane CK: every plane which passes through AB, shall be at right angles to the plane CK.

Let any plane DE pass through AB, and let CE be the common section of the planes DE, CK; take any point F in CE, from which draw FG, in the plane DE, at right† angles to CE: and because AB is perpendicular to the plane CK, therefore* it is also perpendicular to every straight line in that plane meeting it, and consequently it is perpendicular to CE; wherefore ABF is a right angle: but GFB is likewise† a right angle; therefore AB is parallel* to FG: and AB is at right angles to the plane CK; therefore FG is also* at right angles to the same plane. But one plane is at right angles to another plane, when the straight lines drawn in one of the planes at right angles to their common section, are also at right angles* to the other plane; and any straight line FG in the plane DE, which is at right angles to CE, the common section of the planes, has been proved to be perpendicular to the other plane CK; therefore the plane DE is at right angles to the plane CK. In like manner it may be proved, that all planes which pass through AB, are at right angles to the plane CK. Therefore, if a straight line, &c. Q. E. D.

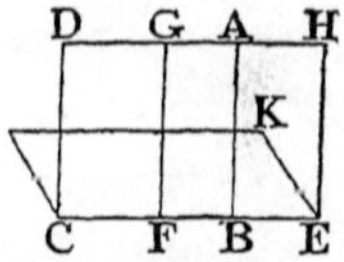

† 11. 1.
* 3 Def. 11.
† Constr.
* 28. 1.
* 8. 11.
* 4 Def. 11.

PROPOSITION XIX.

THEOR.—*If two planes which cut one another be each of them perpendicular to a third plane, their common section shall be perpendicular to the same plane.*

Let the two planes AB, BC be each of them perpendicular to a third plane, and let BD be the common section of the first two: BD shall be perpendicular to the third plane.

† 11. 1. If it be not, from the point D†, draw in the plane AB, the straight line DE at right angles to AD, the common section of the plane AB with the third plane; and in the plane BC, draw DF at right angles to CD, the common section of the plane BC with the third plane. And because the plane AB is perpendicular to the third plane, and DE is drawn in the plane AB at right angles to AD their com-

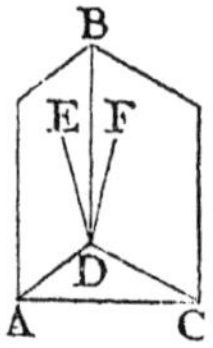

* 4 Def. 11. mon section, DE is perpendicular* to the third plane: in the same manner it may be proved, that DF is perpendicular to the third plane; wherefore, from the point D, two straight lines stand at right angles to the

* 13. 11. third plane, upon the same side of it; which is* impossible: therefore, from the point D, there cannot be any straight line at right angles to the third plane, except BD the common section of the planes AB, BC; therefore BD is perpendicular to the third plane. Wherefore, if two planes, &c. Q. E. D.

PROPOSITION XX.

See N. THEOR.—*If a solid angle be contained by three plane angles, any two of them are greater than the third.*

Let the solid angle at A be contained by the three plane angles BAC, CAD, DAB: any two of them shall be greater than the third.

If the angles BAC, CAD, DAB be all equal, it is evident that any two of them are greater than the third: but if they are not, let BAC be that angle which is not less than either of the other two, and is greater than one of them DAB; and at the point A, in the straight line AB, make, in the plane

* 23. 1. which passes through BA, AC, the angle BAE equal* to the angle DAB; and make AE equal to AD, and through E, draw BEC cutting AB, AC, in the points B, C, and join DB, DC. And because DA is equal to AE, and AB is common, the two DA, AB are equal to the two EA, AB, each to each; and the angle DAB is equal to

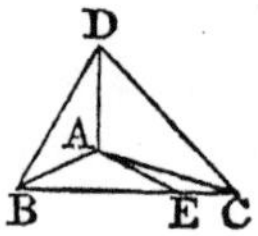

* 4. 1. the angle EAB; therefore the base DB is equal* to the base

BE: and because BD, DC are greater* than CB, and one of them BD has been proved equal to BE a part of CB, therefore the other DC is greater† than the remaining part EC: and because DA is equal to AE, and AC common, but the base DC greater than the base EC, therefore the angle DAC is greater* than the angle EAC; and, by the construction, the angle DAB is equal to the angle BAE; wherefore the angles DAB, DAC are together greater † than BAE, EAC, that is, than the angle BAC: but BAC is not less than either of the angles DAB, DAC; therefore BAC with either of them is greater than the other. Wherefore, if a solid angle, &c. Q. E. D.

* 20. 1.
† 5 Ax.
* 25. 1.
† 4 Ax.

PROPOSITION XXI.

THEOR.—*Every solid angle is contained by plane angles, which together are less than four right angles.*

First, let the solid angle at A be contained by three plane angles BAC, CAD, DAB: these three together, shall be less than four right angles.

Take, in each of the straight lines AB, AC, AD, any points B, C, D, and join BC, CD, DB. Then, because the solid angle at B is contained by the three plane angles CBA, ABD, DBC, any two of them are greater* than the third; therefore the angles CBA, ABD are greater than the angle DBC: for the same reason, the angles BCA, ACD are greater than the angle DCB, and the angles CDA, ADB greater than BDC; wherefore the six angles CBA, ABD, BCA, ACD, CDA, ADB are greater than the three angles DBC, BCD, CDB: but the three angles DBC, BCD, CDB are equal to two right angles*; therefore the six angles CBA, ABD, BCA, ACD, CDA, ADB are greater than two right angles: and because the three angles of each of the triangles ABC, ACD, ADB are equal to two right angles, therefore the nine angles of these three triangles, viz. the angles CBA, BAC, ACB, ACD, CDA, DAC, ADB, DBA, BAD are equal to six right angles: of these, the six angles CBA, ACB, ACD, CDA, ADB, DBA are greater than two right angles; therefore the remaining three angles BAC, CAD, DAB, which contain the solid angle at A, are less than four right angles.

* 20. 11.

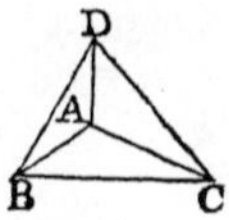

* 32. 1.

Next, let the solid angle at A be contained by any number of plane angles BAC, CAD, DAE, EAF, FAB: these shall together be less than four right angles.

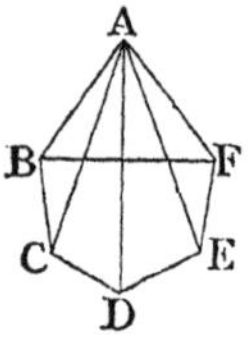

Let the planes in which the angles are, be cut by a plane, and let the common sections of it with those planes be BC, CD, DE, EF, FB. And because the solid angle at B is contained by three plane angles CBA, ABF, FBC, of which any two are greater * than the third, the angles CBA, ABF are greater than the angle FBC: for the same reason, the two plane angles at each of the points C, D, E, F, viz. those angles which are at the bases of the triangles having the common vertex A, are greater than the third angle at the same point, which is one of the angles of the polygon BCDEF; therefore all the angles at the bases of the triangles are together greater than all the angles of the polygon: and because all the angles of the triangles are together equal to twice as many right angles * as there are triangles, that is, as there are sides in the polygon BCDEF; and that all the angles of the polygon, together with four right angles, are likewise equal to twice as many right angles * as there are sides in the polygon; therefore all the angles of the triangles are equal † to all the angles of the polygon together with four right angles: but all the angles at the bases of the triangles are greater than all the angles of the polygon, as has been proved; wherefore the remaining angles of the triangles, viz. those at the vertex, which contain the solid angle at A, are less than four right angles. Therefore, every solid angle, &c. Q. E. D.

* 20. 11.

* 32. 1.

* 1 Cor. 32. 1.

† 1 Ax.

PROPOSITION XXII.

See N.

THEOR.—*If every two of three plane angles be greater than the third, and if the straight lines which contain them be all equal, a triangle may be made of the straight lines that join the extremities of those equal straight lines.*

Let ABC, DEF, GHK be the three plane angles, whereof every two are greater than the third, and let them be contained by the equal straight lines AB, BC, DE, EF, GH, HK: if their extremities be joined by the straight lines AC, DF, GK, a triangle may be made of three straight lines equal to AC,

DF, GK; that is, every two of them shall together be greater than the third.

If the angles at B, E, H, are equal, AC, DF, GK are also equal *, and any two of them greater than the third: but if the angles are not all equal, * 4. 1.

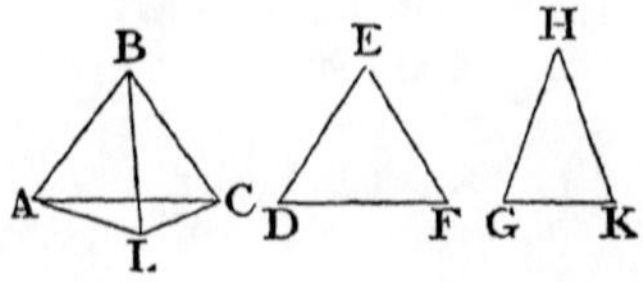

let the angle ABC be not less than either of the two at E, H; therefore the straight line AC is not less * than either of the other two DF, GK; and therefore it is plain that AC, together with either of the other two, must be greater than the third: also DF, with GK shall be greater than AC; for at the point B, in the straight line AB, make * the angle ABL equal to the angle GHK, and make BL equal to one of the straight lines AB, BC, DE, EF, GH, HK, and join AL, LC. Then, because AB, BL are equal to GH, HK, each to each, and the angle ABL to the angle GHK, the base AL is equal † to the base GK: and because the angles at E, H, are greater † than the angle ABC, of which the angle at H is equal to ABL, therefore the remaining angle at E is greater † than the angle LBC: and because the two sides LB, BC are equal to the two DE, EF, each to each, and that the angle DEF is greater than the angle LBC, the base DF is greater * than the base LC: and it has been proved, that GK is equal to AL; therefore DF and GK are greater † than AL and LC: but AL and LC are greater * than AC; much more then are DF and GK greater than AC. Wherefore, every two of these straight lines AC, DF, GK, are greater than the third; and, therefore, a triangle may be made *, the sides of which shall be equal to AC, DF, GK. Q. E. D.

* 4. or 24. 1.
* 23. 1.
† 4. 1.
† Hyp.
† 5 Ax.
* 24. 1.
† 4 Ax.
* 20. 1.
* 22. 1.

PROPOSITION XXIII.

PROB.—*To make a solid angle which shall be contained by three given plane angles, any two of them being greater † than the third, and all three together † less than four right angles.* See N. † 20. 11. † 21. 11.

Let the three given plane angles be ABC, DEF, GHK, any two of which are greater than the third, and all of them together, less than four right angles; it is required to make a solid angle contained by three plane angles equal to ABC, DEF, GHK, each to each.

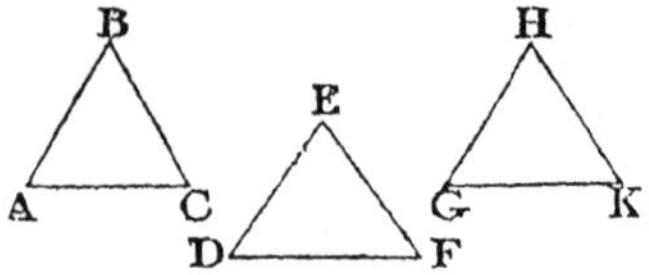

From the straight lines which contain the angles, cut off AB, BC, DE, EF, GH, HK all equal to one another; and join AC,
* 22. 11. DF, GK: then a triangle may be made* of three straight
* 22. 1. lines equal to AC, DF, GK: let this be the triangle LMN*, so that AC be equal to LM, DF to MN, and GK to LN; and
* 5. 4. about the triangle LMN describe* a circle and find† its
† 1. 3. centre X, which will be either within the triangle, or in one of its sides, or without it.

First, let the centre X be within the triangle, and join LX, MX, NX: AB shall be greater than LX. If not, AB must either be equal to, or less than LX: first, let it be equal: then because AB is equal to LX, and that AB is also equal to BC, and LX to XM, AB and BC are equal to LX and XM, each to each; and the base AC is, by construction, equal to the base LM; wherefore the angle ABC is equal to the angle

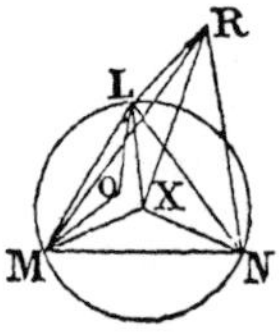

* 8. 1. LXM*: for the same reason, the angle DEF is equal to the angle MXN, and the angle GHK to the angle NXL; therefore the three angles ABC, DEF, GHK are equal to the three angles LXM, MXN, NXL: but the three angles LXM, MXN, NXL are
* 2 Cor. 15. 1. equal* to four right angles; therefore also the three angles ABC, DEF, GHK are equal to four right angles: but, by the hypothesis, they are less than four right angles; which is absurd: therefore AB is not equal to LX. But neither can AB be less than LX: for, if possible, let it be less; and upon the straight line LM, on the side of it on which is the centre X,
† 22. 1. describe† the triangle LOM, of which, two of the sides LO, OM are equal to AB, BC: and because the base LM is equal
* 8. 1. to the base AC, the angle LOM is equal* to the angle ABC: and AB, that is LO, is, by the hypothesis less than LX: wherefore LO, OM fall within the triangle LXM; for, if they
* 21. 1. fell upon its sides, or without it, they would be* equal to, or greater than, LX, XM; therefore the angle LOM, that is the
* 21. 1. angle ABC, is greater* than the angle LXM: in the same manner it may be proved, that the angle DEF is greater than

the angle MXN, and the angle GHK greater than the angle NXL; therefore the three angles ABC, DEF, GHK are greater than the three angles LXM, MXN, NXL, that is †, than four right angles: but the same angles ABC, DEF, GHK are less † than four right angles; which is absurd; therefore AB is not less than LX: and it has been proved, that it is not equal to LX; wherefore AB is greater than LX. † 2 Cor. 15. 1. † Hyp.

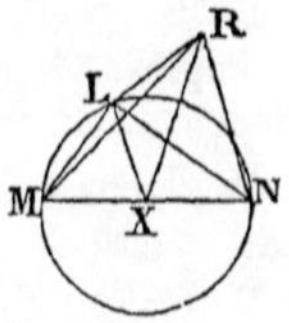

Next, let the centre X of the circle, fall in one of the sides of the triangle, viz. in MN, and join XL: in this case also, AB shall be greater than LX; if not, it is either equal to, or less than it. First, let it be equal to LX; therefore AB, and BC, that is, DE and EF, are equal to MX and XL, that is, to MN: but by the construction, MN is equal to DF; therefore DE, EF are equal to DF; which is † impossible; wherefore AB is not equal to LX: nor is it less; for then, much more, an absurdity would follow; therefore AB is greater than LX. † 20. 1.

But, let the centre X of the circle fall without the triangle LMN, and join LX, MX, NX: in this case likewise, AB shall be greater than LX; if not, it is either equal to or less than LX. First, let it be equal: it may be proved, in the same manner as in the first case, that the angle ABC is equal to the angle MXL, and GHK to LXN; therefore the whole angle MXN is equal to the two angles ABC, GHK: but ABC and GHK are together greater † than the angle DEF; therefore also the angle MXN is greater than DEF: and because DE, EF are equal to MX, XN, each to each, and the base DF to the base MN, the angle MXN is equal * to the angle DEF: but it has been proved, that it is greater than DEF; which is absurd; therefore AB is not equal to LX: neither is it less; for then, as has been proved in the first case, the angle ABC is greater than the angle MXL, and the angle GHK greater than the angle LXN. At the point B, in the straight line CB, make the angle CBP equal to the angle GHK, and make BP equal to HK, and join CP, AP. And because CB is equal to GH; CB, BP are equal to GH, HK, each to each; and they contain equal angles; wherefore the base CP is equal to the base GK, that is, to LN. And in the isosceles triangles ABC, MXL, because the angle ABC is greater than the angle † Hyp. * 8. 1.

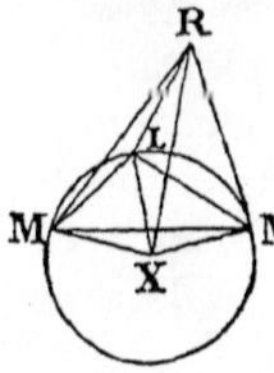

* 32. 1. MXL, therefore the angle MLX at the base, is greater * than the angle ACB at the base: for the same reason, because the angle GHK or CBP, is greater than the angle LXN, the angle XLN is greater than the angle BCP; therefore the whole

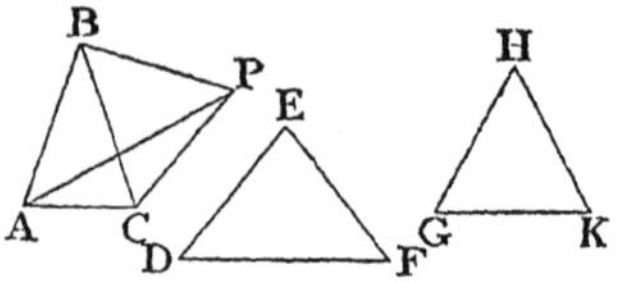

angle MLN is greater than the whole angle ACP. And because ML, LN are equal to AC, CP, each to each, but the angle MLN is greater than the angle ACP, the base MN is
* 24. 1. greater * than the base AP; but MN is equal to DF; therefore also DF is greater than AP. Again, because DE, EF are equal to AB, BP, each to each, but the base DF greater than
* 25. 1. the base AP, the angle DEF is greater * than the angle ABP: but ABP is equal to the two angles ABC, CBP, that is, to the two angles ABC, GHK; therefore the angle DEF is greater
† Hyp. than the two angles ABC, GHK: but it is also less † than these; which is impossible; therefore AB is not less than LX: and it has been proved, that it is not equal to it; therefore AB is greater than LX.

* 12. 11. From the point X, erect * XR at right angles to the plane of the circle LMN. And because it has been proved in all the cases, that AB is greater than LX, find a square equal to the excess of the square of AB above the square of LX, and make RX equal to its side, and join RL, RM, RN: the solid angle at R shall be the angle required.

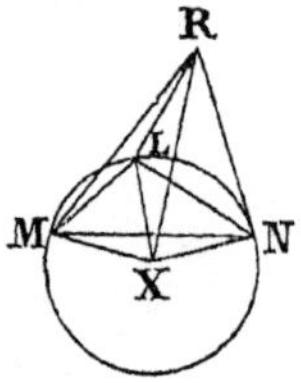

Because RX is perpendicular to the plane
* 3 Def. 11. of the circle LMN, it is * perpendicular to each of the straight lines LX, MX, NX. And because LX is equal to MX, and XR common, and at
† 4. 1. right angles to each of them, the base RL is equal † to the base RM: for the same reason, RN is equal to each of the two RL, RM; therefore the three straight lines RL, RM, RN are all equal. And because the square of XR is equal to the excess of the square of AB above the square of LX, therefore the square of AB is equal to the squares of LX, XR: but the
* 47. 1. square of RL is equal * to the same squares, because LXR is a right angle; therefore the square of AB is equal to the square of RL, and the straight line AB to RL. But each of the straight lines BC, DE, EF, GH, HK, is equal to AB, and

each of the two RM, RN, is equal to RL; therefore AB, BC, DE, EF, GH, HK are each of them equal to each of the straight lines RL, RM, RN. And because RL, RM are equal to AB, BC, each to each, and the base LM to the base AC, the angle LRM is equal* to the angle ABC: for the same reason, the angle MRN is equal to the angle DEF, and NRL to GHK. Therefore, there is made a solid angle at R, which is contained by three plane angles LRM, MRN, NRL, which are equal to the three given plane angles ABC, DEF, GHK, each to each. Which was to be done. * 8. 1.

PROPOSITION A.

THEOR.—*If each of two solid angles be contained by three plane angles, which are equal to one another, each to each, the planes in which the equal angles are, have the same inclination to one another.* See N.

Let there be two solid angles at the points A, B; and let the angle at A be contained by the three plane angles CAD, CAE, EAD, and the angle at B by the three plane angles FBG, FBH, HBG, of which the angle CAD is equal to the angle FBG, and CAE to FBH, and EAD to HBG: the planes in which the equal angles are, shall have the same inclination to one another.

In the straight line AC, take any point K, and from K†, draw, in the plane CAD, the straight line KD at right angles to AC, and in the plane CAE, the straight line KL at right angles to the same AC: therefore the angle DKL is the inclination* of the plane CAD to the plane CAE. In BF, take BM equal to AK, and from the point M, draw in the planes FBG, FBH, the straight lines MG, MN at right angles to BF; therefore the angle GMN is the inclination* of the plane FBG to the plane FBH. Join LD, NG. And because in the triangles KAD, MBG, the angles KAD, MBG are† equal, as also the right angles AKD, BMG, and that the sides AK, BM, adjacent to the equal angles, are equal to one another, therefore KD is equal* to MG, and AD to BG: for the same reason, in the triangles KAL, MBN, KL is

† 11. 1.

* 6 Def. 11.

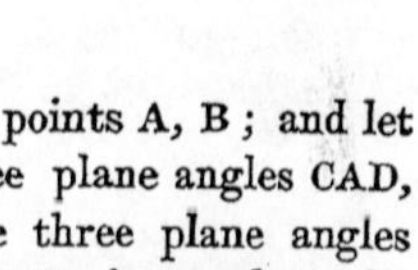

* 6 Def. 11.

† Hyp.

* 26. 1.

equal to MN, and AL to BN; therefore in the triangles LAD, NBG, LA, AD are equal to NB, BG, each to each; and they
* 4. 1. contain equal angles; therefore the base LD is equal* to the base NG. Lastly, in the triangles KLD, MNG, the sides DK, KL are equal to GM, MN each to each, and the base
* 8. 1. LD to the base NG; therefore the angle DKL is equal to* the angle GMN: but the angle DKL is the inclination of the plane CAD to the plane CAE, and the angle GMN is the inclination of the plane FBG to the plane FBH, which planes
* 7 Def. 11. have therefore the same inclination* to one another. And in the same manner it may be demonstrated, that the other planes in which the equal angles are, have the same inclination to one another. Therefore, if each of two solid angles, &c. Q. E. D.

PROPOSITION B.

See N. THEOR.—*If two solid angles be contained, each by three plane angles which are equal to one another, each to each, and alike situated, these solid angles are equal to one another.*

Let there be two solid angles at A and B, of which the solid angle at A is contained by the three plane angles CAD, CAE, EAD, and that at B, by the three plane angles FBG, FBH, HBG, of which CAD is equal to FBG, CAE to FBH, and EAD to HBG: the solid angle at A shall be equal to the solid angle at B.

Let the solid angle at A be applied to the solid angle at B: and first, the plane angle CAD being applied to the plane angle FBG, so that the point A may coincide with the point B, and the straight line AC with BF; then AD coincides with BG, because the angle CAD is equal to the angle FBG: and because the inclination of the plane CAE to

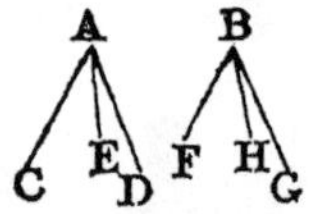

* A. 11. the plane CAD, is equal* to the inclination of the plane FBH to the plane FBG, the plane CAE coincides with the plane FBH, because the planes CAD, FBG coincide with one another: and because the straight lines AC, BF coincide, and that the angle CAE is equal to the angle FBH, therefore AE coincides with BH: and AD coincides with BG; wherefore the plane EAD coin-

cides with the plane HBG: therefore, the solid angle A coincides with the solid angle B, and consequently they are equal* to one another. Q. E. D. * 8 Ax. 1.

PROPOSITION C.

THEOR.—*Solid figures which are contained by the same number of equal and similar planes alike situated, and having none of their solid angles contained by more than three plane angles, are equal and similar to one another.* See N.

Let AG, KQ be two solid figures contained by the same number of similar and equal planes, alike situated, viz. let the plane AC be similar and equal to the plane KM; the plane AF to KP; BG to LQ; GD to QN; DE to NO; and, lastly, FH similar and equal to PR: the solid figure AG shall be equal and similar to the solid figure KQ.

Because the solid angle at A is contained by the three plane angles BAD, BAE, EAD, which, by the hypothesis, are equal to the plane angles LKN, LKO, OKN, which contain the solid angle at K, each to each, therefore the solid angle at A is equal* to the solid angle at K: in the same manner, the other solid angles of the figures are equal to one another. Let then the solid figure AG be applied to the solid figure KQ: first, the plane figure AC being applied to the plane figure KM, so that the straight line AB may coincide with KL, the figure AC must coincide with the figure KM, because they are equal and similar; therefore the straight lines AD, DC, CB coincide with KN, NM, ML, each with each; and the points A, D, C, B, with the points K, N, M, L: and the solid angle at A coincides with* the solid angle at K; wherefore the plane AF coincides with the plane KP, and the figure AF with the figure KP, because they are equal and similar to one another: therefore the straight lines AE, EF, FB coincide with KO, OP, PL; and the points E, F, with the points O, P: in the same manner, the figure AH coincides with the figure KR, and the straight line DH, with NR, and the point H with the point R. And because the solid angle at B is equal to the solid angle at L, it may be proved in the same manner, that the figure BG coincides with * B. 11. * B. 11.

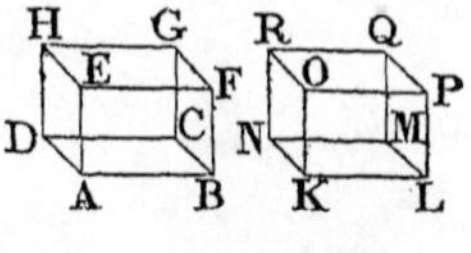

the figure LQ, and the straight line CG with MQ, and the point G with the point Q. Therefore, since all the planes and sides of the solid figure AG, coincide with the planes and sides of the solid figure KQ, AG is equal and similar to KQ. And in the same manner, any other solid figures whatever, contained by the same number of equal and similar planes, alike situated, and having none of their solid angles contained by more than three plane angles, may be proved to be equal and similar to one another. Q. E. D.

PROPOSITION XXIV.

See N. THEOR.—*If a solid be contained by six planes, two and two of which are parallel; the opposite planes are similar and equal parallelograms.*

Let the solid CDGH be contained by the parallel planes AC, GF; BG, CE; FB, AE: its opposite planes shall be similar and equal parallelograms.

Because the two parallel planes BG, CE are cut by the plane
* 16. 11. AC, their common sections AB, CD* are parallel: again, because the two parallel planes BF, AE are cut by the plane
* 16. 11. AC, their common sections AD, BC* are parallel: and AB is parallel to CD; therefore AC is a parallelogram. In like manner it may be proved, that each of the figures CE, FG, GB, BF, AE, is a parallelogram. Join AH, DF: and because AB is parallel to DC, and BH to CF; the two straight lines AB, BH, which meet one another, are parallel to DC and CF, which meet one another, and are not in the same plane with the other two:

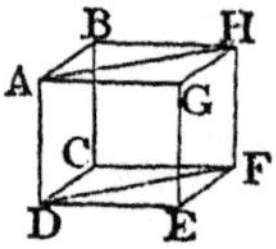

* 10. 11. wherefore they contain* equal angles; therefore the angle ABH is equal to the angle DCF: and because AB, BH are equal to DC, CF, each to each, and the angle ABH equal to
* 4. 1. the angle DCF, therefore the base AH is equal* to the base DF, and the triangle ABH to the triangle DCF: but the
* 34. 1. parallelogram BG is double* of the triangle ABH, and the parallelogram CE double of the triangle DCF; therefore the parallelogram BG is equal and similar to the parallelogram CE. In the same manner it may be proved, that the parallelogram AC is equal and similar to the parallelogram GF, and the parallelogram AE to BF. Therefore, if a solid, &c. Q. E. D.

PROPOSITION XXV.

THEOR.—*If a solid parallelopiped be cut by a plane parallel to two of its opposite planes, it divides the whole into two solids, the base of one of which shall be to the base of the other, as the one solid is to the other.* See N.

Let the solid parallelopiped ABCD be cut by the plane EV, which is parallel to the opposite planes AR, HD, and divides the whole into the two solids ABFV, EGCD: as the base AEFY of the first is to the base EHCF of the other, so shall the solid ABFV be to the solid EGCD.

Produce AH both ways, and take any number of straight lines HM, MN, each equal to EH, and any number AK, KL, each equal to EA, and complete the parallelograms LO, KY, HQ, MS, and the solids LP, KR, HU, MT. Then, because the straight lines LK, KA, AE are all equal, the parallelograms LO, KY, AF are* equal; and likewise the parallelograms KX, KB, AG: also* the parallelograms LZ, KP, AR are equal, because they are opposite planes: for the same reason, the parallelograms EC, HQ, MS are equal*, and the parallelograms HG, HI, IN: as also* HD, MU, NT: therefore three planes of the solid LP are equal and similar to three planes of the solid KR, as also to three planes of the solid AV: but the three planes opposite to these three are equal and similar* to them in the several solids, and none of their solid angles are contained by more than three plane angles; therefore the three solids LP, KR, AV are equal* to one another: for the same reason, the three solids ED, HU, MT are equal to one another: therefore what multiple soever the base LF is of the base AF, the same multiple is the solid LV of the solid AV; and whatever multiple the base NF is of the base HF, the same multiple is the solid NV of the solid ED: and if the base LF be equal to the base NF, the solid LV is equal* to the solid NV; and if the base LF be greater than the base NF, the solid LV is greater than the solid NV; and if less, less. Since then there are four magnitudes, viz. the two bases AF, FH, and the two solids AV, ED; and that of the base AF and solid

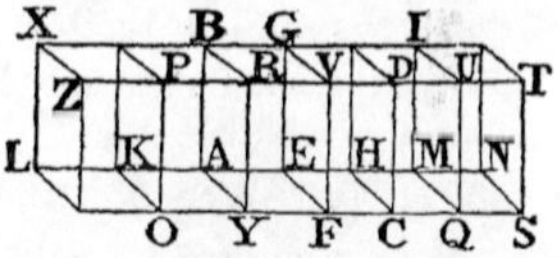

* 36. I.
* 24. 11.
* 36. 1
* 24. 11.
* 24. 11.
* C. 11.
* C. 11.

AV, the base LF and solid LV are any equimultiples whatever; and of the base FH and solid ED, the base FN and solid NV are any equimultiples whatever; and since it has been proved, that if the base LF is greater than the base FN, the solid LV is greater than the solid NV; and if equal, equal;
* 5 Def. 5. and if less, less; therefore* as the base AF is to the base FH, so is the solid AV to the solid ED. Wherefore, if a solid, &c. Q. E. D.

PROPOSITION XXVI.

See N. PROB.—*At a given point in a given straight line, to make a solid angle equal to a given solid angle contained by three plane angles.*

Let AB be a given straight line, A a given point in it, and D a given solid angle contained by the three plane angles EDC, EDF, FDC; it is required to make at the point A, in the straight line AB, a solid angle equal to the solid angle D.

* 11. 11. In the straight line DF take any point F, from which draw* FG perpendicular to the plane EDC, meeting that plane in G,
* 23. 1. and join DG: at the point A, in the straight line AB, make* the angle BAL equal to the angle EDC; and in the plane BAL, make the angle BAK equal to the angle EDG; then make AK
* 12. 11. equal to DG, and from the point K, erect* KH at right angles to the plane BAL, and make KH equal to GF, and join AH: the solid angle at A which is contained by the three plane angles BAL, BAH, HAL, shall be equal to the solid angle at D contained by the three plane angles EDC, EDF, FDC.

Take the equal straight lines AB, DE, and join HB, KB, FE, GE. And because FG is perpendicular to the plane EDC, it
* 3 Def. 11. makes right angles* with every straight line meeting it in that plane; therefore each of the angles FGD, FGE is a right angle: for the same reason, HKA, HKB are right angles. And because KA, AB are equal to GD, DE, each to each, and that they contain equal angles, therefore the base
* 4. 1. BK is equal* to the base EG; and KH
† Constr. is equal † to GF, and HKB, FGE are
* 4. 1. right angles, therefore HB is equal* to FE. Again, because AK, KH are equal to DG, GF, each to each, and contain right angles, the base AH is equal to the

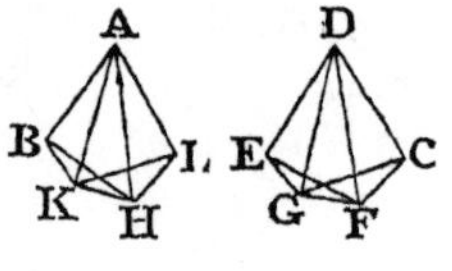

base DF; and AB is equal to DE; therefore, HA, AB are equal to FD, DE, each to each; and the base HB is equal to the base FE; therefore the angle BAH is equal * to the angle EDF: for the same reason, the angle HAL is equal to the angle FDC: because if AL and DC be made equal, and KL, HL, GC, FC be joined; since the whole angle BAL is equal to the whole EDC, and the parts of them BAK, EDG are, by the construction, equal, therefore the remaining angle KAL is equal to the remaining angle GDC: and because KA, AL are equal to GD, DC, each to each, and contain equal angles, the base KL is equal * to the base GC; and KH is equal to GF; so that LK, KH are equal to CG, GF, each to each; and they contain right † angles; therefore the base HL is equal † to the base FC: again, because HA, AL are equal to FD, DC, each to each, and the base HL to the base FC, the angle HAL is equal * to the angle FDC. Therefore, because the three plane angles BAL, BAH, HAL, which contain the solid angle at A, are equal to the three plane angles EDC, EDF, FDC, which contain the solid angle at D, each to each, and are situated in the same order, the solid angle at A is equal * to the solid angle at D. Therefore, at a given point in a given straight line, a solid angle has been made equal to a given solid angle contained by three plane angles. Which was to be done.

* 8. 1.
* 4. 1.
† 3 Def. 11.
† 4. 1.
* 8. 1.
* B. 11.

PROPOSITION XXVII.

THEOR.—*To describe, from a given straight line, a solid parallelopiped similar and similarly situated to one given.* See N.

Let AB be the given straight line, and CD the given solid parallelopiped; it is required from AB, to describe a solid parallelopiped similar and similarly situated to CD.

At the point A of the given straight line AB, make * a solid angle equal to the solid angle at C, and let BAK, KAH, HAB be the three plane angles which contain it, so that BAK be equal to the angle ECG, and KAH to GCF, and HAB to FCE: and as EC to CG, so make * BA to AK; and as GC to CF, so make * KA to AH; wherefore, ex æquali *, as EC to CF, so is BA to AH: com-

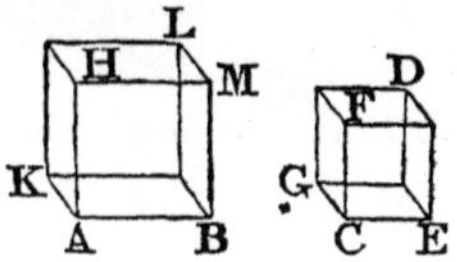

* 26. 11.
* 12. 6.
* 12. 6.
* 22. 5.

plete the parallelogram BH, and the solid AL: AL shall be similar and similarly situated to CD.

Because, as EC to GC, so BA to AK, the sides about the equal angles ECG, BAK, are proportionals; therefore the
† 1 Def. 6. parallelogram BK is similar† to EG: for the same reason, the parallelogram KH is similar to GF, and HB to FE; wherefore three parallelograms of the solid AL are similar to three of the solid CD: and the three opposite ones in each solid are
* 24. 11. equal* and similar to these, each to each. Also, because the plane angles which contain the solid angles of the figures are equal, each to each, and situated in the same order, the solid
* B. 11. angles are equal*, each to each: therefore the solid AL is
* 11 Def. 11. similar* to the solid CD. Wherefore, from a given straight line AB, a solid parallelopiped AL has been described similar and similarly situated to the given one CD. Which was to be done.

PROPOSITION XXVIII.

See N. THEOR.—*If a solid parallelopiped be cut by a plane passing through the diagonals of two of the opposite planes, it shall be cut into two equal parts.*

Let AB be a solid parallelopiped, and DE, CF, the diagonals of the opposite parallelograms AH, GB, viz. those which are drawn betwixt the equal angles in each: and because CD, FE are each of them parallel to GA, and not in the same plane
* 9. 11. with it, CD, FE are* parallel; wherefore the diagonals CF, DE are in the plane in which the
* 16. 11. parallels are, and are themselves* parallels: and the plane CDEF shall cut the solid AB into two equal parts.

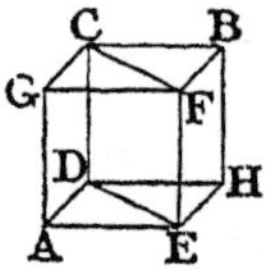

* 34. 1. Because the triangle CGF is equal* to the triangle CBF, and the triangle DAE to DHE;
* 24. 11. and that the parallelogram CA is equal* and similar to the opposite one BE, and the parallelogram GE to CH; therefore the prism contained by the two triangles CGF, DAE, and the
* C. 11. three parallelograms CA, GE, EC, is equal* to the prism contained by the two triangles CBF, DHE, and the three parallelograms BE, CH, EC, because they are contained by the same number of equal and similar planes, alike situated, and none of their solid angles are contained by more than three

plane angles. Therefore, the solid AB is cut into two equal parts by the plane CDEF. Q. E. D.

PROPOSITION XXIX ‡.

THEOR.—*Solid parallelopipeds upon the same base, and of the same altitude, the insisting straight lines of which are terminated in the same straight lines in the plane opposite to the base, are equal to one another.* See N.

Let the solid parallelopipeds AH, AK be upon the same base AB, and of the same altitude, and let their insisting straight lines AF, AG, LM, LN be terminated in the same straight line FN, and CD, CE, BH, BK be terminated in the same straight line DK: the solid AH shall be equal to the solid AK.

First, let the parallelograms DG, HN, which are opposite to the base AB, have a common side HG. Then because the solid AH is cut by the plane AGHC passing through the diagonals AG, CH of the opposite planes ALGF, CBHD, AH is cut into two equal parts * by the plane AGHC; * 28. 11. therefore the solid AH is double of the prism which is contained betwixt the triangles ALG, CBH: for the same reason, because the solid AK is cut by the plane LGHB, through the diagonals LG, BH of the opposite planes ALNG, CBKH, the solid AK is double of the same prism which is contained betwixt the triangles ALG, CBH: therefore the solid AH is equal † to the solid AK. † 6 Ax.

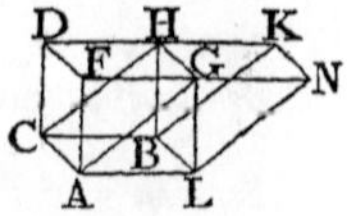

Next, let the parallelograms DM, EN, opposite to the base, have no common side. Then, because CH, CK are parallelograms, CB is equal* * 34. 1. to each of the opposite sides DH, EK; wherefore DH is equal to EK: add, or take away the common part HE; then DE is equal † to HK: † 2 or 3 Ax. wherefore also the triangle CDE is equal * to the triangle BHK, and the parallelogram DG * 38. 1.

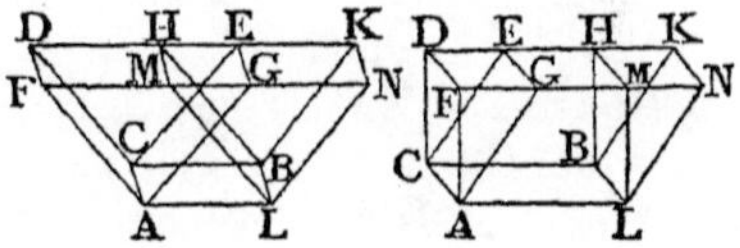

‡ 'N. B. The insisting straight lines of a parallelopiped, mentioned in this 'and some following propositions, are the sides of the parallelograms betwixt the base and the opposite plane parallel to it.'

* 36. 1. is equal* to the parallelogram HN: for the same reason, the triangle AFG is equal to the triangle LMN: and the parallelogram CF is equal* to the parallelogram BM, and CG to BN; for they are opposite. Therefore the prism which is contained by the two triangles AFG, CDE, and the three parallelograms AD, DG, GC, is equal* to the prism contained by the two triangles LMN, BHK, and the three parallelograms BM, MK, KL. If, therefore, the prism LMN, BHK be taken from the solid of which the base is the parallelogram AB, and in which FDKN is the one opposite to it; and if from this same solid there be taken the prism AFG, CDE; the remaining solid, viz. the parallelopiped AH is equal† to the remaining parallelopiped AK. Therefore, solid parallelopipeds, &c. Q. E. D.

* 24. 11.

* C. 11.

† 3 Ax.

PROPOSITION XXX.

See N.

THEOR.—*Solid parallelopipeds upon the same base, and of the same altitude, the insisting straight lines of which are not terminated in the same straight lines in the plane opposite to the base, are equal to one another.*

Let the parallelopipeds CM, CN be upon the same base AB, and of the same altitude, but their insisting straight lines AF, AG, LM, LN, CD, CE, BH, BK not terminated in the same straight lines: the solids CM, CN shall be equal to one another.

Produce FD, MH, and NG, KE, and let them meet one another in the points O, P, Q, R; and join AO, LP, BQ, CR. And because the plane LBHM is parallel to the opposite plane ACDF, and that the plane LBHM is that in which are the parallels LB, MHPQ, in which also is the figure BLPQ; and the plane ACDF is that in which are the parallels AC, FDOR, in which also is the figure CAOR; therefore the figures BLPQ, CAOR are in parallel planes: in like manner, because the plane ALNG is parallel to the opposite plane CBKE, and that the plane ALNG is that in which are the parallels AL, OPGN, in which also is the figure ALPO; and the plane CBKE is that in which are

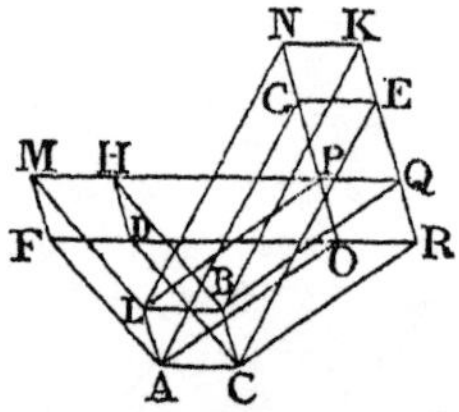

the parallels CB, RQEK, in which also is the figure CBQR; therefore the figures ALPO, CBQR are in parallel planes: and the planes ACBL, ORPQ are† parallel; therefore the solid CP is a parallelopiped: but the solid CM is equal* to the solid CP, because they are upon the same base ACBL, and their insisting straight lines AF, AO, CD, CR; LM, LP, BH, BQ are in the same straight lines FR, MQ: and the solid CP is equal* to the solid CN, for they are upon the same base ACBL, and their insisting straight lines AO, AG, LP, LN; CR, CE, BQ, BK are in the same straight lines ON, RK; therefore the solid CM is equal to the solid CN. Wherefore, solid parallelopipeds, &c. Q. E. D.

† Hyp.
* 29. 11.
* 29. 11.

PROPOSITION XXXI.

THEOR.—*Solid parallelopipeds, which are upon equal bases, and of the same altitude, are equal to one another.* See N.

Let the solid parallelopipeds AE, CF be upon equal bases AB, CD, and be of the same altitude: the solid AE shall be equal to the solid CF.

First, let the insisting straight lines be at right angles to the bases AB, CD, and let the bases be placed in the same plane, and so that the sides CL, LB may be in a straight line; therefore the straight line LM, which is at right angles to the plane in which the bases are, in the point L, is common * to the two solids AE, CF: let the other insisting lines of the solids be AG, HK, BE; DF, OP, CN: and first, let the angle ALB be equal to the angle CLD: then AL, LD are in a straight line*. Produce OD, HB, and let them meet in Q, and complete the solid parallelopiped LR, the base of which is the parallelogram LQ, and of which LM is one of its insisting straight lines. Therefore, because the parallelogram AB is equal to CD, as the base AB is to the base LQ, so is* the base CD to the base LQ. And because the solid parallelopiped AR is cut by the plane LMEB, which is parallel to the opposite planes AK, DR; as the base AB is to the base LQ, so is* the solid AE to the solid LR: for the same reason, because the solid parallelopiped

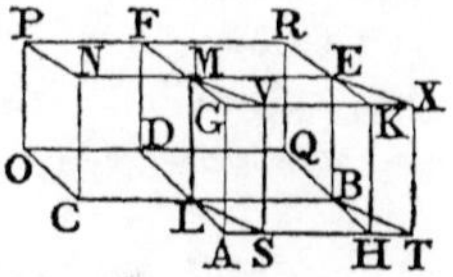

* 13. 11.
* 14. 1. & N. 14. 6.
* 7. 5.
* 25. 11.

CR is cut by the plane LMFD, which is parallel to the opposite planes CP, BR; as the base CD to the base LQ, so is the solid CF to the solid LR: but as the base AB to the base LQ, so the base CD to the base LQ, as before was proved; therefore,
† 11.5. as the solid AE to the solid LR†, so is the solid CF to the solid
* 9. 5. LR: and therefore the solid AE is equal * to the solid CF.

But let the solid parallelopipeds SE, CF be upon equal bases SB, CD, and be of the same altitude, and let their insisting straight lines be at right angles to the bases; and place the bases SB, CD in the same plane, so that CL, LB may be in a straight line; and let the angles SLB, CLD be unequal: the solid SE shall be equal to the solid CF. Produce DL,

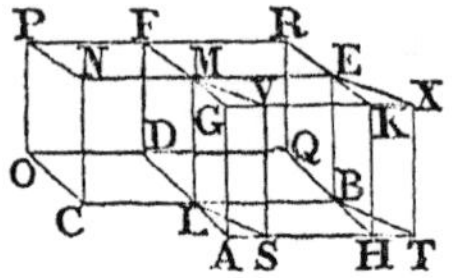

TS until they meet in A; and from B, draw BH parallel to DA; and let HB, OD produced meet in Q, and complete the
* 29. 11. solids AE, LR: therefore the solid AE is equal * to the solid SE, because they are upon the same base LE, and of the same altitude, and their insisting straight lines, viz. LA, LS, BH, BT; MG, MV, EK, EX, are in the same straight lines AT,
* 35. 1. GX: and because the parallelogram AB is equal * to SB, for they are upon the same base LB, and between the same parallels LB, AT: and that the base SB is equal to the base CD; therefore the base AB is equal to the base CD; and the angle ALB is equal to the angle CLD; therefore by the first case, the solid AE is equal to the solid CF: but the solid AE is equal to the solid SE, as was demonstrated; therefore the solid SE is equal to the solid CF.

But if the insisting straight lines AG, HK, BE, LM; CN, RS, DF, OP be not at right angles to the bases AB, CD; in this case likewise, the solid AE shall be equal to the solid CF. From the points G, K, E, M; N, S, F, P, draw the straight

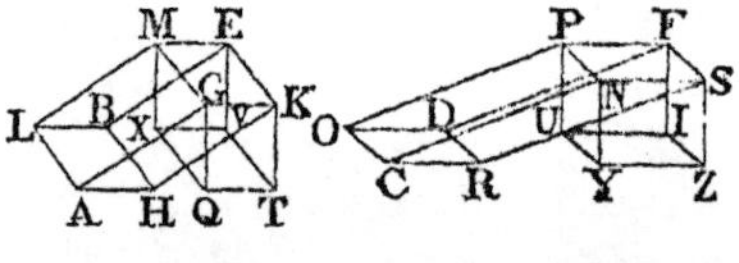

* 11. 11. lines GQ, KT, EV, MX; NY, SZ, FI, PU, perpendicular * to the planes in which are the bases AB, CD; and let them meet them in the points Q, T, V, X; Y, Z, I, U; and join QT, TV, VX, XQ; YZ, ZI, IU, UY. Then, because GQ, KT are
* 6. 11. at right angles to the same plane, they are parallel * to one

another: and MG, EK are parallels; therefore the planes MQ, ET, (of which one passes through MG, GQ, and the other through EK, KT, which are parallel to MG, GQ, and not in the same plane with them,) are parallel * to one another: for the same reason, the planes MV, GT are parallel to one another: therefore the solid QE is a parallelopiped. In like manner it may be proved, that the solid YF is a parallelopiped. But, from what has been demonstrated, the solid EQ is equal to the solid FY, because they are upon equal bases MK, PS, and of the same altitude, and have their insisting straight lines at right angles to the bases: and the solid EQ is equal * to the solid AE, and the solid FY to the solid CF, because they are upon the same bases and of the same altitude; therefore the solid AE is equal to the solid CF. Wherefore, solid parallelopipeds, &c. Q. E. D.

* 15. 11.

* 29. or 30. 11.

PROPOSITION XXXII.

THEOR.—*Solid parallelopipeds which have the same altitude, are to one another as their bases.* See N.

Let AB, CD be solid parallelopipeds of the same altitude: they shall be to one another as their bases; that is, as the base AE to the base CF, so shall the solid AB be to the solid CD.

To the straight line FG, apply the parallelogram FH equal * to AE, so that the angle FGH may be equal to the angle LCG; and upon the base FH, complete the solid parallelopiped GK, one of whose insisting lines is FD, whereby the solids CD, GK must be of the same altitude: therefore the solid AB is equal * to the solid GK, because they are upon equal bases AE, FH, and are of the same altitude: and because the solid parallelopiped CK is cut by the plane DG, which is parallel to its opposite planes, the base HF is * to the base FC, as the solid HD to the solid DC: but the base HF is equal to the base AE, and the solid GK to the solid AB; therefore as the base AE to the base CF, so is the solid AB to the solid CD. Wherefore, solid parallelopipeds, &c. Q. E. D.

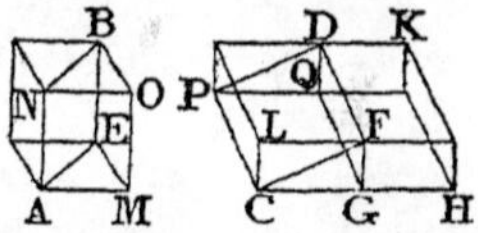

* Cor. 45. 1.

* 31. 11.

* 25. 11.

Cor. From this it is manifest, that prisms upon triangular bases, of the same altitude, are to one another as their bases.

Let the prisms, the bases of which are the triangles AEM, CFG, and NBO, PDQ the triangles opposite to them, have the same altitude: they shall be to one another as their bases. Complete the parallelograms AE, CF, and the solid parallelopipeds AB, CD, in the first of which let MO, and in the other let GQ be one of the insisting lines. And because the solid parallelopipeds AB, CD have the same altitude, they are to one another as the base AE is to the base CF: wherefore the
* 28. 11. prisms, which are their * halves, are to one another, as the base AE to the base CF; that is, as the triangle AEM to the triangle CFG.

PROPOSITION XXXIII.

Theor.—*Similar solid parallelopipeds are one to another in the triplicate ratio of their homologous sides.*

Let AB, CD be similar solid parallelopipeds, and the side AE homologous to the side CF: the solid AB shall have to the solid CD, the triplicate ratio of that which AE has to CF.

Produce AE, GE, HE; and in these produced, take EK equal to CF, EL equal to FN, and EM equal to FR; and complete the parallelogram KL, and the solid KO. Because KE, EL are equal to CF, FN, each to each, and the angle KEL equal to the angle CFN, because it is equal to the angle AEG, which is equal to CFN, by reason that the solids AB, CD are similar; therefore the parallelogram KL is similar and

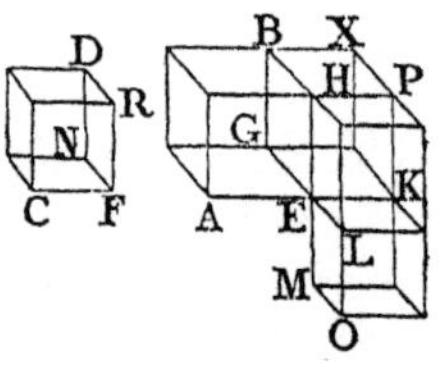

equal to the parallelogram CN: for the same reason the parallelogram MK is similar and equal to CR, and also OE to FD. Therefore three parallelograms of the solid KO are equal and similar to three parallelograms of the solid CD: and the three
* 24. 11. opposite ones in each solid are equal * and similar to these:
* C. 11. therefore the solid KO is equal * and similar to the solid CD. Complete the parallelogram GK; and upon the bases GK, KL, complete the solids EX, LP, so that EH be an insisting straight line in each of them, whereby they must be of the same altitude with the solid AB. And because the solids AB, CD,

are similar, and by permutation, as AE is to CF, so is EG to FN, and so is EH to FR: but FC is equal to EK, and FN to EL, and FR to EM; therefore, as AE to EK, so is EG to EL, and so is HE to EM: but as AE to EK, so * is the parallelogram AG to the parallelogram GK; and as GE to EL, so is * GK to KL; and as HE to EM, so * is PE to KM: therefore as the parallelogram AG to the parallelogram GK, so is GK to KL, and PE to KM: but as AG to GK, so is * the solid AB to the solid EX; and as GK to KL, so * is the solid EX to the solid PL; and as PE to KM, so * is the solid PL to the solid KO: and therefore as the solid AB to the solid EX, so is EX to PL, and PL to KO: but if four magnitudes be continual proportionals, the first is said to have to the fourth, the triplicate † ratio of that which it has to the second; therefore the solid AB has to the solid KO, the triplicate ratio of that which AB has to EX: but as AB is to EX, so is the parallelogram AG to the parallelogram GK, and the straight line AE to the straight line EK; wherefore the solid AB has to the solid KO, the triplicate ratio of that which AE has to EK: but the solid KO is equal to the solid CD, and the straight line EK is equal to the straight line CF; therefore the solid AB has to the solid CD, the triplicate ratio of that which the side AE has to the homologous side CF. Therefore, similar solid parallelopipeds, &c. Q. E. D.

* 1. 6.
* 1. 6.
* 1. 6.
* 25. 11.
* 25. 11.
* 25. 11.
† 11 Def. 5.

COR. From this it is manifest, that, if four straight lines be continual proportionals, as the first is to the fourth, so is the solid parallelopiped described from the first to the similar solid similarly described from the second; because the first straight line has to the fourth, the triplicate ratio of that which it has to the second.

PROPOSITION D.

THEOR.—*Solid parallelopipeds which are contained by parallelograms equiangular to one another, each to each, that is, of which the solid angles are equal, each to each, have to one another the ratio which is the same with the ratio compounded of the ratios of their sides.* See N.

Let AB, CD be solid parallelopipeds, of which AB is contained by the parallelograms AE, AF, AG, which are equiangular, each to each, to the parallelograms CH, CK, CL, which contain the solid CD: the ratio which the solid AB has to the solid CD, shall be the same with that which is compounded of the ratios of the sides AM to DL, AN to DK, and AO to DH.

Produce MA, NA, OA, to P, Q, R, so that AP be equal to DL, AQ to DK, and AR to DH; and complete the solid parallelopiped AX contained by the parallelograms AS, AT, AV, similar and equal to CH, CK, CL, each to each: therefore the solid AX is equal* to the solid CD. Complete likewise the solid AY, the base of which is AS, and AO one of its insisting straight lines. Take any straight line a, and as MA to AP, so make† a to b; and as NA to AQ, so make b to c; and as AO to AR, so c to d. Then, because the parallelogram AE is equiangular to AS, AE is to AS, as the straight line a to c, as is demonstrated in the 23d Prop. Book VI.: and the solids AB, AY, being betwixt the parallel planes BOY, EAS, are of the same altitude; therefore the solid AB is to the solid AY, as* the base AE to the base AS; that is, as the straight line a is to c. And the solid AY is to the solid AX, as* the base OQ is to the base QR; that is, as the straight line OA to AR; that is, as the straight line c to the straight line d. And because the solid AB is to the solid AY, as a is to c, and the solid AY to the solid AX, as c is to d; ex æquali, the solid AB is to the solid AX, or CD which is equal to it, as the straight line a is to d. But the ratio of a to d is said to be compounded* of the ratios of a to b, b to c, and c to d, which are the same with the ratios of the sides MA to AP, NA to AQ, and OA to AR, each to each: and the sides AP, AQ, AR are equal to the sides DL, DK, DH, each to each; therefore the solid AB has to the solid CD, the ratio which is the same with that which is compounded of the ratios of the sides AM to DL, AN to DK, and AO to DH. Q. E. D.

* C. 11. † 12. 6. * 32. 11. * 25. 11. * Def. A. 5.

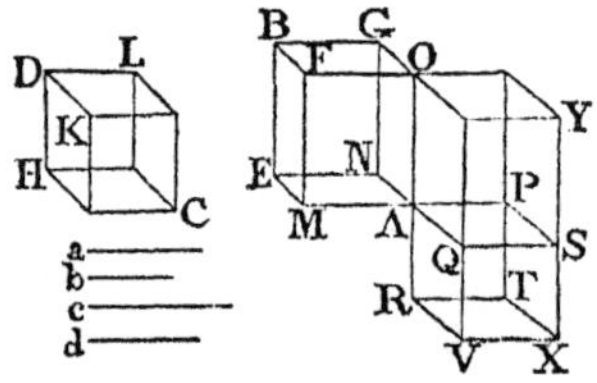

PROPOSITION XXXIV.

See N.

THEOR.—*The bases and altitudes of equal solid parallelopipeds, are reciprocally proportional: and if the bases and altitudes be reciprocally proportional, the solid parallelopipeds are equal.*

Let AB, CD be two solid parallelopipeds: and first, let the insisting straight lines AG, EF, LB, HK; CM, NX, OD, PR, be at right angles to the bases. If the solid AB be equal to the solid CD, their bases shall be reciprocally proportional to their altitudes; that is, as the base EH is to the base NP, so shall CM be to AG. If the base EH be equal to the base NP, then because the solid AB is likewise equal to the solid CD, CM shall be equal to AG: because if the bases EH, NP be equal, but the altitudes AG, CM be not equal, neither shall the solid AB be equal to the solid CD: but the solids are equal, by the hypothesis; therefore the altitude CM is not unequal to the altitude AG; that is, they are equal. Wherefore, as the base EH to the base NP, so is CM to AG.

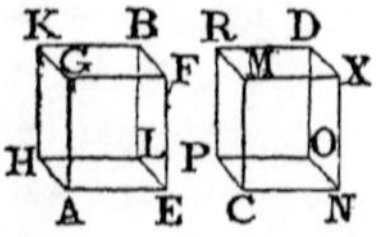

Next, let the bases EH, NP not be equal, but EH greater than the other: then since the solid AB is equal to the solid CD, CM is therefore greater than AG: for if it be not, neither also in this case would the solids AB, CD be equal, which, by the hypothesis, are equal. Make then CT equal to AG, and complete the solid parallelopiped CV, of which the base is NP, and altitude CT. Because the solid AB is equal to the solid CD, therefore the solid AB is to the solid CV, as* the solid CD to the solid CV: * 7. 5.
but as the solid AB to the solid CV, so* is the base EH to the base NP; for the solids AB, CV are of the same altitude: and as the solid CD to CV, so * is the base MP to the base PT, and so is the straight line MC * to CT: and CT is equal to AG; therefore as the base EH to the base NP, so is MC to AG. Wherefore the bases of the solid parallelopipeds AB, CD are reciprocally proportional to their altitudes.

* 32. 11.
* 25. 11.
* 1. 6.

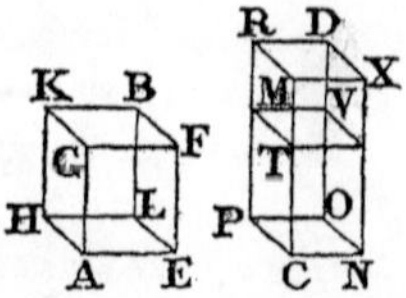

Let now the bases of the solid parallelopipeds AB, CD be reciprocally proportional to their altitudes, viz. as the base EH is to the base NP, so let CM be to AG: the solid AB shall be equal to the solid CD.

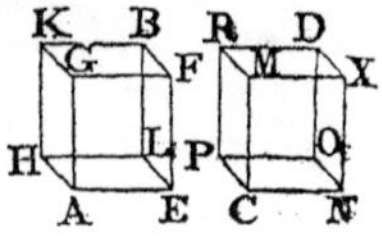

If the base EH be equal to the base NP, then, since EH is to NP as the altitude of the solid CD is to the altitude of the solid AB, therefore the altitude of CD is equal * to the altitude of AB: but * A. 5.

solid parallelopipeds upon equal bases, and of the same
* 31. 11. altitude, are equal * to one another; therefore the solid AB is equal to the solid CD.

But let the bases EH, NP be unequal, and let EH be the greater of the two: therefore, since, as the base EH to the base NP, so is CM the altitude of the solid CD to AG the alti-
* A. 5. tude of AB, CM is greater * than AG. Therefore, as before, take CT equal to AG, and complete the solid CV. And because the base EH is to the base NP, as CM to AG, and that AG is equal to CT, therefore the base EH is to the base NP, as MC to CT. But as the base EH
* 32. 11. is to NP, so * is the solid AB to the solid CV; for the solids AB, CV are of the same altitude: and as MC to CT,
† 1. 6. so † is the base MP to the base PT, and the solid CD to the
* 25. 11. solid * CV: therefore as the solid AB to the solid CV, so is the solid CD to the solid CV; that is, each of the solids AB, CD has the same ratio to the solid CV; and therefore the solid
† 9. 5. AB is equal † to the solid CD.

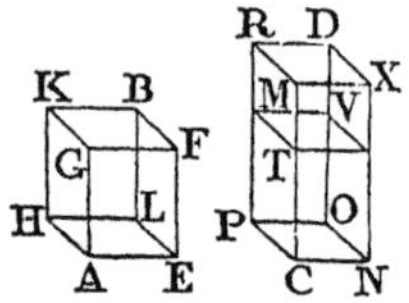

Second general case. Let the insisting straight lines FE, BL, GA, KH; XN, DO, MC, RP not be at right angles to the bases of the solids.

In this case, likewise, if the solids AB, CD be equal, their bases shall be reciprocally proportional to their altitudes, viz. the base EH shall be to the base NP, as the altitude of the solid CD to the altitude of the solid AB.

From the points F, B, K, G; X, D, R, M, draw perpendiculars to the planes in which are the bases EH, NP, meeting those planes in the points S, Y, V, T; Q, I, U, Z; and complete the solids FV, XU, which are parallelopipeds, as was proved in the last part of Prop. 31, of this Book.

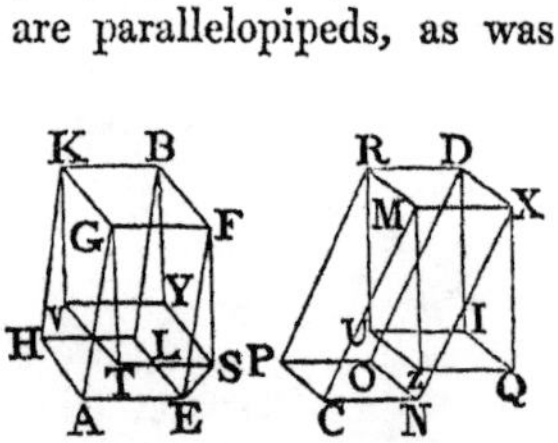

Because the solid AB is equal to the solid CD, and that the
* 29. or 30. 11. solid AB is equal * to the solid BT, for they are upon the same base FK, and of the same altitude; and that the solid CD is
* 29. or 30. 11. equal * to the solid DZ, being upon the same base XR, and of the same altitude; therefore the solid BT is equal to the solid DZ: but the bases are reciprocally proportional to the

altitudes of equal solid parallelopipeds of which the insisting straight lines are at right angles to their bases, as before was proved; therefore as the base FK to the base XR, so is the altitude of the solid DZ to the altitude of the solid BT: and the base FK is equal to the base EH, and the base XR to the base NP; wherefore, as the base EH to the base NP, so is the altitude of the solid DZ to the altitude of the solid BT: but the altitudes of the solids DZ, DC, as also of the solids BT, BA, are the same; therefore as the base EH to the base NP, so is the altitude of the solid CD to the altitude of the solid AB; that is, the bases of the solid parallelopipeds AB, CD are reciprocally proportional to their altitudes.

Next, let the bases of the solids AB, CD be reciprocally proportional to their altitudes, viz. the base EH to the base NP, as the altitude of the solid CD to the altitude of the solid AB: the solid AB shall be equal to the solid CD. The same construction being made; because, as the base EH to the base NP, so is the altitude of the solid CD to the altitude of the solid AB; and that the base EH is equal to the base FK, and NP to XR; therefore the base FK is to the base XR, as the altitude of the solid CD to the altitude of AB: but the altitudes of the solids AB, BT are the same, as also of CD and DZ; therefore as the base FK to the base XR, so is the altitude of the solid DZ to the altitude of the solid BT: wherefore the bases of the solids BT, DZ are reciprocally proportional to their altitudes: and their insisting straight lines are at right angles to the bases; wherefore, as was before proved, the solid BT is equal to the solid DZ: but BT is equal* to the solid BA, and DZ to the solid DC, because they are upon the same bases, and of the same altitude; therefore the solid AB is equal to the solid CD. Therefore, the bases, &c. Q. E. D.

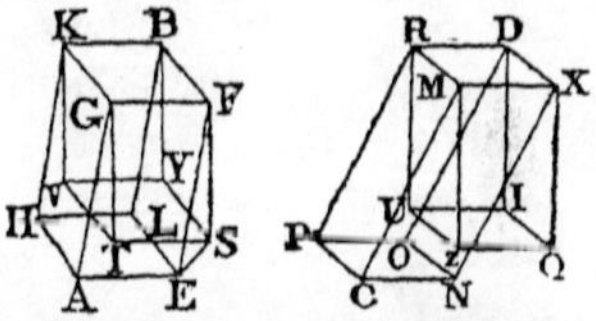

* 29. or 30. 11.

PROPOSITION XXXV.

THEOR.—*If, from the vertices of two equal plane angles, there be drawn two straight lines elevated above the planes in which the angles are, and containing equal angles with the sides of those angles, each to each; and if in the lines above* See N.

the planes there be taken any points, and from them perpendiculars be drawn to the planes in which the first named angles are; and from the points in which they meet the planes, straight lines be drawn to the vertices of the angles first named; these straight lines shall contain equal angles with the straight lines which are above the planes of the angles.

Let BAC, EDF be two equal plane angles: and from the points A, D, let the straight lines AG, DM be elevated above the planes of the angles, making equal angles with their sides, each to each, viz. the angle GAB equal to the angle MDE, and GAC to MDF; and in AG, DM, let any points G, M be taken, and from them, let perpendiculars GL,
† 11. 11. MN be drawn† to the planes BAC, EDF, meeting these planes in the points L, N; and join LA, ND: the angle GAL shall be equal to the angle MDN.

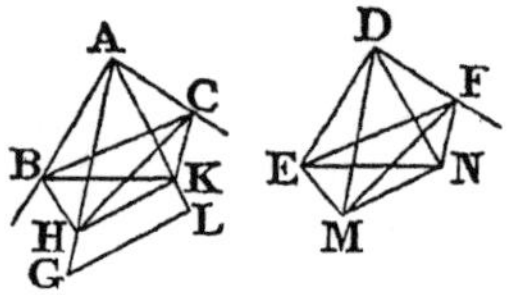

Make AH equal to DM, and through H, draw HK parallel to GL: but GL is perpendicular to the plane BAC; wherefore
* 8. 11. HK is perpendicular* to the same plane. From the points K, N, to the straight lines AB, AC, DE, DF, draw perpendiculars KB, KC, NE, NF; and join HB, BC, ME, EF. Because HK is perpendicular to the plane BAC, the plane HBK
* 18. 11. which passes through HK, is at right angles* to the plane BAC; and AB is drawn in the plane BAC at right angles to the common section BK of the two planes; therefore AB is
* 4 Def. 11. perpendicular* to the plane HBK, and makes right angles*
* 3 Def. 11. with every straight line meeting it in that plane: but BH meets it in that plane; therefore ABH is a right angle: for the same reason, DEM is a right angle, and is therefore equal
† Hyp. to the angle ABH: and the angle HAB is equal† to the angle MDE; therefore in the two triangles HAB, MDE, there are two angles in one, equal to two angles in the other, each to each, and one side equal to one side, opposite to one of the equal angles in each, viz. HA equal to DM; therefore the re-
* 26. 1. maining sides are equal*, each to each; wherefore AB is equal to DE. In the same manner, if HC and MF be joined, it may be demonstrated, that AC is equal to DF: therefore, since AB is equal to DE, BA and AC are equal to ED and DF, each to
† Hyp. each; and the angle BAC is equal† to the angle EDF; where-

fore the base BC is equal * to the base EF, and the remaining angles to the remaining angles; therefore the angle ABC is equal to the angle DEF: and the right angle ABK is equal to the right angle DEN; whence the remaining angle CBK is equal to the remaining angle FEN: for the same reason, the angle BCK is equal to the angle EFN; therefore in the two triangles BCK, EFN, there are two angles in one, equal to two angles in the other, each to each, and one side equal to one side adjacent to the equal angles in each, viz. BC equal to EF; therefore the other sides are equal to the other sides; BK then is equal to EN: and AB is equal to DE; wherefore AB, BK are equal to DE, EN, each to each; and they contain right angles; wherefore the base AK is equal to the base DN. And since AH is equal to DM, the square of AH is equal to the square of DM: but the squares of AK, KH are equal to the square * of AH, because AKH is a right angle; and the squares of DN, NM are equal to the square of DM, for DNM is a right angle; wherefore the squares of AK, KH are equal to the squares of DN, NM: and of these the square of AK is equal to the square of DN; therefore the remaining square of KH is equal to the remaining square of NM; and the straight line KH to the straight line NM; and because HA, AK are equal to MD, DN, each to each, and the base HK to the base MN, as has been proved, therefore the angle HAK is equal * to the angle MDN. Therefore, if from the vertices, &c. Q. E. D.

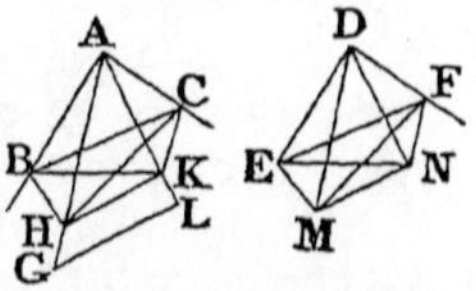

* 4. 1.

* 47. 1.

* 8. 1.

COR. From this it is manifest, that if from the vertices of two equal plane angles, there be elevated two equal straight lines containing equal angles with the sides of the angles, each to each; the perpendiculars drawn from the extremities of the equal straight lines to the planes of the first angles, are equal to one another.

Another Demonstration of the Corollary.

Let the plane angles BAC, EDF be equal to one another, and let AH, DM be two equal straight lines above the planes of the angles, containing equal angles with BA, AC, ED, DF, each to each, viz. the angle HAB equal to MDE, and HAC equal to the angle MDF; and from H, M, let HK, MN be perpendiculars to the planes BAC, EDF: HK shall be equal to MN.

Because the solid angle at A is contained by the three plane angles BAC, BAH, HAC, which are, each to each, equal to the three plane angles EDF, EDM, MDF, containing the solid angle at D; the solid angles at A and D are equal, and therefore coincide with one another; to wit, if the plane angle BAC be applied to the plane angle EDF, the straight line AH coincides with DM, as was shewn in Prop. B. of this book: and because AH is equal to DM, the point H coincides with the point M: wherefore HK, which is perpendicular to the plane BAC,
* 13. 11. coincides with MN* which is perpendicular to the plane EDF, because these planes coincide with one another. Therefore HK is equal to MN. Q. E. D.

PROPOSITION XXXVI.

See N. THEOR.—*If three straight lines be proportionals, the solid parallelopiped described from all three, as its sides, is equal to the equilateral parallelopiped described from the mean proportional, one of the solid angles of which is contained by three plane angles equal, each to each, to the three plane angles containing one of the solid angles of the other figure.*

Let A, B, C be three proportionals, viz. A to B, as B to C: the solid described from A, B, C shall be equal to the equilateral solid described from B, equiangular to the other.

Take a solid angle D contained by three plane angles EDF, FDG, GDE; and make each of the straight lines ED, DF, DG equal to B, and complete the solid parallelopiped DH: make LK equal to A, and at the point K, in the straight line LK,
* 26. 11. make* a solid angle contained by the three plane angles LKM, MKN, NKL, equal to the angles EDF, FDG, GDE, each to each; and make KN equal to B, and KM equal to C; and complete the solid parallelopiped KO. And because, as

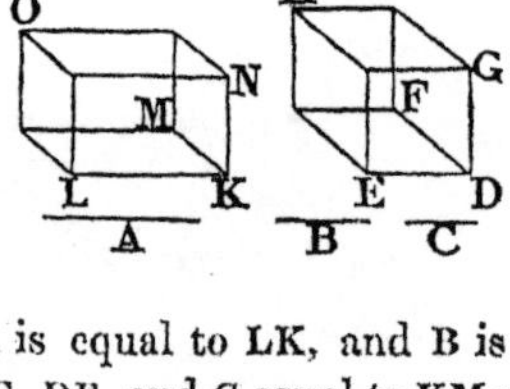

A is to B, so is B to C, and that A is equal to LK, and B is equal to each of the straight lines DE, DF, and C equal to KM; therefore LK is to ED as DF to KM; that is, the sides about the equal angles are reciprocally proportional; therefore the parallelogram LM is equal* to EF: and because EDF, LKM * 14. 6. are two equal plane angles, and the two equal straight lines DG, KN are drawn from their vertices above their planes, and contain equal angles with their sides; therefore the perpendiculars from the points G, N, to the planes EDF, LKM are equal to* one another; therefore the solids KO, DH are of * Cor. 35. 11. the same altitude; and they are upon equal bases LM, EF; and therefore they are equal* to one another: but the solid * 31. 11. KO is described from the three straight lines A, B, C, and the solid DH from the straight line B. Therefore, if three straight lines, &c. *Q. E. D.*

PROPOSITION XXXVII.

THEOR.—*If four straight lines be proportionals, the similar solid parallelopipeds similarly described from them shall also be proportionals: and if the similar parallelopipeds similarly described from four straight lines be proportionals, the straight lines shall be proportionals.* See N.

Let the four straight lines AB, CD, EF, GH be proportionals, viz. as AB to CD, so EF to GH; and let the similar parallelopipeds AK, CL, EM, GN be similarly described from them, AK shall be to CL, as EM to GN.

Make * AB, CD, O, P continual proportionals, as also * 11. 6. EF, GH, Q, R: and because as AB is to CD so EF to GH; and that CD is* to O, as GH to * 11. 5. Q, and O to P, as Q to R; therefore, ex æquali *, AB is to P, as * 22. 5. EF to R: but as AB to P, so* * Cor. 33. 11. is the solid AK, to the solid CL; and as EF to R, so* is the solid * Cor. 33. 11. EM to the solid GN; therefore * * 11. 5. as the solid AK to the solid CL, so is the solid EM to the solid GN.

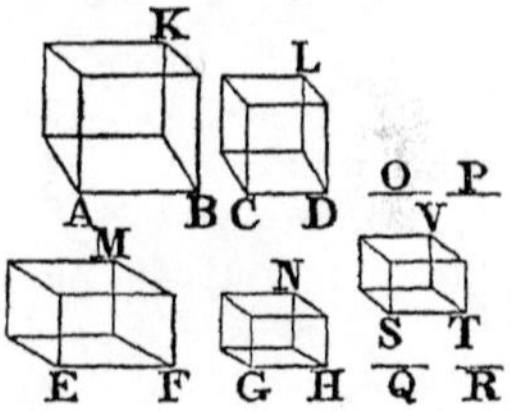

Next let the solid AK be to the solid CL, as the solid EM to the solid GN: the straight line AB shall be to CD, as EF to GH.

Take as AB to CD, so EF to ST, and from ST describe* a * 27. 11. solid parallelopiped SV similar and similarly situated to either

of the solids EM, GN. And because AB is to CD, as EF to ST, and that from AB, CD, the solid parallelopipeds AK, CL are similarly described; and in like manner the solids EM, SV, from the straight lines EF, ST; therefore AK is to CL, as EM to SV; but, by the hypothesis, AK is to CL, as EM
* 9. 5. to GN; therefore GN is equal* to SV: but it is likewise similar and similarly situated to SV; therefore the planes which contain the solids GN, SV are similar and equal, and their homologous sides GH, ST equal to one another: and because as AB to CD, so EF to ST, and that ST is equal to GH, therefore AB is to CD, as EF to GH. Therefore, if four straight lines, &c. Q. E. D.

PROPOSITION XXXVIII.

See N. THEOR.—"*If a plane be perpendicular to another plane, and* "*a straight line be drawn from a point in one of the planes* "*perpendicular to the other plane, this straight line shall* "*fall on the common section of the planes.*"

"Let the plane CD be perpendicular to the plane AB, and "let AD be the common section: if any point E be taken in "the plane CD, the perpendicular drawn from E to the plane "AB, shall fall on AD.

"For if it does not, let it, if possible, fall elsewhere, as "EF; and let it meet the plane AB in the point F; and
* 12. 1. "from F, draw*, in the plane AB, a perpendicular FG to DA, which is also per-
* 4 Def. 11. "pendicular* to the plane CD; and join "EG. Then, because FG is perpendicular "to the plane CD, and the straight line "EG which is in that plane, meets it,

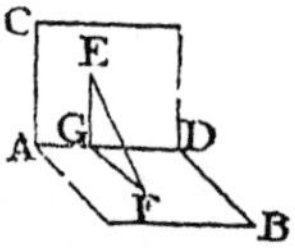

* 3 Def. 11. "therefore FGE is a right angle*: but EF is also at right "angles to the plane AB, and therefore EFG is a right "angle: wherefore two of the angles of the triangle EFG
† 17. 1. "are equal together to two right angles; which is† absurd; "therefore the perpendicular from the point E to the plane "AB, does not fall elsewhere than upon the straight line AD; "it therefore falls upon it. If, therefore, a plane, &c. Q. E. D."

PROPOSITION XXXIX.

THEOR.—*In a solid parallelopiped, if the sides of two of the opposite planes be divided, each into two equal parts, the common section of the planes passing through the points of division, and the diameter of the solid parallelopiped, cut each other into two equal parts.* See N.

Let the sides of the opposite planes CF, AH of the solid parallelopiped AF, be divided each into two equal parts in the points K, L, M, N; X, O, P, R; and join KL, MN, XO, PR: and because DK, CL are equal and parallel, KL is parallel* to DC: for the same reason, MN is parallel to BA: * 33. 1.
and BA is parallel to DC; therefore, because KL, BA are each of them parallel to DC, and not in the same plane with it, KL is parallel* to BA: * 9. 11.
and because KL, MN are each of them parallel to BA, and not in the same plane with it, KL is parallel* * 9. 11.
to MN: wherefore KL, MN are in one plane. In like manner it may be proved, that XO, PR are in one plane. Let YS be the common section of the planes KN, XR; and DG the diameter of the solid parallelopiped AF: YS and DG shall meet, and cut one another into two equal parts.

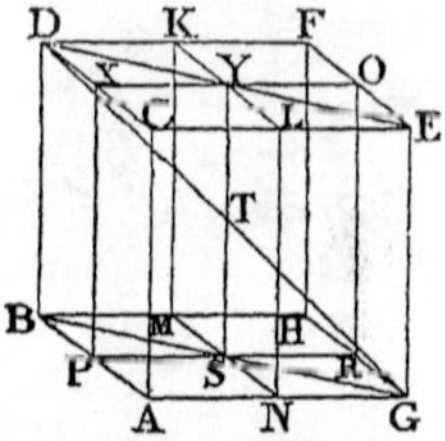

Join DY, YE, BS, SG. Because DX is parallel to OE, the alternate angles DXY, YOE are equal* to one another: and * 29. 1.
because DX is equal to OE, and XY to YO, and that they contain equal angles, the base DY is equal* to the base YE, * 4. 1.
and the other angles are equal; therefore the angle XYD is equal to the angle OYE, and DYE is a straight* line: for the * 14. 1.
same reason, BSG is a straight line, and BS equal to SG. And because CA is equal and parallel to DB, and also equal and parallel to EG, therefore DB is equal and parallel* to * 9. 11.
EG: and DE, BG join their extremities; therefore DE is equal and parallel* to BG: and DG, YS are drawn from * 33. 1.
points in the one, to points in the other, and are therefore in one plane: whence it is manifest, that DG, YS must meet one another: let them meet in T. And because DE is parallel to BG, the alternate angles EDT, BGT are* equal: and * 29. 1.
the angle DTY is equal* to the angle GTS: therefore in the * 15. 1.

triangles DTY, GTS, there are two angles in the one, equal to two angles in the other, and one side equal to one side, opposite to two of the equal angles, viz. DY to GS, for they are the halves of DE, BG; therefore the remaining sides are
* 26. 1. equal*, each to each: wherefore DT is equal to TG, and YT equal to TS. Wherefore, in a solid, &c. Q. E. D.

PROPOSITION XL.

THEOR.—*If there be two triangular prisms of the same altitude, the base of one of which is a parallelogram, and the base of the other a triangle; if the parallelogram be double of the triangle, the prisms shall be equal to one another.*

Let the prisms ABCDEF, GHKLMN be of the same altitude, the first whereof is contained by the two triangles ABE, CDF, and the three parallelograms AD, DE, EC; and the other by the two triangles GHK, LMN, and the three parallelograms LH, HN, NG; and let one of them have a parallelogram AF, and the other a triangle GHK, for its base: if the parallelogram AF be double of the triangle GHK, the prism ABCDEF shall be equal to the prism GHKLMN.

Complete the solids AX, GO: and because the parallelogram AF is double of the triangle GHK, and the parallelo-
* 34. 1. gram HK double* of the same triangle, therefore the parallelogram AF is equal to HK: but solid parallelopipeds upon equal bases, and
* 31. 11. of the same altitude, are equal* to one another; therefore the solid

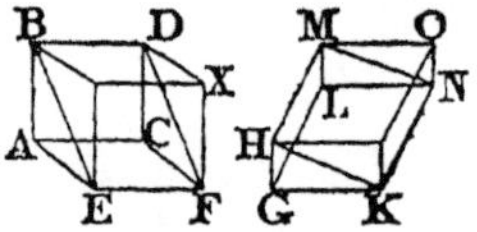

* 28. 11. AX is equal to the solid GO: and the prism ABCDEF is half*
* 28. 11. of the solid AX; and the prism GHKLMN half* of the solid GO: therefore the prism ABCDEF is equal to the prism GHKLMN. Wherefore, if there be two, &c. Q. E. D.

THE

ELEMENTS OF EUCLID.

BOOK XII.

LEMMA I.‡

If from the greater of two unequal magnitudes, there be taken more than its half, and from the remainder more than its half, and so on; there shall at length remain a magnitude less than the least of the proposed magnitudes. See N.

Let AB and C be two unequal magnitudes, of which AB is the greater: if from AB there be taken more than its half, and from the remainder more than its half, and so on; there shall at length remain a magnitude less than C.

For C may be multiplied so as at length to become greater than AB. Let it be so multiplied, and let DE its multiple be greater than AB, and let DE be divided into DF, FG, GE, each equal to C. From AB, take BH greater than its half; and from the remainder AH, take HK greater than its half, and so on, until there be as many divisions in AB as there are in DE: and let the divisions in AB be AK, KH, HB; and the divisions in DE be DF, FG, GE. And because DE is greater than AB, and that EG taken from

‡ This is the first proposition of the tenth book, and is necessary to some of the propositions of this book.

DE is not greater than its half, but BH taken from AB is greater than its half, therefore the remainder GD is greater than the remainder HA. Again, because GD is greater than HA, and that GF is not greater than the half of GD, but HK is greater than the half of HA, therefore the remainder FD is greater than the remainder AK. And FD is equal to C, therefore C is greater than AK; that is, AK is less than C. Q. E. D.

And if only the halves be taken away, the same thing may, in the same way, be demonstrated.

PROPOSITION I.

THEOREM.—*Similar polygons inscribed in circles, are to one another as the squares of their diameters.*

Let ABCDE, FGHKL be two circles, and in them the similar polygons ABCDE, FGHKL: and let BM, GN be the diameters of the circles: as the square of BM is to the square of GN, so shall the polygon ABCDE be to the polygon FGHKL.

Join BE, AM, GL, FN: and because the polygon ABCDE is similar to the polygon FGHKL, the angle BAE is equal to the angle GFL, and as BA to AE, so is GF to FL: therefore the two triangles BAE, GFL, having one angle in one, equal to one angle in the other, and the sides about the equal angles proportionals, are equiangular; and therefore the angle AEB is equal to the angle FLG: but AEB is

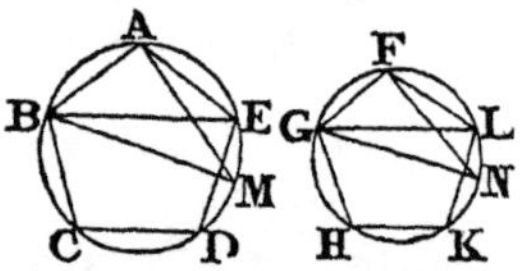

* 21. 3. equal* to AMB, because they stand upon the same circumference: and the angle FLG is, for the same reason, equal to the angle FNG; therefore also the angle AMB is equal to
* 31. 3. FNG: and the right angle BAM is equal to the right* angle GFN; wherefore the remaining angles in the triangles ABM, FGN are equal, and they are equiangular to one another:
* 4. 6. therefore as BM to GN, so* is BA to GF; and therefore the
* 10 Def. 5. & 22. 5. duplicate ratio of BM to GN, is the same* with the duplicate ratio of BA to GF: but the ratio of the square of BM to the
* 20. 6. square of GN, is the duplicate* ratio of that which BM has to GN; and the ratio of the polygon ABCDE to the polygon

FGHKL is the duplicate * of that which BA has to GF: therefore as the square of BM to the square of GN, so is the polygon ABCDE to the polygon FGHKL. Wherefore, similar polygons, &c. Q. E. D. * 20. 6.

PROPOSITION II.

THEOR.—*Circles are to one another as the squares of their diameters.* See N.

Let ABCD, EFGH be two circles, and BD, FH their diameters: as the square of BD to the square of FH, so shall the circle ABCD be to the circle EFGH.

For if it be not so, the square of BD must be to the square of FH, as the circle ABCD is to some space either less than the circle EFGH, or greater than it ‡. First let it be to a space S less than the circle EFGH; and in the circle EFGH † describe the square EFGH. This square is greater than half of the circle EFGH; because, if, through the points E, F, G, H, there be drawn tangents to the circle, the square EFGH is half * of the square described about the circle: and the circle is less than the square described about it; therefore the square EFGH is greater than half of the circle. Divide the circumferences EF, FG, GH, HE, each into two equal parts in the points K, L, M, N, and join EK, KF, FL, LG, GM, MH, HN, NE: therefore each of the triangles EKF, † 6. 4. * 41. 1.

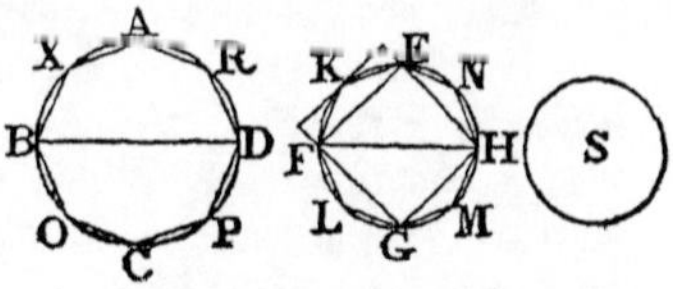

FLG, GMH, HNE, is greater than half of the segment of the circle in which it stands; because, if straight lines touching the circle be drawn through the points K, L, M, N, and the parallelograms upon the straight lines EF, FG, GH, HE be completed, each of the triangles EKF, FLG, GMH, HNE, is the half* of the parallelogram in which it is: but every segment is less than the parallelogram in which it is: wherefore * 41. 1.

‡ For there is some square equal to the circle ABCD; let P be the side of it, and to three straight lines, BD, FH, and P, there can be a fourth proportional; let this be Q: therefore† the squares of these four straight lines are proportionals; that is, to the squares of BD, FH, and the circle ABCD, it is possible there may be a fourth proportional. Let this be S. And in like manner are to be understood some things in some of the following propositions. † 22. 6.

each of the triangles EKF, FLG, GMH, HNE is greater than half the segment of the circle which contains it. Again, if the remaining circumferences be divided each into two equal parts, and their extremities be joined by straight lines, by continuing to do this, there will at length remain segments of the circle, which together are less than the excess of the circle EFGH above the space S; because, by the preceding Lemma, if from the greater of two unequal magnitudes there be taken more than its half, and from the remainder more than its half, and so on, there shall at length remain a magnitude less than the least of the proposed magnitudes. Let then the segments EK, KF, FL, LG, GM, MH, HN, NE be those that remain, and are together less than the excess of the circle EFGH above S: therefore the rest of the circle, viz. the polygon EKFLGMHN is greater than the space S. Describe likewise in the circle ABCD the polygon AXBOCPDR similar to the polygon EKFLGMHN: as therefore the square
* 1. 12. of BD is to the square of FH *, so is the polygon AXBOCPDR to the polygon EKFLGMHN: but the square of BD is also to
† Hyp. the square of FH †, as the circle ABCD is to the space S;
* 11. 5. therefore as the circle ABCD is to the space S, so is* the polygon AXBOCPDR to the polygon EKFLGMHN: but the circle ABCD is greater than the polygon contained in it; where-
* 14. 5. fore the space S is greater * than the polygon EKFLGMHN: but it is likewise less, as has been demonstrated; which is impossible: therefore the square of BD is not to the square of FH, as the circle ABCD is to any space less than the circle EFGH. In the same manner it may be demonstrated, that neither is the square of FH to the square of BD, as the circle EFGH is to any space less than the circle ABCD. Nor is the square of BD to the square of FH, as the circle ABCD

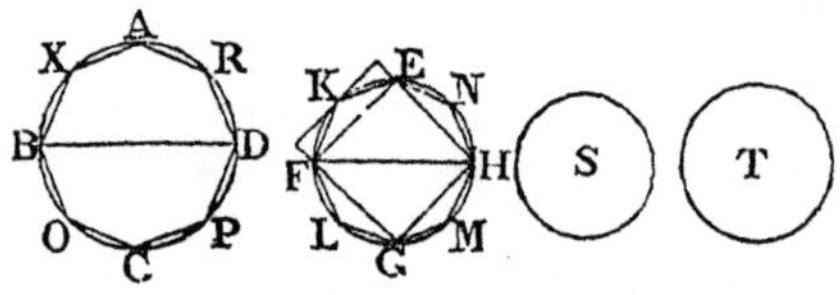

is to any space greater than the circle EFGH. For if possible, let it be so to T, a space greater than the circle EFGH: therefore inversely, as the square of FH to the square of BD, so is the space T to the circle ABCD: but as the space ‡ T is

‡ For as, in the foregoing note, it was explained how it was possible there could be a fourth proportional to the squares of BD, FH, and the circle

to the circle ABCD, so is the circle EFGH to some space, which must be less * than the circle ABCD, because the space T is greater, by hypothesis, than the circle EFGH: therefore as the square of FH is to the square of BD, so is the circle EFGH to a space less than the circle ABCD, which has been demonstrated to be impossible: therefore the square of BD is not to the square of FH, as the circle ABCD is to any space greater than the circle EFGH: and it has been demonstrated that neither is the square of BD to the square of FH, as the circle ABCD to any space less than the circle EFGH: wherefore, as the square of BD to the square of FH, so is the circle ABCD to the circle EFGH ‡. Circles, therefore, are, &c. Q. E. D. * 14. 5.

PROPOSITION III.

THEOR.—*Every pyramid having a triangular base, may be divided into two equal and similar pyramids having triangular bases, and which are similar to the whole pyramid; and into two equal prisms which together are greater than half of the whole pyramid.* See N.

Let there be a pyramid, of which the base is the triangle ABC, and its vertex the point D: the pyramid ABCD may be divided into two equal and similar pyramids having triangular bases, and similar to the whole; and into two equal prisms which together shall be greater than half of the whole pyramid.

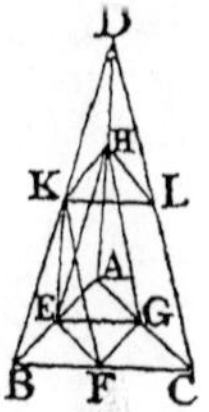

Divide AB, BC, CA, AD, DB, DC, each into two equal parts in the points E, F, G, H, K, L, and join EH, EG, GH, HK, KL, LH, EK, KF, FG. Because AE is equal to EB, and AH to HD, HE is parallel * to DB: for the same reason, HK is parallel to AB; therefore HEBK is a parallelogram, and HK equal * to EB: but EB is equal to AE; therefore also AE is equal to HK: and AH is equal to HD; wherefore EA, AH are equal to KH, HD, each to each; and the angle EAH is equal * * 2. 6. * 34. 1. * 29. 1.

ABCD, which was named S; so, in like manner, there can be a fourth proportional to this other space, named T, and the circles ABCD, EFGH. And the like is to be understood in some of the following propositions.

‡ Because, as a fourth proportional to the squares of BD, FH, and the circle ABCD, is possible, and that it can neither be less nor greater than the circle EFGH, it must be equal to it.

to the angle KHD; therefore the base EH is equal to the base
* 4. 1. KD, and the triangle AEH equal * and similar to the triangle
HKD: for the same reason, the triangle AGH is equal and
similar to the triangle HLD. Again, because the two straight
lines EH, HG, which meet one another, are parallel to KD, DL,
that meet one another, and are not in the same plane with
* 10. 11. them, they contain equal* angles; therefore the angle EHG
is equal to the angle KDL: and because EH,

HG are equal to KD, DL, each to each, and the
angle EHG equal to the angle KDL; therefore
the base EG is equal to the base KL, and the
* 4. 1. triangle EHG equal * and similar to the triangle KDL: for the same reason, the triangle AEG
is also equal and similar to the triangle HKL:
therefore the pyramid, of which the base is the
triangle AEG, and of which the vertex is the
* C. 11. point H, is equal * and similar to the pyramid, the base of
which is the triangle KHL, and vertex the point D. And
because HK is parallel to AB, a side of the triangle ADB, the
triangle ADB is equiangular to the triangle HDK, and their
* 4. 6. sides are* proportionals; therefore the triangle ADB is similar
to the triangle HDK: and for the same reason, the triangle
DBC is similar to the triangle DKL; and the triangle ADC to
the triangle HDL; and also the triangle ABC to the triangle
AEG: but the triangle AEG is similar to the triangle HKL,
* 21. 6. as before was proved; therefore the triangle ABC is similar *
to the triangle HKL: and therefore the pyramid of which the
* B. 11. & base is the triangle ABC, and vertex the point D, is similar*
11 Def. 11. to the pyramid of which the base is the triangle HKL, and
vertex the same point D: but the pyramid of which the base
is the triangle HKL, and vertex the point D, is similar, as
has been proved, to the pyramid the base of which is the
triangle AEG, and vertex the point H; wherefore the pyramid, the base of which is the triangle ABC, and vertex the
point D, is similar to the pyramid of which the base is the
triangle AEG, and vertex H: therefore each of the pyramids
AEGH, HKLD is similar to the whole pyramid ABCD. And
* 41. 1. because BF is equal to FC, the parallelogram EBFG is double*
of the triangle GFC: but when there are two prisms of the
same altitude, of which one has a parallelogram for its base,
and the other a triangle that is half of the parallelogram, these
* 40. 11. prisms are equal * to one another; therefore the prism having

the parallelogram EBFG for its base, and the straight line KH opposite to it, is equal to the prism having the triangle GFC for its base, and the triangle HKL opposite to it; for they are of the same altitude, because they are between the parallel * planes ABC, HKL: and it is manifest that each of these prisms is greater than either of the pyramids of which the triangles AEG, HKL are the bases, and the vertices the points H, D; because, if EF be joined, the prism having the parallelogram EBFG for its base, and KH the straight line opposite to it, is greater than the pyramid of which the base is the triangle EBF, and vertex the point K: but this pyramid is equal* to the pyramid, the base of which is the triangle AEG, and vertex the point H; because they are contained by equal and similar planes: wherefore the prism having the parallelogram EBFG for its base, and opposite side KH, is greater than the pyramid of which the base is the triangle AEG, and vertex the point H: and the prism of which the base is the parallelogram EBFG, and opposite side KH, is equal to the prism having the triangle GFC for its base, and HKL the triangle opposite to it; and the pyramid of which the base is the triangle AEG, and vertex H, is equal to the pyramid of which the base is the triangle HKL, and the vertex D: therefore the two prisms before mentioned are greater than the two pyramids of which the bases are the triangles AEG, HKL, and vertices the points H, D. Therefore, the whole pyramid of which the base is the triangle ABC, and vertex the point D, is divided into two equal pyramids similar to one another, and to the whole pyramid; and into two equal prisms; and the two prisms are together greater than half of the whole pyramid. Q. E. D.

* 15. 11.

* C. 11.

PROPOSITION IV.

THEOR.—*If there be two pyramids of the same altitude upon triangular bases, and each of them be divided into two equal pyramids similar to the whole pyramid, and also into two equal prisms; and if each of these pyramids be divided in the same manner as the first two, and so on; as the base of one of the first two pyramids is to the base of the other, so shall all the prisms in one of them be to all the prisms in the other, that are produced by the same number of divisions.* See N.

Let there be two pyramids of the same altitude upon the triangular bases ABC, DEF, and having their vertices in the points G, H; and let each of them be divided into two equal pyramids similar to the whole, and into two equal prisms; and let each of the pyramids thus made, be conceived to be divided in the like manner, and so on: as the base ABC is to the base DEF, so shall all the prisms in the pyramid ABCG be to all the prisms in the pyramid DEFH made by the same number of divisions.

Make the same construction as in the foregoing proposition: and because BX is equal to XC, and AL to LC, therefore XL
* 2. 6. is parallel* to AB, and the triangle ABC similar to the triangle LXC: for the same reason, the triangle DEF is similar to RVF. And because BC is double of CX, and EF double of
† C. 5. FV, therefore † BC is to CX, as EF to FV: and upon BC, CX, are described the similar and similarly situated rectilineal figures ABC, LXC: and upon EF, FV, in like manner are described the similar figures DEF, RVF: therefore, as the triangle ABC is to the triangle LXC,
* 22. 6. so * is the triangle DEF to the triangle RVF, and, by permutation, as the triangle ABC to the triangle DEF, so is the triangle LXC to the triangle RVF. And because ‡ the planes ABC, OMN,

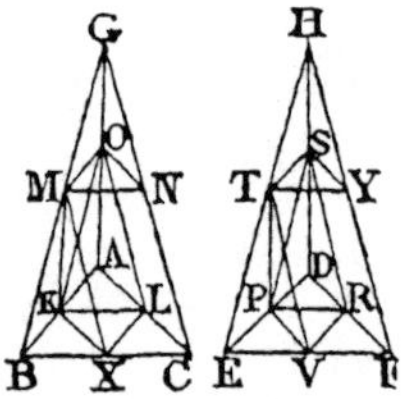

* 15. 11. as also the planes DEF, STY, are parallel*, the perpendiculars drawn from the points G, H, to the bases ABC, DEF, which, by the hypothesis, are equal to one another, shall be cut each
* 17. 11. into two equal* parts by the planes OMN, STY, because the straight lines GC, HF are cut into two equal parts in the points N, Y, by the same planes: therefore the prisms LXCOMN, RVFSTY are of the same altitude; and therefore, as the base LXC to the base RVF, that is, as the triangle ABC to the tri-
* Cor. 32. 11. angle DEF, so* is the prism having the triangle LXC for its base, and OMN the triangle opposite to it, to the prism of which the base is the triangle RVF, and the opposite triangle STY: and because the two prisms in the pyramid ABCG are equal to one another, and also the two prisms in the pyramid DEFH equal to one another; as the prism of which the base

† 2. 6. ‡ Because GO is equal to OA, and GM to MB, therefore † OM is parallel
† 15. 11. to AB; in like manner ON is parallel to AC; therefore † the plane MON is parallel to the plane BAC.

is the parallelogram KBXL and opposite side MO, to the prism having the triangle LXC for its base, and OMN the triangle opposite to it; so is the prism of which the base* is the parallelogram PEVR, and opposite side TS, to the prism of which the base is the triangle RVF, and opposite triangle STY: therefore, componendo, as the prisms KBXLMO, LXCOMN together, are to the prism LXCOMN, so are the prisms PEVRTS, RVFSTY to the prism RVFSTY; and, permutando, as the prisms KBXLMO, LXCOMN are to the prisms PEVRTS, RVFSTY, so is the prism LXCOMN to the prism RVFSTY: but as the prism LXCOMN to the prism RVFSTY, so is, as has been proved, the base ABC to the base DEF; therefore, as the base ABC to the base DEF, so are the two prisms in the pyramid ABCG to the two prisms in the pyramid DEFH: and likewise if the pyramids now made, for example, the two OMNG, STYH, be divided in the same manner; as the base OMN is to the base STY, so are the two prisms in the pyramid OMNG to the two prisms in the pyramid STYH: but the base OMN is to the base STY, as the base ABC to the base DEF; therefore, as the base ABC to the base DEF, so are the two prisms in the pyramid ABCG to the two prisms in the pyramid DEFH; and so are the two prisms in the pyramid OMNG to the two prisms in the pyramid STYH; and so are all four to all four: and the same thing may be shewn of the prisms made by dividing the pyramids AKLO and DPRS, and of all made by the same number of divisions. Q. E. D. * 7. 5.

PROPOSITION V.

THEOR.—*Pyramids of the same altitude which have triangular bases, are to one another as their bases.* See N.

Let the pyramids of which the triangles ABC, DEF are the bases, and of which the vertices are the points G, H, be of the same altitude: as the base ABC to the base DEF, so shall the pyramid ABCG be to the pyramid DEFH.

For if it be not so, the base ABC must be to the base DEF, as the pyramid ABCG to a solid either less than the pyramid DEFH, or greater than it ‡. First, let it be to a solid less than it, viz. to the solid Q; and divide the pyramid DEFH

‡ See note at foot of page 220.

into two equal pyramids, similar to the whole, and into two
* 3. 12. equal prisms; therefore these two prisms are greater * than the half of the whole pyramid. And again, let the pyramids
† Lemma. made by this division be in like manner divided, and so on † until the pyramids which remain undivided in the pyramid DEFH be all of them together, less than the excess of the pyramid DEFH above the solid Q: let these, for example, be the pyramids DPRS, STYH: therefore the prisms, which make the rest of the pyramid DEFH, are greater than the solid Q. Divide likewise the pyramid ABCG in the same manner, and into as many parts as the pyramid DEFH: therefore as the base
* 4. 12. ABC to the base DEF, so * are the prisms in the pyramid ABCG to the prisms in the pyramid DEFH: but as the base ABC to the base DEF, so, by hypothesis, is the pyramid ABCG to the solid Q; and therefore, as the pyramid ABCG to the solid Q, so are the prisms in the pyramid ABCG to the prisms in the pyramid DEFH; but the pyramid ABCG is
* 14. 5. greater than the prisms contained in it; wherefore * also the solid Q is greater than the prisms in the pyramid DEFH: but it is also less, which is impossible: therefore the base ABC is not to the base DEF, as the pyramid ABCG to any solid which is less than the pyramid DEFH. In the same manner it may be demonstrated, that the base DEF is not to the base ABC, as the pyramid DEFH to any solid which is less than the pyramid ABCG. Nor can the base ABC be to the base DEF, as the pyramid ABCG to any solid which is greater than the pyramid DEFH. For if it be possible, let it be so to a greater, viz. the solid Z. And because the base ABC is to the base DEF, as the pyramid ABCG to the solid Z; by inversion, as the base DEF to the base ABC, so is the solid Z to the pyramid ABCG: but as the solid Z is to the pyramid ABCG, so ‡ is
* 14. 5. the pyramid DEFH to some solid, which must be less * than the pyramid ABCG, because the solid Z is greater than the pyramid DEFH; and therefore, as the base DEF to the base ABC, so is the pyramid DEFH to a solid less than the pyramid

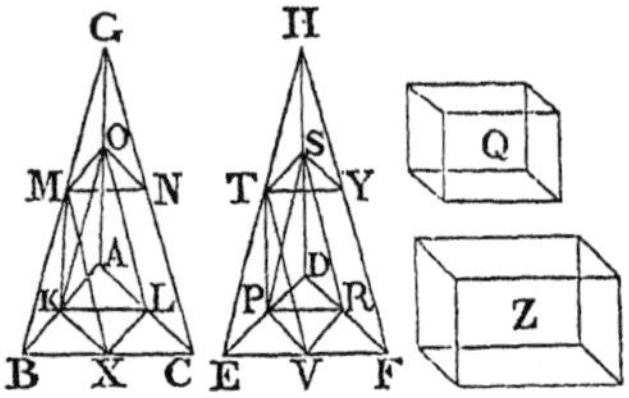

‡ See note at foot of page 220.

ABCG; the contrary to which has been proved: therefore the base ABC is not to the base DEF, as the pyramid ABCG to any solid which is greater than the pyramid DEFH. And it has been proved, that neither is the base ABC to the base DEF, as the pyramid ABCG to any solid which is less than the pyramid DEFH: therefore, as the base ABC is to the base DEF, so is the pyramid ABCG to the pyramid DEFH. Wherefore, pyramids, &c. Q. E. D.

PROPOSITION VI.

THEOR.—*Pyramids of the same altitude which have polygons for their bases, are to one another as their bases.* See N.

Let the pyramids which have the polygons ABCDE, FGHKL for their bases, and their vertices in the points M, N, be of the same altitude: as the base ABCDE to the base FGHKL, so shall the pyramid ABCDEM be to the pyramid FGHKLN.

Divide the base ABCDE into the triangles ABC, ACD, ADE; and the base FGHKL into the triangles FGH, FHK, FKL: and upon the bases ABC, ACD, ADE, let there be as many pyramids of which the common vertex is the point M, and upon the remaining bases as many pyramids having their common vertex in the point N. Therefore, since the triangle ABC is to the triangle FGH, as* the pyramid ABCM to the pyramid FGHN; and the triangle ACD to the triangle FGH, as the pyramid ACDM to the pyramid FGHN; * 5. 12.

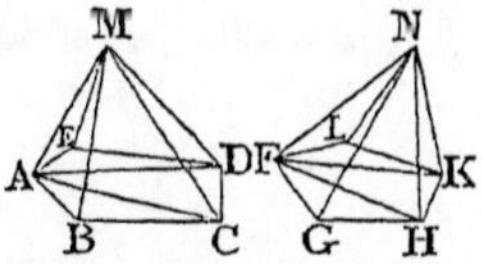

and also the triangle ADE to the triangle FGH, as the pyramid ADEM to the pyramid FGHN; as all the first antecedents to their common consequent, so * are all the other antecedents to their common consequent; that is, as the base ABCDE to the base FGH, so is the pyramid ABCDEM to the pyramid FGHN: and for the same reason, as the base FGHKL to the base FGH, so is the pyramid FGHKLN to the pyramid FGHN; and, by inversion, as the base FGH to the base FGHKL, so is the pyramid FGHN to the pyramid FGHKLN: then, because, as the base ABCDE to the base FGH, so is the pyramid ABCDEM to the pyramid FGHN; and as the base FGH to the base FGHKL, so is the pyramid FGHN to the pyramid FGHKLN; therefore, ex æquali *, as the base ABCDE to the base FGHKL, so the * 2 Cor. 24. 5. * 22. 5.

pyramid ABCDEM to the pyramid FGHKLN. Therefore, pyramids, &c. *Q. E. D.*

PROPOSITION VII.

THEOR.—*Every prism having a triangular base, may be divided into three pyramids that have triangular bases, and are equal to one another.*

Let there be a prism of which the base is the triangle ABC, and DEF the triangle opposite to it: the prism ABCDEF may be divided into three equal pyramids having triangular bases.

Join BD, EC, CD: and because ABED is a parallelogram of
* 34. 1. which BD is the diameter, the triangle ABD is equal* to the triangle EBD; therefore the pyramid of which the base is the
* 5. 12. triangle ABD, and vertex the point C, is equal* to the pyramid of which the base is the triangle EBD, and vertex the point C: but this pyramid is the same with the pyramid the base of which is the triangle EBC, and vertex the point D; for they are contained by the same planes: therefore the pyramid of which the base is the triangle ABD, and vertex the point C, is equal to the pyramid, the base of which is the triangle EBC, and vertex the point D. Again, because FCBE is a parallelogram of which the diameter is CE, the triangle
* 34. 1. ECF is equal* to the triangle ECB; therefore the pyramid of which the base is the triangle ECB, and vertex the point D, is equal to the pyramid the base of which is the triangle ECF, and vertex the point D: but the pyramid of which the base is the triangle ECB, and vertex the point D, has been proved equal to the pyramid of which the base is the triangle ABD, and vertex the point C: therefore the prism ABCDEF is divided into three equal pyramids having triangular bases, viz. into the pyramids ABDC, EBDC, ECFD. And because the pyramid of which the base is the triangle ABD, and vertex the point C, is the same with the pyramid of which the base is the triangle ABC, and vertex the point D, for they are contained by the same planes; and that the pyramid of which the base is the triangle ABD and vertex the point C, has been demonstrated to be a third part of the prism, the base of which is the triangle

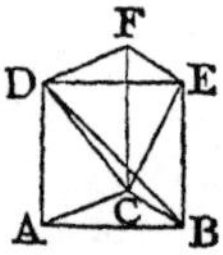

ABC, and DEF the opposite triangle; therefore, the pyramid of which the base is the triangle ABC, and vertex the point D, is the third part of the prism which has the same base, viz. the triangle ABC, and DEF its opposite triangle. Q. E. D.

COR. 1. From this it is manifest, that every pyramid is the third part of a prism which has the same base, and is of an equal altitude with it: for if the base of the prism be any other figure than a triangle, it may be divided into prisms having triangular bases.

COR. 2. Prisms of equal altitudes are to one another as their bases; because the pyramids upon the same bases, and of the same altitude, are* to one another as their bases. * 6.12.

PROPOSITION VIII.

THEOR.—*Similar pyramids having triangular bases, are one to another in the triplicate ratio of that of their homologous sides.*

Let the pyramids having the triangles ABC, DEF for their bases, and the points G, H for their vertices, be similar and similarly situated: the pyramid ABCG shall have to the pyramid DEFH, the triplicate ratio of that which the side BC has to the homologous side EF.

Complete the parallelograms ABCM, GBCN, ABGK, and the solid parallelopiped BGML, contained by these planes and those opposite to them: and, in like manner, complete the solid parallelopiped EHPO contained by the three parallelograms DEFP, HEFR, DEHX, and those opposite to them. And because the pyramid ABCG is similar to the pyramid DEFH, the angle ABC is equal* to the angle DEF, and the angle GBC to the angle HEF, and ABG to DEH: and AB is* to BC, as DE to EF; that is, the sides about the equal angles are proportionals: wherefore the parallelogram BM is similar to EP: for the same reason, the parallelogram BN is similar to ER, and BK to EX: therefore the three parallelograms BM, BN, BK are similar to the three EP, ER, EX: but the three BM, BN, BK are equal and similar* to the three which are opposite to * 24. 11.

* 11 Def. 11.
* 1 Def. 6.

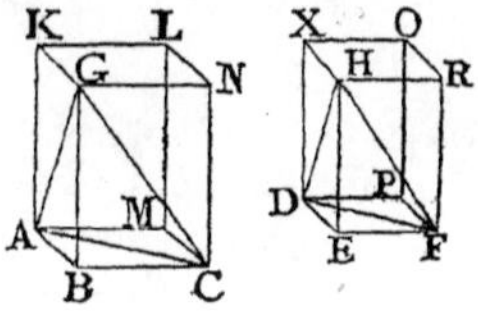

them, and the three EP, ER, EX, equal and similar to the three opposite to them: wherefore the solids BGML, EHPO are contained by the same number of similar planes: and
* B. 11. their solid angles are * equal; and therefore the solid BGML
* 11 Def. 11. is similar * to the solid EHPO: but similar solid parallelo-
* 33. 11. pipeds have the triplicate * ratio of that which their homologous sides have; therefore the solid BGML has to the solid EHPO, the triplicate ratio of that which the side BC has to the homologous side EF: but as the solid BGML is to the
* 15. 5. solid EHPO, so is * the pyramid ABCG to the pyramid DEFH; because the pyramids are the sixth part of the solids, since the
* 28. 11. prism, which is the half * of the solid parallelopiped, is triple *
* 7. 12. of the pyramid: wherefore, likewise, the pyramid ABCG has to the pyramid DEFH, the triplicate ratio of that which BC has to the homologous side EF. Q. E. D.

Cor. From this it is evident, that similar pyramids which have multangular bases, are likewise to one another in the triplicate ratio of their homologous sides: for they may be divided into similar pyramids having triangular bases, because the similar polygons which are their bases, may be divided into the same number of similar triangles homologous to the whole polygons: therefore as one of the triangular pyramids in the first multangular pyramid is to one of the triangular
† 12. 5. pyramids in the other †, so are all the triangular pyramids in the first to all the triangular pyramids in the other; that is, so is the first multangular pyramid to the other: but one triangular pyramid is to its similar triangular pyramid in the triplicate ratio of their homologous sides; and therefore the first multangular pyramid has to the other, the triplicate ratio of that which one of the sides of the first has to the homologous side of the other.

PROPOSITION IX.

Theor.—*The bases and altitudes of equal pyramids having triangular bases are reciprocally proportional; and triangular pyramids, of which the bases and altitudes are reciprocally proportional, are equal to one another.*

Let the pyramids of which the triangles ABC, DEF are the bases, and which have their vertices in the points G, H, be equal to one another: the bases and altitudes of the pyramids ABCG, DEFH, shall be reciprocally proportional, viz. the base ABC shall be to the base DEF, as the altitude of the pyramid DEFH to the altitude of the pyramid ABCG.

Complete the parallelograms AC, AG, GC, DF, DH, HF; and the solid parallelopipeds BGML, EHPO, contained by these planes and those which are opposite to them. And because the pyramid ABCG is equal to the pyramid DEFH, and that the solid BGML is sextuple† of the pyramid ABCG, and the solid EHPO sextuple of the pyramid DEFH; therefore the solid BGML is equal * to the solid EHPO: but the bases and altitudes of equal solid parallelopipeds are reciprocally proportional *; therefore as the base BM to the base EP, so is the altitude of the solid EHPO to the altitude of the solid BGML: but as the base BM to the base EP, so is* the triangle ABC to the triangle DEF; therefore as the triangle ABC to the triangle DEF, so is the altitude of the solid EHPO to the altitude of the solid BGML: but the altitude of the solid EHPO is the same with the altitude of the pyramid DEFH; and the altitude of the solid BGML is the same with the altitude of the pyramid ABCG; therefore, as the base ABC to the base DEF, so is the altitude of the pyramid DEFH to the altitude of the pyramid ABCG: wherefore, the bases and altitudes of the pyramids ABCG, DEFH, are reciprocally proportional.

† 28. 11. & 7. 12.

* 1 Ax. 5.

* 34. 11.

* 15. 5.

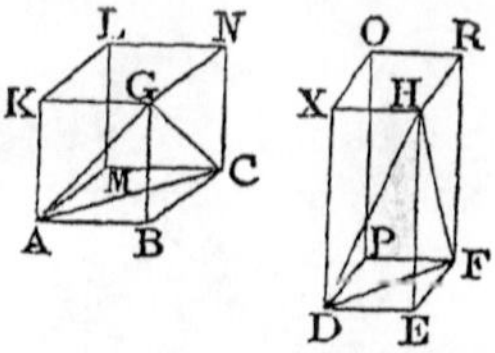

Again, let the bases and altitudes of the pyramids ABCG, DEFH, be reciprocally proportional, viz. the base ABC to the base DEF, as the altitude of the pyramid DEFH to the altitude of the pyramid ABCG: the pyramid ABCG shall be equal to the pyramid DEFH.

The same construction being made; because as the base ABC to the base DEF, so is the altitude of the pyramid DEFH to the altitude of the pyramid ABCG; and as the base ABC to the base DEF, so is the parallelogram BM to the parallelogram EP: therefore the parallelogram BM is to EP, as the altitude of the pyramid DEFH to the altitude of the pyramid ABCG: but the altitude of the pyramid DEFH is the same with the altitude of the solid parallelopiped EHPO; and the altitude of the pyramid ABCG is the same with the altitude

of the solid parallelopiped BGML: therefore as the base BM to the base EP, so is the altitude of the solid parallelopiped EHPO to the altitude of the solid parallelopiped BGML: but solid parallelopipeds having their bases and altitudes recipro-
* 34. 11. cally proportional, are equal * to one another; therefore the solid parallelopiped BGML is equal to the solid parallelopiped EHPO: and the pyramid ABCG is the sixth part of the solid BGML, and the pyramid DEFH is the sixth part of the solid
† 2 Ax. 5. EHPO; therefore the pyramid ABCG is equal † to the pyramid DEFH. Therefore, the bases, &c. Q. E. D.

PROPOSITION X.

THEOR.—*Every cone is the third part of a cylinder which has the same base and is of an equal altitude with it.*

Let a cone have the same base with a cylinder, viz. the circle ABCD, and the same altitude: the cone shall be the third part of the cylinder; that is, the cylinder shall be triple of the cone.

If the cylinder be not triple of the cone, it must either be greater than the triple, or less than it. First, let it be greater than the triple; and inscribe the square
† 2. 12. ABCD in the circle: this square is greater † than the half of the circle ABCD. Upon the square ABCD, erect a prism of the same altitude with the cylinder; this prism shall be greater than half of the cylinder: for let a square be described about the circle, and let a prism be erected upon the square, of the same altitude with the cylinder: then the inscribed square is half of that circumscribed; and upon these square bases are erected solid parallelopipeds, viz. the prisms of the same altitude; therefore the prism upon the square ABCD is the half of the prism upon the square described about the circle; because
* 32. 11. they are to one another * as their bases: and the cylinder is less than the prism upon the square described about the circle ABCD, therefore the prism upon the square ABCD of the same altitude with the cylinder, is greater than half of the cylinder. Bisect the circumferences AB, BC, CD, DA, in the points E, F, G, H; and join AE, EB, BF, FC, CG, GD, DH, HA:

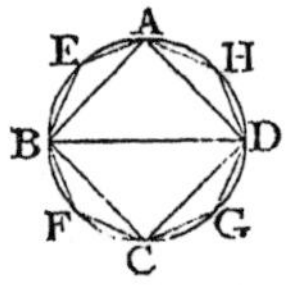

then, each of the triangles AEB, BFC, CGD, DHA is greater than the half of the segment of the circle in which it stands, as was shewn in Prop. 2, of this Book. Erect prisms upon each of these triangles, of the same altitude with the cylinder; each of these prisms shall be greater than half of the segment of the cylinder in which it is; because, if through the points E, F, G, H, parallels be drawn to AB, BC, CD, DA, and parallelograms be completed upon the same AB, BC, CD, DA, and solid parallelopipeds be erected upon the parallelograms; the prisms upon the triangles AEB, BFC, CGD, DHA, are the halves of the solid parallelopipeds *; and the segments of the cylinder which are upon the segments of the circle cut off by AB, BC, CD, DA, are less than the solid parallelopipeds which contain them; therefore the prisms upon the triangles AEB, BFC, CGD, DHA, are greater than half of the segments of the cylinder in which they are: therefore, if each of the circumferences be divided into two equal parts, and straight lines be drawn from the points of division to the extremities of the circumferences, and upon the triangles thus made, prisms be erected of the same altitude with the cylinder, and so on, there must at length remain some segments of the cylinder which together are less * than the excess of the cylinder above the triple of the cone: let them be those upon the segments of the circle, AE, EB, BF, FC, CG, GD, DH, HA; therefore the rest of the cylinder, that is, the prism of which the base is the polygon AEBFCGDH, and of which the altitude is the same with that of the cylinder, is greater than the triple of the cone: but this prism is triple * of the pyramid upon the same base, of which the vertex is the same with the vertex of the cone; therefore the pyramid upon the base AEBFCGDH, having the same vertex with the cone, is greater than the cone of which the base is the circle ABCD: but it is also less, for the pyramid is contained within the cone; which is impossible; therefore the cylinder is not greater than the triple of the cone.

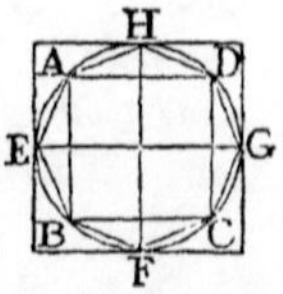

* 2 Cor. 7. 12.

* Lemma.

* 1 Cor. 7. 12.

Nor can the cylinder be less than the triple of the cone. Let it be less, if possible; therefore, inversely, the cone is greater than the third part of the cylinder. In the circle ABCD, inscribe a square: this square is greater than the half of the circle: and upon the square ABCD erect a pyramid,

having the same vertex with the cone; this pyramid is greater than the half of the cone; because, as was before demonstrated, if a square be described about the circle, the square ABCD is the half of it: and if upon these squares there be erected solid parallelopipeds of the same altitude with the cone, which are also prisms, the prism upon the square ABCD is the half of that which is upon the square described about the circle; for they are to one another as their
* 32. 11. bases *; as are also the third parts of them: therefore the paramid, the base of which is the square ABCD, is half of the pyramid upon the square described about the circle: but this last pyramid is greater than the cone which it contains; therefore the pyramid upon the square ABCD, having the same vertex with the cone, is greater than the half of the cone. Bisect the circumferences AB, BC, CD, DA, in the points E, F, G, H, and join AE, EB, BF, FC, CG, GD, DH, HA: therefore each of the triangles AEB, BFC, CGD, DHA, is greater than half of the segment of the circle in which it is: upon each of these triangles erect pyramids having the same vertex with the cone: therefore each of those pyramids is greater than the half of the segment of the cone in which it is, as was before demonstrated of the prisms and segments of the cylinder: and thus dividing each of the circumferences into two equal parts, and joining the points of division and their ex-

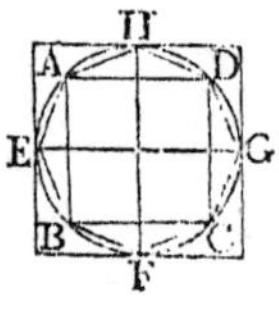

tremities by straight lines, and upon the triangles erecting pyramids having their vertices the same with that of the cone, and so on, there must at length remain some segments of the
† Lemma. cone, which together are less † than the excess of the cone above the third part of the cylinder: let these be the segments upon AE, EB, BF, FC, CG, GD, DH, HA: therefore the rest of the cone, that is, the pyramid of which the base is the polygon AEBFCGDH, and of which the vertex is the same with that of the cone, is greater than the third part of the cylinder: but this pyramid is the third part of the prism upon the same base AEBFCGDH, and of the same altitude with the cylinder; therefore this prism is greater than the cylinder of which the base is the circle ABCD: but it is also less, for it is contained within the cylinder; which is impossible: therefore the cylinder is not less than the triple of the cone. And it has been demonstrated, that neither is it greater

than the triple; therefore the cylinder is triple of the cone, or, the cone is the third part of the cylinder. Wherefore, every cone, &c. Q. E. D.

PROPOSITION XI.

THEOR.—*Cones and cylinders of the same altitude, are to one another as their bases.* See N.

Let the cones and cylinders, of which the bases are the circles ABCD, EFGH, and the axes KL, MN, and AC, EG the diameters of their bases, be of the same altitude: as the circle ABCD to the circle EFGH, so shall the cone AL be to the cone EN.

If it be not so, the circle ABCD must be to the circle EFGH, as the cone AL to some solid either less than the cone EN, or greater than it. First, let it be to a solid less than EN, viz. to the solid X; and let Z be the solid which is equal to the excess of the cone EN above the solid X; therefore the cone EN is equal to the solids X, Z together. In the circle EFGH, inscribe the square EFGH; therefore this square is greater than the half of the circle: upon the square EFGH, erect a pyramid of the same altitude with the cone; this pyramid shall be greater than half of the cone: for, if a square be described about the circle, and a pyramid be erected upon it, having the same vertex with the cone ‡, the pyramid inscribed in the cone is half of the pyramid circumscribed about it, because they are to one another as their bases *: but the cone is less than the circumscribed pyramid; therefore the pyramid of which the base is the square EFGH, and its vertex the same with that of the cone, is greater than half of the cone.

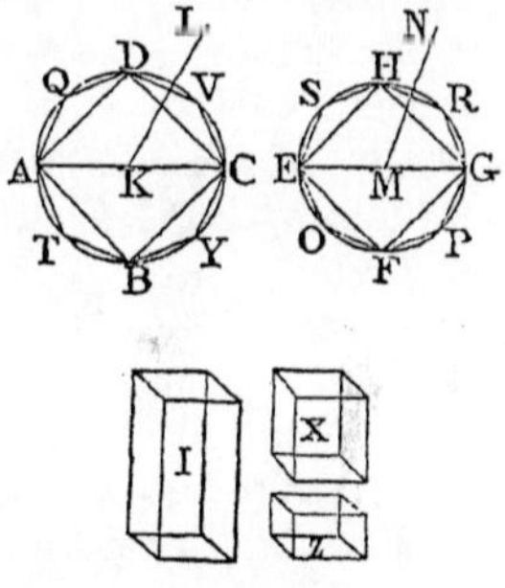

* 6. 12.

‡ Vertex is put in place of altitude, which is in the Greek, because the pyramid, in what follows, is supposed to be circumscribed about the cone, and so must have the same vertex. And the same change is made in some places following.

Divide the circumferences EF, FG, GH, HE, each into two equal parts in the points O, P, R, S, and join EO, OF, FP, PG, GR, RH, HS, SE: therefore each of the triangles EOF, FPG, GRH, HSE, is greater than half of the segment of the circle in which it is: upon each of these triangles, erect a pyramid having the same vertex with the cone; each of these pyramids is greater than the half of the segment of the cone in which it is: and thus dividing each of these circumferences into two equal parts, and, from the points of division, drawing straight lines to the extremities of the circumferences, and upon each of the triangles, thus made, erecting pyramids having the same vertex with the cone, and so on, there must at length remain
* Lemma. some segments of the cone which are together less * than the solid Z; let these be the segments upon EO, OF, FP, PG, GR, RH, HS, SE: therefore the remainder of the cone, viz. the pyramid of which the base is the polygon EOFPGRHS, and its vertex the same with that of the cone, is greater than the solid X. In the circle ABCD, inscribe the polygon ATBYCVDQ similar to the polygon EOFPGRHS, and upon it erect a pyramid having the same vertex with the cone AL: and because
* 1. 12. as the square of AC is to the square of EG, so * is the polygon ATBYCVDQ to the polygon EOFPGRHS; and as the
* 2. 12. square of AC to the square of EG, so is * the circle ABCD to
* 11. 5. the circle EFGH; therefore the circle ABCD is * to the circle EFGH, as the polygon ATBYCVDQ to the polygon EOFPGRHS: but as the circle ABCD to the circle EFGH, so is the cone AL to the solid X; and as the polygon ATBYCVDQ to the polygon EOFPGRHS, so
* 6. 12. is * the pyramid of which the base is the first of these polygons, and vertex L, to the pyramid of which the base is the other polygon, and its vertex N: therefore, as the cone AL to the solid X, so is the pyramid of which the base is the polygon ATBYCVDQ, and vertex L, to the pyramid the base of which is the pylygon EOFPGRHS, and vertex N: but the cone AL is greater than the pyramid
* 14. 5. contained in it; therefore the solid X is greater * than the pyramid in the cone EN: but it is less, as was shewn; which

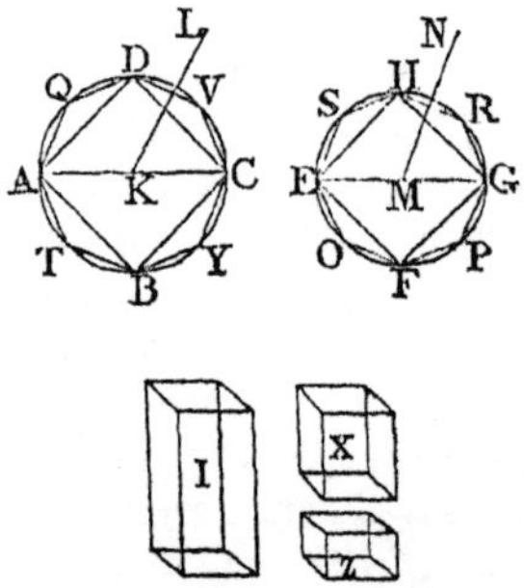

is absurd: therefore the circle ABCD is not to the circle EFGH, as the cone AL to any solid which is less than the cone EN. In the same manner it may be demonstrated, that the circle EFGH is not to the circle ABCD, as the cone EN to any solid less than the cone AL. Nor can the circle ABCD be to the circle EFGH, as the cone AL, to any solid greater than the cone EN. For if it be possible, let it be so to the solid I, which is greater than the cone EN: therefore, by inversion, as the circle EFGH to the circle ABCD, so is the solid I to the cone AL: but as the solid I to the cone AL, so is the cone EN to some solid, which must be less * than the cone AL, * 14. 5
because the solid I is greater than the cone EN; therefore, as the circle EFGH is to the circle ABCD, so is the cone EN to a solid less than the cone AL, which was shewn to be impossible; therefore the circle ABCD is not to the circle EFGH, as the cone AL is to any solid greater than the cone EN. And it has been demonstrated, that neither is the circle ABCD to the circle EFGH, as the cone AL to any solid less than the cone EN; therefore the circle ABCD is to the circle EFGH, as the cone AL to the cone EN: but as the cone is to the cone, so * is the cylinder to the cylinder, because the cylin- * 15. 5.
ders are triple * of the cones, each of each: therefore as the * 10. 12.
circle ABCD to the circle EFGH, so are the cylinders upon them of the same altitude. Wherefore cones and cylinders, &c. Q. E. D.

PROPOSITION XII.

THEOR.—*Similar cones and cylinders have to one another, the triplicate ratio of that which the diameters of their bases have.* See N.

Let the cones and cylinders of which the bases are the circles ABCD, EFGH, and the diameters of the bases AC, EG, and KL, MN the axes of the cones or cylinders, be similar: the cone of which the base is the circle ABCD and vertex the point L, shall have to the cone of which the base is the circle EFGH and vertex N, the triplicate ratio of that which AC has to EG.

For if the cone ABCDL has not to the cone EFGHN, the triplicate ratio of that which AC has to EG, the cone ABCDL must have the triplicate of that ratio to some solid which is less or greater than the cone EFGHN. First, let it have it to a less, viz. to the solid X. Make the same construction as in the preceding proposition, and it may be demonstrated the very same way as in that proposition, that the pyramid of which the base is the polygon EOFPGRHS, and vertex N, is greater than the solid X. Inscribe also in the circle ABCD, the polygon ATBYCVDQ similar to the polygon EOFPGRHS, upon which, erect a pyramid having the same vertex with the cone: and let LAQ be one of the triangles containing the pyramid upon the polygon ATBYCVDQ, the vertex of which is L; and let NES be one of the triangles containing the pyramid upon the polygon EOFPGRHS of which the vertex is N;

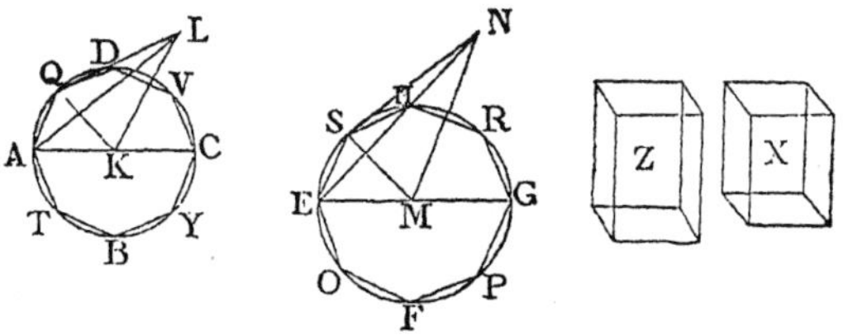

and join KQ, MS. Then, because the cone ABCDL is similar
*24 Def.11. to the cone EFGHN, AC is * to EG as the axis KL to the axis
* 15. 5. MN; and as AC to EG, so * is AK to EM; therefore as AK to EM, so is KL to MN; and alternately, AK to KL, as EM to MN: and the right angles AKL, EMN are equal: therefore, the sides about these equal angles being proportionals,
* 6. 6. the triangle AKL is similar * to the triangle EMN. Again, because AK is to KQ, as EM to MS, and that these sides are about equal angles AKQ, EMS, because these angles are, each of them, the same part of four right angles at the centres K,
* 6. 6. M, therefore the triangle AKQ is similar * to the triangle EMS. And because it has been shewn, that as AK to KL, so is EM to MN, and that AK is equal to KQ, and EM to MS, therefore as QK to KL, so is SM to MN; and therefore, the sides about the right angles QKL, SMN, being proportionals, the triangle LKQ is similar to the triangle NMS. And because of the similarity of the triangles AKL, EMN, as LA is to AK, so is NE to EM; and by the similarity of the triangles
* 22. 5. AKQ, EMS, as KA to AQ, so ME to ES: therefore, ex æquali *, LA is to AQ, as NE to ES. Again, because of the similarity of the triangles LQK, NSM, as LQ to QK, so NS to SM; and

from the similarity of the triangles KAQ, MES, as KQ to QA, so MS to SE: therefore, ex æquali*, LQ is to QA, as NS to SE: * 22. 5. and it was proved, that QA is to AL, as SE to EN: therefore, again, ex æquali, as QL to LA, so is SN to NE: wherefore the triangles LQA, NSE, having the sides about all their angles proportionals, are equiangular* and similar to one another: * 5. 6. and therefore the pyramid of which the base is the triangle AKQ, and vertex L, is similar to the pyramid the base of which is the triangle EMS, and vertex N, because their solid angles are equal* to one another, and they are contained by * B. 11. the same number of similar planes: but similar pyramids which have triangular bases, have to one another the triplicate* ratio of that which their homologous sides have; there- * 8. 12. fore the pyramid AKQL has to the pyramid EMSN, the triplicate ratio of that which AK has to EM. In the same manner, if straight lines be drawn from the points D, V, C, Y, B, T, to K, and from the points H, R, G, P, F, O, to M, and pyramids be erected upon the triangles, having the same vertices with the cones, it may be demonstrated, that each pyramid in the first cone has to each in the other, taking them in the same order, the triplicate ratio of that which the side AK has to the side EN; that is, which AC has to EG: but as one antecedent to its consequent, so are all the antecedents to all the consequents*, therefore as the pyramid AKQL to the py- * 12. 5. ramid EMSN, so is the whole pyramid the base of which is the polygon DQATBYCV, and vertex L, to the whole pyramid of which the base is the polygon HSEOFPGR, and vertex N: wherefore also the first of these two last-named pyramids has to the other, the triplicate ratio of that which AC has to EG; but, by the hypothesis, the cone of which the base is the circle ABCD, and vertex L, has to the solid X, the triplicate

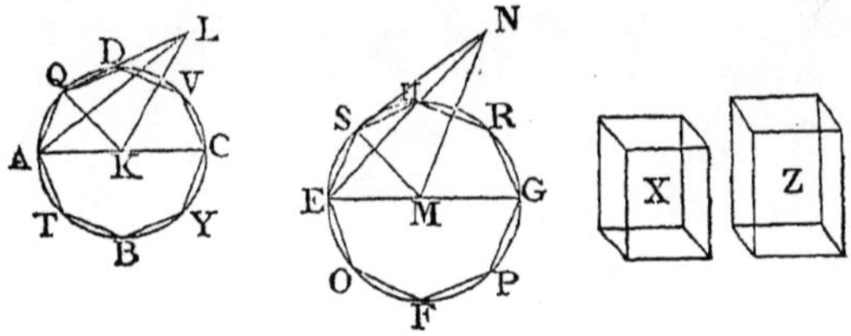

ratio of that which AC has to EG; therefore, as the cone of which the base is the circle ABCD, and vertex L, is to the

solid X, so is the pyramid the base of which is the polygon DQATBYCV, and vertex L, to the pyramid the base of which is the polygon HSEOFPGR, and vertex N: but the said cone is greater than the pyramid contained in it; therefore the
* 14. 5. solid X is greater* than the pyramid, the base of which is the polygon HSEOFPGR, and vertex N: but it is also less; which is impossible: therefore the cone, of which the base is the circle ABCD and vertex L, has not to any solid which is less than the cone of which the base is the circle EFGH and vertex N, the triplicate ratio of that which AC has to EG. In the same manner it may be demonstrated, that neither has the cone EFGHN to any solid which is less than the cone ABCDL, the triplicate ratio of that which EG has to AC. Nor can the cone ABCDL have to any solid which is greater than the cone EFGHN, the triplicate ratio of that which AC has to EG. For if it be possible, let it have it to a greater, viz. to the solid Z: therefore, inversely, the solid Z has to the cone ABCDL, the triplicate ratio of that which EG has to AC: but as the solid Z is to the cone ABCDL, so is the cone EFGHN
* 14. 5. to some solid, which must be less* than the cone ABCDL, because the solid Z is greater than the cone EFGHN; therefore the cone EFGHN has to a solid which is less than the cone ABCDL, the triplicate ratio of that which EG has to AC, which was demonstrated to be impossible: therefore the cone ABCDL has not to any solid greater than the cone EFGHN, the triplicate ratio of that which AC has to EG: and it was demonstrated, that it could not have that ratio to any solid less than the cone EFGHN: therefore the cone ABCDL has to the cone EFGHN, the triplicate ratio of that which AC has to
* 15. 5. EG, but as the cone is to the cone, so* the cylinder to the
† 10. 12. cylinder; for every cone is the third† part of the cylinder upon the same base, and of the same altitude: therefore also the cylinder has to the cylinder, the triplicate ratio of that which AC has to EG. Wherefore, similar cones, &c. *Q. E. D.*

PROPOSITION XIII.

See N. THEOR.—*If a cylinder be cut by a plane parallel to its opposite planes or bases, it divides the cylinder into two cylinders, one of which is to the other, as the axis of the first to the axis of the other.*

Let the cylinder AD be cut by the plane GH parallel to the opposite planes AB, CD, meeting the axis EF in the point K; and let the line GH be the common section of the plane GH, and the surface of the cylinder AD. Let AEFC be the parallelogram in any position of it, by the revolution of which about the straight line EF, the cylinder AD is described; and let GK be the common section of the plane GH, and the plane AEFC. And because the parallel planes AB, GH are cut by the plane AEKG, AE, KG, their common sections with it, are parallel*: wherefore AK is a parallelogram, and GK equal to EA, the straight line from the centre of the circle AB: for the same reason, each of the straight lines drawn from the point K to the line GH, may be proved to be equal to those which are drawn from the centre of the circle AB to its circumference, and are therefore all equal to one another; therefore the line GH is the circumference of a circle * of which the centre is the point K: therefore the plane GH divides the cylinder AD into the cylinders AH, GD; for they are the same which would be described by the revolution of the parallelograms AK, GF, about the straight lines EK, KF: and it is to be shewn, that the cylinder AH is to the cylinder HC, as the axis EK to the axis KF.

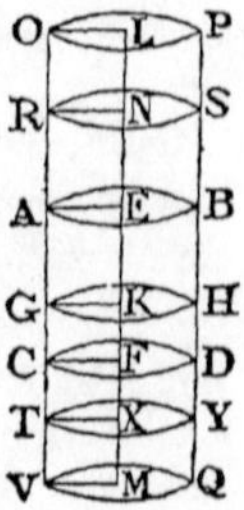

* 16. 11.

* 15 Def. 1.

Produce the axis EF both ways: and take any number of straight lines EN, NL, each equal to EK; and any number FX, XM, each equal to FK; and let planes parallel to AB, CD, pass through the points L, N, X, M: therefore the common sections of these planes with the cylinder produced, are circles, the centres of which are the points L, N, X, M, as was proved of the plane GH: and these planes cut off the cylinders PR, RB, DT, TQ. And because the axes LN, NE, EK are all equal, therefore the cylinders PR, RB, BG are* to one another as their bases: but their bases are equal, and therefore the cylinders PR, RB, BG are equal: and because the axes LN, NE, EK are equal to one another, as also the cylinders PR, RB, BG, and that there are as many axes as cylinders; therefore whatever multiple the axis KL is of the axis KE, the same multiple is the cylinder PG of the cylinder GB: for the same reason, whatever multiple the axis MK is of the axis KF, the same multiple is the cylinder QG of the cylinder GD: and if

* 11. 12.

the axis KL be equal to the axis KM, the cylinder PG is equal to the cylinder GQ; and if the axis KL be greater than the axis KM, the cylinder PG is greater than the cylinder GQ; and if less, less: therefore, since there are four magnitudes, viz. the axes EK, KF, and the cylinders BG, GD; and that of the axis EK and cylinder BG there have been taken any equimultiples whatever, viz. the axis KL and cylinder PG, and of the axis KF and cylinder GD, any equimultiples whatever, viz. the axis KM and cylinder GQ; and since it has been demonstrated, that if the axis KL be greater than the axis KM, the cylinder PG is greater than the cylinder GQ;

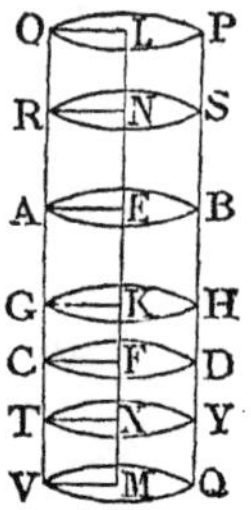

* 5 Def. 5. and if equal, equal; and if less, less; therefore* as the axis EK is to the axis KF, so is the cylinder BG to the cylinder GD. Wherefore, if a cylinder, &c. Q. E. D.

PROPOSITION XIV.

See N. THEOR.—*Cones and cylinders upon equal bases, are to one another as their altitudes.*

Let the cylinders EB, FD be upon the equal bases AB, CD: as the cylinder EB to the cylinder FD, so shall the axis GH be to the axis KL.

Produce the axis KL to the point N, and make LN equal to the axis GH; and let CM be a cylinder of which the base is CD, and axis LN. Then, because the cylinders EB, CM have

* 11. 12. the same altitude, they are to one another as their bases*: but their bases are equal, therefore also the cylinders EB, CM are equal: and because the cylinder FM is cut by the plane CD parallel to its opposite planes, as the

* 13. 12. cylinder CM to the cylinder FD, so is* the axis LN to the axis KL: but the cylinder CM is equal to the cylinder EB, and the axis LN to the axis GH; therefore as the cylinder EB to the cylinder FD, so is the axis GH to the axis

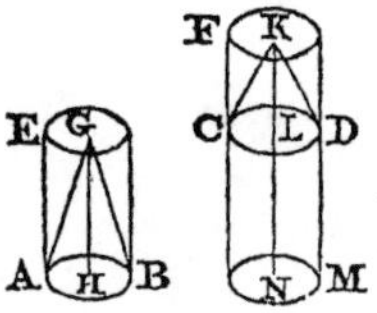

* 15. 5. KL: and as the cylinder EB to the cylinder FD, so is* the cone ABG to the cone CDK, because the cylinders are

* 10. 12. triple* of the cones: therefore also the axis GH is to the axis

KL, as the cone ABG to the cone CDK, and as the cylinder EB to the cylinder FD. Wherefore, cones, &c. Q. E. D.

PROPOSITION XV.

THEOR.—*The bases and altitudes of equal cones and cylinders are reciprocally proportional; and if the bases and altitudes be reciprocally proportional, the cones and cylinders are equal to one another.* See N.

Let the circles ABCD, EFGH, the diameters of which are AC, EG, be the bases, and KL, MN, the axes, as also the altitudes, of equal cones and cylinders; and let ALC, ENG be the cones, and AX, EO, the cylinders: the bases and altitudes of the cylinders AX, EO, shall be reciprocally proportional; that is, as the base ABCD to the base EFGH, so shall the altitude MN be to the altitude KL.

Either the altitude MN is equal to the altitude KL, or these altitudes are not equal. First let them be equal; and the cylinders AX, EO being also equal, and cones and cylinders of the same altitude being to one another as their bases*, * 11. 12.
therefore the base ABCD is equal* to the base EFGH; and * A. 5.
as the base ABCD is to the base EFGH, so is the altitude MN to the altitude KL. But let the altitudes KL, MN be unequal, and MN the greater of the two, and from MN, take MP equal to KL, and through the point P, cut the cylinder EO by the plane TYS, parallel to the opposite planes of the circles EFGH,

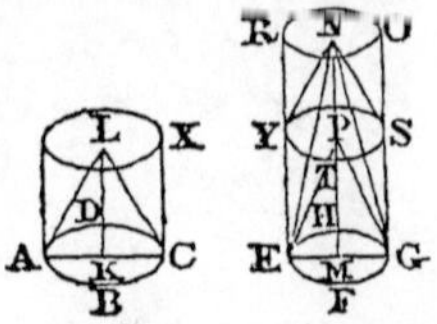

RO: therefore the common section of the plane TYS and the cylinder EO is a circle, and consequently ES is a cylinder, the base of which is the circle EFGH, and altitude MP: and because the cylinder AX is equal to the cylinder EO, as AX is to the cylinder ES, so* is the cylinder EO to the same ES: * 7. 5.
but as the cylinder AX to the cylinder ES, so* is the base * 11. 12.
ABCD to the base EFGH; for the cylinders AX, ES are of the same altitude; and as the cylinder EO to the cylinder ES, so* * 13. 12.
is the altitude MN to the altitude MP, because the cylinder EO is cut by the plane TYS parallel to its opposite planes; therefore as the base ABCD to the base EFGH, so is the altitude MN to the altitude MP: but MP is equal to the altitude

KL: wherefore as the base ABCD to the base EFGH, so is the altitude MN to the altitude KL; that is, the bases and altitudes of the equal cylinders AX, EO, are reciprocally proportional.

But let the bases and altitudes of the cylinders AX, EO, be reciprocally proportional, viz. the base ABCD to the base EFGH, as the altitude MN to the altitude KL: the cylinder AX shall be equal to the cylinder EO.

First, let the base ABCD be equal to the base EFGH: then, because as the base ABCD is to the base EFGH, so is
* A. 5. the altitude MN to the altitude KL; MN is equal * to
* 11. 12. KL; and therefore the cylinder AX is equal * to the cylinder EO.

But let the bases ABCD, EFGH be unequal, and let ABCD be the greater; and because as ABCD is to the base EFGH, so is the altitude MN to the altitude KL; therefore MN is
* A. 5. greater * than KL. Then, the same construction being made as before, because as the base ABCD to the base EFGH, so is the altitude MN to the altitude KL; and because the altitude
* 11. 12. KL is equal to the altitude MP; therefore the base ABCD is* to the base EFGH, as the cylinder AX to the cylinder ES; and as the altitude MN to the altitude MP or KL, so is the cylinder EO to the cylinder ES: therefore the cylinder AX is to the cylinder ES, as the cylinder EO is to the same ES: whence, the cylinder AX is equal to the cylinder EO: and the same reasoning holds in cones. Q. E. D.

PROPOSITION XVI.

PROB.—*In the greater of two circles that have the same centre, to inscribe a polygon of an even number of equal sides, that shall not meet the lesser circle.*

Let ABCD, EFGH be two given circles having the same centre K: it is required to inscribe in the greater circle ABCD, a polygon of an even number of equal sides, that shall not meet the lesser circle.

Through the centre K, draw the straight line BD, and from the point G, where it meets the circumference of the lesser circle, draw GA at right angles to BD, and produce it to C;
* Cor. 16. 3. therefore AC touches* the circle EFGH: then, if the circum-

ference BAD be bisected, and the half of it be again bisected, and so on, there must at length remain a circumference less* than AD: let this be LD; and from the point L, draw LM perpendicular to BD, and produce it to N; and join LD, DN; therefore LD is equal to DN: and because LN is parallel to AC, and that AC touches the circle EFGH; therefore LN does not meet the circle EFGH; and much less shall the straight lines LD, DN meet the circle EFGH: so that if straight lines equal to LD, be applied in the circle ABCD from the point L around to N, there shall be inscribed in the circle a polygon of an even number of equal sides not meeting the lesser circle. Which was to be done.

* Lemma.

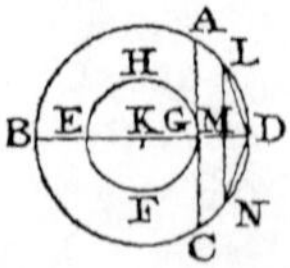

LEMMA II.

If two trapeziums ABCD, EFGH *be inscribed in the circles, the centres of which are the points* K, L; *and if the sides* AB, DC *be parallel, as also* EF, HG; *and the other four sides* AD, BC, EH, FG *be all equal to one another, but the side* AB *greater than* EF, *and* DC *greater than* HG; *the straight line* KA, *from the centre of the circle in which the greater sides are, is greater than the straight line* LE, *drawn from the centre to the circumference of the other circle.*

If it be possible, let KA be not greater than LE; then KA must be either equal to it, or less than it. First, let KA be equal to LE: therefore, because in two equal circles, AD, BC in the one, are equal to EH, FG in the other, the circumferences AD, BC are equal* to the circumferences EH, FG; but because the straight lines AB, DC are respectively greater than EF, GH, the circumferences AB, DC are greater than EF, HG; therefore the whole circumference ABCD is greater than the whole EFGH; but it is also equal to it, which is impossible; therefore the straight line KA is not equal to LE.

* 28. 3.

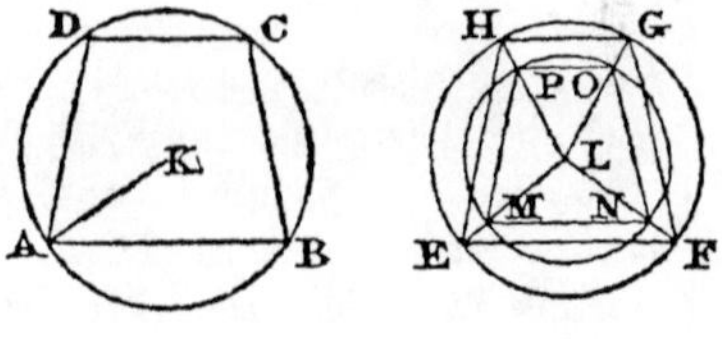

But let KA be less than LE, and make LM equal to KA; and from the centre L, and distance LM, describe the circle MNOP, meeting the straight lines LE, LF, LG, LH, in M, N, O, P; and join MN, NO, OP, PM which are respectively
* 2. 6. parallel * to, and less than, EF, FG, GH, HE: then because EH is greater than MP, AD is greater than MP; and the circles ABCD, MNOP are equal; therefore the circumference AD is greater than MP: for the same reason, the circumference BC is greater than NO: and because the straight line AB is greater than EF, which is greater than MN, much more is AB greater than MN: therefore the circumference AB is greater than MN; and for the same reason, the circumference DC is greater than PO: therefore the whole circumference ABCD is greater than the whole MNOP: but it is likewise equal to it, which is impossible; therefore KA is not less than LE: nor is it equal to it; therefore, the straight line KA must be greater than LE. Q. E. D.

COR. And if there be an isosceles triangle, the sides of which are equal to AD, BC, but its base less than AB the greater of the two sides AB, DC; the straight line KA may, in the same manner, be demonstrated to be greater than the straight line drawn from the centre to the circumference of the circle described about the triangle.

PROPOSITION XVII.

See N. PROB.—*In the greater of two spheres which have the same centre, to inscribe a solid polyhedron, the superficies of which shall not meet the lesser sphere.*

Let there be two spheres about the same centre A; it is required to inscribe in the greater, a solid polyhedron, the superficies of which shall not meet the lesser sphere.

Let the spheres be cut by a plane passing through the centre; the common sections of it with the spheres shall be circles, because the sphere is described by the revolution of a semicircle about the diameter remaining unmoveable; so that in whatever position the semicircle be conceived, the common section of the plane in which it is with the superficies of the sphere is the circumference of a circle; and this is a great circle of the sphere, because the diameter of the sphere,
* 15. 3. which is likewise the diameter of the circle, is greater* than

any straight line in the circle or sphere. Let then the circle made by the section of the plane with the greater sphere be BCDE, and with the lesser sphere be FGH; and draw the two diameters BD, CE, at right angles to one another; and in BCDE, the greater of the two circles, inscribe * a polygon of an even number of equal sides not meeting the lesser circle FGH; and let its sides, in BE the fourth part of the circle, be BK, KL, LM, ME; join KA, and produce it to N; and from A, draw † AX at right angles to the plane of the circle BCDE, meeting the superficies of the sphere in the point X: and let planes pass through AX, and each of the straight lines BD, KN, which, from what has been said, shall produce great circles on the superficies of the sphere, and let BXD, KXN be the semicircles thus made upon the diameters BD, KN: therefore, because XA is at right angles to the plane of the circle BCDE, every plane which passes through XA is at right * angles to the plane of the circle BCDE; wherefore the semicircles BXD, KXN are at right angles to that plane: and because the semicircles BED, BXD, KXN upon the equal diameters BD, KN, are equal to one another, their halves BE, BX, KX are equal to one another; therefore as many sides of the polygon as are in BE, so many are there in BX, KX, equal to the sides BK, KL, LM, ME: let these polygons be described, and their sides be BO, OP, PR, RX; KS, ST, TY, YX; and join OS, PT, RY; and from the points O, S, draw OV, SQ perpendiculars to AB, AK: and because the plane BOXD is at right angles to the plane BCDE, and in one of them BOXD, OV is drawn perpendicular to AB the common section of the planes, therefore OV is perpendicular * to the plane BCDE: for the same reason, SQ is perpendicular to the same plane, because the plane KSX is at right angles to the plane BCDE. Join VQ and because in the equal semicircles BXD, KXN, the circumferences BO, KS are equal, and OV, SQ are perpendicular to

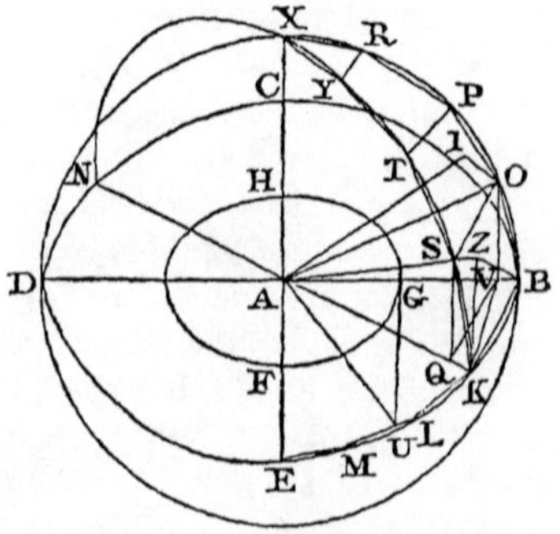

* 16. 12.

† 12. 11.

* 18. 11.

* 4 Def. 11.

* 26. 1. their diameters, therefore* OV is equal to SQ, and BV equal to KQ: but the whole BA is equal to the whole KA, therefore the remainder VA is equal to the remainder QA: therefore
* 2. 6. as BV is to VA, so is KQ to QA; wherefore VQ is parallel* to BK: and because OV, SQ are each of them at right angles
* 6. 11. to the plane of the circle BCDE, OV is parallel* to SQ; and it has been proved, that it is also equal to it; therefore QV,
* 33. 1. SO are equal* and parallel: and because QV is parallel to
* 9. 11. SO, and also to KB, OS is parallel* to BK; and therefore BO, KS, which join them, are in the same plane in which these parallels are, and the quadrilateral figure KBOS is in one plane: and if PB, TK be joined, and perpendiculars be drawn from the points P, T, to the straight lines AB, AK, it may be demonstrated, that TP is parallel to KB in the very same way
* 9. 11. that SO was shewn to be parallel to the same KB; wherefore* TP is parallel to SO, and the quadrilateral figure SOPT is in one plane: for the same reason, the quadrilateral TPRY is in
* 2. 11. one plane: and the figure YRX* is also in one plane: therefore, if from the points O, S, P, T, R, Y, there be drawn straight lines to the point A, there will be formed a solid polyhedron between the circumferences BX, KX, composed of pyramids, the bases of which are the quadrilaterals KBOS, SOPT, TPRY, and the triangle YRX, and of which the common vertex is the point A: and if the same construction be made upon each of the sides KL, LM, ME, as has been done upon BK, and the like be done also in the other three quadrants, and in the other hemisphere, there will be formed a solid polyhedron inscribed in the sphere, composed of pyramids, the bases of which are the aforesaid quadrilateral figures, and the triangle YRX, and those formed in the like manner in the rest of the sphere, the common vertex of them all being the point A.

Also the superficies of this solid polyhedron, shall not meet the lesser sphere in which is the circle FGH. For, from
* 11. 11. the point A, draw* AZ perpendicular to the plane of the quadrilateral KBOS, meeting it in Z, and join BZ, ZK: and because AZ is perpendicular to the plane KBOS, it makes right angles with every straight line meeting it in that plane; therefore AZ is perpendicular to BZ and ZK: and because AB is equal to AK, and that the squares of AZ, ZB are equal to the square
* 47. 1. of AB, and the squares of AZ, ZK, to the square of AK*; therefore the squares of AZ, ZB are equal to the squares of AZ,

ZK: take from these equals the square of AZ, and the remaining square of BZ is equal to the remaining square of ZK; and therefore the straight line BZ is equal to ZK: in the like manner it may be demonstrated, that the straight lines drawn from the point Z to the points O, S, are equal to BZ or ZK; therefore the circle described from the centre Z, and distance ZB, will pass through the points K, O, S, and KBOS will be a quadrilateral figure in the circle: and because KB is greater than QV, and QV equal to SO, therefore KB is greater than SO: but KB is equal to each of the straight lines BO, KS; wherefore each of

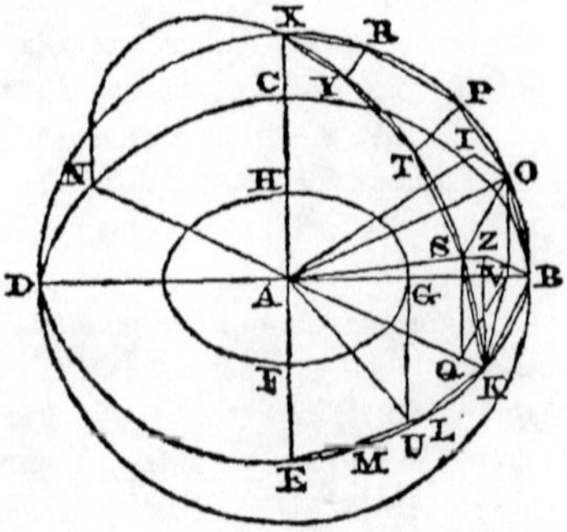

the circumferences cut off by KB, BO, KS, is greater than that cut off by OS; and these three circumferences, together with a fourth equal to one of them, are greater than the same three together with that cut off by OS; that is, than the whole circumference of the circle; therefore the circumference subtended by KB is greater than the fourth part of the whole circumference of the circle KBOS, and consequently, the angle BZK at the centre is greater than a right angle: and because the angle BZK is obtuse, the square of BK is greater* than the squares of BZ, ZK; that is, greater than twice the square of BZ. Join KV: and because, (in the triangles KBV, OBV,) KB, BV are equal to OB, BV, and that they contain equal angles, the angle KVB is equal* to the angle OVB: and OVB is a right angle; therefore also KVB is a right angle: and because BD is less than twice DV, the rectangle contained by BD, BV, is less than twice the rectangle DVB; that is*, the square of KB is less than twice the square of KV: but the square of KB is greater than twice the square of BZ; therefore the square of KV is greater than the square of BZ: and because BA is equal to AK, and that the squares of BZ, ZA are equal together to the square of BA, and the squares of KV, VA to the square of AK; therefore the squares of BZ, ZA are equal to the squares of KV, VA; and

* 12. 2.
* 4. 1.
* 8. 6.

of these the square of KV is greater than the square of BZ; therefore the square of VA is less than the square of ZA, and the straight line AZ greater than VA: much more then is AZ greater than AG, because, in the preceding proposition, it was shewn that KV falls without the circle FGH: and AZ is perpendicular to the plane KBOS, and is therefore the shortest of all the straight lines that can be drawn from A, the centre of the sphere, to that plane: therefore the plane KBOS does not meet the lesser sphere.

And that the other planes between the quadrants BX, KX, fall without the lesser sphere, is thus demonstrated. From the point A, draw AI perpendicular to the plane of the quadrilateral SOPT, and join IO; and, as was demonstrated of the plane KBOS and the point Z, in the same way it may be shewn, that the point I is the centre of a circle described about SOPT; and that OS is greater than PT; and PT was shewn to be parallel to OS: therefore, because the two trapeziums KBOS, SOPT inscribed in circles, have their sides BK, OS parallel, as also OS, PT; and their other sides BO, KS, OP, ST all equal to one another, and that BK is greater than OS, and OS greater than PT, therefore the straight line ZB is greater* than IO. Join AO which will be equal to AB; and because AIO, AZB are right angles, the squares of AI, IO are equal to the square of AO or of AB; that is, to the squares of AZ, ZB; and the square of ZB is greater than the square of IO, therefore the square of AZ is less than the square of AI; and the straight line AZ less than the straight line AI: and it was proved, that AZ is greater than AG; much more then is AI greater than AG: therefore the plane SOPT falls wholly without the lesser sphere. In the same manner it may be demonstrated, that the plane TPRY falls without the same sphere*, as also the triangle YRX. And after the same way it may be demonstrated, that all the planes which contain the solid polyhedron, fall without the lesser sphere. Therefore, in the greater of two spheres which have the same centre, a solid polyhedron is inscribed, the superficies of which does not meet the lesser sphere. Which was to be done.

* 2 Lem. 12.

* Cor. Lem. 2. 12.

But the straight line AZ may be demonstrated to be greater than AG otherwise, and in a shorter manner, without the help of Prop. 16, as follows. From the point G, draw GU at right angles to AG, and join AU. If then the circumference BE be bisected, and its half again bisected, and so on, there will at length be left a circumference less than the circumference which is subtended by a straight line equal to GU, inscribed in the circle BCDE: let this be the circumference KB; therefore the straight line KB is less than GU: and because the angle BZK is obtuse, as was proved in the preceding, therefore BK is greater than BZ: but GU is greater than BK; much more then is GU greater than BZ, and the square of GU than the square of BZ: and AU is equal to AB; therefore the square of AU, that is, the squares of AG, GU are equal to the square of AB, that is, to the squares of AZ, ZB: but the square of BZ is less than the square of GU; therefore the square of AZ is greater than the square of AG, and the straight line AZ consequently greater than the straight line AG.

COR. And if in the lesser sphere there be inscribed a solid polyhedron, by drawing straight lines betwixt the points in which the straight lines from the centre of the sphere, drawn to all the angles of the solid polyhedron in the greater sphere, meet the superficies of the lesser; in the same order in which are joined the points in which the same lines from the centre meet the superficies of the greater sphere; the solid polyhedron in the sphere BCDE shall have to this other solid polyhedron, the triplicate ratio of that which the diameter of the sphere BCDE has to the diameter of the other sphere. For if these two solids be divided into the same number of pyramids, and in the same order, the pyramids shall be similar to one another, each to each: because they have the solid angles at their common vertex, the centre of the sphere, the same in each pyramid, and their other solid angles at the bases, equal to one another, each to each*, because they are (* B. 11.) contained by three plane angles, each equal to each; and the pyramids are contained by the same number of similar planes; and are therefore similar * to one another, each to each: but (* 11 Def. 11.) similar pyramids have to one another, the triplicate* ratio of (* Cor. 8. 12.) their homologous sides: therefore the pyramid of which the base is the quadrilateral KBOS, and vertex A, has to the pyramid in the other sphere of the same order, the triplicate ratio of their homologous sides, that is, of that ratio which AB from the centre of the greater sphere, has to the straight line from the same centre to the superficies of the lesser sphere. And in like manner, each pyramid in the greater sphere has to each of the same order in the lesser, the triplicate ratio of that which AB has to the semi-diameter of the lesser sphere. And as one antecedent is to its consequent, so are all the antecedents to all the consequents. Wherefore, the whole solid polyhedron in the greater sphere has to the whole solid polyhedron in the other, the triplicate ratio of that which AB the semi-diameter of the first has to the semi-diameter of the other; that is, which the diameter BD of the greater has to the diameter of the other sphere.

PROPOSITION XVIII.

THEOR.—*Spheres have to one another, the triplicate ratio of that which their diameters have.*

Let ABC, DEF be two spheres, of which the diameters are BC, EF: the sphere ABC shall have to the sphere DEF, the triplicate ratio of that which BC has to EF.

For if it has not, the sphere ABC must have to a sphere either less or greater than DEF, the triplicate ratio of that which BC has to EF. First, let it have that ratio to a less, viz. to the sphere GHK; and let the sphere DEF have the same

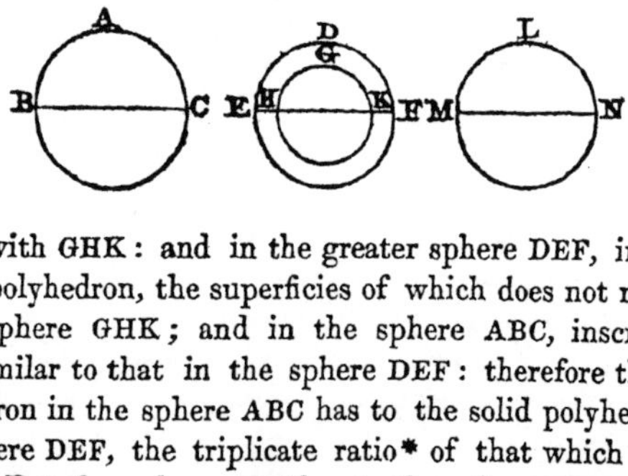

* 17. 12. centre with GHK: and in the greater sphere DEF, inscribe* a solid polyhedron, the superficies of which does not meet the lesser sphere GHK; and in the sphere ABC, inscribe another similar to that in the sphere DEF: therefore the solid polyhedron in the sphere ABC has to the solid polyhedron in
* Cor. 17. 12. the sphere DEF, the triplicate ratio* of that which BC has to EF. But the sphere ABC has to the sphere GHK, the triplicate ratio of that which BC has to EF; therefore as the sphere ABC to the sphere GHK, so is the solid polyhedron in the sphere ABC to the solid polyhedron in the sphere DEF: but the sphere ABC is greater than the solid polyhedron in it;
* 14. 5. therefore* also the sphere GHK is greater than the solid polyhedron in the sphere DEF: but it is also less, because it is contained within it, which is impossible: therefore the sphere ABC has not to any sphere less than DEF, the triplicate ratio

of that which BC has to EF. In the same manner it may be demonstrated, that the sphere DEF has not to any sphere less than ABC, the triplicate ratio of that which EF has to BC. Nor can the sphere ABC have to any sphere greater than DEF, the triplicate ratio of that which BC has to EF: for if it can, let it have that ratio to a greater sphere LMN: therefore by inversion, the sphere LMN has to the sphere ABC, the triplicate ratio of that which the diameter EF has to the diameter BC. But as the sphere LMN to ABC, so is the sphere DEF to some sphere which must be less * than the sphere ABC, * 14. 5.
because the sphere LMN is greater than the sphere DEF; therefore the sphere DEF has to a sphere less than ABC, the triplicate ratio of that which EF has to BC; which was shewn to be impossible: therefore the sphere ABC has not to any sphere greater than DEF, the triplicate ratio of that which BC has to EF: and it was demonstrated, that neither has it that ratio to any sphere less than DEF. Therefore, the sphere ABC has to the sphere DEF, the triplicate ratio of that which BC has to EF. *Q. E. D.*

END OF THE ELEMENTS.

NOTES,

CRITICAL AND GEOMETRICAL.

The following Notes, by Dr. Simson, contain Observations on some of the foregoing Propositions, with an Account of those things in which this Edition of the Elements differs from the Greek Text, and the reasons of the alterations that have been made.

BOOK I.

DEFINITIONS.

I. to VI. It is necessary to consider a solid, that is, a magnitude which has length, breadth, and thickness, in order to understand aright the definitions of a point, line, and superficies; for these all arise from a solid, and exist in it. The boundary, or boundaries, which contain a solid are called superficies; or the boundary which is common to two solids which are contiguous, or which divides one solid into two contiguous parts, is called a superficies: Thus, if BCGF be one of the boundaries which contain the solid ABCDEFGH, or which is the common boundary of this solid and the solid BKLCFNMG, and is therefore in the one as well as in the other solid, it is called a superficies, and has no thickness; for if it have any, this thickness must either be a part of the thickness of the solid AG, or of the solid BM, or a part of the thickness of each of them. It cannot be a part of the thickness of the solid BM; because

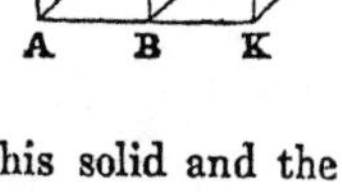

if this solid be removed from the solid AG, the superficies BCGF, the boundary of the solid AG, remains still the same as it was. Nor can it be a part of the thickness of the solid AG; because if this be removed from the solid BM, the superficies BCGF, the boundary of the solid BM, does nevertheless remain; therefore the superficies BCGF has no thickness, but only length and breadth.

The boundary of a superficies is called a line, or a line is the common boundary of two superficies that are contiguous, or which divides one superficies into two contiguous parts; thus, if BC be one of the boundaries which contain the superficies ABCD, or which is the common boundary of this superficies and of the superficies KBCL, which is contiguous to it, this boundary BC is called a line, and has no breadth: for if it have any, this must be part either of the breadth of the superficies ABCD, or of the superficies KBCL, or part of each of them. It is not part of the breadth of the superficies KBCL; for, if this superficies be removed from the superficies ABCD, the line BC, which is the boundary of the superficies ABCD, remains the same as it was: nor can the breadth that BC is supposed to have, be a part of the breadth of the superficies ABCD; because, if this be removed from the superficies KBCL, the line BC, which is the boundary of the superficies KBCL, does nevertheless remain: therefore the line BC has no breadth: and because the line BC is in a superficies, and that a superficies has no thickness, as was shewn, therefore a line has neither breadth nor thickness, but only length.

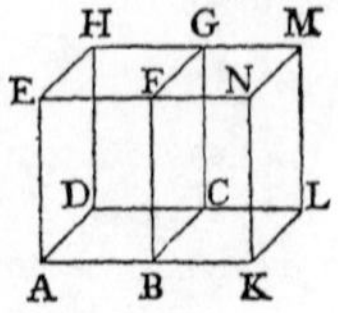

The boundary of a line is called a point, or a point is the common boundary or extremity of two lines, that are contiguous: thus, if B be the extremity of the line AB, or the common extremity of the two lines AB, KB, this extremity is called a point, and has no length. For if it have any, this length must either be part of the length of the line AB, or of the line KB. It is not part of the length of KB; for if the line KB be removed from AB, the point B, which is the extremity of the line AB, remains the same as it was: nor is it part of the length of the line AB; for, if AB be removed from the line KB, the point B, which is the extremity of the line KB, does nevertheless remain: therefore the point B has no length: and because a point is in a line, and a line has neither breadth nor thickness, therefore a point has no length, breadth, nor thickness. And in this manner the definitions of a point, line, and superficies, are to be understood.

VII. Instead of this definition as it is in the Greek copies, a more distinct one is given from a property of a plane superficies, which is manifestly supposed in the Elements, viz. that a straight line drawn from any point in a plane to any other in it, is wholly in that plane.

VIII. It seems that he who made this definition designed that it should comprehend not only a plane angle contained by two straight lines, but likewise the angle which some conceive to be made by a straight line and a curve, or by two curve lines which meet one another in a plane: but, though the meaning of the words ἐπ' εὐθείας, that is, in a straight line, or in the same direction, be plain, when two straight lines are said to be in a straight line, it does not appear what ought to be understood by these words, when a straight line and a curve, or two curve lines, are said to be in the same direction; at least it cannot be explained in this place: which makes it probable that this definition, and that of the angle of a segment, and what is said of the angle of a semicircle, and the angles of segments, in the 16th and 31st Propositions of Book III., are the additions of some less skilful editor: on which account, especially since they are quite useless, these definitions are distinguished from the rest by inverted double commas.

XVII. The words "which also divides the circle into two equal parts" are added at the end of this definition in all the copies, but are now left out as not belonging to the definition, being only a corollary from it. Proclus demonstrates it, by conceiving one of the parts into which the diameter divides the circle, to be applied to the other; for it is plain they must coincide, else the straight lines from the centre to the circumference would not be at all equal: the same thing is easily deduced from the 31st Proposition of Book III. and the 24th of the same; from the first of which it follows, that semicircles are similar segments of a circle; and from the other, that they are equal to one another.

XXXIII. This definition has one condition more than is necessary; because every quadrilateral figure which has its opposite sides equal to one another, has likewise its opposite angles equal; and on the contrary.

Let ABCD be a quadrilateral figure of which the opposite sides AB, CD are equal to one another, as also AD and BC; join BD; the two sides AD, DB are equal to the two CB, BD, and the base AB is equal to the base CD; therefore * the angle ADB is equal to the angle CBD; and * the angle BAD is equal to the angle DCB, and ABD to BDC; and therefore also the angle ADC is equal to the angle ABC. * 8. 1. * 4. 1.

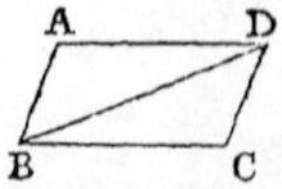

And if the angle BAD be equal to the opposite angle BCD, and the angle ABC to ADC; the opposite sides are equal; because * the angles of the quadrilateral figure ABCD are together equal to four right angles, and the two angles BAD, ADC are together equal to the two angles BCD, ABC: wherefore BAD, ADC are the half of all the four angles; that is, BAD and ADC are equal to two right angles; and therefore AD, CD * are parallels. In the same manner, AD, BC are parallels: therefore ABCD is a parallelogram, and its opposite sides * are equal. * 32. 1. * 28. 1. * 34. 1.

PROPOSITION VII.

There are two cases of this proposition, one of which is not in the Greek text, but is as necessary as the other: and that the case left out has been formerly in the text, appears plainly from this, that the second part of Prop. 5. which is necessary to the demonstration of this case, can be of no use at all in the Elements, or any where else, but in this demonstration; because the second part of Prop. 5. clearly follows from the first part, and Prop. 13. Book I. This part must therefore have been added to Prop. 5. on account of some proposition betwixt the 5th and 13th, but none of these stand in need of it except the 7th Proposition, on account of which it has been added: besides, the translation from the Arabic has this case explicitly demonstrated. And Proclus acknowledges that the second part of Prop. 5. was added on account of Prop 7., but gives a ridiculous reason for it, "that it might afford an answer to objections made against the 7th," as if the case of the 7th, which is left out, were, as he expressly makes it, an objection against the proposition itself. Whoever is curious may read what Proclus says of this in his commentary on the 5th and 7th Propositions; for it is not worth while to relate his trifles at full length.

It was thought proper to change the enunciation of this 7th Prop. so as to preserve the very same meaning; the literal translation from the Greek being extremely harsh and difficult to be understood by beginners.

PROPOSITION XI.

A corollary is added to this proposition, which is necessary to Prop. 1. B. 11. and otherwise.

PROPOSITIONS XX. XXI.

Proclus, in his commentary, relates, that the Epicureans derided Prop. 20, as being manifest even to asses, and needing no demonstration; and his answer is, that though the truth of it be manifest to our senses, yet it is science which must give the reason why two sides of a triangle are greater than the third: but the right answer to this objection against this and the 21st, and some other plain propositions, is, that the number of axioms ought not to be increased without necessity, as it must be if these propositions be not demonstrated. Mons. Clairault in the Preface to his Elements of Geometry, published in French at Paris, anno 1741, says "That Euclid has been at the pains to prove, that the two sides of a triangle which is included within another, are together less than the two sides of the triangle which includes it:" but he has forgot to add this condition, viz. that the triangles must be upon the same base: because unless this be added, the sides of the included triangle may be greater than the sides of the triangle which includes it, in any ratio which is less than that of two to one, as Pappus Alexandrinus has demonstrated in Prop. 3. B. 3. of his Mathematical Collections.

PROPOSITION XXII.

Some authors blame Euclid because he does not demonstrate that the two circles made use of in the construction of this problem must cut one another: but this is very plain from the determination he has given, viz. that any two of the straight lines DF, FG, GH, must be greater than the third. For who is so dull, though only beginning to learn the Elements, as not to perceive that the circle described from the centre F, at the distance FD, must meet FH betwixt F and H, because FD is less than FH; and that for the like reason, the circle described from the centre G, at the distance GH or GM, must meet DG betwixt D and G; and that these circles must meet one another, because FD and GH are together greater than FG? And this determination is easier to be understood than that which Mr. Thomas Simpson derives from it, and puts instead of Euclid's, in the 49th page of his Elements of Geometry, that he may supply the omission he blames Euclid for, which determination is, that any of the three straight lines must be less than the sum, but greater than the difference, of the other two: from this he shews, that the circles must meet one another, in one case; and says, that it may be proved after the same manner in any other case: but the straight line GM, which he bids take from GF, may be greater than it, as in the figure here annexed; in which case his demonstration must be changed into another.

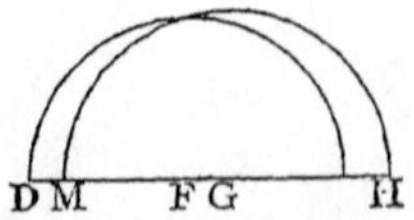

PROPOSITION XXIV.

To this is added, "of the two sides DE, DF, let DE be that which is not greater than the other," that is, take that side of the two DE, DF, which is not greater than the other, in order to make with it the angle EDG equal to BAC; because without this restriction there might be three different cases of the proposition, as Campanus and others make.

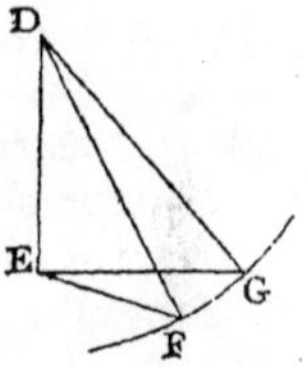

Mr. Thomas Simpson, in p. 262, of the second edition of his Elements of Geometry, printed anno 1760, observes in his notes, that it ought to have been shewn, that the point F falls below the line EG. This probably Euclid omitted, as it is very easy to perceive, that DG being equal to DF, the point G is in the circumference of a circle described from the centre D at the distance DF, and must be in that part of it which is above the straight line EF, because DG falls above DF, the angle EDG being greater than the angle EDF.

PROPOSITION XXIX.

The proposition which is usually called the 5th postulate or 11th axiom, by some the 12th, on which this 29th depends, has given a great deal to do, both to ancient and modern geometers; it seems not to be properly placed among the axioms, as indeed it is not self-evident; but it may be demonstrated thus:

DEFINITIONS.

I. The distance of a point from a straight line, is the perpendicular drawn to it from the point.

II. One straight line is said to go nearer to, or further from, another straight line, when the distances of the points of the first from the other straight line become less or greater than they were; and two straight lines are said to keep the same distance from one another, when the distance of the points of one of them from the other is always the same.

AXIOM.

A straight line cannot first come nearer to another straight line, and then go further from it, before it cuts it; and, in like manner, a straight line cannot go further from another straight line, and then come nearer to it; nor can a straight line keep the same distance from another straight line, and then come nearer to it, or go further from it; for a straight line keeps always the same direction.

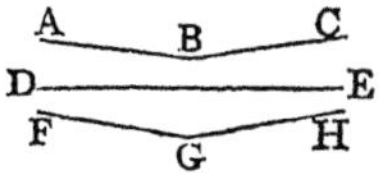

See the preceding figure.

For example, the straight line ABC cannot first come nearer to the straight line DE, as from the point A to the point B, and then, from the point B to the point C, go further from the same DE: and, in like manner, the straight line FGH cannot go further from DE, as from F to G, and then, from G to H, come nearer to the same DE: and so in the last case, as in fig. 2.

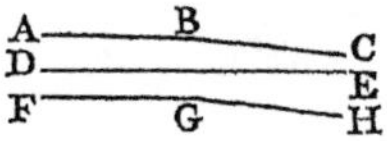

PROPOSITION I.

If two equal straight lines AC, BD *be each at right angles to the same straight line* AB; *if the points* C, D *be joined by the straight line* CD, *the straight line* EF, *drawn from any point* E *in* AB *to* CD, *at right angles to* AB, *shall be equal to* AC, *or* BD.

If EF be not equal to AC, one of them must be greater than the other; let AC be the greater: then, because FE is less than CA, the straight line CFD is nearer to the straight line AB at the point F than at the point C, that is, CF comes nearer to AB from the point C to F: but because DB is greater than FE, the straight line CFD is further from AB at the point D than at F, that is, FD goes further from AB from F to D: therefore the straight line CFD first comes nearer to the straight line AB, and then goes further from it, before it cuts it, which is impossible: if FE be said to be greater than CA, or DB, the straight line CFD first goes further from the straight line AB, and then comes nearer to it, which is also impossible: therefore FE is not unequal to AC, that is, it is equal to it.

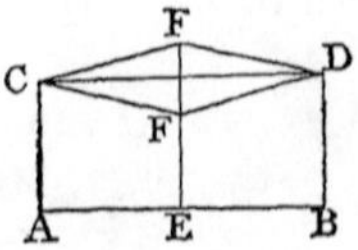

PROPOSITION II.

If two equal straight lines AC, BD *be each at right angles to the same straight line* AB, *the straight line* CD *which joins their extremities, makes right angles with* AC *and* BD.

Join AD, BC; and because, in the triangles CAB, DBA, the sides CA, AB are equal to DB, BA, and the angle CAB equal to the angle DBA, the base BC is equal* to the base AD: and in the triangles ACD, BDC, the sides AC, CD are equal to BD, DC, and the base AD is equal to the base BC; therefore the angle ACD is equal * to the angle BDC: from any point E in AB, draw EF to CD, at right angles to AB; therefore, (by Prop. 1.) EF is equal to AC, or BD; wherefore, as has been just shewn, the angle ACF is equal to the angle EFC: in the same manner, the angle BDF is equal to the angle EFD; but the angles ACD, BDC are equal; therefore the angles EFC and EFD are equal, and they are right angles*; wherefore, also the angles ACD, BDC are right angles.

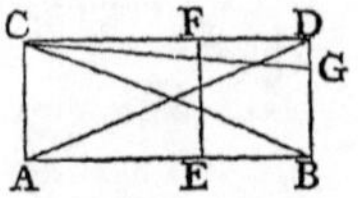

* 4. 1.

* 8. 1.

* 10 Def. 1.

Corollary. Hence, if two straight lines AB, CD be at right angles to the same straight line AC, and if betwixt them a straight line BD be drawn at right angles to either of them, as to AB, then BD is equal to AC, and BDC is a right angle.

If AC be not equal to BD, take BG equal to AC, and join CG: therefore, by this proposition, the angle ACG is a right angle; but ACD is also a right angle; wherefore the angles ACD, ACG are equal to one another, which is impossible: therefore BD is equal to AC; and by this proposition BDC is a right angle.

PROPOSITION III.

If two straight lines which contain an angle be produced, there may be found in either of them a point from which the perpendicular drawn to the other shall be greater than any given straight line.

Let AB, AC be two straight lines which make an angle with one another, and let AD be the given straight line; a point may be found either in AB or AC, as in AC, from which, the perpendicular drawn to the other AB, shall be greater than AD.

In AC take any point E, and draw EF perpendicular to AB: produce AE to G, so that EG be equal to AE; and produce FE to H, and make EH equal to FE, and join HG. Because, in the triangles AEF, GEH, the two sides AE, EF are equal to the two GE, EH, each

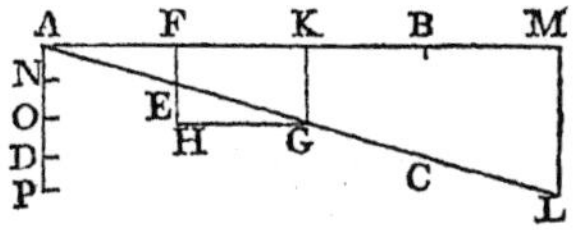

* 15. 1. to each, and contain equal *
* 4. 1. angles, the angle GHE is therefore equal to * the angle AFE, which is a right angle: draw GK perpendicular to AB; and because the straight lines FK, HG are at right angles to FH, and KG at right angles to FK, KG is equal to FH, (by Cor. Prop. 2.) that is, to the double of FE. In the same manner if AG be produced to L, so that GL be equal to AG, and LM be drawn perpendicular to AB, then LM is double of GK, and so on. In AD take AN equal to FE, and AO equal to KG, that is, to the double of FE, or AN; also take AP equal to LM, that is, to the double of KG, or AO; and let this be done till the straight line taken be greater than AD: let this straight line so taken be AP, and because AP is equal to LM, therefore LM is greater than AD. Which was to be done.

PROPOSITION IV.

If two straight lines AB, CD *make equal angles* EAB, ECD *with another straight line* EAC *towards the same parts of it,* AB *and* CD *are at right angles to some straight line.*

Bisect AC in F, and draw FG perpendicular to AB; take

CH in the straight line CD equal to AG, and on the contrary side of AC to that on which AG is, and join FH; therefore, in the triangles AFG, CFH, the sides FA, AG are equal to FC, CH, each to each, and the angle FAG, that is * EAB, is equal to the angle FCH; wherefore *, the angle AGF is equal to CHF, and AFG to the angle CFH: to these last add the common angle AFH; therefore the two angles AFG, AFH are equal to the two angles CFH, HFA, which two last are together equal to two right angles *: therefore, also AFG, AFH are equal to two right angles, and consequently *, GF and FH are in one straight line. And because AGF is a right angle, CHF which is equal to it is also a right angle: therefore the straight lines AB, CD are at right angles to GH.

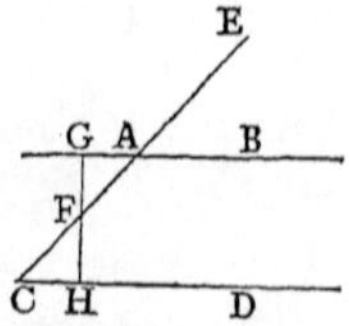

* 15. 1.
* 4. 1.
* 13. 1.
* 14. 1.

PROPOSITION V.

If two straight lines AB, CD *be cut by a third* ACE, *so as to make the interior angles* BAC, ACD, *on the same side of it, together less than two right angles*, AB *and* CD *being produced, shall meet one another towards the parts on which are the two angles, which are less than two right angles.*

At the point C in the straight line CE, make * the angle ECF equal to the angle EAB, and draw to AB, the straight line CG at right angles to CF: then, because the angles ECF, EAB are equal to one another, and that the angles ECF, FCA are together equal * to two right angles, the angles EAB, FCA are equal to two right angles. But, by the hypothesis, the angles EAB, ACD are together less than two right angles; therefore the angle FCA is greater than ACD, and CD falls between CF and AB: and because CF and CD make an angle with one another, (by Prop. 3.) a point may be found in either of them CD, from which the perpendicular drawn to CF shall be greater than the straight line CG. Let this point be H, and draw HK perpendicular to CF, meeting AB in L: and because AB, CF contain equal angles with AC on the same side of it, (by Prop. 4.) AB and CF are at right angles to the straight line

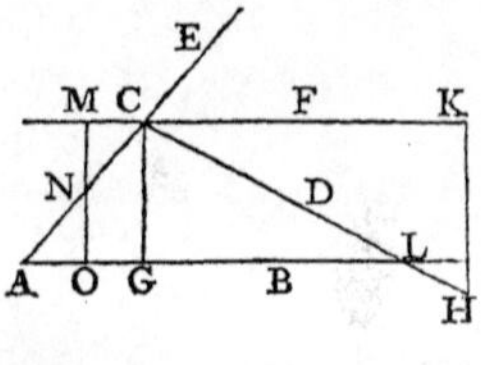

* 23. 1.
* 13. 1.

MNO, which bisects AC in N, and is perpendicular to CF: therefore (by Cor. Prop. 2.) CG and KL, which are at right angles to CF, are equal to one another: and HK is greater than CG, and therefore is greater than KL, and consequently the point H is in KL produced. Wherefore the straight line CDH, drawn betwixt the points C, H, which are on contrary sides of AL, must necessarily cut the straight line AB.

PROPOSITION XXXV.

The demonstration of this proposition is changed, because, if the method which is used in it was followed, there would be three cases to be separately demonstrated, as is done in the translation from the Arabic; for, in the Elements, no case of a proposition that requires a different demonstration ought to be omitted. On this account we have chosen the method which Mons. Clairault has given, the first of any, as far as I know, in his Elements, page 21, and which afterwards Mr. Simpson gives in his, page 32. But whereas Mr. Simpson makes use of Prop. 26. Book 1., from which the equality of the two triangles does not immediately follow, because to prove that, the 4th of Book 1. must likewise be made use of, as may be seen in the very same case in the 34th Prop. Book 1., it was thought better to make use only of the 4th of Book 1.

PROPOSITION XLV.

The straight line KM is proved to be parallel to FL from the 33d Prop., whereas KH is parallel to FG by construction, and KHM, FGL have been demonstrated to be straight lines. A corollary is added from Commandine, as being often used.

BOOK II.

PROPOSITION XIII.

In this proposition only acute-angled triangles are mentioned, whereas it holds true of every triangle; and the demonstrations of the cases omitted are added. Commandine and Clavius have likewise given their demonstrations of these cases.

PROPOSITION XIV.

In the demonstration of this, some Greek editor has ignorantly inserted the words, "but if not, one of the two BE, ED, is the greater: let BE be the greater, and produce it to F;" as if it was of any consequence whether the greater or lesser be produced: therefore, instead of these words, there ought to be read only, "but if not, produce BE to F."

BOOK III.

PROPOSITION I.

Several authors, especially among the modern mathematicians and logicians, inveigh too severely, and sometimes ignorantly enough, against indirect or apogogic demonstrations, not being aware that there are some things that cannot be demonstrated any other way: of this the present proposition is a very clear instance, as no direct demonstration can be given of it: because, besides the definition of a circle, there is no principle or property relating to a circle antecedent to this problem, from which either a direct or indirect demonstration can be deduced: wherefore it is necessary that the point found by the construction of the problem be proved to be the centre of the circle, by the help of this definition, and some of the preceding propositions: and because, in the demonstration, this proposition must be brought in, viz. straight lines from the centre of a circle to the circumference are equal, and that the point found by the construction cannot be assumed as the centre, for this is the thing to be demonstrated; it is manifest some other point must be assumed as the centre: and if from this assumption an absurdity follows, as Euclid demonstrates there must, then it is not true that the point assumed is the centre; and as any point whatever was assumed, it follows that no point, except that found by the construction, can be the centre, from which the necessity of an indirect demonstration in this case is evident.

PROPOSITION XIII.

As it is much easier to imagine that two circles may touch one another within in more points than one, upon the same side, than upon opposite sides, the figure of that case ought not to have been omitted; but the construction in the Greek text would not have suited with this figure so well, because the centres of the circles must have been placed near to the circumferences; on which account another construction and demonstration is given, which is the same with the second part of that which Campanus has translated from the Arabic, where, without any reason, the demonstration is divided into two parts.

PROPOSITION XV.

The converse of the second part of this proposition is wanting, though in the preceding, the converse is added, in a like case, both in the enunciation and demonstration; and it is now added in this. Besides, in the demonstration of the first part of this fifteenth, the diameter AD (see Commandine's figure) is proved to be greater than the straight line BC by means of another straight line MN; whereas it may be better done without it: on which accounts, we have given a different demonstration, like to that which Euclid gives in the preceding 14th, and to that which Theodosius gives in Prop. 6. B. I. of his Spherics, in this very affair.

PROPOSITION XVI.

In this we have not followed the Greek nor the Latin translation literally, but have given what is plainly the meaning of the proposition, without mentioning the angle of the semicircle, or that which some call the cornicular angle, which they conceive to be made by the circumference and the straight line which is at right angles to the diameter, at its extremity; which angles have furnished matter of great debate between some of the modern geometers, and given occasion of deducing strange consequences from them, which are quite avoided by the manner in which we have expressed the proposition. And in like manner, we have given the true meaning of Prop. 31. B. III. without mentioning the angles of the greater or lesser segments. These passages, Vieta, with good reason, suspects to be adulterated in the 386th page of his Oper. Math.

PROPOSITION XX.

The first words of the second part of this demonstration,

"κεκλάσθω δὴ πάλιν" are wrong translated by Mr. Briggs and Dr. Gregory, "Rursus inclinetur"; for the translation ought to be "Rursus inflectatur"; as Commandine has it. A straight line is said to be inflected either to a straight or curve line, when a straight line is drawn to this line from a point, and from the point in which it meets it, a straight line making an angle with the former, is drawn to another point, as is evident from the 90th Prop. of Euclid's Data: for thus the whole line betwixt the first and last points is inflected or broken at the point of inflection, where the two straight lines meet. And in the like sense, two straight lines are said to be inflected from two points to a third point, when they make an angle at this point; as may be seen in the description given by Pappus Alexandrinus of Apollonius's Books de Locis planis, in the preface to the 7th Book: we have made the expression fuller from the 90th Prop. of the Data.

PROPOSITION XXI.

There are two cases of this proposition, the second of which, viz. when the angles are in a segment not greater than a semicircle, is wanting in the Greek: and of this a more simple demonstration is given than that which is in Commandine, as being derived only from the first case without the help of triangles.

PROPOSITIONS XXIII. XXIV.

In Prop. 24th it is demonstrated that the segment AEB must coincide with the segment CFD, (see Commandine's figure,) and that it cannot fall otherwise, as CGD, so as to cut the other circle in a third point G, from this, that if it did, a circle could cut another in more points than two; but this ought to have been proved to be impossible in the 23d Prop., as well as that one of the segments cannot fall within the other. This part, then, is left out in the 24th, and put in its proper place, the 23d Proposition.

PROPOSITION XXV.

This proposition is divided into three cases, of which two have the same construction and demonstration; therefore it is now divided only into two cases.

PROPOSITION XXXIII.

This also in the Greek is divided into three cases, of which two, viz. one in which the given angle is acute, and the other in which it is obtuse, have exactly the same construction and demonstration; on which account, the demonstration of the last case is left out, as quite superfluous, and the addition of some unskilful editor; besides the demonstration of the case when the angle given is a right angle, is done in a roundabout way, and is therefore changed to a more simple one, as was done by Clavius.

PROPOSITION XXXV.

As the 25th and 33d Propositions are divided into more cases, so this 35th is divided into fewer cases than are necessary. Nor can it be supposed that Euclid omitted them because they are easy; as he has given the case, which by far is the easiest of them all, viz. that in which both the straight lines pass through the centre: and in the following proposition he separately demonstrates the case in which the straight line passes through the centre, and that in which it does not pass through the centre: so that it seems Theon, or some other, has thought them too long to insert: but cases that require different demonstrations, should not be left out in the Elements, as was before taken notice of: these cases are in the translation from the Arabic, and are now put into the text.

PROPOSITION XXXVII.

At the end of this, the words "in the same manner it may "be demonstrated, if the centre be in AC", are left out, being the addition of some ignorant editor.

BOOK IV.

DEFINITIONS.

When a point is in a straight line, or any other line, this point is by the Greek geometers said ἅπτεσθαι, to be upon or in that line; and when a straight line or circle meets a circle any way, the one is said ἅπτεσθαι, to meet the other: but when a straight line or circle meets a circle so as not to cut it, it is said ἐφάπτεσθαι, to touch the circle; and these two terms are never promiscuously used by them: therefore, in the 5th Definition of Book IV. the compound ἐφάπτηται must be read, instead of the simple ἅπτηται: and in the 1st, 2d, 3d, and 6th Definitions in Commandine's translation, " tangit", must be read instead of " contingit": and in the 2d and 3d Definitions of Book 3. the same change must be made: but in the Greek text of Propositions 11th, 12th, 13th, 18th, 19th, Book 3, the compound verb is to be put for the simple.

PROPOSITION IV.

In this, as also in the 8th and 13th Propositions of this Book, it is demonstrated indirectly, that the circle touches a straight line; whereas in the 17th, 33d, and 37th Propositions of Book 3. the same thing is directly demonstrated: and this way we have chosen to use in the propositions of this book, as it is shorter.

PROPOSITION V.

The demonstration of this has been spoiled by some unskilful hand: for he does not demonstrate, as is necessary, that the two straight lines which bisect the sides of the triangle at right angles, must meet one another; and, without any reason, he divides the proposition into three cases: whereas, one and the same construction and demonstration serves for them all, as Campanus has observed; which useless repetitions are now left out: the Greek text also in the corollary is manifestly vitiated, where mention is made of a given angle, though there neither is, nor can be, any thing in the proposition relating to a given angle.

PROPOSITIONS XV. XVI.

In the corollary of the first of these, the words "equilateral and equiangular" are wanting in the Greek; and in Prop. 16. instead of "the circle ABCD," ought to be read "the circumference ABCD;" where mention is made of its containing fifteen equal parts.

BOOK V.

DEFINITIONS.

III. Many of the modern mathematicians reject this definition: the very learned Dr. Barrow has explained it at large at the end of his third lecture of the year 1666, in which also he answers the objections made against it as well as the subject would allow: and at the end gives his opinion upon the whole as follows:

"I shall only add, that the author had, perhaps, no other "design in making this definition, than (that he might more "fully explain and embellish his subject) to give a general "and summary idea of ratio to beginners, by premising this "metaphysical definition, to the more accurate definition of "ratios that are the same to one another, or one of which is "greater or less than the other: I call it a metaphysical, for "it is not properly a mathematical, definition, since nothing "in mathematics depends on it, or is deduced, nor, as I "judge, can be deduced from it: and the definition of ana- "logy, which follows, viz. Analogy is the similitude of ratios, "is of the same kind, and can serve for no purpose in mathe- "matics, but only to give beginners some general, though "gross and confused, notion of analogy: but the whole of the "doctrine of ratios, and the whole of mathematics, depend "upon the accurate mathematical definitions which follow "this: to these we ought principally to attend, as the doc- "trine of ratios is more perfectly explained by them; this "third, and others like it, may be entirely spared without "any loss to geometry; as we see in the 7th book of the Ele- "ments, where the proportion of numbers to one another is "defined, and treated of, yet without giving any definition "of the ratio of numbers; though such a definition was as ne- "cessary and useful to be given in that book as in this: but in- "deed there is scarce any need of it in either of them: though "I think that a thing of so general and abstracted a nature, "and thereby the more difficult to be conceived and explained, "cannot be more commodiously defined than as the author "has done: upon which account I thought fit to explain it at "large, and defend it against the captious objections of those "who attack it." To this citation from Dr. Barrow I have nothing to add, except that I fully believe the 3d and 8th Definitions are not Euclid's, but added by some unskilful editor.

XI. It was necessary to add the word "continual" before "proportionals" in this definition; and thus it is cited in the 33d Prop. of Book 11.

After this definition, ought to have followed the definition of compound ratio, as this was the proper place for it; duplicate and triplicate ratio being species of compound ratio: but Theon has made it the 5th Def. of B. 6. where he gives an absurd and entirely useless definition of compound ratio: for this reason we have placed another definition of it betwixt the 11th and 12th of this book, which, no doubt, Euclid gave; for he cites it expressly in Prop. 23. B. 6. and which Clavius, Herigon, and Barrow, have likewise given; but they retain also Theon's, which they ought to have left out of the Elements.

XIII. This, and the rest of the definitions following, contain the explication of some terms which are used in the 5th and following books; which, except a few, are easily enough understood from the propositions of this book, where they are first mentioned: they seem to have been added by Theon, or some other. However it be, they are explained something more distinctly for the sake of learners.

PROPOSITION IV.

In the construction preceding the demonstration of this, the words ἃ ἔτυχε "any whatever," are twice wanting in the Greek, as also in the Latin translations, and are now added, as being wholly necessary.

Ibid. in the demonstration; in the Greek, and in the Latin translation of Commandine, and in that of Mr. Henry Briggs, which was published in London in 1620, together with the Greek text of the first six books, which translation in this place is followed by Dr. Gregory in his edition of Euclid, there is this sentence following, viz. "and of A and C have "been taken equimultiples K, L; and of B and D, any equi-"multiples whatever (ἃ ἔτυχε) M, N;" which is not true, the words "any whatever" ought to be left out: and it is strange that neither Mr. Briggs, who did right to leave out these words in one place of Prop. 13. of this Book, nor Dr.

Gregory, who changed them into the word "some" in three places, and left them out in a fourth of that same Prop. 13., did not also leave them out in this place of Prop. 4. and in the second of the two places where they occur in Prop. 17. of this book, in neither of which they can stand consistent with truth: and in none of all these places, even in those which they corrected in their Latin translation, have they cancelled the words ἃ ἔτυχε in the Greek text, as they ought to have done.

The same words ἃ ἔτυχε are found in four places of Prop. 11. of this Book, in the first and last of which they are necessary, but in the second and third, though they are true, they are quite superfluous; as they likewise are in the second of the two places in which they are found in the 12th Prop. and in the like places of Prop. 22, 23, of this Book; but are wanting in the last place of Prop. 23, as also in Prop. 25, Book 11.

COROLLARY PROPOSITION IV.

This corollary has been unskilfully annexed to this proposition, and has been made instead of the legitimate demonstration, which, without doubt, Theon, or some other editor, has taken away, not from this, but from its proper place in this book: the author of it designed to demonstrate, that if four magnitudes E, G, F, H be proportionals, they are also proportionals inversely; that is, G is to E, as H to F; which is true; but the demonstration of it does not in the least depend upon this 4th Prop. or its demonstration: for, when he says, "because it is demonstrated, that if K be greater than "M, L is greater than N," &c. this indeed is shewn in the demonstration of the 4th Prop. but not from this, that E, G, F, H are proportionals; for this last is the conclusion of the proposition. Wherefore these words, "because it is demon-"strated," &c. are wholly foreign to his design: and he should have proved, that if K be greater than M, L is greater than N, from this, that E, G, F, H are proportionals, and from the 5th Def. of this Book, which he has not; but is done in Prop. B, which we have given in its proper place, instead of this corollary; and another corollary is placed after the 4th Prop. which is often of use; and is necessary to the demonstration of Prop. 18. of this book.

PROPOSITION V.

In the construction which precedes the demonstration of this proposition, it is required that EB may be the same multiple of CG, that AE is of CF; that is, that EB be divided into as many equal parts as there are parts in AE equal to CF: from which it is evident, that this construction is not Euclid's, for he does not shew the way of dividing straight lines, and far less other magnitudes, into any number of equal parts, until the 9th Prop. of B. 6. and he never requires any thing to be done in the construction of which he had not before given the method of doing: for this reason, we have changed the construction to one, which, without doubt, is Euclid's, in which nothing is required but to add a magnitude to itself a certain number of times; and this is to be found in the translation from the Arabic, though the enunciation of the proposition and the demonstration are there very much spoiled. Jacobus Peletarius, who was the first, as far as I know, who took notice of this error, gives also the right construction in his edition of Euclid, after he had given the other which he blames. He says, he would not leave it out, because it was fine, and might sharpen one's genius to invent others like it: whereas there is not the least difference between the two demonstrations, except a single word in the construction, which very probably has been owing to an unskilful librarian. Clavius likewise gives both the ways; but neither he nor Peletarius takes notice of the reason why the one is preferable to the other.

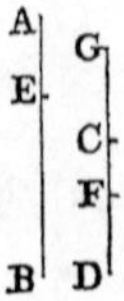

PROPOSITION VI.

There are two cases of this proposition, of which only the first and simplest is demonstrated in the Greek. And it is probable Theon thought it was sufficient to give this one, since he was to make use of neither of them in his mutilated edition of the 5th Book; and he might as well have left out the other, as also the 5th Prop. for the same reason. The demonstration of the other case is now added, because both of them, as also the 5th Prop. are necessary to the demonstration of the 18th Prop. of this Book. The translation from the Arabic gives both cases briefly.

PROPOSITION A.

This proposition is frequently used by geometers, and it is necessary in the 25th Prop. of this Book, 31st of the 6th, and 34th of the 11th, and 15th of the 12th Book: it seems to have been taken out of the Elements by Theon, because it appeared evident enough to him, and others, who substitute the confused and indistinct idea the vulgar have of proportionals, in place of that accurate idea which is to be got from the 5th Def. of this Book. Nor can there be any doubt that Eudoxus or Euclid gave it a place in the Elements, when we see the 7th and 9th of the same Book demonstrated, though they are quite as easy and evident as this. Alphonsus Borellus takes occasion from this proposition to censure the 5th Definition of this Book very severely, but most unjustly. In p. 126. of his Euclid Restored, printed at Pisa in 1658, he says, "Nor can even this least degree of knowledge be ob-"tained from the aforesaid property," viz. that which is contained in 5th Def. 5. "That, if four magnitudes be propor-"tionals, the third must necessarily be greater than the "fourth, when the first is greater than the second: as Cla-"vius acknowledges in the 16th Prop. of the 5th Book of the "Elements." But though Clavius makes no such acknowledgment expressly, he has given Borellus a handle to say this of him; because when Clavius, in the above-cited place, censures Commandine, and that very justly, for demonstrating this proposition by help of the 16th of the 5th; yet he himself gives no demonstration of it, but thinks it plain from the nature of proportionals, as he writes in the end of the 14th and 16th Prop. B. 5. of his edition, and is followed by Herigon in Schol. 1. Prop. 4. B. 5. as if there was any nature of proportionals antecedent to that which is to be derived and understood from the definition of them: and, indeed, though it is very easy to give a right demonstration of it, nobody, as far as I know, has given one, except the learned Dr. Barrow, who, in answer to Borellus's objection, demonstrates it indirectly, but very briefly and clearly, from the 5th Def. in the 322d page of his Lect. Mathem. from which definition it may also be easily demonstrated directly: on which account we have placed it next to the propositions concerning equimultiples.

PROPOSITION B.

This also is easily deduced from the 5th Def. B. 5. and therefore is placed next to the other; for it was very ignorantly made a corollary from the 4th Prop. of this Book. See the note on that corollary.

PROPOSITION C.

This is frequently made use of by geometers, and is necessary to the 5th and 6th Propositions of the 10th Book. Clavius, in his notes subjoined to the 8th Def. of Book 5. demonstrates it only in numbers, by help of some of the propositions of the 7th Book; in order to demonstrate the property contained in the 5th Def. of the 5th Book, when applied to numbers, from the property of proportionals contained in the 20th Def. of the 7th Book: and most of the commentators judge it difficult to prove that four magnitudes which are proportionals according to the 20th Def. of 7th Book, are also proportionals according to the 5th Def. of 5th Book. But this is easily made out as follows:

First, if A, B, C, D be four magnitudes, such that A is the same multiple, or the same part of B, which C is of D: A, B, C, D are proportionals: This is demonstrated in proposition C.

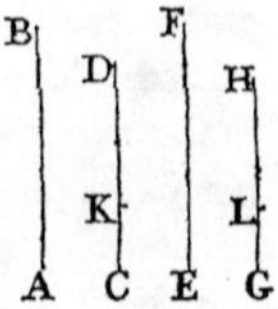

Secondly, if AB contain the same parts of CD that EF does of GH; in this case likewise AB is to CD, as EF to GH.

Let CK be a part of CD, and GL the same part of GH; and let AB be the same multiple of CK, that EF is of GL: therefore*, AB is to CK, as EF to GL: and CD, GH are equimultiples of CK, GL, the second and fourth; wherefore*, AB is to CD, as EF to GH.

* C. 5.

* Cor. 4. 5.

And if four magnitudes be proportionals according to the 5th Def. of Book 5. they are also proportionals according to the 20th Def. of Book 7.

First, if A be to B, as C to D; then if A be any multiple or part of B*, C is the same multiple or part of D.

* D, 5.

Next, if AB be to CD, as EF to GH: then if AB contain any parts of CD, EF contains the same parts of GH: for let CK be

a part of CD, and GL the same part of GH, and let AB be a multiple of CK: EF is the same multiple of GL: take M the same multiple of GL that AB is of CK; there-
* C. 5. fore *, AB is to CK, as M to GL: and CD, GH, are
* Cor. 4. 5. equimultiples of CK, GL; wherefore *, AB is to CD, as M to GH. And, by the hypothesis, AB is to CD,
* 9. 5. as EF to GH; therefore, M is equal * to EF, and consequently EF is the same multiple of GL that AB is of CK.

PROPOSITION D.

This is not unfrequently used in the demonstration of other propositions, and is necessary in that of Prop. 9. B. 6. It seems Theon has left it out for the reasons mentioned in the notes at Prop. A.

PROPOSITION VIII.

In the demonstration of this, as it is now in the Greek, there are two cases, (see the demonstration in Hervagius, or Dr. Gregory's edition,) of which the first is that in which AE is less than EB; and in this it necessarily follows, that HΘ the multiple of EB is greater than ZH the same multiple of AE, which last multiple, by the construction, is greater than Δ; whence also HΘ must be greater than Δ: but in the second case, viz. that in which EB is less than AE, though ZH be greater than Δ, yet HΘ may be less than the same Δ; so that there cannot be taken a multiple of Δ which is the first that is greater than K or HΘ, because Δ itself is greater than it: upon this account, the author of this demonstration found it necessary to change one part of the construction that was made use of in the first case: but he has, without any necessity, changed also another part of it, viz. when he orders to take N that multiple of Δ which is the first that is greater than ZH; for he might have taken that multiple of Δ which is the first that is greater than HΘ, or K, as was done in the first case: he likewise brings in this K into the demonstration of both cases, without any reason; for it serves to no purpose but to lengthen the demonstration. There is also a

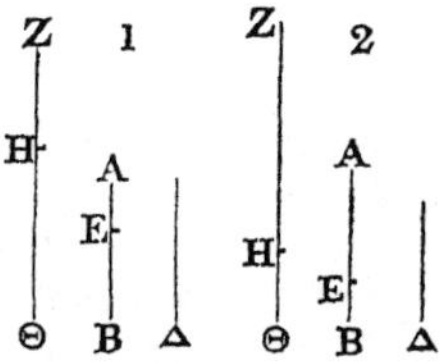

third case which is not mentioned in this demonstration, viz. that in which AE in the first, or EB in the second of the two other cases is greater than Δ: and in this any equimultiples, as the doubles of AE, EB, are to be taken, as is done in this edition, where all the cases are at once demonstrated: and from this it is plain that Theon, or some other unskilful editor, has vitiated this proposition.

PROPOSITION IX.

Of this there is given a more explicit demonstration than that which I found in the Elements.

PROPOSITION X.

It was necessary to give another demonstration of this proposition, because that which is in the Greek and Latin, or other editions, is not legitimate: for the words *greater, the same,* or *equal, lesser,* have a quite different meaning when applied to magnitudes and ratios, as is plain from the 5th and 7th definitions of Book 5. By the help of these let us examine the demonstration of the 10th Prop. which proceeds thus: "Let A have to C a greater ratio than B to C: I say that A "is greater than B: for if it be not greater, it is either equal "or less. But A cannot be equal to B, because then each of "them would have the same ratio to C; but they have not. "Therefore A is not equal to B." The force of which reasoning is this: if A had to C the same ratio that B has to C, then if any equimultiples whatever of A and B be taken, and any multiple whatever of C; if the multiple of A be greater than the multiple of C, then* the multiple of B is also greater than that of C; but from the hypothesis, that A has a greater ratio to C, than B has to C, there must † be certain equimultiples of A and B, and some multiple of C, such that the multiple of A is greater than the multiple of C, but the multiple of B is not greater than the same multiple of C, and this proposition directly contradicts the preceding; wherefore A is not equal to B. The demonstration of the 10th Prop. goes on thus: "But neither is A less than B; because then A would have a "less ratio to C than B has to it: but it has not a less ratio, "therefore A is not less than B", &c. Here it is said, that "A would have a less ratio to C than B has to C", or, which is the same thing, that B would have a greater ratio to C than

* 5 Def. 5.

† 7 Def. 5.

* 7 Def. 5. A to C; that is *, there must be some equimultiples of B and A, and some multiple of C, such that the multiple of B is greater than the multiple of C, but the multiple of A is not greater than it: and it ought to have been proved, that this can never happen if the ratio of A to C be greater than the ratio of B to C; that is, it should have been proved, that, in this case, the multiple of A is always greater than the multiple of C, whenever the multiple of B is greater than the multiple of C; for when this is demonstrated, it will be evident that B cannot have a greater ratio to C, than A has to C, or, which is the same thing, that A cannot have a less ratio to C than B has to C. But this is not at all proved in the 10th Prop.: but if the 10th were once demonstrated, it would immediately follow from it, but cannot without it be easily demonstrated, as he that tries to do it will find. Wherefore the 10th Prop. is not sufficiently demonstrated. And it seems that he who has given the demonstration of the 10th Prop. as we now have it instead of that which Eudoxus or Euclid had given, has been deceived in applying what is manifest, when understood of magnitudes, unto ratios, viz. that a magnitude cannot be both greater and less than another. That those things which are equal to the same are equal to one another, is a most evident axiom when understood of magnitudes; yet Euclid does not make use of it to infer, that those ratios, which are the same to the same ratio, are the same to one another, but explicitly demonstrates this in Prop. 11. of Book 5. The demonstration we have given of the 10th Prop. is no doubt the same with that of Eudoxus or Euclid, as it is immediately and directly derived from the definition of a greater ratio, viz. the 7th of the 5th.

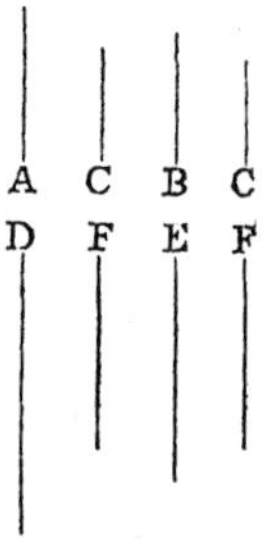

The above-mentioned proposition, viz. If A have to C a greater ratio than B to C; and if of A and B there be taken certain equimultiples, and some multiple of C; then if the multiple of B be greater than the multiple of C, the multiple of A is also greater than the same, is thus demonstrated.

Let D, E be equimultiples of A, B, and F a multiple of C, such, that E the multiple of B is greater than F; D the multiple of A is also greater than F.

† 10. 5. Because A has a greater ratio to C, than B to C, A † is greater

than B, therefore D the multiple of A is greater than E the same multiple of B: and E is greater than F; much more therefore D is greater than F.

PROPOSITION XIII.

In Commandine's, Brigg's, and Gregory's translations at the beginning of this demonstration, it is said, "and the multiple "of C is greater than the multiple of D; but the multiple of "E is not greater than the multiple of F": which words are a literal translation from the Greek: but the sense evidently requires that it be read, "so that the multiple of C be greater "than the multiple of D; but the multiple of E be not greater "than the multiple of F." And thus this place was restored to the true reading in the first editions of Commandine's Euclid, printed in 8vo. at Oxford; but in the later editions, at least in that of 1747, the error of the Greek text was kept in.

There is a corollary added to Prop. 13, as it is necessary to the 20th and 21st Prop. of this Book, and is as useful as the proposition.

PROPOSITION XIV.

The two cases of this, which are not in the Greek, are added; the demonstration of them not being exactly the same with that of the first case.

PROPOSITION XVII.

The order of the words in a clause of this is changed to one more natural: as was also done in Prop. 11.

PROPOSITION XVIII.

The demonstration of this is none of Euclid's, nor is it legitimate; for it depends upon this hypothesis, that to any three magnitudes, two of which, at least, are of the same kind, there may be a fourth proportional: which, if not proved, the demonstration now in the text is of no force; but this is assumed without any proof; nor can it, as far as I am able to discern, be demonstrated by the propositions preceding this: so far is it from deserving to be reckoned an axiom, as Clavius, after other commentators, would have it, at the end of the definitions of the 5th Book. Euclid does not demonstrate it, nor does he show how to find the fourth proportional, before the 12th Prop. of the 6th Book: and he never assumes any thing in the demonstration of a proposition which he had not before demonstrated; at least, he assumes nothing the existence of which is not evidently possible; for a certain conclusion can never be deduced by the means of an uncertain proposition: upon this account we have given a legitimate demonstration of this proposition instead of that in the Greek and other editions, which very probably Theon, at least some other, has put in the place of Euclid's, because he thought it too prolix: and as the 17th Prop. of which this 18th is the converse, is demonstrated by help of the first and second propositions of this book; so, in the demonstration now given of the 18th, the 5th Prop. and both cases of the 6th are necessary, and these two propositions are the converses of the 1st and 2d. Now the 5th and 6th do not enter into the demonstration of any proposition in this book as we now have it: nor can they be of use in any proposition of the Elements, except in this 18th, and this is a manifest proof, that Euclid made use of them in his demonstration of it, and that the demonstration now given, which is exactly the converse of that of the 17th, as it ought to be, differs nothing from that of Eudoxus or Euclid: for the 5th and 6th have undoubtedly been put into the 5th Book for the sake of some propositions in it, as all the other propositions about equimultiples have been.

Hieronymus Saccherius, in his book named "Euclides ab "omni nævo vindicatus", printed at Milan, ann. 1733, in 4to. acknowledges this blemish in the demonstration of the 18th; and that he may remove it, and render the demonstration we now have of it legitimate, he endeavours to demonstrate the following proposition, which is in page 115 of this book, viz.

"Let A, B, C, D be four magnitudes of which the two first "are of one kind, and also the two others either of the same "kind with the two first, or of some other the same kind with "one another. I say the ratio of the third C to the fourth "D, is either equal to, or greater, or less, than the ratio of "the first A to the second B."

And after two propositions premised as lemmas, he proceeds thus:

"Either among all the possible equimultiples of the first

" A, and of the third C, and, at the same time, among all
" the possible equimultiples of the second B, and of the
" fourth D, there can be found some one multiple EF of the
" first A, and one IK of the second B, that are equal to one
" another; and also (in the same case) some one multiple GH
" of the third C, equal to LM the multiple of the fourth D, or
" such equality is *no where* to be found. If the first case
" happen [i. e. if such equali-
" ty be to be found], it is
" manifest from what is before
" demonstrated, that A is to B,
" as C to D; but if such si-
" multaneous equality be not
" to be found upon both

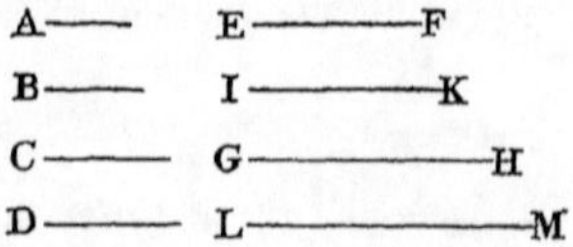

" sides, it will be found either upon one side, as upon the
" side of A [and B]; or it will be found upon neither side:
" if the first happen; therefore (from Euclid's definition
" of greater and lesser ratio foregoing) A has to B a greater
" or less ratio than C to D; according as GH the multiple
" of the third C is less, or greater, than LM the multiple
" of the fourth D: but if the second case happen; therefore
" upon the one side, as upon the side of A the first and B the
" second, it may happen that the multiple EF [viz. of the
" first] may be less than IK the multiple of the second, while,
" on the contrary, upon the other side [viz. of C and D], the
" multiple GH [of the third C] is greater than the other mul-
" tiple LM [of the fourth D]: and then (from the same defi-
" nition of Euclid) the ratio of the first A to the second B is
" less than the ratio of the third C to the fourth D; or on the
" contrary."

" Therefore the axiom [i. e. the proposition before set
" down] remains demonstrated," &c.

Not in the least; but it still remains undemonstrated: for what he says may happen, may, in innumerable cases, never happen; and therefore his demonstration does not hold: for example, if A be the side, and B the diameter of a square; and C the side, and D the diameter of another square; there can in no case be any multiple of A equal to any of B; nor any one of C equal to one of D, as is well known; and yet it can never happen that when any multiple of A is greater than a multiple of B, the multiple of C can be less than the multiple of D, nor when the multiple of A is less than that of B,

the multiple of C can be greater than that of D, viz. taking equimultiples of A and C, and equimultiples of B and D: for A, B, C, D are proportionals, and so if the multiple of A be
* 5 Def. 5. greater, &c. than that of B, so must that of C * be greater, &c. than that of D.

The same objection holds good against the demonstration which some give of the first Prop. of the 6th Book, which we have made against this of the 18th Prop. because it depends upon the same insufficient foundation with the other.

PROPOSITION XIX.

A corollary is added to this, which is as frequently used as the proposition itself. The corollary which is subjoined to it in the Greek, plainly shews that the 5th Book has been vitiated by editors who were not geometers: for the conversion of ratios does not depend upon this 19th, and the demonstration which several of the commentators on Euclid give of conversion is not legitimate, as Clavius has rightly observed, who has given a good demonstration of it which we have put in proposition E; but he makes it a corollary from the 19th, and begins it with the words, "Hence it easily follows," though it does not at all follow from it.

PROPOSITIONS XX. XXI. XXII. XXIII. XXIV.

The demonstrations of the 20th and 21st Propositions are shorter than those Euclid gives of easier propositions, either in the preceding or following books. Wherefore it was proper to make them more explicit, and the 22d and 23d Propositions are, as they ought to be, extended to any number of magnitudes: and in like manner may the 24th be, as is taken notice of in the corollary; and another corollary is added, as useful as the proposition, and the words "any whatever" are applied near the end of Prop. 23, which are wanting in the Greek text and the translations from it.

In a paper written by Philippus Naudæus, and published, after his death, in the History of the Royal Academy of Sciences of Berlin, anno 1745, page 50, the 23d Prop. of the 5th Book is censured as being obscurely enunciated, and, because of this, prolixly demonstrated: the enunciation there given is not Euclid's but Tacquet's, as he acknowledges, which, though not so well expressed, is, upon the matter, the same with that which is now in the Elements. Nor is there any thing obscure in it, though the author of the paper has set down the proportionals in a disadvantageous order, by which it appears to be obscure: but no doubt Euclid enunciated this 23d, as well as the 22d, so as to extend it to any number of magnitudes, which, taken two and two, are proportionals, and not of six only: and to this general case the enunciation which Naudæus gives cannot be well applied.

The demonstration which is given of this 23d, in that paper, is quite wrong; because, if the proportional magnitudes be plane or solid figures, there can be no rectangle (which he improperly calls a *product*) conceived to be made by any two of them: and if it should be said, that in this case straight lines are to be taken which are proportional to the figures, the demonstration would this way become much longer than Euclid's: but even though his demonstration had been right, who does not see that it could not be made use of in the 5th Book?

PROPOSITIONS F. G. H. K.

These propositions are annexed to the 5th book, because they are frequently made use of by both ancient and modern geometers: and in many cases, compound ratios cannot be brought into demonstration, without making use of them.

Whoever desires to see the doctrine of ratios delivered in this 5th Book solidly defended, and the arguments brought against it by And. Tacquet, Alph. Borellus, and others, fully refuted, may read Dr. Barrow's Mathematical Lectures, viz. the 7th and 8th of the year 1666.

The 5th book being thus corrected, I most readily agree to what the learned Dr. Barrow says *, " That there is nothing " in the whole body of the Elements of a more subtile invention, nothing more solidly established, and more accurately " handled, than the doctrine of proportionals." And there is some ground to hope, that geometers will think that this could not have been said with as good reason, since Theon's time, till the present.

* Page 336.

BOOK VI.

DEFINITIONS.

II. and V. The 2d definition does not seem to be Euclid's, but some unskilful editor's: for there is no mention made by Euclid, nor, as far as I know, by any other geometer, of reciprocal figures: it is obscurely expressed, which made it proper to render it more distinct: it would be better to put the following definition in place of it, viz.

DEF. II.—*Two magnitudes are said to be reciprocally proportional to two others, when one of the first is to one of the other magnitudes as the remaining one of the last two is to the remaining one of the first.*

But the 5th definition, which, since Theon's time, has been kept in the Elements, to the great detriment of learners, is now justly thrown out of them, for the reasons given in the notes on the 23d Prop. of this Book.

PROPOSITIONS I. and II.

To the first of these a corollary is added, which is often used: and the enunciation of the second is made more general.

PROPOSITION III.

A second case of this, as useful as the first, is given in Prop. A.; viz. the case in which the exterior angle of a triangle is bisected by a straight line: the demonstration of it is very like to that of the first case, and upon this account may, probably, have been left out, as also the enunciation, by some unskilful editor. At least it is certain, that Pappus makes use of this case, as an elementary proposition, without a demonstration of it, in Prop. 39 of his 7th Book of Mathematical Collections.

PROPOSITION VI.

To this a case is added which occurs not unfrequently in demonstrations.

PROPOSITION VIII.

It seems plain that some editor has changed the demonstration that Euclid gave of this proposition: for, after he has demonstrated that the triangles are equiangular to one another, he particularly shews that their sides about the equal angles are proportionals, as if this had not been done in the demonstration of the 4th Prop. of this Book: this superfluous part is not found in the translation from the Arabic, and is now left out.

PROPOSITION IX.

This is demonstrated in a particular case, viz. that in which the third part of a straight line is required to be cut off; which is not at all like Euclid's manner: besides, the author of the demonstration, from four magnitudes being proportionals, concludes that the third of them is the same multiple of the fourth, which the first is of the second; now, this is no where demonstrated in the 5th Book, as we now have it; but the editor assumes it from the confused notion which the vulgar have of proportionals: on this account it was necessary to give a general and legitimate demonstration of this proposition.

PROPOSITION XVIII.

The demonstration of this seems to be vitiated. For the proposition is demonstrated only in the case of quadrilateral figures, without mentioning how it may be extended to figures of five or more sides: besides, from two triangles being equiangular, it is inferred, that a side of the one is to the homologous side of the other, as another side of the first is to the side homologous to it of the other, without permutation of the proportionals; which is contrary to Euclid's manner, as is clear from the next proposition; and the same fault occurs again in the conclusion, where the sides about the equal angles are not shewn to be proportionals by reason of again neglecting permutation. On these accounts, a demonstration is given in Euclid's manner, like to that he makes use of in the 20th Prop. of this Book; and it is extended to five-sided figures, by which it may be seen how to extend it to figures of any number of sides.

PROPOSITION XXIII.

Nothing is usually reckoned more difficult in the elements of geometry by learners, than the doctrine of compound ratio, which Theon has rendered absurd and ungeometrical by substituting the 5th Definition in the 6th Book in place of the right Definition, which, without doubt, Eudoxus or Euclid gave, in its proper place, after the definition of triplicate ratio, &c. in the 5th Book. Theon's definition is this: a ratio is said to be compounded of ratios ὅταν αἱ τῶν λόγων πηλικότητες ἐφ᾽ ἑαυτὰς πολλαπλασιασθεῖσαι ποιῶσι τινα: which Commandine thus translates: "Quando rationum quantitates inter se "multiplicatæ aliquam efficient rationem;" that is, when the quantities of the ratios being multiplied by one another make a certain ratio. Dr. Wallis translates the word πηλικότητες "rationum exponentes," the exponents of the ratios: and Dr. Gregory renders the last words of the definition by "illius facit quantitatem," makes the quantity of that ratio: but in whatever sense the "quantities" or "exponents of the ratios," and their "multiplication," be taken, the definition will be ungeometrical and useless: for there can be no multiplication but by a number: now the quantity or exponent of a ratio (according to Eutochius in his Comment. on Prop. 4. Book 2. of Arch. de Sph. et Cyl. and as the moderns explain that term) is the number which multiplied into the consequent term of a ratio, produces the antecedent, or, which is the same thing, the number which arises by dividing the antecedent by the consequent; but there are many ratios, such, that no number can arise from the division of the antecedent by the consequent; ex. gr. the ratio which the diameter of a square has to the side of it; and the ratio which the circumference of a circle has to its diameter, and such like. Besides, there is not the least mention made of this definition in the writings of Euclid, Archimedes, Apollonius, or other ancients, though they frequently make use of compound ratio: and in this 23d Prop. of the 6th Book, where compound ratio is first mentioned, there is not one word which can relate to this definition, though here, if in any place, it was necessary to be brought in; but the right definition is expressly cited in these words: "but the ratio of K to M is "compounded of the ratio of K to L, and of the ratio of L to "M." This definition therefore of Theon is quite useless and absurd: for that Theon brought it into the Elements can scarce be doubted; as it is to be found in his Commentary upon Ptolemy's Μεγάλη Σύνταξις, page 62, where he also, gives a childish explication of it, as agreeing only to such ratios as can be expressed by numbers; and from this place the definition and explication have been exactly copied and prefixed to the definitions of the 6th Book, as appears from Hervagius's edition: but Zambertus and Commandine, in their Latin translations, subjoin the same to these definitions. Neither Campanus, nor, as it seems, the Arabic manuscripts from which he made his translation, have this definition. Clavius, in his observations upon it, rightly judges that the definition of compound ratio might have been made after the same manner in which the definitions of duplicate and triplicate ratio are given, viz. "that as in several magnitudes that "are continual proportionals, Euclid named the ratio of the "first to the third, the duplicate ratio of the first to the se-"cond; and the ratio of the first to the fourth, the triplicate "ratio of the first to the second; that is, the ratio com-"pounded of two or three intermediate ratios that are equal "to one another, and so on; so, in like manner, if there be "several magnitudes of the same kind, following one another, "which are not continual proportionals, the first is said to "have to the last the ratio compounded of all the intermediate "ratios,—only for this reason, that these intermediate ratios "are interposed betwixt the two extremes, viz. the first and "last magnitudes; even, as, in the 10th Def. of the 5th "Book, the ratio of the first to the third was called the du-"plicate ratio, merely upon account of two ratios being in-"terposed betwixt the extremes, that are equal to one an-"other: so that there is no difference betwixt the compound-"ing of ratios, and the duplication or triplication of them "which are defined in the 5th Book, but that in the duplica-"tion, triplication, &c. of ratios, all the interposed ratios are "equal to one another; whereas, in the compounding of "ratios, it is not necessary that the intermediate ratios, "should be equal to one another." Also Mr. Edmund Scarburg, in his English translation of the first six books, pages 238, 266, expressly affirms, that the 5th Def. of the 6th Book is supposititious, and that the true definition of compound ratio is contained in the 10th Def. of the 5th Book, viz. the definition of duplicate ratio, or to be understood from it, to wit, in the same manner as Clavius has explained it in the preceding citation. Yet these, and the rest of the moderns, do notwithstanding retain this 5th Def. of the 6th Book, and illustrate and explain it by long commentaries, when they ought rather to have taken it quite away from the Elements.

For, by comparing Def. 5. Book 6. with Prop. 5. Book 8. it will clearly appear that this definition has been put into the Elements in place of the right one, which has been taken out of them: because, in Prop. 5. Book 8. it is demonstrated, that the plane number of which the sides are C, D, has to the plane number of which the sides are E, Z, (see Hervagius's or Gregory's edition,) the ratio which is compounded of the ratios of their sides; that is, of the ratios of C to E, and D to Z; and by Def. 5. Book 6. and the explication given of it by all the commentators, the ratio which is compounded of the ratios of C to E, and D to Z, is the ratio of the product made by the multiplication of the antecedents C, D, to the product by the consequents E, Z, that is, the ratio of the plane number of which the sides are C, D, to the plane number of which the sides are E, Z. Wherefore the proposition which is the 5th Def. of Book 6. is the very same with the 5th Prop. of Book 8. and therefore it ought necessarily to be cancelled in one of these places; because it is absurd, that the same proposition should stand as a definition in one place of the Elements, and be demonstrated in another place of them. Now, there is no doubt that Prop. 5. Book 8. should have a place in the Elements, as the same thing is demonstrated in it concerning plane numbers, which is demonstrated in Prop. 23. Book 6. of equiangular parallelograms; wherefore Def. 5. Book 6. ought not to be in the Elements. And from this it is evident, that this definition is not Euclid's, but Theon's, or some other unskilful geometer's.

But nobody, as far as I know, has hitherto shewn the true use of compound ratio, or for what purpose it has been introduced into geometry; for every proposition in which compound ratio is made use of, may without it be both enunciated and demonstrated. Now the use of compound ratio consists wholly in this, that by means of it circumlocutions may be avoided, and thereby propositions may be more briefly either enunciated or demonstrated, or both may be done; for instance, if this 23d Prop. of the 6th Book were to be enunciated, without mentioning compound ratio, it might be done as follows: if two parallelograms be equiangular, and if as a side of the first to a side of the second, so any assumed straight line be made to a second straight line; and as the other side of the first to the other side of the second, so the second straight line be made to a third: the first parallelogram is to the second, as the first straight line to the third. And the demonstration would be exactly the same as we now have it. But the ancient geometers when they observed this enunciation could be made shorter, by giving a name to the ratio which the first straight line has to the last, by which name the intermediate ratios might likewise be signified, of the first to the second, and of the second to the third, and so on, if there were more of them, they called this ratio of the first to the last, the ratio compounded of the ratios of the first to the second, and of the second to the third straight line; that is, in the present example, of the ratios which are the same with the ratios of the sides, and by this they expressed the proposition more briefly thus: if there be two equiangular parallelograms, they have to one another the ratio which is the same with that which is compounded of ratios that are the same with the ratios of the sides; which is shorter than the preceding enunciation, but has precisely the same meaning. Or yet shorter thus: equiangular parallelograms have to one another the ratio which is the same with that which is compounded of the ratios of their sides. And these two enunciations, the first especially, agree to the demonstration which is now in the Greek. The proposition may be more briefly demonstrated, as Candalla does, thus: let ABCD, CEFG be two equiangular parallelograms, and complete the parallelogram CDHG: then, because there are three parallelograms AC, CH, CF, the first AC (by the definition of compound ratio) has to the third CF, the ratio which is compounded of the ratio of the first AC to the second CH, and of the ratio of CH to the third CF; but the parallelogram AC is to the parallelogram CH, as the straight line BC to CG: and the parallelogram CH is to CF, as the straight line CD is to CE; therefore the parallelogram AC has to CF the ratio which is compounded of ratios that are the same with the ratios of the sides. And to this demonstration agrees the enunciation which is at present in the text, viz. equiangular parallelograms

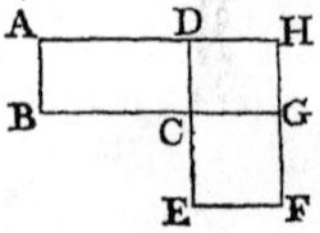

have to one another the ratio which is compounded of the ratios of the sides; for the vulgar reading, "which is compounded of their sides", is absurd. But, in this edition, we have kept the demonstration which is in the Greek text, though not so short as Candalla's; because the way of finding the ratio which is compounded of the ratios of the sides, that is, of finding the ratio of parallelograms, is shewn in that, but not in Candalla's demonstration; whereby beginners may learn, in like cases, how to find the ratio which is compounded of two or more given ratios.

From what has been said, it may be observed, that in any magnitudes whatever of the same kind A, B, C, D, &c. the ratio compounded of the ratios of the first to the second, of the second to the third, and so on to the last, is only a name or expression by which the ratio which the first A has to the last D is signified, and by which, at the same time, the ratios of all the magnitudes A to B, B to C, C to D, from the first to the last, to one another, whether they be the same, or be not the same, are indicated; as in magnitudes which are continual proportionals, A, B, C, D, &c. the duplicate ratio of the first to the second is only a name, or expression, by which the ratio of the first A to the third C is signified, and by which, at the same time, is shewn, that there are two ratios of the magnitudes from the first to the last, viz. of the first A to the second B, and of the second B to the third or last C, which are the same with one another; and the triplicate ratio of the first to the second is a name or expression by which the ratio of the first A to the fourth D is signified, and by which, at the same time, is shewn, that there are three ratios of the magnitudes from the first to the last, viz. of the first A to the second B, and of B to the third C, and of C to the fourth or last D, which are all the same with one another; and so in the case of any other multiplicate ratios. And that this is the right explication of the meaning of these ratios is plain from the definitions of duplicate and triplicate ratio, in which Euclid makes use of the word λέγεται, "is said to be", or "is called"; which word, he, no doubt, made use of also in the definition of compound ratio, which Theon, or some other, has expunged from the Elements; for the very same word is still retained in the wrong definition of compound ratio, which is now the 5th of the 6th Book: but in the citation of these definitions it is sometimes retained, as in the demonstration of

Prop. 19. B. 6. "the first is said to have, ἔχειν λέγεται, to "the third the duplicate ratio", &c. which is wrong translated by Commandine and others, "has", instead of "is said to "have", and sometimes it is left out, as in the demonstration of Prop. 33. of the 11th Book, in which we find, "the first "has, ἔχει, to the third the triplicate ratio"; but without doubt ἔχει, "has", in this place signifies the same as ἔχειν λέγεται, "is said to have": so likewise in Prop. 23. B. 6. we find this citation, "but the ratio of K to M is compounded, "σύγκειται, of the ratio of K to L, and the ratio of L to M", which is a shorter way of expressing the same thing, which, according to the definition, ought to have been expressed by σύγκεισθαι λέγεται, is said to be compounded.

From these remarks, together with the propositions subjoined to the 5th Book, all that is found concerning compound ratio, either in the ancient or modern geometers, may be understood and explained.

PROPOSITION XXIV.

It seems that some unskilful editor has made up this demonstration, as we now have it, out of two others; one of which may be made from the 2d Prop. and the other from the 4th of this Book. For after he has, from the 2d of this Book, and composition and permutation, demonstrated, that the sides about the angle common to the two parallelograms, are proportionals, he might have immediately concluded, that the sides about the other equal angles were proportionals, viz. from Prop. 34. B. 1. and Prop. 7. B. 5. This he does not, but proceeds to shew, that the triangles and parallelograms are equiangular: and in a tedious way, by help of Prop. 4, of this Book, and the 22d of Book 5. deduces the same conclusion: From which, it is plain, that this ill-composed demonstration is not Euclid's: these superfluous things are now left out, and a more simple demonstration is given from the 4th Prop. of this Book, the same which is in the translation from the Arabic, by help of the 2d Prop. and composition; but in this the author neglects permutation, and does not shew the parallelograms to be equiangular, as it is proper to do for the sake of beginners.

PROPOSITION XXV.

It is very evident, that the demonstration which Euclid had given of this proposition has been vitiated by some unskilful hand: for, after this editor had demonstrated, that "as the "rectilineal figure ABC is to the rectilineal figure KGH, so is "the parallelogram BE to the parallelogram EF"; nothing more should have been added but this, "and the rectilineal "figure ABC is equal to the parallelogram BE: therefore the "rectilineal KGH is equal to the parallelogram EF", viz. from Prop. 14. Book 5. But betwixt these two sentences he has inserted this, "wherefore, by permutation, as the recti-"lineal figure ABC to the parallelogram BE, so is the recti-"lineal KGH to the parallelogram EF"; by which it is plain, he thought it was not so evident to conclude, that the second of four proportions is equal to the fourth from the equality of the first and third, which is a thing demonstrated in the 14th Prop. of B. 5. as to conclude that the third is equal to the fourth, from the equality of the first and second, which is no where demonstrated in the Elements as we now have them: but though this proposition, viz. the third of four proportionals, is equal to the fourth, if the first be equal to the second, had been given in the Elements by Euclid, as very probably it was, yet he would not have made use of it in this place, because, as was said, the conclusion would have been immediately deduced without this superfluous step by permutation: this we have shewn at the greater length, both because it affords a certain proof of the vitiation of the text of Euclid; for the very same blunder is found twice in the Greek text of Prop. 23. Book 11., and twice in Prop. 2. Book 12., and in the 5th, 11th, 12th, and 18th of that Book; in which places of Book 12., except the last of them, it is rightly left out in the Oxford edition of Commandine's translation: and also that geometers may beware of making use of permutation in the like cases; for the moderns not unfrequently commit this mistake, and among others Commandine himself in his commentary on Prop. 5. Book 3. p. 6. b. of Pappus Alexandrinus, and in other places: the vulgar notion of proportionals has, it seems, pre-occupied many so much, that they do not sufficiently understand the true nature of them.

Besides, though the rectilineal figure ABC, to which another is to be made similar, may be of any kind whatever: yet in the demonstration the Greek text has "triangle" instead of "rectilineal figure," which error is corrected in the above-named Oxford edition.

PROPOSITION XXVII.

The second case of this has ἄλλως, "otherwise," prefixed to it, as if it were a different demonstration, which probably has been done by some unskilful librarian. Dr. Gregory has rightly left it out: the scheme of this second case ought to be marked with the same letters of the alphabet which are in the scheme of the first, as is now done.

PROPOSITIONS XXVIII. XXIX.

These two problems, to the first of which the 27th Prop. is necessary, are the most general and useful of all in the Elements, and are most frequently made use of by the ancient geometers in the solution of other problems; and therefore are very ignorantly left out by Tacquet and Dechales in their editions of the Elements, who pretend that they are scarce of any use: the cases of these problems, wherein it is required to apply a rectangle which shall be equal to a given square, to a given straight line, either deficient or exceeding by a square; as also to apply a rectangle which shall be equal to another given, to a given straight line, deficient or exceeding by a square, are very often made use of by geometers: and on this account, it is thought proper, for the sake of beginners, to give their constructions as follow:

1. *To apply a rectangle, which shall be equal to a given square, to a given straight line, deficient by a square: but the given square must not be greater than that upon the half of the given line.*

Let AB be the given straight line, and let the square upon the given straight line C, be that to which the rectangle to be applied must be equal, and this square by the determination is not greater than that upon half of the straight line AB.

Bisect AB in D, and if the square upon AD be equal to the square upon C, the thing required is done: but if it be not equal to it, AD must be greater than C, according to the determination: draw DE at right angles to AB, and make it equal to C; produce ED to F, so that EF be equal to AD or DB; and from the centre E, at the distance EF, describe a circle meeting AB in G; and upon GB, describe the square GBKH, and complete the rectangle AGHL; also join EG: and because AB is bisected in D, the rectangle AG, GB together with the square of DG, is equal* to (the square of DB, that is, of EF or EG, that is, to) the squares of ED, DG: take away the square of DG from each of these equals; therefore the remaining rectangle AG, GB is equal to the square of ED, that is, of C: but the rectangle AG, GB is the rectangle AH, because GH is equal to GB; therefore the rectangle AH is equal to the given square upon the straight line C. Wherefore the rectangle AH, equal to the given square upon C, has been applied to the given straight line AB, deficient by the square GK. Which was to be done.

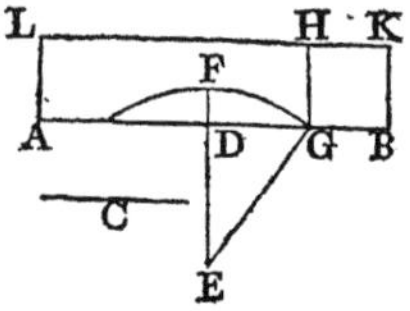

* 5. 2.

2. *To apply a rectangle which shall be equal to a given square, to a given straight line, exceeding by a square.*

Let AB be the given straight line, and let the square upon the given straight line C, be that to which the rectangle to be applied must be equal.

Bisect AB in D, and draw BE at right angles to it, so that BE be equal to C; and having joined DE, from the centre D, at the distance DE, describe a circle meeting AB produced in G; upon BG describe the square BGHK, and complete the rectangle AGHL. And because AB is bisected in D, and produced to G, the rectangle AG, GB together with the square of DB, is equal* to (the square of DG, or DE, that is to) the squares of EB, BD. From each of these equals take the square of DB; therefore, the remaining rectangle AG, GB is equal to the square of BE, that is, to the square upon C. But the rectangle AG, GB is the rectangle AH, because GH is equal to GB: therefore the rectangle AH is equal to the square upon C. Wherefore the rectangle AH, equal to the given square upon C, has been applied to the given straight line AB, exceeding by the square GK. Which was to be done.

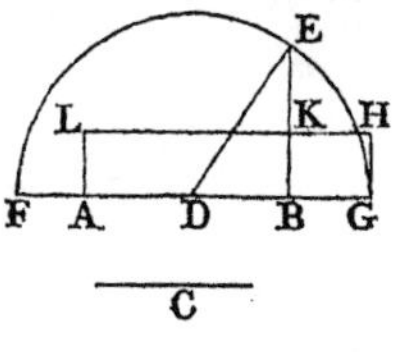

* 6. 2.

3. *To apply a rectangle to a given straight line which shall be equal to a given rectangle, and be deficient by a square: but the given rectangle must not be greater than the square upon the half of the given straight line.*

Let AB be the given straight line, and let the given rectangle be that which is contained by the straight lines C, D, which is not greater than the square upon the half of AB; it is required to apply to AB a rectangle equal to the rectangle C, D, deficient by a square.

Draw AE, BF at right angles to AB, upon the same side of it, and make AE equal to C, and BF to D; join EF, and bisect it in G; and from the centre G, at the distance GE, describe a circle meeting AE again in H: join HF, and draw GK parallel to it, and GL parallel to AE, meeting AB in L.

Because the angle EHF in a semicircle is equal to the right angle EAB, AB and HF are parallels, and AH and BF are parallels; wherefore AH is equal to BF, and the rectangle EA, AH equal to the rectangle EA, BF, that is, to the rectangle C, D: and because EG, GF are equal to one another, and AE, LG, BF parallels; therefore AL and LB are equal, also EK is equal to KH * and the rectangle C, D, from the determination, is not greater than the square of AL, the half of AB; wherefore the rectangle EA, AH is not greater than the square of AL, that is, of KG: add to each the square of KE; therefore the square * of AK is not greater than the squares of EK, KG, that is, than the square of EG, and consequently, the straight line AK or GL is not greater than GE. Now, if GE be equal to GL, the circle EHF touches AB in L, and therefore the square of AL is * equal to the rectangle EA, AH, that is, to the given rectangle C, D, and that which was required is done: but if EG, GL be unequal, EG must be the greater; and therefore the circle EHF cuts the straight line AB: let it cut it in the points M, N; and upon NB, describe the square NBOP, and complete the rectangle ANPQ: because ML is equal to * LN, and it has been proved, that AL is equal to LB, therefore AM is equal to NB, and the rectangle AN, NB, equal to the rectangle NA, AM, that is, to the rectangle * EA, AH,

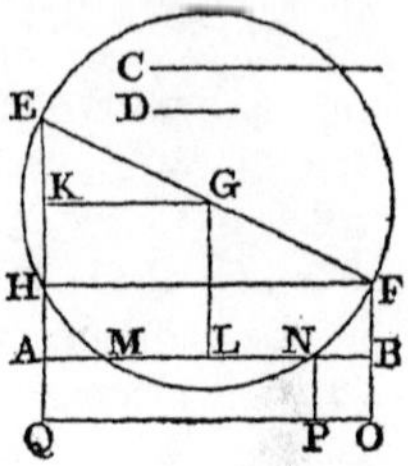

* 3. 3.

* 6. 2.

* 36. 3.

* 3. 3.

* Cor. 36. 3.

or the rectangle C, D: but the rectangle AN, NB is the rectangle AP, because PN is equal to NB; therefore the rectangle AP is equal to the rectangle C, D; and the rectangle AP, equal to the given rectangle C, D, has been applied to the given straight line AB, deficient by the square BP. Which was to be done.

4. *To apply a rectangle to a given straight line, that shall be equal to a given rectangle, exceeding by a square.*

Let AB be the given straight line, and C, D, the given rectangle; it is required to apply a rectangle to AB, equal to C, D, exceeding by a square.

Draw AE, BF at right angles to AB, on the contrary sides of it, and make AE equal to C, and BF equal to D: join EF and bisect it in G; and from the centre G, at the distance GE, describe a circle meeting AE again in H: join HF, and draw GL parallel to AE; let the circle meet AB produced in M, N; and upon BN, describe the square NBOP, and complete the rectangle ANPQ: because the angle EHF in a semicircle is equal to the right angle EAB, AB and HF are parallels, and therefore AH and BF are equal, and the rectangle EA, AH equal to the rectangle EA, BF, that is, to the rectangle C, D: and because ML is equal to LN, and AL to LB, therefore MA is equal to BN, and the rectangle AN, NB, to MA, AN, that is* to the rectangle EA, AH, or the rectangle C, D: therefore the rectangle AN, NB, that is, AP, is equal to the rectangle C, D; and to the given straight line AB, the rectangle AP has been applied equal to the given rectangle C, D, exceeding by the square BP. Which was to be done.

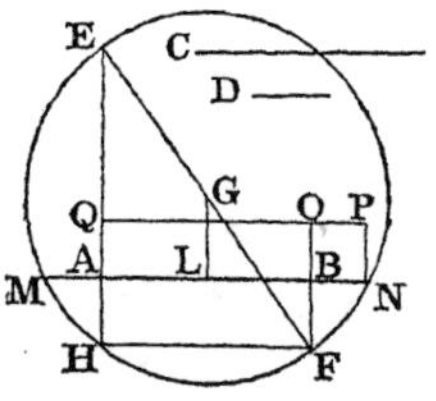

* 35. 3.

Willebrordus Snellius was the first, as far as I know, who gave these constructions of the 3d and 4th Problems, in his Apollonius Batavus: and afterwards the learned Dr. Halley gave them in the Scholium of the 18th Prop. of the 8th Book of Apollonius's Conics restored by him.

The 3d Problem is otherwise enunciated thus: to cut a given straight line AB in the point N, so as to make the rectangle AN, NB, equal to a given space: or, which is the same thing, having given AB the sum of the sides of a rectangle, and the magnitude of it being likewise given, to find its sides.

And the 4th Problem is the same with this: to find a point N in the given straight line AB produced, so as to make the rectangle AN, NB, equal to a given space: or, which is the same thing, having given AB the difference of the sides of a rectangle, and the magnitude of it, to find the sides.

PROPOSITION XXXI.

In the demonstration of this, the inversion of proportionals is twice neglected, and is now added, that the conclusion may be legitimately made by help of the 24th Prop. of Book 5. as Clavius had done.

PROPOSITION XXXII.

The enunciation of the preceding 26th Prop. is not general enough; because not only two similar parallelograms that have an angle common to both, are about the same diameter, but likewise two similar parallelograms that have vertically opposite angles, have their diameters in the same straight line: but there seems to have been another, and that a direct, demonstration of these cases, to which this 32d Prop. was needful: and the 32d may be otherwise, and something more briefly demonstrated, as follows:

PROPOSITION XXXII.

If two triangles which have two sides of the one, &c.

Let GAF, HFC be two triangles which have two sides AG, GF, proportional to the two sides FH, HC, viz. AG to GF, as FH to HC; and let AG be parallel to FH, and GF to HC: AF and FC are in a straight line.

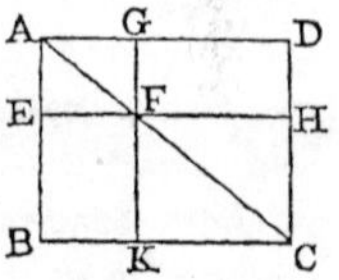

Draw CK parallel * to FH, and let it meet GF produced in K: because AG, KC are each of them parallel to FH, they are parallel * to one another, and therefore the alternate angles AGF, FKC are equal: and AG is to GF, as (FH to HC, that is *) CK to KF; wherefore the triangles AGF, CKF are equiangular *, and the angle AFG equal to the angle CFK: but GFK is a straight line, therefore AF and FC are in a straight line *.

* 31. 1.
* 30. 1.
* 34. 1.
* 6. 6.
* 14. 1.

The 26th Prop. is demonstrated from the 22d, as follows:

If two similar and similarly placed parallelograms have an angle common to both, or vertically opposite angles, their diameters are in the same straight line.

First, let the parallelograms ABCD, AEFG have the angle BAD common to both, and be similar, and similarly placed: ABCD, AEFG are about the same diameter.

Produce EF, GF, to H, K, and join FA, FC; then because the parallelograms ABCD, AEFG are similar, DA is to AB, as GA to AE: wherefore the remainder DG
* Cor.19.5. is * to the remainder EB, as GA to AE: but DG is equal to FH, EB to HC, and AE to GF: therefore as FH to HC, so is AG to GF; and FH, HC are parallel to AG, GF; and the triangles AGF, FHC are joined at one angle in the point F; wherefore
* 32. 6. AF, FC are in the same straight line *.

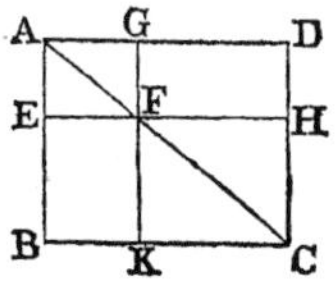

Next, let the parallelograms KFHC, GFEA, which are similar and similarly placed, have their angles KFH, GFE vertically opposite; their diameters AF, FC are in the same straight line.

Because AG, GF are parallel to FH, HC; and that AG is to GF, as FH to HC; therefore AF, FC are in the same straight
* 32. 6. line *.

PROPOSITION XXXIII.

The words "because they are at the centre," are left out, as the addition of some unskilful hand.

In the Greek, as also in the Latin translation, the words, ἃ ἔτυχε, "any whatever", are left out in the demonstrations of both parts of the proposition, and are now added as quite necessary; and in the demonstration of the second part, where the triangle BGC is proved to be equal to CGK, the illative particle ἄρα, in the Greek text ought to be omitted.

The second part of the proposition is an addition of Theon's, as he tells us in his Commentary on Ptolemy's Μεγάλη Σύνταξις, p. 50.

PROPOSITIONS B. C. D.

These three propositions are added, because they are frequently made use of by geometers.

BOOK XI.

DEFINITIONS.

IX. and XI. The similitude of plane figures is defined from the equality of their angles, and the proportionality of the sides about the equal angles; for from the proportionality of the sides only, or only from the equality of the angles, the similitude of the figures does not follow, except in the case when the figures are triangles: the similar position of the sides which contain the figures, to one another, depending partly upon each of these: and for the same reason, those are similar solid figures which have all their solid angles equal, each to each, and are contained by the same number of similar plane figures: for there are some solid figures contained by similar plane figures, of the same number, and even of the same magnitude, that are neither similar nor equal, as shall be demonstrated after the notes on the 10th Def.: upon this account it was necessary to amend the definition of similar solid figures, and to place the definition of a solid angle before it: and from this and the 10th Def. it is sufficiently plain how much the Elements have been spoiled by unskilful editors

X. Since the meaning of the word "equal" is known and established before it comes to be used in this definition; therefore the proposition which is the 10th Def. of this Book, is a theorem, the truth or falsehood of which ought to be demonstrated, not assumed; so that Theon, or some other editor, has ignorantly turned a theorem, which ought to be demonstrated, into this 10th Def. That figures are similar, ought to be proved from the definition of similar figures; that they are equal, ought to be demonstrated from the axiom, "Magnitudes that wholly coincide are equal to one another;" or from Prop. A. of Book 5. or the 9th Prop. or the 14th of the same Book, from one of which the equality of all kinds of figures must ultimately be deduced. In the preceding books, Euclid has given no definition of equal figures, and it is certain he did not give this: for what he called the first Def. of the third Book, is really a theorem in which those circles are said to be equal, that have the straight lines from the centres to the circumferences equal, which is plain from the definition of a circle; and therefore has by some editor been improperly placed among the definitions. The equality of figures ought not to be defined, but demonstrated: therefore, though it were true, that solid figures contained by the same number of similar and equal plane figures are equal to one another, yet he would justly deserve to be blamed who would make a definition of this proposition, which ought to be demonstrated. But if this proposition be not true, must it not be confessed that geometers have, for these thirteen hundred years, been mistaken in this elementary matter? And this should teach us modesty, and to acknowledge how little, through the weakness of our minds, we are able to prevent mistakes, even in the principles of sciences which are justly reckoned amongst the most certain; for that the proposition is not universally true, can be shewn by many examples: the following is sufficient:

Let there be any plane rectilineal figure, as the triangle
* 12. 11. ABC, and from a point D within it, draw * the straight line DE at right angles to the plane ABC: in DE, take DE, DF, equal to one another, upon the opposite sides of the plane, and let G be any point in EF; join DA, DB, DC; EA, EB, EC; FA, FB, FC; GA, GB, GC: because the straight line EDF is at right angles to the plane ABC, it makes right angles with DA, DB, DC, which it meets in that plane; and in the triangles EDB, FDB, the sides ED

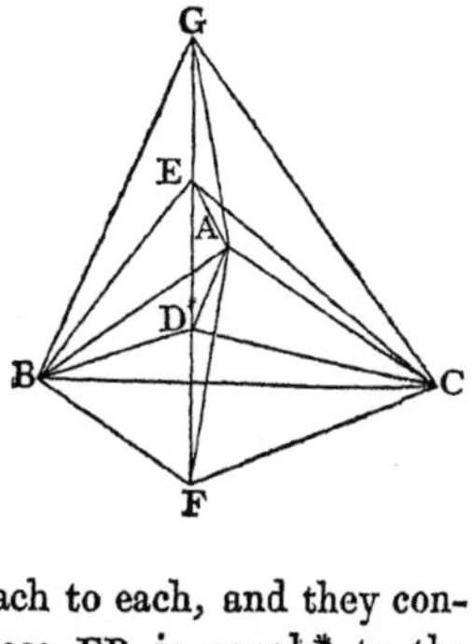

and DB are equal to FD and DB, each to each, and they con-
* 4. 1. tain right angles; therefore the base EB is equal * to the base FB; in the same manner EA is equal to FA, and EC to FC: and in the triangles EBA, FBA, the sides EB, BA are equal to FB, BA, and the base EA is equal to the base
* 8. 1. FA; wherefore the angle EBA is equal * to the angle FBA,
* 4. 1. and the triangle EBA equal * to the triangle FBA, and the other angles equal to the other angles; therefore these tri-

angles are similar *: in the same manner the triangle EBC is similar to the triangle FBC, and the triangle EAC to FAC; therefore there are two solid figures, each of which is contained by six triangles; one of them by three triangles, the common vertex of which is the point G, and their bases the straight lines AB, BC, CA, and by three other triangles, the common vertex of which is the point E, and their bases the same lines AB, BC, CA: the other solid is contained by the same three triangles, the common vertex of which is G, and their bases AB, BC, CA, and by three other triangles of which the common vertex is the point F, and their bases the same straight lines AB, BC, CA: now the three triangles GAB, GBC, GCA are common to both solids, and the three others EAB, EBC, ECA, of the first solid, have been shewn equal and similar to the three others FAB, FBC, FCA, of the other solid, each to each; therefore these two solids are contained by the same number of equal and similar planes: but that they are not equal, is manifest, because the first of them is contained in the other: therefore it is not universally true that solids are equal which are contained by the same number of equal and similar planes.

* 4. 6. & 1 Def. 6.

Corollary. From this it appears that two unequal solid angles may be contained by the same number of equal plane angles.

For the solid angle at B, which is contained by the four plane angles EBA, EBC, GBA, GBC, is not equal to the solid angle at the same point B which is contained by the four plane angles FBA, FBC, GBA, GBC; for this last contains the other: and each of them is contained by four plane angles, which are equal to one another, each to each, or are the self-same, as has been proved: and indeed there may be innumerable solid angles all unequal to one another, which are each of them contained by plane angles that are equal to one another, each to each: it is likewise manifest, that the before-mentioned solids are not similar, since their solid angles are not all equal.

And that there may be innumerable solid angles all unequal to one another, which are each of them contained by the same plane angles disposed in the same order, will be plain from the three following propositions.

PROPOSITION I.

PROBLEM.—*Three magnitudes,* A, B, C, *being given, to find a fourth such, that every three shall be greater than the remaining one.*

Let D be the fourth: therefore D must be less than A, B, C, together: of the three A, B, C, let A be that which is not less than either of the two B and C: and first, let B and C together be not less than A; therefore B, C, D, together are greater than A: and because A is not less than B; A, C, D, together are greater than B: in the like manner A, B, D, together are greater than C; wherefore in the case in which B and C together are not less than A, any magnitude D which is less than A, B, C together, will answer the problem.

But if B, and C together, be less than A: then, because it is required that B, C, D together, be greater than A, from each of these taking away B, C, the remaining one D must be greater than the excess of A above B and C: take therefore any magnitude D which is less than A, B, C together, but greater than the excess of A above B and C: then B, C, D together, are greater than A; and because A is greater than either B or C, much more will A and D, together with either of the two B, C, be greater than the other: and by the construction, A, B, C are together greater than D.

COR. If besides it be required, that A and B together, shall not be less than C and D together; the excess of A and B together, above C, must not be less than D, that is D must not be greater than that excess.

PROPOSITION II.

PROB.—*Four magnitudes* A, B, C, D *being given, of which* A *and* B *together, are not less than* C *and* D *together, and such that any three of them whatever are greater than the fourth; to find a fifth magnitude* E *such that any two of the three* A, B, E, *shall be greater than the third, and also that any two of the three* C, D, E, *shall be greater than the third.*

Let A be not less than B, and C not less than D.

First, let the excess of C above D be not less than the excess of A above B: it is plain that a magnitude E can be taken which is less than the sum of C and D, but greater than the excess of C above D; let it be taken; then E is greater likewise than the excess of A above B; wherefore E and B together are greater than A; and A is not less than B; therefore A and B together are greater than B: and by the hypothesis, A and E together are not less than C and D together, and C and D together are greater than E; therefore likewise A and B are greater than E.

But let the excess of A above B be greater than the excess of C above D: and because, by the hypothesis, the three B, C, D are together greater than the fourth A, C and D together are greater than the excess of A above B: therefore a magnitude may be taken which is less than C and D together, but greater than the excess of A above B. Let this magnitude be E: and because E is greater than the excess of A above B, B together with E is greater than A: and as, in the preceding case, it may be shewn that A together with E is greater than B, and that A together with B is greater than E: therefore in each of the cases it has been shewn, that any two of the three, A, B, E, are greater than the third.

And because in each of the cases, E is greater than the excess of C above D, E together with D is greater than C; and by the hypothesis, C is not less than D; therefore E together with C is greater than D; and by the construction, C and D together are greater than E: therefore any two of the three C, D, E, are greater than the third.

PROPOSITION III.

Theorem.—*There may be innumerable solid angles, all unequal to one another, each of which is contained by the same four plane angles placed in the same order.*

Take three plane angles A, B, C, of which A is not less than either of the other two, and such, that A and B together are less than two right angles; and, by Prob. 1, and its corollary, find a fourth angle D such, that any three whatever of the angles A, B, C, D, be greater than the remaining angle, and such, that A and B together be not less than C and D together: and, by Prob. 2, find a fifth angle E such, that any two of the angles A, B, E, be greater than the third, and also that any two of the angles C, D, E, be greater than the third: and because A and B together are less than two right angles, the double of A and B together is less than four right angles: but A and B together are greater than the angle E; wherefore the double of A, B together, is greater than the three angles A, B, E together, which three are consequently less than four right angles; and every two of the

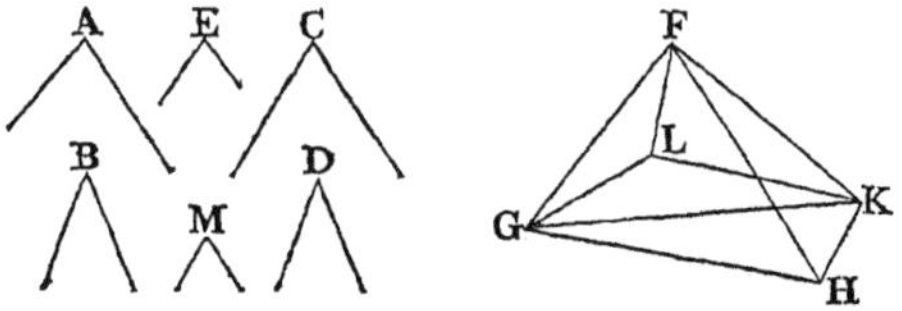

† 23. 11. same angles A, B, E, are greater than the third; therefore† a solid angle may be made, contained by three plane angles, equal to the angles A, B, E, each to each. Let this be the angle F, contained by the three plane angles GFH, HFK, GFK, which are equal to the angles A, B, E, each to each: and because the angles C, D together, are not greater than the angles A, B together, therefore the angles C, D, E are not greater than the angles A, B, E: but these last three are less than four right angles, as has been demonstrated; wherefore also the angles C, D, E are together less than four right angles; and every two of them are greater than the third; therefore a solid angle may be made, which shall be contained by three
* 23. 11. plane angles equal to the angles C, D, E, each to each*; and†,
† 26. 11. at the point F, in the straight line FG, a solid angle may be made equal to that which is contained by the three plane angles that are equal to the angles C, D, E: let this be made, and let the angle GFK, which is equal to E, be one of the three; and let KFL, GFL be the other two which are equal to the angles C, D, each to each: thus, there is a solid angle constituted at the point F, contained by the four plane angles GFH, HFK, KFL, GFL, which are equal to the angles A, B, C, D, each to each.

Again: find another angle M such, that every two of the three angles A, B, M, be greater than the third, and also every two of the three C, D, M, be greater than the third: and, it may be demonstrated, as in the preceding part, that the three A, B, M are less than four right angles, as also that the three C, D, M are less than four right angles. Make
* 23. 11. therefore * a solid angle at N contained by the three plane angles ONP, PNQ, ONQ which are equal to A, B, M, each to

each: and make† at the point N, in the straight line ON, a solid angle contained by three plane angles, of which one is the angle ONQ equal to M, and the other two are the angles QNR, ONR, which are equal to the angles C, D, each to each: thus, at the point N, there is a solid angle contained by the four plane angles ONP, PNQ, QNR, ONR, which are equal to the angles A, B, C, D, each to each. And that the two solid angles at the points F, N, each of which is contained by the above-named four plane angles, are not equal to one another, or, that they cannot coincide, will be plain by considering that the angles GFK, ONQ, that is, the angles E, M, are unequal by the construction; and therefore the straight lines GF, FK cannot coincide with ON, NQ, nor consequently can the solid angles, which therefore are unequal. † 26. 11.

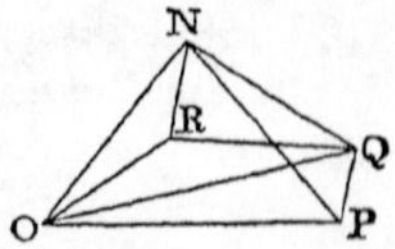

And because from the four plane angles A, B, C, D, there can be found innumerable other angles that will serve the same purpose with the angles E and M, it is plain that innumerable other solid angles may be constituted which are each contained by the same four plane angles, and all of them unequal to one another. Q. E. D.

And from this it appears, that Clavius and other authors are mistaken, who assert, that those solid angles are equal which are contained by the same number of plane angles that are equal to one another, each to each. Also it is plain, that the 26th Prop. of Book 11. is by no means sufficiently demonstrated, because the equality of two solid angles, whereof each is contained by three plane angles which are equal to one another, each to each, is only assumed and not demonstrated.

PROPOSITION I.

The words at the end of this, "for a straight line cannot "meet a straight line in more than one point," are left out, as an addition by some unskilful hand; for this is to be demonstrated, not assumed.

Mr. Thomas Simpson, in his notes at the end of the second edition of his Elements of Geometry, p. 262, after repeating the words of this note, adds: "Now can it possibly show any "want of skill in an editor" he means Euclid or Theon, "to "refer to an axiom which Euclid himself hath laid down, "Book 1. No. 14," he means Barrow's Euclid, for it is the 10th in the Greek, "and not to have demonstrated, what no "man can demonstrate?" But all that in this case can follow from that axiom is, that, if two straight lines could meet each other in two points, the parts of them betwixt these points must coincide, and so they would have a segment betwixt these points common to both. Now, as it has not been shewn in Euclid, that they cannot have a common segment, this does not prove that they cannot meet in two points, from which their not having a common segment is deduced in the Greek edition: but, on the contrary, because they cannot have a common segment, as is shewn in Cor. of 11th Prop. Book 1. of 4to edition, it follows plainly, that they cannot meet in two points, which the remarker says no man can demonstrate.

Mr. Simpson, in the same notes, p. 265, justly observes, that in the corollary of Prop. 11. Book 1. 4to edition, the straight lines AB, BD, BC, are supposed to be all in the same plane, which cannot be assumed in 1st Prop. Book 11. This, soon after the 4to edition was published, I observed and corrected as it is now in this edition: he is mistaken in thinking the 10th axiom he mentions here to be Euclid's; it is none of Euclid's, but is the 10th in Dr. Barrow's edition, who had it from Herigon's Cursus, vol. 1. and in place of it the corollary of 11th Prop. Book 1. was added.

PROPOSITION II.

This proposition seems to have been changed and vitiated by some editor; for all the figures defined in the 1st Book of the Elements, and among them triangles, are, by the hypothesis, plane figures; that is, such as are described in a plane; wherefore the second part of the enunciation needs no demonstration. Besides, a convex superficies may be terminated by three straight lines meeting one another: the thing that should have been demonstrated is, that two or three straight lines, that meet one another, are in one plane. And as this is not sufficiently done, the enunciation and demonstration are changed into those now put into the text.

PROPOSITION III.

In this proposition the following words near to the end of it are left out, viz. "therefore DEB, DFB are not straight "lines; in the like manner, it may be demonstrated, that "there can be no other straight line between the points D, B;" because from this, that two lines include a space, it only follows that one of them is not a straight line: and the force of the argument lies in this, viz. if the common section of the planes be not a straight line, then two straight lines could include a space, which is absurd: therefore the common section is a straight line.

PROPOSITION IV.

The words "and the triangle AED to the triangle BEC" are omitted, because the whole conclusion of the 4th Prop. Book 1. has been so often repeated in the preceding books, it was needless to repeat it here.

PROPOSITION V.

In this, near to the end, ἐπιπέδῳ ought to be left out in the Greek text: and the word "plane" is rightly left out in the Oxford edition of Commandine's translation.

PROPOSITION VII.

This proposition has been put into this book by some unskilful editor, as is evident from this, that straight lines which are drawn from one point to another in a plane, are, in the preceding books, supposed to be in that plane: and if they were not, some demonstrations in which one straight line is supposed to meet another would not be conclusive, because these lines would not meet one another: for instance, in Prop. 30. Book 1. the straight line GK would not meet EF, if GK were not in the plane in which are the parallels AB, CD, and in which, by hypothesis, the straight line EF is: besides, this 7th Prop. is demonstrated by the preceding 3d; in which the very thing which is proposed to be demonstrated in the 7th is twice assumed, viz. that the straight line drawn from one point to another in a plane, is in that plane: and the same thing is assumed in the preceding 6th Prop. in which the straight line which joins the points B, D, that are in the plane to which AB and CD are at right angles, is supposed to be in that plane: and the 7th, of which another demonstration is given, is kept in the book merely to preserve the number of the propositions; for it is evident, from the 7th and 35th Definitions of the 1st Book, though it had not been in the Elements.

PROPOSITION VIII.

In the Greek, and in Commandine's and Dr. Gregory's translations, near to the end of this proposition, are the following words: "but DC is in the plane through BA, AD," instead of which, in the Oxford edition of Commandine's translation is rightly put, "but DC is in the plane through "BD, DA." But all the editions have the following words, viz. "because AB, BD are in the plane through BD, DA, and "DC is in the plane in which are AB, BD," which are manifestly corrupted, or have been added to the text; for there was not the least necessity to go so far about to shew that DC is in the same plane in which are BD, DA, because it immediately follows, from Prop. 7. preceding, that BD, DA are in the plane in which are the parallels AB, CD: therefore instead of these words, there ought only to be, "because all "three are in the plane in which are the parallels AB, CD."

PROPOSITION XV.

After the words "and because BA is parallel to GH," the following are added, "for each of them is parallel to DE, and "are not both in the same plane with it," as being manifestly forgotten to be put into the text.

PROPOSITION XVI.

In this, near to the end, instead of the words, "but straight "lines which meet neither way," ought to be read, "but "straight lines in the same plane which produced meet "neither way." Because, though in citing this Def. in Prop. 27. Book 1. it was not necessary to mention the words "in "the same plane," all the straight lines in the books preceding this, being in the same plane; yet here it was quite necessary.

PROPOSITION XX.

In this, near the beginning, are the words, "but if not, "let BAC be the greater:" but the angle BAC may happen to be equal to one of the other two: wherefore this place should be read thus, "but if not, let the angle BAC be not "less than either of the other two, but greater than DAB."

At the end of this proposition it is said, "in the same "manner it may be demonstrated," though there is no need of any demonstration; because the angle BAC being not less than either of the other two, it is evident that BAC together with one of them, is greater than the other.

PROPOSITION XXII.

And likewise in this, near the beginning, it is said, "but "if not, let the angles at B, E, H, be unequal, and let the angle "at B be greater than either of those at E, H:" which words manifestly shew this place to be vitiated, because the angle at B may be equal to one of the other two. They ought therefore to be read thus, "but if not, let the angles at B, E, H, be "unequal, and let the angle at B be not less than either of the "other two at E, H: therefore the straight line AC is not less "than either of the two DF, GK."

PROPOSITION XXIII.

The demonstration of this is made something shorter, by not repeating in the third case the things which were demonstrated in the first; and by making use of the construction which Campanus has given; but he does not demonstrate the second and third cases: the construction and demonstration of the third case are made a little more simple than in the Greek text.

PROPOSITION XXIV.

The word "similar" is added to the enunciation of this proposition, because the planes containing the solids which are to be demonstrated to be equal to one another, in the 25th Prop. ought to be similar and equal, that the equality of the solids may be inferred from Prop. C. of this Book: and in the Oxford edition of Commandine's translation, a corollary is added to Prop. 24. to shew that the parallelograms mentioned in this proposition are similar, that the equality of the solids in Prop. 25. may be deduced from the 10th Def. of Book 11.

PROPOSITIONS XXV. XXVI.

In the 25th Prop. solid figures which are contained by the same number of similar and equal plane figures, are supposed to be equal to one another. And it seems that Theon or some other editor, that he might save himself the trouble of demonstrating the solid figures mentioned in this proposition to be equal to one another, has inserted the 10th Def. of this Book, to serve instead of a demonstration; which was very ignorantly done.

Likewise in the 26th Prop. two solid angles are supposed to be equal, if each of them be contained by three plane angles which are equal to one another, each to each. And it is strange enough, that none of the commentators on Euclid have, as far as I know, perceived, that something is wanting in the demonstrations of these two propositions. Clavius, indeed, in a note upon the 11th Def. of this Book, affirms, that it is evident that those solid angles are equal which are contained by the same number of plane angles, equal to one another, each to each, because they will coincide, if they be conceived to be placed within one another; but this is said without any proof, nor is it always true, except when the solid angles are contained by three plane angles only which are equal to one another, each to each; and in this case the proposition is the same with this; that two spherical triangles that are equilateral to one another are also equiangular to one another, and can coincide: which ought not to be granted without a demonstration. Euclid does not assume this in the case of rectilineal triangles but demonstrates it in Prop. 8. Book 1, that triangles which are equilateral to one another, are also equiangular to one another; and from this their total equality appears by Prop. 4. Book 1. And Menelaus, in the 4th Prop. of his first Book of Spherics, explicitly demonstrates, that spherical triangles which are mutually equilateral, are also equiangular to one another; from which it is easy to shew that they must coincide, providing they have their sides disposed in the same order and situation.

To supply these defects, it was necessary to add the three propositions marked A, B, C, to this Book. For the 25th, 26th, and 28th propositions of it, and consequently eight others, viz. the 27th, 31st, 32d, 33d, 34th, 36th, 37th, and 40th of the same, which depend upon them, have hitherto stood upon an infirm foundation; as also the 8th, 12th, Cor. of 17th and 18th of 12th Book, which depend upon the 9th Def. For it has been shewn in the notes on Def. 10th of this Book, that solid figures which are contained by the same number of similar and equal plane figures, as also solid angles that are contained by the same number of equal plane angles, are not always equal to one another.

It is to be observed that Tacquet, in his Euclid, defines equal solid angles to be such, "as being put within one another do coincide:" but this is an axiom, not a definition; for it is true of all magnitudes whatever. He made this useless definition, that by it he might demonstrate the 36th Prop. of this Book, without the help of the 35th of the same: concerning which demonstration, see the note upon Prop. 36.

PROPOSITION XXVIII.

In this it ought to have been demonstrated, not assumed, that the diagonals are in one plane. Clavius has supplied this defect.

PROPOSITION XXIX.

There are three cases of this proposition: the first is, when the two parallelograms opposite to the base AB have a side common to both: the second is, when these parallelograms are separated from one another: and the third, when there is a part of them common to both; and to this last only, the demonstration that has hitherto been in the Elements does agree. The first case is immediately deduced from the preceding 28th Prop., which seems for this purpose to have been premised to this 29th, for it is necessary to none but to it, and to the 40th of this Book, as we now have it, to which last it would, without doubt, have been premised, if Euclid had not made use of it in the 29th; but some unskilful editor has taken it away from the Elements, and has mutilated

Euclid's demonstration of the two other cases, which is now restored, and serves for both at once.

PROPOSITION XXX.

In the demonstration of this, the opposite planes of the solid CP, in the figure in this edition, that is, of the solid CO in Commandine's figure, are not proved to be parallel; which it is proper to do for the sake of learners.

PROPOSITION XXXI.

There are two cases of this proposition: the first is when the insisting straight lines are at right angles to the basis; the other, when they are not: the first case is divided again into two others, one of which is, when the bases are equiangular parallelograms: the other, when they are not equiangular. The Greek editor makes no mention of the first of these two last cases, but has inserted the demonstration of it as a part of that of the other: and therefore should have taken notice of it in a corollary; but we thought it better to give these two cases separately: the demonstration also is made something shorter by following the way Euclid has made use of in Prop. 14. Book 6. Besides, in the demonstration of the case in which the insisting straight lines are not at right angles to the bases, the editor does not prove that the solids described in the construction are parallelopipeds, which it is not to be thought that Euclid neglected: also the words, "of which the insisting straight lines are not in "the same straight lines," have been added by some unskilful hand: for they may be in the same straight lines.

PROPOSITION XXXII.

The editor has forgot to order the parallelogram FH to be applied in the angle FGH equal to the angle LCG, which is necessary. Clavius has supplied this.

Also, in the construction, it is required to complete the solid of which the base is FH, and altitude the same with that of the solid CD: but this does not determine the solid to be completed, since there may be innumerable solids upon the same base, and of the same altitude: it ought therefore to be said, "complete the solid of which the base is FH, and "one of its insisting straight lines is FD:" The same correction must be made in Prop. 33.

PROPOSITION D.

It is very probable that Euclid gave this proposition a place in the Elements, since he gave the like proposition concerning equiangular parallelograms in the 23d Prop. B. 6.

PROPOSITION XXXIV.

In this the words, ὧν αἱ ἐφεστῶσαι οὐκ εἰσὶν ἐπὶ τῶν αὐτῶν εὐθειῶν, "of which the insisting straight lines are not in the "same straight lines," are thrice repeated: but these words ought either to be left out, as they are by Clavius, or, in place of them, ought to be put, "whether the insisting straight "lines be, or be not, in the same straight lines:" for the other case is, without any reason, excluded; also the words ὧν τὰ ὕψη, "of which the altitudes," are twice put for ὧν αἱ ἐφεστῶσαι "of which the insisting straight lines;" which is a plain mistake: for the altitude is always at right angles to the base.

PROPOSITION XXXV.

The angles ABH, DEM, are demonstrated to be right angles in a shorter way than in the Greek; and in the same way ACH, DFM may be demonstrated to be right angles: also the repetition of the same demonstration, which begins with "in "the same manner", is left out, as it was probably added to the text by some editor: for the words, "in like manner we "may demonstrate", are not inserted except when the demonstration is not given, or when it is something different from the other if it be given, as in Prop. 26. of this Book. Campanus has not this repetition.

We have given another demonstration of the corollary, besides the one in the original, by help of which the 36th Prop. may be demonstrated without the 35th.

PROPOSITION XXXVI.

Tacquet in his Euclid demonstrates this proposition without the help of the 35th; but it is plain, that the solids mentioned in the Greek text in the enunciation of the proposition as equiangular, are such that their solid angles are contained by three plane angles equal to one another, each to each; as is evident from the construction. Now Tacquet does not demonstrate, but assumes, these solid angles to be equal to one another; for he supposes the solids to be already made, and does not give the construction by which they are made; but, by the second demonstration of the preceding corollary, his demonstration is rendered legitimate likewise in the case where the solids are constructed as in the text.

PROPOSITION XXXVII.

In this it is assumed, that the ratios which are triplicate of those ratios which are the same with one another, are likewise the same with one another: and that those ratios are the same with one another, of which the triplicate ratios are the same with one another: but this ought not to be granted without a demonstration; nor did Euclid assume the first and easiest of these two propositions, but demonstrated it in the case of duplicate ratios, in the 22d Prop. Book 6. On this account, another demonstration is given of this Proposition, like to that which Euclid gives in Prop. 22. Book 6. as Clavius has done.

PROPOSITION XXXVIII.

When it is required to draw a perpendicular from a point in one plane which is at right angles to another plane, unto this last plane, it is done by drawing a perpendicular from the point to the common section of the planes; for this perpendicular will be perpendicular to the plane by Def. 4. of this Book: and it would be foolish in this case to do it by the 11th Prop. of the same: but Euclid*, Apollonius, and other geometers, when they have occasion for this problem, direct a perpendicular to be drawn from the point to the plane, and conclude that it will fall upon the common section of the planes, because this is the very same thing as if they had made use of the construction above mentioned, and then concluded that the straight line must be perpendicular to the plane; but is expressed in fewer words. Some editor, not perceiving this, thought it was necessary to add this proposition, which can never be of any use to the 11th Book; and its being near to the end among propositions with which it has no connexion, is a mark of its having been added to the text.

* 17. 12. in other editions.

PROPOSITION XXXIX.

In this it is supposed, that the straight lines which bisect the sides of the opposite planes, are in one plane, which ought to have been demonstrated; as is now done.

BOOK XII.

The learned Mr. Moore, professor of Greek in the University of Glasgow, observed to me that it plainly appears from Archimedes's Epistle to Dositheus, prefixed to his books of the Sphere and Cylinder, which epistle he has restored from ancient manuscripts, that Eudoxus was the author of the chief propositions in this 12th Book.

PROPOSITION II.

At the begining of this it is said, "if it be not so, the "square of BD shall be to the square of FH, as the circle "ABCD is to some space either less than the circle EFGH, or "greater than it." And the like is to be found near to the end of this proposition, as also in Prop. 5, 11, 12, 18. of this Book: concerning which it is to be observed, that in the demonstration of theorems, it is sufficient, in this and the like cases, that a thing made use of in the reasoning can possibly exist, provided this be evident, though it cannot be exhibited or found by a geometrical construction: so, in this place, it is assumed, that there may be a fourth proportional to these three magnitudes, viz. the squares of BD, FH, and the circle ABCD; because it is evident that there is some square equal to the circle ABCD, though it cannot be found geometrically; and to the three rectilineal figures, viz. the squares of BD, FH, and the square which is equal to the circle ABCD, there is a fourth square proportional; because to the three straight lines which are their sides, there is a fourth straight line proportional*, * 12. 6.
and this fourth square, or a space equal to it, is the space which in this proposition is denoted by the letter S: and the like is to be understood in the other places above cited: and it is probable that this has been shewn by Euclid, but left out by some editor; for the lemma which

some unskilful hand has added to this proposition explains nothing of it.

PROPOSITION III.

In the Greek text and the translations, it is said, "and be-"cause the two straight lines BA, AC, which meet one an-"other," &c. Here the angles BAC, KHL, are demonstrated to be equal to one another, by 10th Prop. B. 11., which had been done before: because the triangle EAG was proved to be similar to the triangle KHL: this repetition is left out, and the triangles BAC, KHL, are proved to be similar in a shorter way by Prop. 21. B. 6.

PROPOSITION IV.

A few things in this are more fully explained, than in the Greek text.

PROPOSITION V.

In this, near to the end, are the words, ὡς ἔμπροσθεν ἐδείχθη "as was before shewn;" and the same are found again in the end of Prop. 18. of this Book; but the demonstration referred to, except it be in the useless lemma annexed to the 2d Prop. is no where in these Elements, and has been perhaps left out by some editor, who has forgot to cancel those words also.

PROPOSITION VI.

A shorter demonstration is given of this; and that which is in the Greek text may be made shorter by a step than it is: for the author of it makes use of the 22d Prop. of B. 5. twice: whereas once would have served his purpose; because that proposition extends to any number of magnitudes which are proportionals taken two and two, as well as to three which are proportional to other three.

COROLLARY PROPOSITION VIII.

The demonstration of this is imperfect, because it is not shewn, that the triangular pyramids into which those upon multangular bases are divided, are similar to one another, as ought necessarily to have been done, and is done in the like case in Prop. 12. of this Book: the full demonstration of the corollary is as follows:

Upon the polygonal bases ABCDE, FGHKL, let there be similar and similarly situated pyramids which have the points M, N, for their vertices: the pyramid ABCDEM has to the pyramid FGHKLN, the triplicate ratio of that which the side AB has to the homologous side FG.

Let the polygons be divided into the triangles ABE, EBC, ECD; FGL, LGH, LHK, which are similar *, each to each: * 20. 6.
and because the pyramids are similar, therefore * the triangle * 11 Def. 11.
EAM is similar to the triangle LFN, and the triangle ABM to FGN: wherefore * ME is to EA, as NL to LF; and as AE to * 4. 6.
EB, so is FL to LG, because the triangles EAB, LFG are similar; therefore, ex æquali, as ME to EB, so is NL to LG: in like manner it may be shewn, that EB is to BM, as LG to GN; therefore, again, ex æquali, as EM to MB, so is LN to NG: wherefore the triangles EMB, LNG, having their sides proportionals, are * equiangular, and similar to one another: there- * 5. 6.

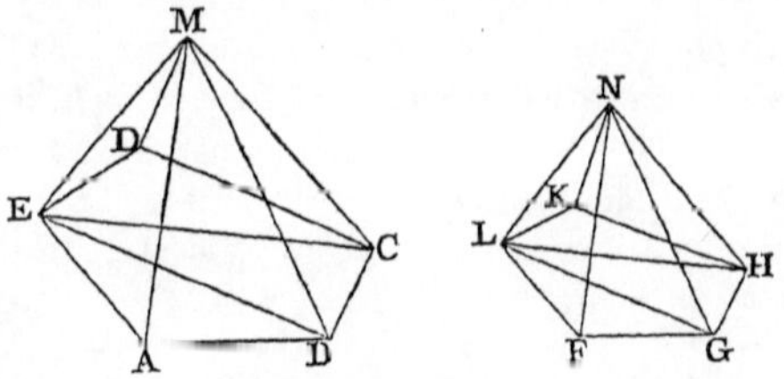

fore the pyramids which have the triangles EAB, LFG for their bases, and the points M, N for their vertices, are similar * to one another, for their solid angles are * equal, and the * 11 Def. 11. * B. 11.
solids themselves are contained by the same number of similar planes: in the same manner the pyramid EBCM may be shewn to be similar to the pyramid LGHN, and the pyramid ECDM to LHKN: and because the pyramids EABM, LFGN are similar, and have triangular bases, the pyramid EABM has * to * 8. 12.
LFGN, the triplicate ratio of that which EB has to the homologous side LG. And, in the same manner, the pyramid EBCM has to the pyramid LGHN, the triplicate ratio of that which EB has to LG: therefore as the pyramid EABM is to the pyramid LFGN, so is the pyramid EBCM to the pyramid LGHN: in like manner, as the pyramid EBCM is to LGHN, so is the pyramid ECDM to the pyramid LHKN: and as one of the antecedents is to one of the consequents, so are all the antecedents to all the consequents: therefore as the pyramid

EABM to the pyramid LFGN, so is the whole pyramid ABCDEM to the whole pyramid FGHKLN: and the pyramid EABM has to the pyramid LFGN, the triplicate ratio of that which AB has to FG; therefore the whole pyramid has to the whole pyramid, the triplicate ratio of that which AB has to the homologous side FG. Q. E. D.

PROPOSITIONS XI. XII.

The order of the letters of the alphabet is not observed in these two propositions, according to Euclid's manner, and is now restored: by which means, the first part of Prop. 12. may be demonstrated in the same words with the first part of Prop. 11.; on this account the demonstration of that first part is left out, and assumed from Prop. 11.

PROPOSITION XIII.

In this proposition, the common section of a plane parallel to the bases of a cylinder, with the cylinder itself, is supposed to be a circle, and it was thought proper briefly to demonstrate it; whence it is sufficiently manifest, that this plane divides the cylinder into two others: and the same thing is understood to be supplied in Prop. 14.

PROPOSITION XV.

"And complete the cylinders AX, EO." Both the enunciation and exposition of the proposition represent the cylinders as well as the cones, as already described: wherefore the reading ought rather to be, "and let the cones be ALC, ENG; "and the cylinders AX, EO."

The first case in the second part of the demonstration is wanting; and something also in the second case of that part, before the repetition of the construction is mentioned; which are now added.

PROPOSITION XVII.

In the enunciation of this proposition the Greek words εἰς τὴν μείζονα σφαῖραν στερεὸν πολύεδρον ἐγγράψαι μὴ ψαῦον τῆς ἐλάσσονος σφαίρας κατὰ τὴν ἐπιφάνειαν are thus translated by Commandine and others, "in majori solidum polyhedrum de-"scribere quod minoris sphæræ superficiem non tangat;" that is, "to describe in the greater sphere a solid polyhedron "which shall not meet the superficies of the lesser sphere:" whereby they refer the words κατὰ τὴν ἐπιφάνειαν to these next to them τῆς ἐλάσσονος σφαίρας: but they ought by no means to be thus translated: for the solid polyhedron doth not only meet the superficies of the lesser sphere, but pervades the whole of that sphere: therefore the aforesaid words are to be referred to τὸ στερεὸν πολύεδρον, and ought thus to be translated, viz. "to describe in the greater sphere a solid polyhedron whose superficies shall not meet the lesser sphere;" as the meaning of the proposition necessarily requires.

The demonstration of the proposition is spoiled and mutilated: for some easy things are very explicitly demonstrated, while others not so obvious are not sufficiently explained; for example, when it is affirmed, that the square of KB is greater than the double of the square of BZ, in the first demonstration; and that the angle BZK is obtuse, in the second: both which ought to have been demonstrated: besides, in the first demonstration, it is said, "draw KΩ from the point K, perpendicular to BD;" whereas it ought to have been said, "join KV," and it should have been demonstrated, that KV is perpendicular to BD: for it is evident from the figure in Hervagius's and Gregory's editions, and from the words of the demonstration, that the Greek editor did not perceive that the perpendicular drawn from the point K to the straight line BD, must necessarily fall upon the point V, for in the figure it is made to fall upon the point Ω, a different point from V, which is likewise supposed in the demonstration. Commandine seems to have been aware of this; for in this figure he marks one and the same point with two letters V, Ω; and before Commandine, the learned John Dee, in the commentary he annexes to this proposition in Henry Billingsley's translation of the Elements, printed at London, ann. 1570, expressly takes notice of this error, and gives a demonstration suited to the construction in the Greek text, by which he shews that the perpendicular drawn from the point K to BD, must necessarily fall upon the point V.

Likewise it is not demonstrated, that the quadrilateral figures SOPT, TPRY, and the triangle YRX, do not meet the lesser sphere, as was necessary to have been done; only Clavius, as far as I know, has observed this, and demonstrated it by a lemma, which is now premised to this proposition, something altered, and more briefly demonstrated.

In the corollary of this proposition, it is supposed that a solid polyhedron is described in the other sphere similar to that which is described in the sphere BCDE; but, as the construction by which this may be done is not given, it was thought proper to give it, and to demonstrate that the pyramids in it are similar to those of the same order in the solid polyhedron described in the sphere BCDE.

From the preceding notes, it is sufficiently evident how much the Elements of Euclid, who was a most accurate geometer, have been vitiated and mutilated by ignorant editors. The opinion which the greatest part of learned men have entertained concerning the present Greek edition, viz. that it is very little or nothing different from the genuine work of Euclid, has without doubt deceived them, and made them less attentive and accurate in examining that edition; whereby several errors, some of them gross enough, have escaped their notice, from the age in which Theon lived, to this time. Upon which account there is some ground to hope, that the pains we have taken in correcting those errors, and freeing the Elements as far as we could from blemishes, will not be unacceptable to good judges, who can discern when demonstrations are legitimate, and when they are not.

The objections which since the first edition have been made against some things in the notes, especially against the doctrine of proportionals, have either been fully answered in Dr. Barrow's Lect. Mathemat. and in these notes; or are such, except one which has been taken notice of in the note on Prop. 1. Book 11., as shew that the person who made them has not sufficiently considered the things against which they are brought; so that it is not necessary to make any further answer to these objections and others like them against Euclid's definition of proportionals, of which definition Dr. Barrow justly says in page 297 of the above-named book, that "Nisi "machinis impulsa validioribus, æternùm persistet incon-"cussa."

END OF THE NOTES.

THE

BOOK

OF

EUCLID'S DATA.

DEFINITIONS.

I. SPACES, lines, and angles, are said to be given in magnitude, when equals to them can be found.

II. A ratio is said to be given, when a ratio of a given magnitude to a given magnitude which is the same ratio with it can be found.

III. Rectilineal figures are said to be given in species, which have each of their angles given, and the ratios of their sides given.

IV. Points, lines, and spaces, are said to be given in position, which have always the same situation, and which are either actually exhibited, or can be found.

A. An angle is said to be given in position, which is contained by straight lines given in position.

V. A circle is said to be given in magnitude, when a straight line from its centre to the circumference is given in magnitude.

VI. A circle is said to be given in position and magnitude, the centre of which is given in position, and a straight line from it to the circumference is given in magnitude.

VII. Segments of circles are said to be given in magnitude, when the angles in them, and their bases, are given in magnitude.

VIII. Segments of circles are said to be given in position and magnitude, when the angles in them are given in magnitude, and their bases are given both in position and magnitude.

IX. A magnitude is said to be greater than another by a given magnitude, when this given magnitude being taken from it, the remainder is equal to the other magnitude.

X. A magnitude is said to be less than another by a given magnitude, when this given magnitude being added to it, the whole is equal to the other magnitude.

(1.‡)

PROPOSITION I.

See N. *The ratio of given magnitudes to one another is given.*

Let A, B be two given magnitudes: the ratio of A to B is given.

* 1 Def. Dat. Because A is a given magnitude, there may* be found one equal to it; let this be C: and because B is given, one equal to it may be found; let it be D: and since A is equal to C, * 7. 5. and B to D; therefore * A is to B, as C to D; and consequently the ratio of A to B is given, because the ratio of the given magnitudes C, D, which is the same with it, has been found.

(2.)

PROPOSITION II.

See N. *If a given magnitude have a given ratio to another magnitude, " and if unto the two magnitudes by which the given " ratio is exhibited, and the given magnitude, a fourth pro- " portional can be found"; the other magnitude is given.*

Let the given magnitude A have a given ratio to the magnitude B: if a fourth proportional can be found to the three magnitudes above named, B is given in magnitude.

‡ The figures between parentheses in the margin, opposite the head of each proposition, shew the number of that proposition in the other editions.

Because A is given, a magnitude may be found equal to it*; let this be C: and because the ratio of A to B is given, a ratio which is the same with it may be found; let this be the ratio of the given magnitude E to the given magnitude F: unto the magnitudes E, F, C, find a fourth proportional D, which, by the hypothesis, can be done. Wherefore, because A is to B, as E to F: and as E to F, so is C to D; A is* to B, as C to D. But A is equal to C; therefore* B is equal to D. The magnitude B is therefore given*, because a magnitude D equal to it has been found.

* 1 Def.
* 11. 5.
* 14. 5.
* 1 Def.

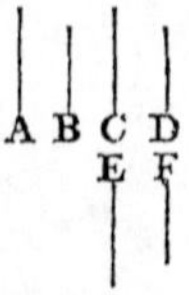

The limitation within the inverted commas is not in the Greek text, but is now necessarily added; and the same must be understood in all the propositions of the book which depend upon this second proposition, where it is not expressly mentioned. See the note upon it.

PROPOSITION III. (3.)

If any given magnitudes be added together, their sum shall be given.

Let any given magnitudes AB, BC be added together: their sum AC is given.

Because AB is given, a magnitude equal to it may* be found; let this be DE: and because BC is given, one equal to it may be found; let this be EF: wherefore, because AB is equal to DE, and BC equal to EF; the whole AC is equal to the whole DF; AC is therefore given, because DF has been found which is equal to it.

* 1 Def.

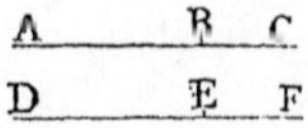

PROPOSITION IV. (4.)

If a given magnitude be taken from a given magnitude, the remaining magnitude shall be given.

From the given magnitude AB, let the given magnitude AC be taken: the remaining magnitude CB is given.

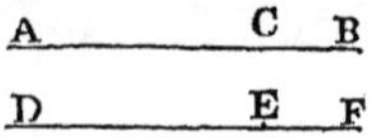

Because AB is given, a magnitude equal to it may* be found; let this be DF: and because AC is given, one equal to it may be found; let this be DE: wherefore, because AB is equal to DF, and

* 1 Def.

AC to DE, the remainder CB is equal to the remainder FE. * 1 Def. CB is therefore given *, because FE, which is equal to it, has been found.

(12.)

PROPOSITION V.

See N. *If of three magnitudes, the first together with the second be given, and also the second together with the third; either the first is equal to the third, or one of them is greater than the other by a given magnitude.*

Let AB, BC, CD be three magnitudes, of which AB together with BC, that is, AC, is given; and also BC together with CD, that is, BD, is given: either AB is equal to CD, or one of them is greater than the other by a given magnitude.

Because AC, BD are each of them given, they are either equal to one another, or not equal.

First, let them be equal, and because AC is equal to BD, take away the common part BC; therefore the remainder AB is equal to the remainder CD.

But if they be unequal, let AC be greater than BD, and make CE equal to BD. Therefore, CE is given, because BD is given. And the whole AC is given; * 4 Dat. therefore * AE the remainder is given.

And because EC is equal to BD, by taking BC from both, the remainder EB is equal to the remainder CD. And AE is given; wherefore AB exceeds EB, that is, CD, by the given magnitude AE.

(5.)

PROPOSITION VI.

See N. *If a magnitude have a given ratio to a part of it, it shall also have a given ratio to the remaining part of it.*

Let the magnitude AB have a given ratio to AC a part of it: it has also a given ratio to the remainder BC.

Because the ratio of AB to AC is given, a ratio may be * 2 Def. found * which is the same to it: let this be the ratio of DE, a given magnitude to the given magnitude DF.

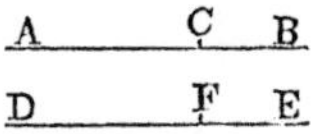

And because DE, * 4 Dat. DF are given, the remainder FE is * given: * E. 5. and because AB is to AC, as DE to DF, by conversion * AB is to BC, as DE to EF. Therefore, the ratio of AB to BC is given, because the ratio of the given magnitudes DE, EF, which is the same with it, has been found.

COR. From this it follows, that the parts AC, CB have a given ratio to one another: because as AB to BC, so is DE to EF; by division* AC is to CB, as DF to FE; and DF, FE are given; therefore* the ratio of AC to CB is given. * 17. 5. * 2 Def.

PROPOSITION VII. (6.)

If two magnitudes which have a given ratio to one another be added together, the whole magnitude shall have to each of them a given ratio. See N.

Let the magnitudes AB, BC, which have a given ratio to one another, be added together: the whole AC has to each of the magnitudes AB, BC, a given ratio.

Because the ratio of AB to BC is given, a ratio may be found* which is the same with it; let this be the ratio of the given magnitudes DE, EF: and because DE, EF are given, the whole DF is given*: and because as AB to BC, so is DE to EF; by composition* AC is to CB as DF to FE; and by conversion*, AC is to AB, as DF to DE: wherefore because AC is to each of the magnitudes AB, BC, as DF to each of the others DE, EF; the ratio of AC to each of the magnitudes AB, BC is given*. * 2 Def. * 3 Dat. * 18. 5. * E. 5. * 2 Def.

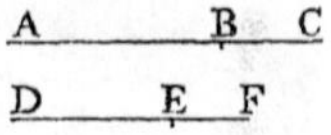

PROPOSITION VIII. (7.)

If a given magnitude be divided into two parts which have a given ratio to one another, and if a fourth proportional can be found to the sum of the two magnitudes by which the given ratio is exhibited, one of them, and the given magnitude; each of the parts is given. See N.

Let the given magnitude AB be divided into the parts AC, CB, which have a given ratio to one another: if a fourth proportional can be found to the above-named magnitudes, AC and CB are each of them given.

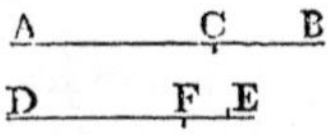

Because the ratio of AC to CB is given, the ratio of AB to BC is given*, therefore a ratio, which is the same with it, can be found*; let this be the ratio of the given magnitudes DE, EF: and because the given magnitude AB has to BC the * 7 Dat. * 2 Def.

given ratio of DE to EF, if unto DE, EF, AB, a fourth pro-
* 2 Dat. portional can be found, this which is BC is given*; and be-
* 4 Dat. cause AB is given, the other part AC is given*.

In the same manner, and with the like limitation, if the difference AC of two magnitudes AB, BC, which have a given ratio be given, each of the magnitudes AB, BC, is given.

(8.)

PROPOSITION IX.

Magnitudes which have given ratios to the same magnitude, have also a given ratio to one another.

Let A, C have each of them a given ratio to B: A has a given ratio to C.

Because the ratio of A to B is given, a ratio which is the
* 2 Def. same to it may be found*; let this be the ratio of the given magnitudes D, E: and because the ratio of B to C is given, a ratio which is the
* 2 Def. same with it may be found*: let this be the ratio of the given magnitudes F, G: to F, G, E, find a fourth proportional H, if it can be done; and because as A is to B, so is D to E: and as B to C, so is (F to G, and so is) E to H; ex æquali, as A to C, so is D to H: therefore the ratio of A to
* 2 Def. C is given*, because the ratio of the given magnitudes D and H, which is the same with it, has been found: but if a fourth proportional to F, G, E, cannot be found, then it can only be said that the ratio of A to C is compounded of the ratios of A to B, and B to C, that is, of the given ratios of D to E, and F to G.

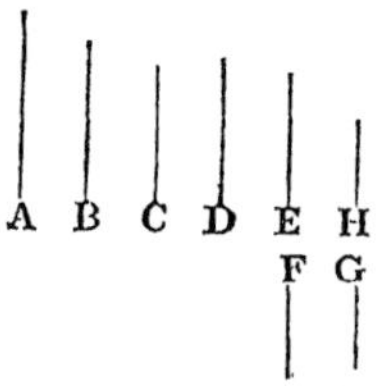

(9.)

PROPOSITION X.

If two or more magnitudes have given ratios to one another, and if they have given ratios, though they be not the same, to some other magnitudes; these other magnitudes shall also have given ratios to one another.

Let two or more magnitudes A, B, C, have given ratios to one another; and let them have given ratios, though they be not the same, to some other magnitudes D, E, F: the magnitudes D, E, F have given ratios to one another.

Because the ratio of A to B is given, and likewise the ratio of A to D: therefore the ratio of D to B is given*; but * 9 Dat.
the ratio of B to E is given; therefore* the ratio of D to E is given: * 9 Dat.
and because the ratio of B to C is given, and also the ratio of B to E; the ratio of E to C is given*: and * 9 Dat.
the ratio of C to F is given; wherefore the ratio of E to F is given; D, E, F have therefore given ratios to one another.

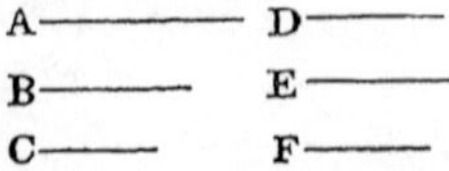

PROPOSITION XI. (22.)

If two magnitudes have each of them a given ratio to another magnitude, both of them together shall have a given ratio to that other.

Let the magnitudes AB, BC have a given ratio to the magnitude D: AC has a given ratio to the same D.

Because AB, BC have each of them a given ratio to D, the ratio of AB to BC is given*: and by composition, the ratio of * 9 Dat.
AC to CB is given*: but the ratio of BC to D is given: therefore* the ratio of AC to D is given. * 7 Dat. * 9 Dat.

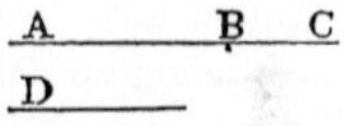

PROPOSITION XII. (23.)

If the whole have to the whole a given ratio, and the parts have to the parts, given, but not the same, ratios; every one of them, whole or part, shall have to every one a given ratio. See N.

Let the whole AB have a given ratio to the whole CD, and the parts AE, EB have given, but not the same, ratios to the parts CF, FD: every one shall have to every one, whole or part, a given ratio.

Because the ratio of AE to CF is given; as AE to CF, so make AB to CG; the ratio, therefore, of AB to CG is given: wherefore the ratio of the remainder EB to the remainder FG is given, because it is the same* with the ratio of AB to CG: and the ratio of EB to FD is given, wherefore the ratio of FD to FG is given*; and, by conversion, the * 19. 5. * 9 Dat.
ratio of FD to DG is given*: and because AB has to each of the magnitudes CD, CG, a given ratio, the ratio of CD to CG * 6 Dat.

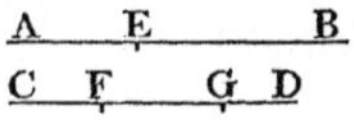

* 9 Dat. is given*, and therefore* the ratio of CD to DG is given:
* 6 Dat. but the ratio of GD to DF is given, wherefore* the ratio of
* 9 Dat. * Cor. 6. CD to DF is given, and consequently* the ratio of CF to FD
Dat. is given; but the ratio of CF to AE is given, as also the ratio
* 10 Dat. of FD to EB; wherefore* the ratio of AE to EB is given; as
* 7 Dat. also the ratio of AB to each of them*. The ratio, therefore, of every one to every one is given.

(24.)

PROPOSITION XIII.

See N. *If the first of three proportional straight lines have a given ratio to the third, the first shall also have a given ratio to the second.*

Let A, B, C be three proportional straight lines; that is, as A to B, so is B to C: if A has to C a given ratio, A shall also have to B a given ratio.

Because the ratio of A to C is given, a ratio which is the
* 2 Def. same with it may be found*: let this be the ratio of the given
* 13. 6. straight lines D, E; and between D and E, find a* mean proportional F; therefore the rectangle contained by D and E is equal to the square of F, and the rectangle D, E is given, because its sides D, E are given; wherefore the square of F, and the straight line F is given: and because, as A is to C,
* 2 Cor. 20. 6. so is D to E; but as A to C, so is* the square of
* 2 Cor. 20. 6. A to the square of B; and as D to E, so is* the square of D to the square of F: therefore the
* 11. 5. square* of A is to the square of B, as the square of
* 22. 6. D to the square of F: as therefore* the straight line A to the straight line B, so is the straight line D to the straight line
* 2 Def. F; therefore the ratio of A to B is given*, because the ratio of the given straight lines D, F, which is the same with it, has been found.

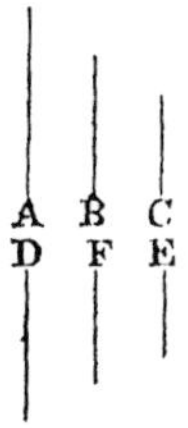

(A.)

PROPOSITION XIV.

See N. *If a magnitude, together with a given magnitude, have a given ratio to another magnitude; the excess of this other magnitude above a given magnitude, has a given ratio to the first magnitude: and if the excess of a magnitude above a given magnitude have a given ratio to another magnitude; this other magnitude, together with a given magnitude, has a given ratio to the first magnitude.*

Let the magnitude AB, together with the given magnitude BE, that is AE, have a given ratio to the magnitude CD: the excess of CD above a given magnitude has a given ratio to AB.

Because the ratio of AE to CD is given; as AE to CD, so make BE to FD; therefore the ratio of BE to FD is given, and BE is given; wherefore FD is given*: and * 2 Dat.
because as AE to CD, so is BE to FD, the remainder AB is* to the remainder CF, as * 19. 5.
AE to CD: but the ratio of AE to CD is given; therefore the ratio of AB to CF is given; that is, CF, the excess of CD above the given magnitude FD, has a given ratio to AB.

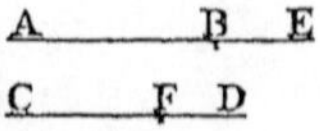

Next, let the excess of the magnitude AB above the given magnitude BE, that is, let AE have a given ratio to the magnitude CD; CD together with a given magnitude has a given ratio to AB.

Because the ratio of AE to CD is given; as AE to CD so make BE to FD; therefore the ratio of BE to FD is given; and BE is given; wherefore FD is given*: * 2 Dat.
and because, as AE to CD, so is BE to FD, AB is to CF, as* AE to CD: but the ratio * 12. 5.
of AE to CD is given, therefore the ratio of AB to CF is given; that is, CF, which is equal to CD together with the given magnitude DF, has a given ratio to AB.

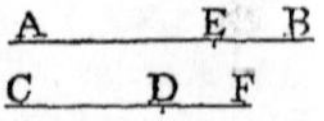

PROPOSITION XV. (B.)

If a magnitude, together with that to which another magnitude has a given ratio, be given; the sum of this other, and that to which the first magnitude has a given ratio, is given. See N.

Let AB, CD be two magnitudes, of which AB, together with BE, to which CD has a given ratio, is given: CD is given, together with that magnitude to which AB has a given ratio.

Because the ratio of CD to BE is given; as BE to CD so make AE to FD; therefore the ratio of AE to FD is given, and AE is given, wherefore* FD is given: and because as BE to CD, so * 2 Dat.
is AE to FD: AB is* to FC, as BE to CD: and the ratio of BE * Cor. 19. 5.

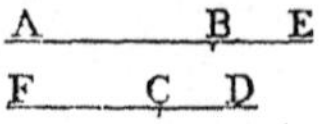

to CD is given: wherefore the ratio of AB to FC is given: and FD is given, that is, CD together with FC, to which AB has a given ratio, is given.

(10.) PROPOSITION XVI.

See N. *If the excess of a magnitude above a given magnitude, have a given ratio to another magnitude, the excess of both together above a given magnitude, shall have to that other a given ratio: and if the excess of two magnitudes together above a given magnitude, have to one of them a given ratio, either the excess of the other above a given magnitude has to that one a given ratio, or the other is given together with the magnitude to which that one has a given ratio.*

Let the excess of the magnitude AB above a given magnitude, have a given ratio to the magnitude BC: the excess of AC both of them together, above the given magnitude, has a given ratio to BC.

Let AD be the given magnitude, the excess of AB above which, viz. DB, has a given ratio to BC: and because DB has a given ratio to BC,

* 7 Dat. the ratio of DC to CB is given*, and AD is given; therefore DC, the excess of AC above the given magnitude AD, has a given ratio to BC.

Next, let the excess of two magnitudes AB, BC together, above a given magnitude, have to one of them BC a given ratio; either the excess

of the other of them AB above the given magnitude, shall have to BC a given ratio; or AB is given, together with the magnitude to which BC has a given ratio.

Let AD be the given magnitude, and first let it be less than AB; and because DC the excess of AC above AD has a given ratio to BC, DB has* a given ratio to BC; that is, DB the excess of AB above the given magnitude AD has a given ratio to BC.

* Cor. 6. Dat.

But let the given magnitude be greater than AB, and make AE equal to it; and because EC, the excess of AC above AE, has to BC a given ratio, BC has* a given ratio to BE; and because AE is given, AB together with BE, to which BC has a given ratio, is given.

* 6 Dat.

PROPOSITION XVII. (11.)

If the excess of a magnitude above a given magnitude, have a given ratio to another magnitude, the excess of the same first magnitude above a given magnitude, shall have a given ratio to both the magnitudes together. And if the excess of either of two magnitudes above a given magnitude, have a given ratio to both magnitudes together, the excess of the same above a given magnitude, shall have a given ratio to the other. See N.

Let the excess of the magnitude AB above a given magnitude, have a given ratio to the magnitude BC: the excess of AB above a given magnitude, has a given ratio to AC.

Let AD be the given magnitude; and because DB, the excess of AB above AD, has a given ratio to BC; the ratio of DC to DB is given*; make the ratio of AD to DE the same with this ratio; therefore the ratio of AD to DE is given; and AD is given, wherefore* DE and the remainder AE are given. And because as DC to DB, so is AD to DE, AC is* to EB, as DC to DB; and the ratio of DC to DB is given; wherefore the ratio of AC to EB is given; and because the ratio of EB to AC is given, and that AE is given, therefore EB the excess of AB above the given magnitude AE, has a given ratio to AC. * 7 Dat. * 2 Dat. * 12. 5.

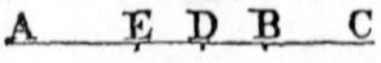

Next, let the excess of AB above a given magnitude have a given ratio to AB and BC together, that is, to AC; the excess of AB above a given magnitude has a given ratio to BC.

Let AE be the given magnitude; and because EB the excess of AB above AE has to AC a given ratio, as AC to EB so make AD to DE; therefore the ratio of AD to DE is given, as also* the ratio of AD to AE: and AE is given, wherefore* AD is given: and because, as the whole AC to the whole EB, so is AD to DE, the remainder DC is* to the remainder DB, as AC to EB; and the ratio of AC to EB is given; wherefore the ratio of DC to DB is given, as also* the ratio of DB to BC: and AD is given; therefore DB, the excess of AB above a given magnitude AD, has a given ratio to BC. * 6 Dat. * 2 Dat. * 19. 5. * Cor. 6. Dat.

(14.)

PROPOSITION XVIII.

If to each of two magnitudes which have a given ratio to one another, a given magnitude be added, the wholes shall either have a given ratio to one another, or the excess of one of them above a given magnitude, shall have a given ratio to the other.

Let the two magnitudes AB, CD have a given ratio to one another, and to AB let the given magnitude BE be added, and the given magnitude DF to CD: the wholes AE, CF either have a given ratio to one another, or the excess of one of them above a given magnitude, has a given ratio to the other.

* 1 Dat. Because BE, DF are each of them given, their ratio * is given, and if this ratio be the same with the ratio of AB to
* 12. 5. CD, the ratio of AE to CF, which is the same* with the given ratio of AB to CD, shall be given.

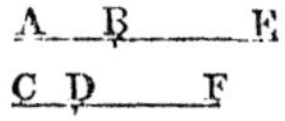

But if the ratio of BE to DF be not the same with the ratio of AB to CD, either it is greater than the ratio of AB to CD, or, by inversion, the ratio of DF to BE is greater than the ratio of CD to AB: first, let the ratio of BE to DF be greater than the ratio of AB to CD; and as AB to CD, so make BG to DF; therefore the ratio of BG to DF is given;
* 2 Dat. and DF is given, therefore * BG is given: and because BE has a greater ratio to DF
* 10. 5. than (AB to CD, that is than) BG to DF, BE is greater * than BG: and because as AB to CD, so is BG to DF; therefore AG
* 12. 5. is * to CF, as AB to CD: but the ratio of AB to CD is given, wherefore the ratio of AG to CF is given; and because BE, BG are each of them given, GE is given: therefore AG, the excess of AE above a given magnitude GE, has a given ratio to CF. The other case is demonstrated in the same manner.

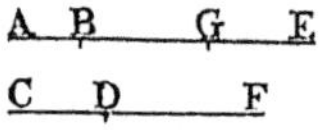

(15.)

PROPOSITION XIX.

If from each of two magnitudes which have a given ratio to one another, a given magnitude be taken, the remainders shall either have a given ratio to one another, or the excess of one of them above a given magnitude, shall have a given ratio to the other.

Let the magnitudes AB, CD have a given ratio to one another, and from AB let the given magnitude AE be taken, and from CD the given magnitude CF: the remainders EB, FD shall either have a given ratio to one another, or the excess of one of them above a given magnitude shall have a given ratio to the other.

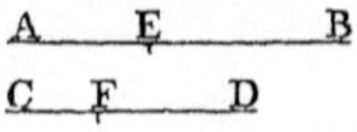

Because AE, CF are each of them given, their ratio is given *; and if this ratio be the same with the ratio of AB to CD, the ratio of the remainder EB to the remainder FD, which is the same * with the given ratio of AB to CD, shall be given. * 1 Dat. * 19. 5.

But if the ratio of AB to CD be not the same with the ratio of AE to CF, either it is greater than the ratio of AE to CF or, by inversion, the ratio of CD to AB is greater than the ratio of CF to AE: first, let the ratio of AB to CD be greater than the ratio of AE to CF, and as AB to CD, so make AG to CF: therefore the ratio of AG to CF is given, and CF is given, wherefore * AG is given: and because the ratio of AB to CD, that is, the ratio of AG to CF, is greater than the ratio of AE to CF; AG is greater * than AE: and AG, AE are given, therefore the remainder EG is given; and as AB to CD, so is AG to CF, and so is * the remainder GB to the remainder FD; and the ratio of AB to CD is given: wherefore the ratio of GB to FD is given; therefore GB, the excess of EB above a given magnitude EG, has a given ratio to FD. In the same manner the other case is demonstrated. * 2 Dat. * 10. 5. * 19. 5.

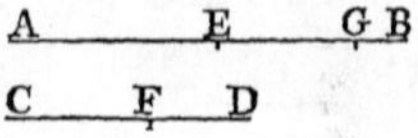

PROPOSITION XX. (16.)

If to one of two magnitudes which have a given ratio to one another, a given magnitude be added, and from the other a given magnitude be taken; the excess of the sum above a given magnitude, shall have a given ratio to the remainder.

Let the two magnitudes AB, CD have a given ratio to one another, and to AB let the given magnitude EA be added, and from CD let the given magnitude CF be taken: the excess of the sum EB above a given magnitude has a given ratio to the remainder FD.

Because the ratio of AB to CD is given, make as AB to CD, so AG to CF: therefore the ratio of AG to CF is given, and
* 2 Dat. CF is given, wherefore * AG is given: and EA is given, therefore the whole EG is given; and because as AB to CD, so is
* 19. 5. AG to CF, and so is * the remainder GB to the remainder FD; the ratio of GB to FD is given. And EG is given, therefore GB, the excess of the sum EB above the given magnitude EG, has a given ratio to the remainder FD.

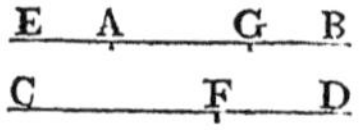

(C.) PROPOSITION XXI.

See N. *If two magnitudes have a given ratio to one another, if a given magnitude be added to one of them, and the other be taken from a given magnitude; the sum, together with the magnitude to which the remainder has a given ratio, is given: and the remainder is given together with the magnitude to which the sum has a given ratio.*

Let the two magnitudes AB, CD have a given ratio to one another; and to AB let the given magnitude BE be added, and let CD be taken from the given magnitude FD: the sum AE is given, together with the magnitude to which the remainder FC has a given ratio.

Because the ratio of AB to CD is given, make as AB to CD, so GB to FD: therefore the ratio of GB to FD is given, and FD
* 2 Dat. is given, wherefore GB is given *; and BE is given, the whole GE is therefore given: and because as AB to CD, so is
* 19. 5. GB to FD, and so is * GA to FC; the ratio of GA to FC is given: and AE together with GA is given, because GE is given; therefore, the sum AE, together with GA, to which the remainder FC has a given ratio, is given. The second part is manifest from Prop. 15.

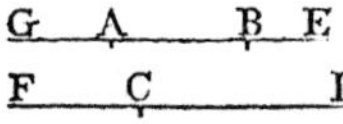

(D.) PROPOSITION XXII.

See N. *If two magnitudes have a given ratio to one another, if from one of them a given magnitude be taken, and the other be taken from a given magnitude; each of the remainders is given, together with the magnitude to which the other remainder has a given ratio.*

Let the two magnitudes AB, CD have a given ratio to one another, and from AB let the given magnitude AE be taken, and let CD be taken from the given magnitude CF: the remainder EB is given, together with the magnitude to which the other remainder DF has a given ratio.

Because the ratio of AB to CD is given, make as AB to CD, so AG to CF: the ratio of AG to CF is therefore given, and CF is given, wherefore* AG is given; and * 2 Dat.
AE is given, and therefore the remainder EG is given: and because as AB to CD, so is AG to CF: and so is* the remainder BG * 19. 5.
to the remainder DF; the ratio of BG to DF is given: and EB together with BG is given, because EG is given: therefore the remainder EB, together with BG, to which DF, the other remainder has a given ratio, is given. The second part is plain from this and Prop. 15.

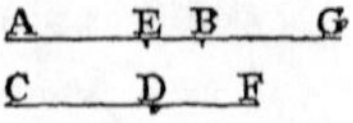

PROPOSITION XXIII. (20.)

If from two given magnitudes there be taken magnitudes which have a given ratio to one another, the remainders shall either have a given ratio to one another, or the excess of one of them above a given magnitude shall have a given ratio to the other. See N.

Let AB, CD be two given magnitudes, and from them let the magnitudes AE, CF which have a given ratio to one another, be taken: the remainders EB, FD, either have a given ratio to one another, or the excess of one of them above a given magnitude has a given ratio to the other.

Because AB, CD are each of them given, the ratio of AB to CD is given: and if this ratio be the same with the ratio of AE to CF, then the remainder EB has* the same given ratio to the remainder FD. * 19. 5.

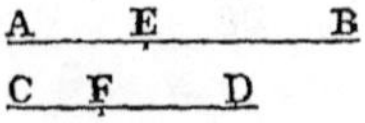

But if the ratio of AB to CD be not the same with the ratio of AE to CF, it is either greater than it, or, by inversion, the ratio of CD to AB is greater than the ratio of CF to AE: first, let the ratio of AB to CD be greater than the ratio of AE to CF; and as AE to CF, so make AG to CD; therefore the ratio of AG to CD is given, because the ratio of AE to CF is given; and CD is given; wherefore* AG is given; and because the * 2 Dat.

ratio of AB to CD is greater than the ratio of (AE to CF that
* 10. 5. is, than the ratio of) AG to CD; AB is greater * than AG: and AB, AG, are given; therefore the remainder BG is given: and because as AE
* 19. 5. to CF, so is AG to CD, and so is * EG to FD; the ratio of EG to FD is given: and GB is given; therefore EG, the excess of EB above a given magnitude GB, has a given ratio to FD. The other case is shewn in the same way.

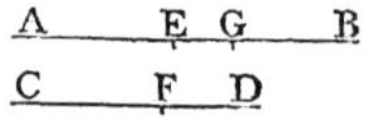

(13.)

PROPOSITION XXIV.

See N. *If there be three magnitudes, the first of which has a given ratio to the second, and the excess of the second above a given magnitude has a given ratio to the third; the excess of the first above a given magnitude shall also have a given ratio to the third.*

Let AB, CD, E be the three magnitudes of which AB has a given ratio to CD, and the excess of CD above a given magnitude has a given ratio to E: the excess of AB above a given magnitude has a given ratio to E.

Let CF be the given magnitude, the excess of CD above which, viz. FD has a given ratio to E: and because the ratio of AB to CD is given, as AB to CD, so make AG to CF; therefore the ratio of AG to CF, is given: and CF is given, where-
* 2 Dat. fore * AG is given: and because as AB to CD, so
* 19. 5. is AG to CF, and so is * GB to FD; the ratio of GB to FD is given. And the ratio of FD to E is given,
* 9 Dat. wherefore* the ratio of GB to E is given, and AG is given; therefore GB the excess of AB above a given magnitude AG has a given ratio to E.

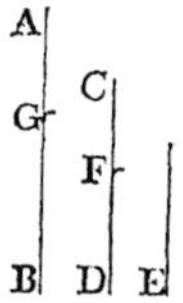

Cor. 1. And if the first have a given ratio to the second, and the excess of the first above a given magnitude, have a given ratio to the third; the excess of the second above a given magnitude, shall have a given ratio to the third. For, if the second be called the first, and the first the second, this corollary will be the same with the proposition.

Cor. 2. Also, if the first have a given ratio to the second, and the excess of the third above a given magnitude, have also a given ratio to the second, the same excess shall have a given ratio to the first; as is evident from the 9th Dat.

PROPOSITION XXV. (17.)

If there be three magnitudes, the excess of the first whereof above a given magnitude has a given ratio to the second, and the excess of the third above a given magnitude has a given ratio to the same second; the first shall either have a given ratio to the third, or the excess of one of them above a given magnitude shall have a given ratio to the other.

Let AB, C, DE be three magnitudes, and let the excesses of each of the two AB, DE, above given magnitudes, have given ratios to C: AB, DE, either have a given ratio to one another, or the excess of one of them above a given magnitude, has a given ratio to the other.

Let FB the excess of AB above the given magnitude AF, have a given ratio to C; and let GE, the excess of DE above the given magnitude DG, have a given ratio to C; and because FB, GE have each of them a given ratio to C, they have a given * ratio to one another. But to FB, GE the given magnitudes AF, DG are added; therefore * the whole magnitudes AB, DE have either a given ratio to one another, or the excess of one of them above a given magnitude, has a given ratio to the other.

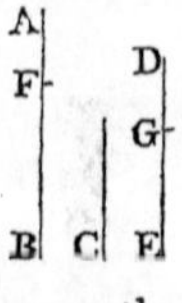

* 9 Dat.

* 18 Dat.

PROPOSITION XXVI. (18.)

If there be three magnitudes, the excesses of one of which above given magnitudes, have given ratios to the other two magnitudes; these two shall either have a given ratio to one another, or the excess of one of them above a given magnitude, shall have a given ratio to the other.

Let AB, CD, EF be three magnitudes, and let GD, the excess of one of them CD above the given magnitude CG, have a given ratio to AB: and also let KD, the excess of the same CD above the given magnitude CK, have a given ratio to EF: either AB has a given ratio to EF, or the excess of one of them above a given magnitude has a given ratio to the other.

Because GD has a given ratio to AB, as GD to AB, so make CG to HA; therefore the ratio of CG to HA is given; and CG is given, wherefore * HA is given: and because as GD to AB, * 2 Dat.

* 12. 5. so is CG to HA, and so is * CD to HB: the ratio of CD to HB is given: also because KD has a given ratio to EF, as KD to EF, so make CK to LE: therefore the ratio of CK to LE is given; and CK is given,
* 2 Dat. wherefore LE * is given; and because as KD to
* 12. 5. EF, so is CK to LE, and so * is CD to LF; the ratio of CD to LF is given: but the ratio of CD
* 9 Dat. to HB is given; wherefore * the ratio of HB to LF is given; and from HB, LF the given magnitudes HA, LE being taken, the remainders AB, EF shall either have a given ratio to one another, or the excess of one of them above a given
* 19 Dat. magnitude has a given ratio to the other *.

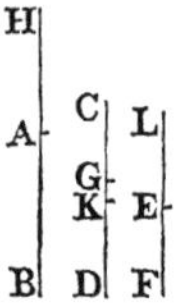

Another Demonstration.

Let AB, C, DE be three magnitudes, and let the excesses of one of them C above given magnitudes, have given ratios to AB, and DE: either AB, DE have a given ratio to one another, or the excess of one of them above a given magnitude, has a given ratio to the other.

Because the excess of C above a given magnitude has a given
* 14 Dat. ratio to AB: therefore * AB together with a given magnitude has a given ratio to C: let this given magnitude be AF, wherefore FB has a given ratio to C: also because the excess of C above a given magnitude
* 14 Dat. has a given ratio to DE; therefore * DE together with a given magnitude has a given ratio to C: let this given magnitude be DG, wherefore GE has a
* 9 Dat. given ratio to C: and FB has a given ratio to C, therefore * the ratio of FB to GE is given: and from FB, GE, the given magnitudes AF, DG being taken, the remainders AB, DE, either have a given ratio to one another, or the excess of one of them above a given magnitude, has a given ratio to the
* 19 Dat. other *.

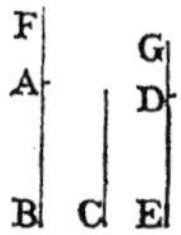

(19.)

PROPOSITION XXVII.

If there be three magnitudes, the excess of the first of which above a given magnitude has a given ratio to the second, and the excess of the second above a given magnitude has also a given ratio to the third; the excess of the first above a given magnitude shall have a given ratio to the third.

Let AB, CD, E be three magnitudes, the excess of the first of which AB above the given magnitude AG, viz. GB, has a given ratio to CD: and FD the excess of CD above the given magnitude CF, has a given ratio to E: the excess of AB above a given magnitude has a given ratio to E.

Because the ratio of GB to CD is given, as GB to CD, so make GH, to CF; therefore the ratio of GH to CF is given; and CF is given, wherefore * GH is given; and AG * 2 Dat.
is given, wherefore the whole AH is given: and because as GB to CD, so is GH to CF, and so is * * 19. 5.
the remainder HB to the remainder FD; the ratio of HB to FD is given: and the ratio of FD to E is given, wherefore * the ratio of HB to E is given: * 9 Dat.
and AH is given; therefore HB the excess of AB above a given magnitude AH, has a given ratio to E.

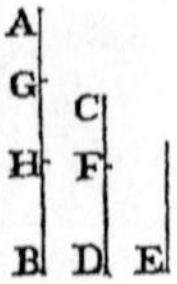

"*Otherwise*,

"Let AB, C, D be three magnitudes, the excess EB of the "first of which AB above the given magnitude AE has a given "ratio to C, and the excess of C above a given magni- "tude has a given ratio to D: the excess of AB "above a given magnitude has a given ratio to D.

"Because EB has a given ratio to C, and the "excess of C above a given magnitude has a given "ratio to D, therefore * the excess of EB above * 24 Dat.
"a given magnitude has a given ratio to D: let "this given magnitude be EF; therefore FB the excess of EB "above EF has a given ratio to D: and AF is given, because "AE, EF are given: therefore FB the excess of AB above a "given magnitude AF, has a given ratio to D."

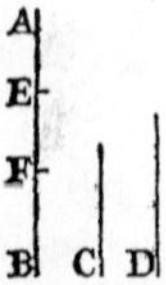

PROPOSITION XXVIII. (25.)

If two lines given in position cut one another, the point or points in which they cut one another are given. See N.

Let two lines AB, CD, given in position, cut one another in the point E: the point E is given.

Because the lines AB, CD are given in position, they have always the same situation *, and therefore the point, or points, * 4 Def.
in which they cut one another have always the same situation: and because the lines AB, CD can be found *, the point, or points, in which they cut one another, are * 4 Def.
likewise found; and therefore are given in position *. * 4 Def.

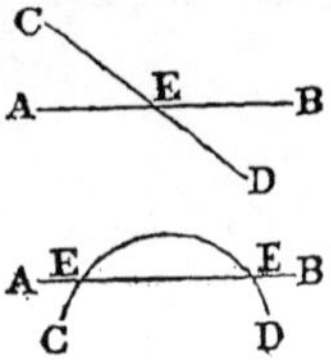

(26.) PROPOSITION XXIX.

If the extremities of a straight line be given in position, the straight line is given in position and magnitude.

Because the extremities of the straight line are given, * 4 Def. they can be found *: let these be the points A, B, between * 1 Postulate. which a straight line AB can be drawn *; this has an invariable position, because between two given points there can be drawn but one straight line: and when the straight line AB is drawn, its magnitude is at the same time exhibited, or given: therefore the straight line AB is given in position and magnitude.

(27.) PROPOSITION XXX.

If one of the extremities of a straight line given in position and magnitude be given, the other extremity shall also be given.

Let the point A be given, to wit, one of the extremities of a straight line given in magnitude, and which lies in the straight line AC given in position: the other extremity is also given.

Because the straight line is given in magnitude, one equal * 1 Def. to it can be found *: let this be the straight line D: from the greater straight line AC, cut off AB equal to the lesser D: therefore the other extremity B of the straight line AB is found: and the point B has always the same situation; because any other point in AC, upon the same side of A, cuts off between it and the point A, a greater or less straight line * 4 Def. than AB, that is, than D: therefore the point B is given *: and it is plain another such point can be found in AC, produced upon the other side of the point A.

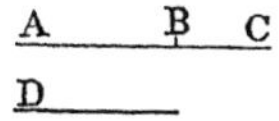

(28.) PROPOSITION XXXI.

If a straight line be drawn through a given point, parallel to a straight line given in position, that straight line is given in position.

Let A be a given point, and BC a straight line given in position: the straight line drawn through A, parallel to BC, is given in position.

Through A, draw* the straight line DAE parallel to BC; the straight line DAE has always the same position, because no other straight line can be drawn through A parallel to BC: therefore, the straight line DAE, which has been found, is given* in position. * 31. 1.

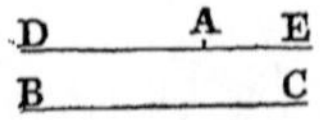

* 4 Def.

PROPOSITION XXXII. (29.)

If a straight line be drawn to a given point in a straight line given in position, and make a given angle with it, that straight line is given in position.

Let AB be a straight line given in position, and C a given point in it: the straight line drawn to C, which makes a given angle with CB, is given in position.

Because the angle is given, one equal to it can be found*; let this be the angle at D. At the given point C, in the given straight line AB, make* the angle ECB equal to the angle at D: therefore the straight line EC has always the same situation, because any other straight line FC, drawn to the point C, makes with CB a greater or less angle than the angle EBC, or the angle at D: therefore the straight line EC, which has been found, is given in position.

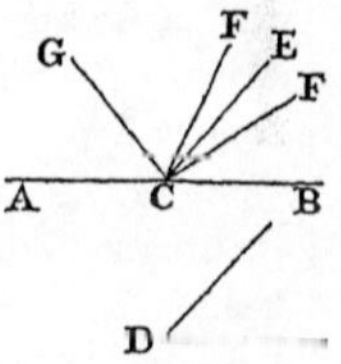

* 1 Def.

* 23. 1.

It is to be observed, that there are two straight lines EC, GC, upon one side of AB, that make equal angles with it, and which make equal angles with it when produced to the other side.

PROPOSITION XXXIII. (30.)

If a straight line be drawn from a given point to a straight line given in position, and make a given angle with it, that straight line is given in position.

From the given point A, let the straight line AD be drawn to the straight line BC given in position, and make with it a given angle ADC: AD is given in position.

* 31. 1. Through the point A, draw* the straight line EAF parallel to BC; and because through the given point A, the straight line EAF is drawn parallel to BC, which is given in position, EAF is therefore given
* 31 Dat. in position*. And because the straight line AD meets the parallels BC, EF, the
* 29. 1. angle EAD is equal* to the angle ADC;

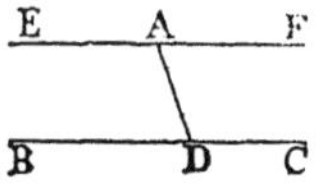

and ADC is given, wherefore also the angle EAD is given: therefore, because the straight line DA is drawn to the given point A, in the straight line EF given in position, and makes
* 32 Dat. with it a given angle EAD, AD is given* in position.

(31.)

PROPOSITION XXXIV.

See N. *If from a given point to a straight line given in position, a straight line be drawn which is given in magnitude, the same is also given in position.*

Let A be a given point, and BC a straight line given in position: a straight line given in magnitude, drawn from the point A to BC, is given in position.

Because the straight line is given in magnitude, one equal
* 1 Def. to it can be found*; let this be the straight line D: from the point A, draw AE perpendicular to BC: and because AE is the shortest of all the straight lines which can be drawn from the point A to BC, the straight line D, to which one equal is to be drawn from the point A to BC, cannot be less than AE. If therefore D be equal to

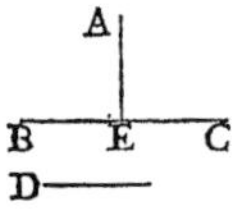

AE, AE is the straight line given in magnitude drawn from the given point A to BC: and it is evident that AE is given
* 33 Dat. in position*, because it is drawn from the given point A to BC, which is given in position, and makes with BC the given angle AEC.

But if the straight line D be not equal to AE, it must be greater than it: produce AE, and make AF equal to D; and from the centre A, at the distance AF, describe the circle GFH, and join AG, AH: because the circle GFH is given
* 6 Def. in position*, and the straight line BC is also given in position; therefore their

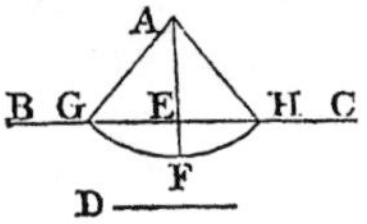

* 28 Dat. intersection G is given*; and the point A is given; where-

fore AG is given in position *; that is, the straight line AG given in magnitude (for it is equal to D), and drawn from the given point A, to the straight line BC given in position, is also given in position; and in like manner AH is given in position: therefore in this case there are two straight lines AG, AH, of the same given magnitude which can be drawn from a given point A, to a straight line BC given in position. * 29 Dat.

PROPOSITION XXXV. (32.)

If a straight line be drawn between two parallel straight lines given in position, and make given angles with them, the straight line is given in magnitude.

Let the straight line EF be drawn between the parallels AB, CD, which are given in position, and make the given angles BEF, EFD: EF is given in magnitude.

In CD take the given point G; and through G, draw * GH parallel to EF: and because CD meets the parallels GH, EF, the angle EFD is equal * to the angle HGD: and EFD is a given angle; wherefore the angle HGD is given: and because HG is drawn to the given point G, in the straight line CD, given in position, and makes a given angle HGD, the straight line HG is given in position *: and AB is given in position; therefore the point H is given *: and the point G is also given, wherefore GH is given in magnitude *: and EF is equal to it, therefore EF is given in magnitude. * 31. 1. * 29. 1. * 32 Dat. * 28 Dat. * 29 Dat.

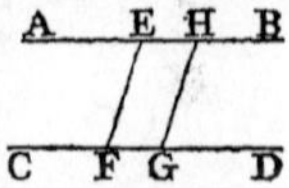

PROPOSITION XXXVI. (33.)

If a straight line given in magnitude, be drawn between two parallel straight lines given in position, it shall make given angles with the parallels. See N.

Let the straight line EF given in magnitude be drawn between the parallel straight lines AB, CD, which are given in position: the angles AEF, EFC shall be given.

Because EF is given in magnitude, a straight line equal to it can be found *; let this be G: in AB take a given point H, and from it draw * HK perpendicular to CD: therefore the straight line G, that is, EF cannot be less than HK: and if G be equal to HK, EF also is * 1 Def. * 12. 1.

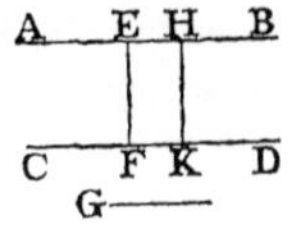

equal to it; wherefore EF is at right angles to CD; for if it be not, EF would be greater than HK, which is absurd. Therefore the angle EFD is a right, and consequently a given angle.

But if the straight line G be not equal to HK, it must be greater than it; produce HK, and take HL equal to G; and from the centre H, at the distance HL, describe the circle
* 6 Def. MLN, and join HM, HN: and because the circle* MLN, and the straight line CD, are given in position, the points M, N
* 28 Dat. are* given: and the point H is given, wherefore the straight lines HM, HN
* 29 Dat. are given in position*: and CD is given in position; therefore the angles HMN, HNM are given in posi-
* A. Def. tion*: of the straight lines HM, HN,

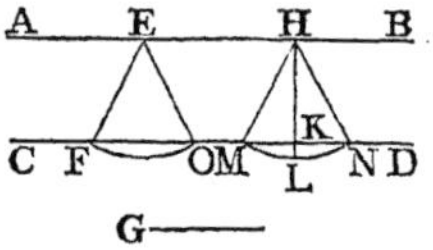

let HN be that which is not parallel to EF, for EF cannot be parallel to both of them; and draw EO parallel to HN: EO
* 34. 1. therefore is equal* to HN, that is, to G; and EF is equal to G; wherefore EO is equal to EF, and the angle EFO to the angle
* 29. 1. EOF, that is*, to the given angle HNM, and because the angle HNM, which is equal to the angle EFO, or EFD, has been found; therefore the angle EFD, that is, the angle AEF, is
* 1 Def. given in magnitude*; and consequently the angle EFC.

(E.)

PROPOSITION XXXVII.

See N. *If a straight line given in magnitude be drawn from a point to a straight line given in position, in a given angle, the straight line drawn through that point parallel to the straight line given in position, is given in position.*

Let the straight line AD given in magnitude be drawn from the point A, to the straight line BC given in position, in the given angle ADC: the straight line EAF drawn through A, parallel to BC, is given in position.

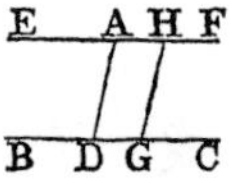

In BC take a given point G, and draw GH parallel to AD: and because HG is drawn to a given point G in the straight line BC given
* 29. 1. in position, in a given angle HGC, for it is equal* to the
* 32 Dat. given angle ADC, HG is given in position*: but it is given
* 34. 1. also in magnitude, because it is equal to* AD, which is given in magnitude: therefore because G, one of the extremities of the straight line GH, given in position and magnitude, is given,

the other extremity H is given*; and the straight line EAF, * 30 Dat.
which is drawn through the given point H, parallel to BC given in position, is therefore given* in position. * 31 Dat.

PROPOSITION XXXVIII. (34.)

If a straight line be drawn from a given point to two parallel straight lines given in position, the ratio of the segments between the given point and the parallels shall be given.

Let the straight line EFG be drawn from the given point E, to the parallels AB, CD: the ratio of EF to EG is given.

From the point E, draw EHK perpendicular to CD; and because from a given point E, the straight line EK is drawn to CD which is given in position, in a given angle EKC, EK is given in position*; and AB, * 33 Dat.
CD are given in position; therefore * the points H, K are given: * 28 Dat.

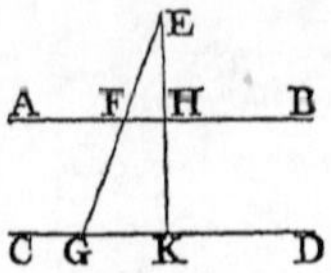

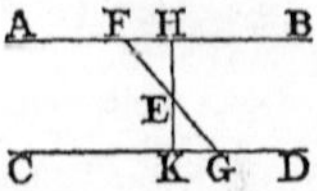

and the point E is given; wherefore * EH, EK are given in * 29 Dat.
magnitude, and the ratio * of them is therefore given. But * 1 Dat.
as EH to EK, so is EF to EG, because AB, CD are parallels; therefore the ratio of EF to EG is given.

PROPOSITION XXXIX. (35, 36.)

If the ratio of the segments of a straight line between a given point in it and two parallel straight lines be given; if one of the parallels be given in position, the other is also given in position. See N.

From the given point A, let the straight line AED be drawn to the two parallel straight lines FG, BC, and let the ratio of

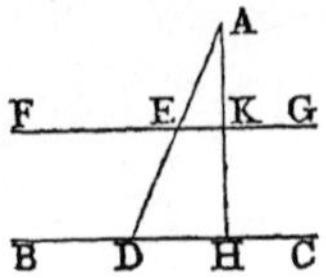

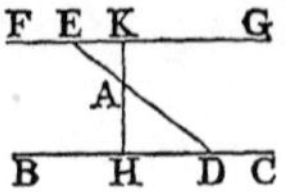

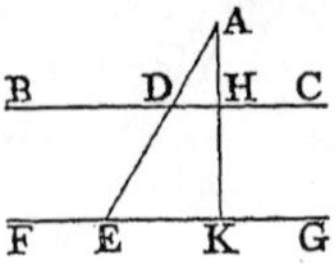

the segments AE, AD, be given: if one of the parallels BC be given in position, the other FG is also given in position.

From the point A, draw AH perpendicular to BC, and let it meet FG in K; and because AH is drawn from the given point A, to the straight line BC given in position, and makes
* 33 Dat. a given angle AHD, AH is given * in position; and BC is like-
* 28 Dat. wise given in position, therefore the point H is given *: the
* 29 Dat. point A is also given; wherefore AH is given in magnitude *, and because FG, BC are parallels, as AE to AD, so is AK to AH; and the ratio of AE to AD is given, wherefore the ratio of AK to AH is given; but AH is given in magnitude, there-
* 2 Dat. fore * AK is given in magnitude; and it is also given in posi-
* 30 Dat. tion, and the point A is given: wherefore * the point K is given. And because the straight line FG is drawn through the given point K, parallel to BC which is given in position,
* 31 Dat. therefore * FG is given in position.

(37, 38.) PROPOSITION XL.

See N. *If the ratio of the segments of a straight line into which it is cut by three parallel straight lines be given; if two of the parallels be given in position, the third is also given in position.*

Let AB, CD, HK be three parallel straight lines, of which AB, CD are given in position; and let the ratio of the segments GE, GF, into which the straight line GEF is cut by the three parallels, be given: the third parallel HK is given in position.

In AB take a given point L, and draw LM perpendicular to CD, meeting HK in N: because LM is drawn from the given point L, to CD which is given in position, and makes a given angle LMD, LM is given in
* 33 Dat. position *; and CD is given in position, where-

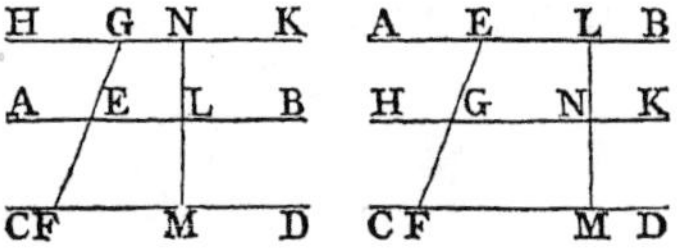

* 28 Dat. fore the point M is given *; and the point L is given; LM is
* 29 Dat. therefore given in magnitude *; and because the ratio of GE to GF is given, and as GE to GF, so is NL to NM; the ratio
* Cor. 6. or 7 Dat. of NL to NM is given; and therefore * the ratio of ML to
* Cor. 6. or 7 Dat. LN is given; but ML is given in magnitude *, wherefore *
* 2 Dat. LN is given in magnitude: and it is also given in position,
* 30 Dat. and the point L is given, wherefore * the point N is given: and because the straight line HK is drawn through the given point

N, parallel to CD which is given in position, therefore HK is given in position*. * 31 Dat.

PROPOSITION XLI. (F.)

If a straight line meet three parallel straight lines which are given in position, the segments into which they cut it have a given ratio. See N.

Let the parallel straight lines AB, CD, EF, given in position, be cut by the straight line GHK: the ratio of GH to HK is given.

In AB take a given point L, and draw LM perpendicular to CD, meeting EF in N; therefore* LM is given in position: and CD, EF are given in position, wherefore the points M, N are given: and the point L is given; therefore* the straight lines LM, MN are given in magnitude; and the ratio of LM to MN is therefore given*: but as LM to MN, so is GH to HK; wherefore the ratio of GH to HK is given.

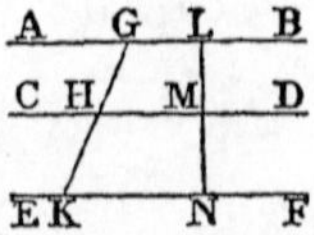

* 33 Dat.
* 29 Dat.
* 1 Dat.

PROPOSITION XLII. (39.)

If each of the sides of a triangle be given in magnitude, the triangle is given in species. See N.

Let each of the sides of the triangle ABC be given in magnitude: the triangle ABC is given in species.

Make a triangle* DEF, the sides of which are equal, each to each, to the given straight lines AB, BC, CA; which can be done, because any two of them must be greater than the third; and let DE be equal to AB, EF to BC, and FD to CA; and because the two sides ED, DF are equal to the two BA, AC, each to each, and the base EF equal to the base BC; the angle EDF is equal* to the angle BAC; therefore, because the angle EDF, which is equal to the angle BAC, has been found, the angle BAC is given*: in like manner the angles at B, C, are given. And because the sides AB, BC, CA are given, their ratios to one another are given*; therefore the triangle ABC is given* in species.

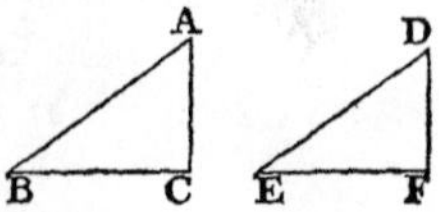

* 22. 1.
* 8. 1.
* 1 Def.
* 1 Dat.
* 3 Def.

(40.)

PROPOSITION XLIII.

If each of the angles of a triangle be given in magnitude, the triangle is given in species.

Let each of the angles of the triangle ABC be given in magnitude: the triangle ABC is given in species.

Take a straight line DE given in position and magnitude,
* 23. 1. and at the points D, E, make* the angle EDF equal to the angle BAC, and the angle DEF equal to ABC; therefore the other angles EFD, BCA are equal, and each of the angles at the points A, B, C, is given, wherefore each of those at the points DEF is given.

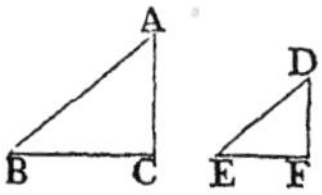

And because the straight line FD is drawn to the given point D, in DE which is given in position, making the given angle
* 32 Dat. EDF, therefore DF is given in position*; in like manner EF also is given in position; wherefore the point F is given: and the points D, E are given; therefore each of the straight
* 29 Dat. lines DE, EF, FD, is given* in magnitude; wherefore the
* 42 Dat. * 4. 6. triangle DEF is given in species*; and it is similar* to the
1 Def. 6. triangle ABC; which therefore is given in species.

(41.)

PROPOSITION XLIV.

If one of the angles of a triangle be given, and if the sides about it have a given ratio to one another, the triangle is given in species.

Let the triangle ABC have one of its angles BAC given, and let the sides BA, AC, about it, have a given ratio to one another: the triangle ABC is given in species.

Take a straight line DE given in position and magnitude, and at the point D, in the given straight line DE, make the angle EDF equal to the given angle BAC: wherefore the angle EDF is given; and because the straight line FD is drawn to the given point D in ED, which is given in position, making the given angle EDF;

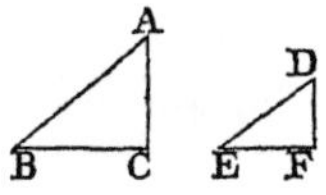

* 32 Dat. therefore FD is given in position*. And because the ratio of BA to AC is given, make the ratio of ED to DF the same with it, and join EF; and because the ratio of ED to DF is given,

and ED is given, therefore* DF is given in magnitude: and * 2 Dat.
it is given also in position, and the point D is given, where-
fore the point F is given*; and the points D, E are given, * 30 Dat.
wherefore DE, EF, FD are given * in magnitude; and the * 29 Dat.
triangle DEF is therefore given* in species; and because the * 42 Dat.
triangles ABC, DEF have one angle BAC equal to one angle
EDF, and the sides about these angles proportionals; the
triangles are * similar, but the triangle DEF is given in * 6. 6.
species, and therefore also the triangle ABC.

PROPOSITION XLV. (42.)

If the sides of a triangle have to one another given ratios, the triangle is given in species. See N.

Let the sides of the triangle ABC have given ratios to one another: the triangle ABC is given in species.

Take a straight line D given in magnitude; and because
the ratio of AB to BC is given, make the ratio of D to E the
same with it; and D is given, therefore* E is given. And * 2 Dat.
because the ratio of BC to CA is given, to this make the ratio
of E to F the same; and E is given, and therefore* F. And * 2 Dat.
because as AB to BC, so is D to E; by
composition AB and BC together are to
BC, as D and E to F: but as BC to CA,
so is E to F; therefore, ex æquali*, as * 22. 5.
AB and BC are to CA, so are D and E
to F, and AB and BC are greater* than * 20. 1.
CA; therefore D and E are greater* * A. 5.
than F. In the same manner, any two of the three D, E, F
are greater than the third. Make* the triangle GHK, whose * 22. 1.
sides are equal to D, E, F, so that GH be equal to D, HK to
E, and KG to F; and because D, E, F are each of them given,
therefore GH, HK, KG are each of them given in magnitude;
therefore the triangle GHK is given* in species: but as AB * 42 Dat.
to BC, so is (D to E, that is) GH to HK; and as BC to CA, so
is (E to F, that is) HK to KG; therefore, ex æquali, as AB
to AC, so is GH to GK. Wherefore*, the triangle ABC is * 5. 6.
equiangular and similar to the triangle GHK; and the triangle
GHK is given in species; therefore also the triangle ABC is
given in species.

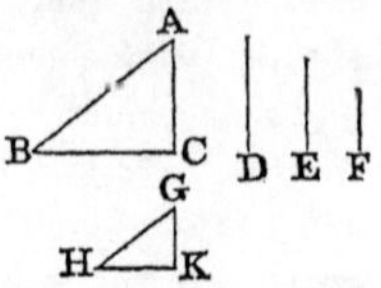

COR. If a triangle be required to be made, the sides of

which shall have the same ratios which three given straight lines D, E, F have to one another; it is necessary that every two of them be greater than the third.

(43.) PROPOSITION XLVI.

If the sides of a right-angled triangle about one of the acute angles, have a given ratio to one another, the triangle is given in species.

Let the sides AB, BC about the acute angle ABC of the triangle ABC, which has a right angle at A, have a given ratio to one another: the triangle ABC is given in species.

Take a straight line DE given in position and magnitude; and because the ratio of AB to BC is given, make as AB to BC, so DE to EF; and because DE has a given ratio to EF,
* 2 Dat. and DE is given, therefore* EF is given; and because as AB
* 19. 1. to BC, so is DE to EF; and AB is less* than BC, therefore
* A. 5. DE is less* than EF. From the point D, draw DG at right angles to DE, and from the centre E, at the distance EF, describe a circle which shall meet DG in two points; let G be either of them, and join EG: therefore the circumference

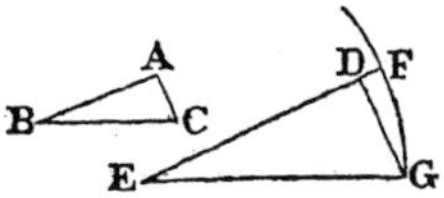

* 6 Def. of the circle is given* in position;
* 32 Dat. and the straight line DG is given* in position, because it is drawn to the given point D in DE given in position, in a
* 28 Dat. given angle; therefore* the point G is given, and the points
* 29 Dat. D, E are given, wherefore DE, EG, GD are given* in magni-
* 42 Dat. tude, and the triangle DEG in species*. And because the triangles ABC, DEG have the angle BAC equal to the angle EDG, and the sides about the angles ABC, DEG proportionals, and each of the other angles BCA, EGD, less than a right
* 7. 6. angle; the triangle ABC is equiangular* and similar to the triangle DEG; but DEG is given in species; therefore the triangle ABC is given in species: and in the same manner, the triangle made by drawing a straight line from E to the other point in which the circle meets DG, is given in species.

PROPOSITION XLVII. (44.)

If a triangle have one of its angles which is not a right angle given, and if the sides about another angle have a given ratio to one another, the triangle is given in species. See N.

Let the triangle ABC have one of its angles ABC a given, but not a right angle, and let the sides BA, AC, about another angle BAC, have a given ratio to one another: the triangle ABC is given in species.

First, let the given ratio be the ratio of equality, that is, let the sides BA, AC, and consequently the angles ABC, ACB, be equal; and because the angle ABC is given, the angle ACB, and also the remaining* angle BAC is given; therefore the triangle ABC is given* in species; and it is evident that in this case the given angle ABC must be acute.

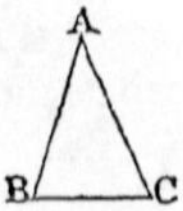

* 32. 1.
* 43 Dat.

Next, let the given ratio be the ratio of a less to a greater, that is, let the side AB adjacent to the given angle be less than the side AC: take a straight line DE given in position and magnitude, and make the angle DEF equal to the given angle ABC: therefore EF is given* in position; and because the ratio of BA to AC is given, as BA to AC, so make ED to DG; and because the ratio of ED to DG is given, and ED is given, the straight line DG is given*, and BA is less than AC, therefore ED is less* than DG. From the centre D, at the distance DG, describe the circle GF meeting EF in F, and join DF; and because the circle is given* in position, as also the straight line EF, the point F is given*; and the points D, E are given; wherefore the straight lines DE, EF, FD are given* in magnitude, and the triangle DEF in species*. And because BA is less than AC, the angle ACB is less* than the angle ABC, and therefore ACB is less* than a right angle. In the same manner because ED is less than DG or DF, the angle DFE is less than a right angle: and because the triangles ABC, DEF have the angle ABC equal to the angle DEF, and the sides about the angles BAC, EDF, proportionals, and each of the other angles ACB, DFE, less than a right angle; the triangles

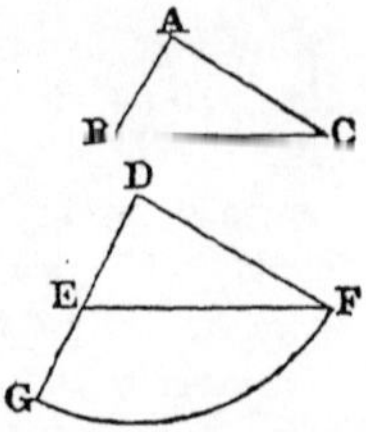

* 32 Dat.
* 2 Dat.
* A. 5.
* 6 Dat.
* 28 Dat.
* 29 Dat.
* 42 Dat.
* 18. 1.
* 17. 1.

* 7. 6. ABC, DEF are* similar, and DEF is given in species, wherefore the triangle ABC is also given in species.

Thirdly, let the given ratio be the ratio of a greater to a less, that is, let the side AB adjacent to the given angle be greater than AC; and, as in the last case, take a straight line DE given in position and magnitude, and make the angle DEF equal to
* 32 Dat. the given angle ABC; therefore EF is given* in position: also draw DG perpendicular to EF; therefore if the ratio of BA to AC be the same with the ratio of ED to the perpendicular DG, the triangles ABC, DEG are simi-
* 7. 6. lar*, because the angles ABC, DEG are equal, and DGE is a right angle: therefore the angle ACB is a right
* 43 Dat. angle, and the triangle ABC is given* in species.

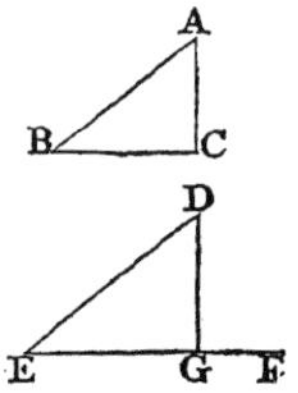

But if, in this last case, the given ratio of BA to AC be not the same with the ratio of ED to DG, that is, with the ratio of BA to the perpendicular AM drawn from A to BC; the
* 8. 5. ratio of BA to AC must be less* than the ratio of BA to AM, because AC is greater than AM. Make as BA to AC, so ED to DH; therefore the ratio of ED to DH is less than the ratio of (BA to AM, that is, than the ratio of)
* 10. 5. ED to DG: and consequently DH is greater* than DG; and because BA is greater than
* A. 5. AC, ED is greater* than DH. From the centre D, at the distance DH, describe the circle KHF which necessarily meets the straight line EF in two points, because DH is greater than DG, and less than DE. Let the circle meet EF in the points F, K, which are given, as was shewn in the preceding case; and DF, DK being joined, the triangles DEF, DEK are given in species, as was there shewn. From the centre A, at the distance AC, describe a circle meeting BC again in L: and if the angle ACB be less than a right angle, ALB must be greater than a right angle; and on the contrary. In the same manner, if the angle DFE be less than a right angle, DKE must be greater than one; and on the contrary. Let each of the angles ACB, DFE be either less or greater than a right angle; and because in the triangles ABC, DEF, the angles ABC, DEF are equal, and the sides BA, AC, and ED, DF, about two
* 7. 6. of the other angles proportionals, the triangle ABC is similar*

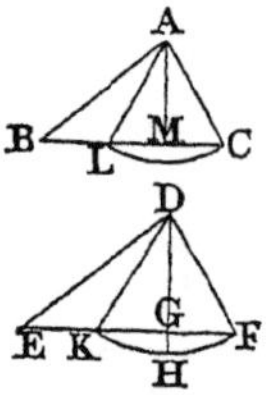

to the triangle DEF. In the same manner the triangle ABL is similar to DEK. And the triangles DEF, DEK are given in species; therefore also the triangles ABC, ABL are given in species. And from this it is evident, that in this third case, there are always two triangles of a different species, to which the things mentioned as given in the proposition can agree.

PROPOSITION XLVIII. (45.)

If a triangle have one angle given, and if both the sides together about that angle have a given ratio to the remaining side, the triangle is given in species.

Let the triangle ABC have the angle BAC given, and let the sides BA, AC together, about that angle, have a given ratio to BC: the triangle ABC is given in species.

Bisect* the angle BAC by the straight line AD; therefore the angle BAD is given. And because as BA to AC, so is* BD to DC, by permutation, as AB to BD, so is AC to CD; and as BA and AC together to BC, so is* AB to BD. But the ratio of BA and AC together to BC is given, wherefore the ratio of AB to BD is given, and the angle BAD is given; therefore* the triangle ABD is given in species, and the angle ABD is therefore given; the angle BAC is also given, wherefore the triangle ABC is given in species*. * 9. 1. * 3. 6. * 12. 5. * 47 Dat. * 43 Dat.

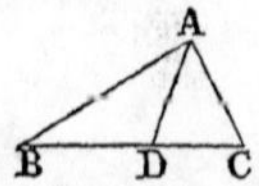

A triangle which shall have the things that are mentioned in the proposition to be given, can be found in the following manner. Let EFG be the given angle, and let the ratio of H to K be the given ratio which the two sides about the angle EFG must have to the third side of the triangle; therefore, because two sides of a triangle are greater than the third side, the ratio of H to K must be the ratio of a greater to a less. Bisect * the angle EFG by the straight line FL, and by the 47th Prop. find a triangle of which EFL is one of the angles, and in which the ratio of the sides about the angle opposite to FL is the same with the ratio of H to K: to do which, take FE given in position and magnitude, and draw EL perpendicular to FL: then if the ratio of H to K be the same with the ratio of FE to EL, produce EL, and let it meet FG in P; the triangle FEP is that which was to be found: for it has the given angle EFG; and because this angle is bi- * 9. 1.

* 3. 6. sected by FL, the sides EF, EP, together are to EP as* FE to EL, that is, as H to K.

But if the ratio of H to K be not the same with the ratio of FE to EL, it must be less than it, as was shewn in Prop. 47., and in this case there are two triangles, each of which has the given angle EFL, and the ratio of the sides about the angle opposite to FL the same with the ratio of H to K. By Prop. 47. find these triangles EFM, EFN, each of which has the angle EFL for one of its angles, and the ratio of the side FE to EM or EN the same with the ratio of H to K; and let the angle EMF be greater, and ENF less, than a right angle. And because H is greater than K, EF is greater than EN, and therefore the angle EFN, that is, the angle NFG,

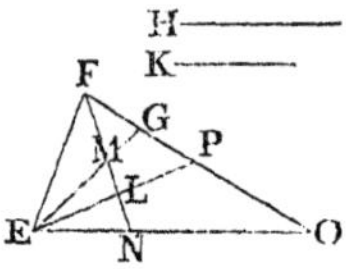

* 18. 1. is less* than the angle ENF. To each of these add the angles NEF, EFN; therefore the angles NEF, EFG are less than the angles NEF, EFN, FNE, that is, than two right angles; therefore the straight lines, EN, FG, must meet together when produced; let them meet in O, and produce EM to G. Each of the triangles EFG, EFO, has the things mentioned to be given in the proposition: for each of them has the given angle EFG; and because this angle is bisected by the straight line FMN, the sides EF, FG together, have to EG the third side, the ratio of FE to EM, that is, of H to K. In like manner, the sides EF, FO together, have to EO, the ratio which H has to K.

(46.)

PROPOSITION XLIX.

If a triangle have one angle given, and if the sides about another angle both together have a given ratio to the third side, the triangle is given in species.

Let the triangle ABC have one angle ABC given, and let the two sides BA, AC, about another angle BAC have a given ratio to BC: the triangle ABC is given in species.

Suppose the angle BAC to be bisected by the straight line AD; BA and AC together are to BC, as AB to BD, as was shewn in the preceding proposition. But the ratio of BA and AC together to BC is given; therefore, also, the ratio of

* 44 Dat. AB to BD is given. And the angle ABD is given, wherefore* the triangle ABD is given in species; and consequently the an-

gle BAD, and its double the angle BAC, are given: and the angle ABC is given; therefore the triangle ABC is given in species*. * 43 Dat.

A triangle which shall have the things mentioned in the proposition to be given may be thus found. Let EFG be the given angle, and the ratio of H to K the given ratio; and by Prop. 44. find the triangle EFL, which has the angle EFG for one of its angles, and the ratio of the sides EF, FL, about this angle the same with the ratio of H to K; and make the angle LEM equal to the angle FEL. And because the ratio of H to K is the ratio which two sides of a triangle have to the third, H must be greater than K. And because EF is to FL, as H to K, therefore EF is greater than FL, and the angle FEL, that is, LEM, is therefore less than the angle ELF. Wherefore the angles LFE, FEM are less than two right angles, as was shewn in the foregoing proposition, and the straight lines FL, EM must meet if produced; let them meet in G, EFG is the triangle which was to be found: for EFG is one of its angles, and because the angle FEG is bisected by EL, the two sides FE, EG together, have to the third side FG, the ratio of EF to FL, that is, the given ratio of H to K.

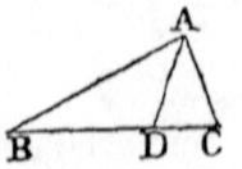

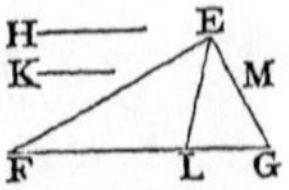

PROPOSITION L. (70.)

If from the vertex of a triangle given in species, a straight line be drawn to the base in a given angle, it shall have a given ratio to the base.

From the vertex A, of the triangle ABC which is given in species, let AD be drawn to the base BC in a given angle ADB: the ratio of AD to BC is given.

Because the triangle ABC is given in species, the angle ABD is given, and the angle ADB is given, therefore the triangle ABD is given* * 43 Dat.
in species: wherefore the ratio of AD to AB is given. And the ratio of AB to BC is given; and therefore* * 9 Dat.
the ratio of AD to BC is given.

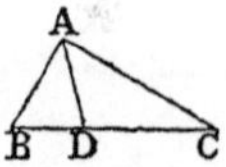

PROPOSITION LI. (47.)

Rectilineal figures given in species, are divided into triangles which are given in species.

Let the rectilineal figure ABCDE be given in species: ABCDE may be divided into triangles given in species.

Join BE, BD: and because ABCDE is given in species, the
* 3 Def. angle BAE is given*, and the ratio of BA to AE is given*;
* 3 Def. wherefore the triangle BAE is given in species*, and the angle AEB is therefore given*.
* 44 Dat. * 3 Def. But the whole angle AED is given, and therefore the remaining angle BED is given, and the ratio of AE to EB is given, as also the ratio of AE to ED; therefore the ratio of BE
* 9 Dat. to ED is given*. And the angle BED is given, wherefore the
* 44 Dat. triangle BED is given* in species. In the same manner, the triangle BDC is given in species; therefore rectilineal figures which are given in species, are divided into triangles given in species.

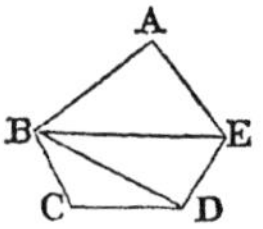

(48.)

PROPOSITION LII.

If two triangles given in species be described upon the same straight line, they shall have a given ratio to one another.

Let the triangles ABC, ABD, given in species, be described upon the same straight line AB: the ratio of the triangle ABC to the triangle ABD is given.

Through the point C, draw CE parallel to AB, and let it meet DA produced in E, and join BE. Because the triangle ABC is given in species, the angle BAC, that is, the angle ACE, is given; and because the triangle ABD is given in species, the angle DAB, that is, the angle AEC, is given. Therefore the triangle ACE is given in

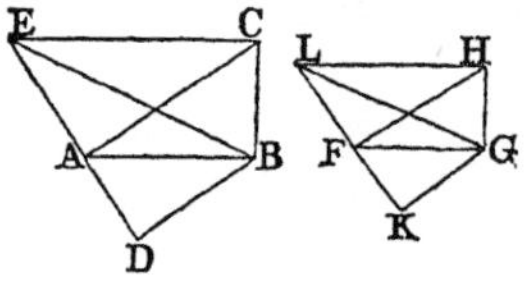

* 3 Def. species; wherefore the ratio of EA to AC is given*, and the ratio of CA to AB is given, as also the ratio of BA to AD;
* 9 Dat. therefore the ratio of* EA to AD is given, and the triangle
* 37. 1. ACB is equal* to the triangle AEB, and as the triangle AEB,
* 1. 6. or ACB, is to the triangle ADB, so is* the straight line EA to AD: but the ratio of EA to AD is given; therefore the ratio of the triangle ACB to the triangle ADB is given.

PROBLEM.—*To find the ratio of two triangles* ABC, ABD, *given in species, and which are described upon the same straight line* AB.

Take a straight line FG given in position and magnitude, and because the angles of the triangles ABC, ABD are given, at the points F, G, of the straight line FG, make the angles GFH, GFK* equal to the angles BAC, BAD, and the angles FGH, FGK, equal to the angles ABC, ABD, each to each. Therefore the triangles ABC, ABD are equiangular to the triangles FGH, FGK, each to each. Through the point H, draw HL parallel to FG, meeting KF produced in L. And because the angles BAC, BAD are equal to the angles GFH, GFK, each to each; therefore the angles ACE, AEC are equal to FHL, FLH, each to each, and the triangle AEC equiangular to the triangle FLH. Therefore as EA to AC, so is LF to FH, and as CA to AB, so HF to FG; and as BA to AD, so is GF to FK; wherefore, ex æquali, as EA to AD, so is LF to FK. But, as was shewn, the triangle ABC is to the triangle ABD, as the straight line EA to AD, that is, as LF to FK. The ratio, therefore, of LF to FK has been found, which is the same with the ratio of the triangle ABC to the triangle ABD.

* 23. 1.

PROPOSITION LIII. (49.)

If two rectilineal figures given in species be described upon the same straight line, they shall have a given ratio to one another. Sec N.

Let any two rectilineal figures ABCDE, ABFG, which are given in species, be described upon the same straight line AB: the ratio of them to one another is given.

Join AC, AD, AF; each of the triangles AED, ADC, ACB, AGF, ABF is given* in species. And because the triangles ADE, ADC, given in species, are described upon the same straight line AD, the ratio of EAD to DAC is given*; and by composition, the ratio of EACD to DAC is given*. And the ratio of DAC to CAB is given*, because they are described upon the same straight line AC; therefore the ratio of EACD to ACB is given*; and, by composition, the ratio of ABCDE to ABC is given. In the same manner, the ratio of ABFG, to ABF is given. But the ratio of the triangle ABC to the triangle ABF is given; wherefore* because the ratio of ABCDE to ABC is given, as also the ratio of ABC to

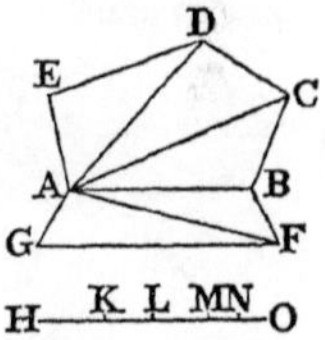

* 15 Dat.
* 52 Dat.
* 7 Dat.
* 52 Dat.
* 9 Dat.
* 52 Dat.

ABF, and the ratio of ABF to ABFG; the ratio of the recti-
* 9 Dat. lineal ABCDE to the rectilineal ABFG is given*.

PROB.—*To find the ratio of two rectilineal figures given in species, and described upon the same straight line.*

Let ABCDE, ABFG be two rectilineal figures given in species, and described upon the same straight line AB, and join AC, AD, AF. Take a straight line HK given in position and magnitude, and by the 52d Dat. find the ratio of the triangle ADE to the triangle ADC, and make the ratio of HK to KL the same with it. Find also the ratio of the triangle ACD to the triangle ACB, and make the ratio of KL to LM the same. Also, find the ratio of the triangle ABC to the triangle ABF, and make the ratio of LM to MN the same. And, lastly, find the ratio of the triangle AFB to the triangle AFG, and make the ratio of MN to NO the same. Then the ratio of ABCDE to ABFG is the same with the ratio of HM to MO.

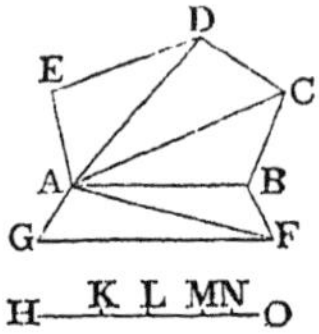

Because the triangle EAD is to the triangle DAC, as the straight line HK to KL; and as the triangle DAC to CAB, so is the straight line KL to LM; therefore by using composition as often as the number of triangles requires, the rectilineal ABCDE is to the triangle ABC, as the straight line HM to ML. In like manner, because the triangle GAF is to FAB as ON to NM, by composition, the rectilineal ABFG is to the triangle ABF as MO to NM, and by inversion, as ABF to ABFG so is NM to MO. And the triangle ABC is to ABF, as LM to MN. Wherefore, because as ABCDE to ABC, so is HM to ML: and as ABC to ABF, so is LM to MN; and as ABF to ABFG, so is MN to MO; ex æquali, as the rectilineal ABCDE to ABFG, so is the straight line HM to MO.

(50.)

PROPOSITION LIV.

If two straight lines have a given ratio to one another, the similar rectilineal figures described upon them similarly, shall have a given ratio to one another.

Let the straight lines AB, CD have a given ratio to one another, and let the similar and similarly placed rectilineal figures E, F, be described upon them: the ratio of E to F is given.

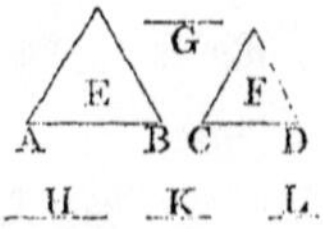

To AB, CD, let G be a third proportional; therefore as AB to CD, so is CD to G. And the ratio of AB to CD is given; wherefore the ratio of CD to G is given; and consequently the ratio of AB to G is also given*. * 9 Dat.
But as AB to G, so is the figure E to the figure* F. Therefore the ratio of E to F is given. * 2 Cor. 20. 6.

Prob.—*To find the ratio of two similar rectilineal figures* E, F, *similarly described upon straight lines* AB, CD, *which have a given ratio to one another.*

Let G be a third proportional to AB, CD.

Take a straight line H given in magnitude; and because the ratio of AB to CD is given, make the ratio of H to K the same with it; and because H is given, K is given. As H is to K, so make K to L; then the ratio of E to F is the same with the ratio of H to L: for AB is to CD, as H to K, wherefore CD is to G, as K to L; and, ex æquali, as AB to G, so is H to L; but the figure E is to* the figure F, as AB to G, that is, as H to L. * 2 Cor. 20. 6.

PROPOSITION LV. (51.)

If two straight lines have a given ratio to one another, the rectilineal figures given in species, described upon them, shall have to one another a given ratio.

Let AB, CD be two straight lines which have a given ratio to one another: the rectilineal figures E, F, given in species and described upon them, have a given ratio to one another.

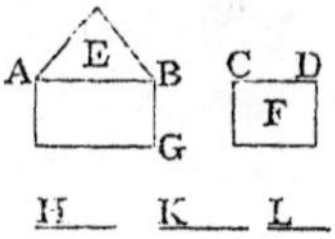

Upon the straight line AB, describe the figure AG similar and similarly placed to the figure F; and because F is given in species, AG is also given in species: therefore, since the figures E, AG, which are given in species, are described upon the same straight line AB, the ratio of E to AG is given*, and because the ratio of AB to CD is given, and upon them are described the similar and similarly placed rectilineal figures AG, F, the ratio of AG to F is given*; and the ratio of AG to E is given; therefore the ratio of E to F is given*. * 53 Dat. * 54 Dat. * 9 Dat.

Prob.—*To find the ratio of two rectilineal figures* E, F, *given in species and described upon the straight lines* AB, CD, *which have a given ratio to one another.*

Take a straight line H given in magnitude; and because the rectilineal figures E, AG, given in species, are described upon the same straight line AB, find their ratio by the 53d Dat. and make the ratio of H to K the same, K is therefore given. And because the similar rectilineal figures AG, F are described upon the straight lines AB, CD, which have a given ratio, find their ratio by the 54th Dat. and make the ratio of K to L the same: the figure E has to F the same ratio which H has to L: for by the construction, as E is to AG, so is H to K; and as AG to F, so is K to L: therefore, ex æquali, as E to F, so is H to L.

(52.)

PROPOSITION LVI.

If a rectilineal figure given in species be described upon a straight line given in magnitude, the figure is given in magnitude.

Let the rectilineal figure ABCDE given in species, be described upon the straight line AB given in magnitude: the figure ABCDE is given in magnitude.

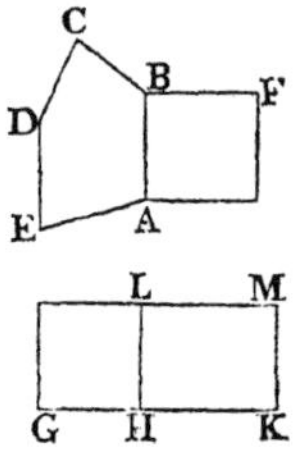

Upon AB let the square AF be described; therefore AF is given in species and magnitude, and because the rectilineal figures ABCDE, AF, given in species, are described upon the same straight line AB, the ratio

* 53 Dat. of ABCDE to AF is given *: but the square

* 2 Dat. AF is given in magnitude, therefore * also the figure ABCDE is given in magnitude.

Prob.—*To find the magnitude of a rectilineal figure given in species described upon a straight line given in magnitude.*

Take the straight line GH equal to the given straight line AB, and by the 53d Dat. find the ratio which the square AF upon AB has to the figure ABCDE; and make the ratio of GH to HK the same; and upon GH describe the square GL and complete the parallelogram LHKM; the figure ABCDE is equal to LHKM. Because AF is to ABCDE, as the straight line GH to HK, that is, as the figure GL to HM; and AF is equal to GL; therefore ABCDE is equal to HM *. * 14. 5.

PROPOSITION LVII. (53.)

If two rectilineal figures be given in species, and if a side of one of them have a given ratio to a side of the other, the ratios of the remaining sides to the remaining sides shall be given.

Let AC, DF be two rectilineal figures given in species, and let the ratio of the side AB to the side DE be given: the ratios of the remaining sides to the remaining sides are also given.

Because the ratio of AB to DE is given, as also * the ratios of AB to BC, and of DE to EF, the ratio of BC to EF is given*. In the same manner the ratios of the other sides to the other sides are given. * 3 Def. * 10 Dat.

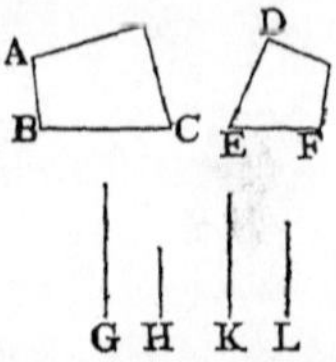

The ratio which BC has to EF may be found thus: take a straight line G given in magnitude, and because the ratio of BC to BA is given, make the ratio of G to H the same; and because the ratio of AB to DE is given, make the ratio of H to K the same; and make the ratio of K to L the same with the given ratio of DE to EF. Since therefore as BC to BA, so is G to H; and as BA to DE, so is H to K; and as DE to EF, so is K to L; ex æquali, BC is to EF, as G to L; therefore the ratio of G to L has been found, which is the same with the ratio of BC to EF.

PROPOSITION LVIII. (G.)

If two similar rectilineal figures have a given ratio to one another, their homologous sides have also a given ratio to one another. See N.

Let the two similar rectilineal figures A, B have a given ratio to one another: their homologous sides have also a given ratio.

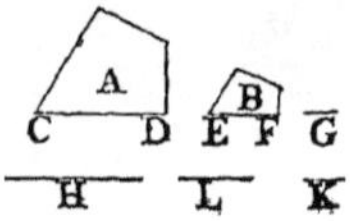

Let the side CD be homologous to EF; and to CD, EF, let the straight line G be a third proportional. As therefore* CD to G, so is the figure A to B; and the ratio of A to B is given, therefore the ratio of CD to G is * 2 Cor. 20. 6.

* 13 Dat. given; and CD, EF, G are proportionals; wherefore* the ratio of CD to EF is given.

The ratio of CD to EF may be found thus: take a straight line H given in magnitude; and because the ratio of the figure A to B is given, make the ratio of H to K the same with it: and, as the 13th Dat. directs to be done, find a mean proportional L between H and K; the ratio of CD to EF is the same with that of H to L. Let G be a third proportional to CD, EF; therefore as CD to G, so is (A to B, and so is) H to K; and as CD to EF, so is H to L, as is shewn in the 13th Dat.

(54.)

PROPOSITION LIX.

See N. *If two rectilineal figures given in species have a given ratio to one another, their sides shall likewise have given ratios to one another.*

Let the two rectilineal figures A, B, given in species, have a given ratio to one another: their sides shall also have given ratios to one another.

If the figure A be similar to B, their homologous sides shall have a given ratio to one another, by the preceding proposition; and because the figures are given in species, the sides

* 3 Def. of each of them have given ratios* to one another; therefore
* 9 Dat. each side of one of them has* to each side of the other a given ratio.

But if the figure A be not similar to B, let CD, EF be any two of their sides; and upon EF, conceive the figure EG to be described similar and similarly placed to the figure A, so that CD, EF be homologous sides: therefore EG is given in species: and the figure B is

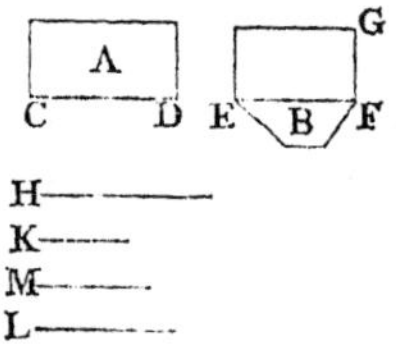

* 53 Dat. given in species; wherefore* the ratio of B to EG is given:
* 9 Dat. and the ratio of A to B is given, therefore* the ratio of the
* 58 Dat. figure A to EG is given: and A is similar to EG; therefore*
* 9 Dat. the ratio of the side CD to EF is given; and consequently* the ratios of the remaining sides to the remaining sides are given.

The ratio of CD to EF may be found thus: take a straight line H given in magnitude, and because the ratio of the figure A to B is given, make the ratio of H to K the same with it. And by the 53d Dat. find the ratio of the figure B to EG, and make the ratio of K to L the same: between H and L find a mean proportional M, the ratio of CD to EF is the same with the ratio of H to M. Because the figure A is to B as H to K, and as B to EG, so is K to L; ex æquali, as A to EG, so is H to L: and the figures A, EG are similar, and M is a mean proportional between H and L; therefore, as was shewn in the preceding proposition, CD is to EF as H to M.

PROPOSITION LX. (55.)

If a rectilineal figure be given in species and magnitude, the sides of it shall be given in magnitude.

Let the rectilineal figure A be given in species and magnitude: its sides are given in magnitude.

Take a straight line BC given in position and magnitude, and upon BC describe * the figure D similar, and similarly placed, to the figure A, and let EF be the side of the figure A homologous to BC the side of D; therefore the figure D is given in species. And because upon the given straight line BC the figure D given in species is described, D is given* in magnitude, and the figure A is given in magnitude, therefore the ratio of A to D is given: and the figure A is similar to D: therefore the ratio of the side EF to the homologous side BC is given*; and BC is given, wherefore * EF is given: and the ratio of EF to EG is given*, therefore EG is given. And, in the same manner, each of the other sides of the figure A can be shewn to be given.

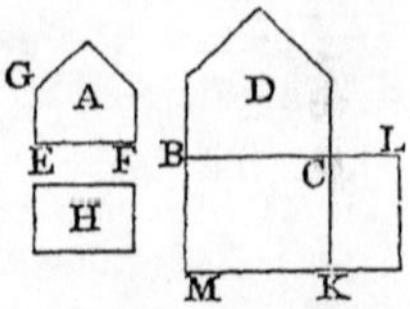

* 18. 6.
* 56 Dat.
* 58 Dat.
* 2 Dat.
* 3 Def.

* Prob.—*To describe a rectilineal figure* A *similar to a given figure* D, *and equal to another given figure* H.

* This is Prop. 25. B. 6.

Because each of the figures D, H is given, their ratio is given, which may be found by making * upon the given straight line BC the parallelogram BK equal to D, and upon its side CK making * the parallelogram KL equal to H in the angle KCL equal to the angle MBC; therefore the ratio of D to H, that is, of BK to KL, is the same with the ratio of BC to CL: and because the figures D, A are similar, and that the ratio of D to A, or H, is the same with the ratio of BC to CL, the ratio * of the homologous sides BC, EF, is the same with

* Cor. 45. 1.
* Cor. 45. 1.
* 58 Dat.

the ratio of BC to the mean proportional between BC and CL. Find EF the mean proportional; then EF is the side of the figure to be described, homologous to BC the side of D, and the figure itself can be described by the 18th Prop. B. 6., which, by the construction, is similar to D; and because D is
* 2 Cor. 20. 6. to A, as * BC to CL, that is, as the figure BK to KL; and that
* 14. 5. D is equal to BK, therefore A * is equal to KL, that is, to H.

(57.)

PROPOSITION LXI.

See N. *If a parallelogram given in magnitude have one of its sides and one of its angles given in magnitude, the other side also is given.*

Let the parallelogram ABDC given in magnitude, have the side AB and the angle BAC given in magnitude: the other side AC is given.

Take a straight line EF given in position and magnitude; and because the parallelogram AD is given in magnitude, a
* 1 Def. rectilineal figure equal to it can be found*. And a parallelo-
* Cor. 45. 1. gram equal to this figure can be applied* to the given straight line EF in an angle equal to the given angle BAC. Let this be the parallelogram EFHG, having the angle FEG equal to the angle BAC. And because the parallelograms AD, EH, are equal, and have the angles at A and E equal; the sides about them are recipro-
* 14. 6. cally proportional*; therefore as AB to EF, so is EG to AC: and AB, EF, EG are given,
* 12. 6. therefore also AC is given*. Whence the way of finding AC is manifest.

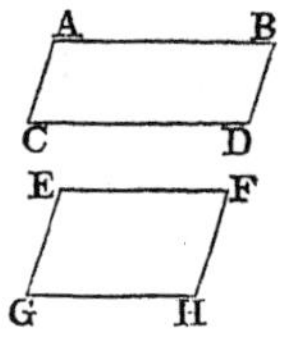

(H.)

PROPOSITION LXII.

If a parallelogram have a given angle, the rectangle contained by the sides about that angle has a given ratio to the parallelogram.

Let the parallelogram ABCD have the given angle ABC: the rectangle AB, BC has a given ratio to the parallelogram AC.

From the point A, draw AE perpendicular to BC; because the angle ABC is given, as also the angle AEB, the triangle
* 43 Dat. ABE is given* in species; therefore the ratio of BA to AE is

given. But as BA to AE, so is* the rectangle AB, BC, to the rectangle AE, BC; therefore the ratio of the rectangle AB, BC, to AE, BC, that is*, to the parallelogram AC, is given. * 1. 6. * 35. 1.

And it is evident how the ratio of the rectangle to the parallelogram may be found, by making the angle FGH equal to the given angle ABC, and drawing from any point F in one of its sides, FK perpendicular to the other GH: for GF is to FK, as BA to AE, that is, as the rectangle AB, BC, to the parallelogram AC.

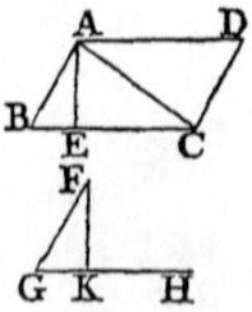

COR. And if a triangle ABC has a given angle ABC, the rectangle AB, BC, contained by the sides about that angle, shall have a given ratio to the triangle ABC. (66.)

Complete the parallelogram ABCD; therefore by this proposition, the rectangle AB, BC, has a given ratio to the parallelogram AC; and AC has a given ratio to its half the triangle* ABC; therefore the rectangle AB, BC has a given* ratio to the triangle ABC. * 41. 1. * 9 Dat.

And the ratio of the rectangle to the triangle is found thus: make the triangle FGK as was shewn in the proposition: the ratio of GF to the half of the perpendicular FK, is the same with the ratio of the rectangle AB, BC, to the triangle ABC. Because, as was shewn, GF is to FK, as AB, BC, to the parallelogram AC; and FK is to its half, as AC is to its half, which is the triangle ABC: therefore, ex æquali, GF is to the half of FK, as the rectangle AB, BC, is to the triangle ABC.

PROPOSITION LXIII. (56.)

If two parallelograms be equiangular, as the side of the first to a side of the second, so is the other side of the second to the straight line to which the other side of the first has the same ratio which the first parallelogram has to the second. And consequently, if the ratio of the first parallelogram to the second be given, the ratio of the other side of the first to that straight line is given; and if the ratio of the other side of the first to that straight line be given, the ratio of the first parallelogram to the second is given.

Let AC, DF be two equiangular parallelograms: as BC, a side of the first, is to EF, a side of the second, so is DE, the other side of the second, to the straight line to which AB, the other side of the first, has the same ratio which AC has to DF.

Produce the straight line AB, and make as BC to EF so DE to BG, and complete the parallelogram BGHC; therefore, because BC or GH, is to EF, as DE to BG, the sides about the equal angles BGH, DEF, are reciprocally proportional; wherefore* the parallelogram BH is equal to DF; and AB is to BG, as the parallelogram AC is to BH, that is, to DF; as therefore BC is to EF, so is DE to BG, which is the straight line to which AB has the same ratio that AC has to DF.

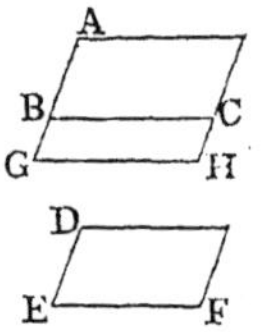

* 14. 6.

And if the ratio of the parallelogram AC to DF be given, then the ratio of the straight line AB to BG is given; and if the ratio of AB to the straight line BG be given, the ratio of the parallelogram AC to DF is given.

(74. 73.)

PROPOSITION LXIV.

See N.

If two parallelograms have unequal, but given angles, and if as a side of the first to a side of the second, so the other side of the second be made to a certain straight line; if the ratio of the first parallelogram to the second be given, the ratio of the other side of the first to that straight line shall be given. And if the ratio of the other side of the first to that straight line be given, the ratio of the first parallelogram to the second shall be given.

Let ABCD, EFGH be two parallelograms which have the unequal but given angles ABC, EFG; and as BC to FG, so make EF to the straight line M: if the ratio of the parallelogram AC to EG be given, the ratio of AB to M is given.

At the point B of the straight line BC, make the angle CBK equal to the angle EFG, and complete the parallelogram KBCL. And because the ratio of AC to EG is given, and that AC is equal* to the parallelogram KC, therefore the ratio of KC to EG is given; and KC, EG, are equiangular, therefore as BC to FG, so is* EF to the straight line to which KB has a given ratio, viz. the same which the parallelogram KC has to EG; but as BC to FG, so is EF to the straight line M; therefore KB has a given ratio to M; and the ratio of AB to

* 35. 1.

* 63 Dat.

BK is given, because the triangle ABK is given in species*; * 43 Dat.
therefore the ratio of AB to M is given*. * 9 Dat.

And if the ratio of AB to M be given, the ratio of the parallelogram AC to EG is given; for since the ratio of KB to BA is given, as also the ratio of AB to M, the ratio of KB to M is given*; and because the parallelograms KC, EG are equiangular, as BC to FG, so is* EF to the straight line to which KB has the same ratio which the parallelogram KC has to EG: but as BC to FG, so is EF to M; therefore KB is to M as the parallelogram KC is to EG; and the ratio of KB to M is given, therefore the ratio of the parallelogram KC, that is, of AC to EG is given. * 9 Dat. * 63 Dat.

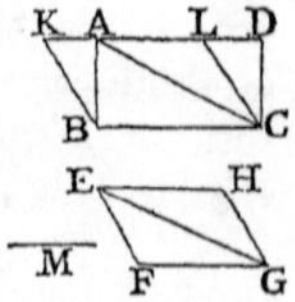

COR. And if two triangles ABC, EFG, have two equal angles, or two unequal but given angles ABC, EFG, and if as BC a side of the first to FG a side of the second, so the other side of the second EF be made to a straight line M; if the ratio of the triangles be given, the ratio of the other side of the first to the straight line M is given. (75.)

Complete the parallelograms ABCD, EFGH; and because the ratio of the triangle ABC to the triangle EFG is given, the ratio of the parallelogram AC to EG is given*, because the parallelograms are double* of the triangles; and because BC is to FG, as EF to M, the ratio of AB to M is given by the 63 Dat. if the angles ABC, EFG, are equal; but if they be unequal but given angles, the ratio of AB to M is given by this proposition. * 15. 5. * 41. 1.

And if the ratio of AB to M be given, the ratio of the parallelogram AC to EG is given by the same propositions; and therefore the ratio of the triangle ABC to EFG is given.

PROPOSITION LXV. (68.)

If two equiangular parallelograms have a given ratio to one another, and if one side have to one side a given ratio, the other side shall also have to the other side a given ratio.

Let the two equiangular parallelograms AB, CD have a given ratio to one another, and let the side EB have a given ratio to the side FD: the other side AE has also a given ratio to the other side CF.

Because the two equiangular parallelograms AB, CD have a given ratio to one another; as EB, a side of the first, is to
* 63 Dat. FD, a side of the second, so is* FC, the other side of the second, to the straight line to which AE, the other side of the first, has the same given ratio which the first parallelogram AB has to the other CD. Let this straight line be EG; therefore the ratio of AE to EG is given; and EB is to FD, as FC to EG, therefore the ratio of FC to EG is given, because the ratio of EB to FD is given; and because the ratio of AE to EG, as also the ratio of FC to EG is given; the ratio of AE to CF is
* 9 Dat. given*.

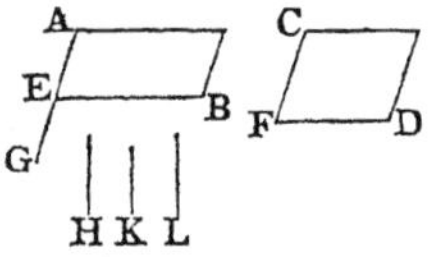

The ratio of AE to CF may be found thus: take a straight line H given in magnitude: and because the ratio of the parallelogram AB to CD is given, make the ratio of H to K the same with it. And because the ratio of FD to EB is given, make the ratio of K to L the same: the ratio of AE to CF is the same with the ratio of H to L. Make as EB to FD, so FC to EG: therefore, by inversion, as FD to EB, so is EG to
* 63 Dat. FC; and as AE to EG, so is* (the parallelogram AB to CD, and so is) H to K: but as EG to FC, so is (FD to EB, and so is) K to L; therefore, ex æquali, as AE to FC, so is H to L.

(69.)

PROPOSITION LXVI.

If two parallelograms have unequal but given angles, and a given ratio to one another; if one side have to one side a given ratio, the other side has also a given ratio to the other side.

Let the two parallelograms ABCD, EFGH, which have the given unequal angles ABC, EFG, have a given ratio to one another, and let the ratio of BC to FG be given: the ratio also of AB to EF is given.

At the point B of the straight line BC, make the angle CBK equal to the given angle EFG, and complete the parallelogram BKLC: and because each of the angles BAK, AKB
* 43 Dat. is given, the triangle ABK is given* in species; therefore the ratio of AB to BK is given; and because, by the hypothesis, the ratio of the parallelogram AC to EG is given, and that AC

is equal* to BL; therefore the ratio of BL to EG is given: * 35. 1.
and because BL is equiangular to EG, and, by the hypothesis, the ratio of BC to FG is given; therefore* * 65 Dat.
the ratio of KB to EF is given, and the ratio of KB to BA is given; the ratio therefore* of * 9 Dat.
AB to EF is given.

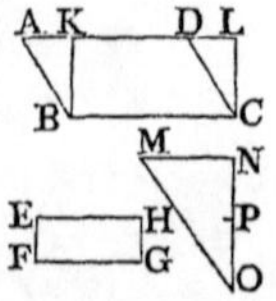

The ratio of AB to EF may be found thus: take the straight line MN given in position and magnitude; and make the angle NMO equal to the given angle BAK, and the angle MNO equal to the given angle EFG or AKB: and because the parallelogram BL is equiangular to EG, and has a given ratio to it, and that the ratio of BC to FG is given; find * the ratio of KB to EF; * 65 Dat.
and make the ratio of NO to OP the same with it: then the ratio of AB to EF is the same with the ratio of MO to OP: for since the triangle ABK is equiangular to MON, as AB to BK, so is MO to ON: and as KB to EF, so is NO to OP; therefore, ex æquali, as AB to EF, so is MO to OP.

PROPOSITION LXVII. (70.)

If the sides of two equiangular parallelograms have given ratios to one another, the parallelograms shall have a given ratio to one another. See N.

Let ABCD, EFGH be two equiangular parallelograms, and let the ratio of AB to EF, as also the ratio of BC to FG, be given: the ratio of the parallelogram AC to EG is given.

Take a straight line K given in magnitude, and because the ratio of AB to EF is given, make the ratio of K to L the same with it; therefore L is given*: and * 2 Dat.
because the ratio of BC to FG is given, make the ratio of L to M the same: therefore M is given *, and K is given: * 2 Dat.
wherefore * the ratio of K to M is given: * 1 Dat.
but the parallelogram AC is to the parallelogram EG, as the straight line K to the straight line M, as is demonstrated in the 23d Prop. of B. 6. Elem.; therefore the ratio of AC to EG is given.

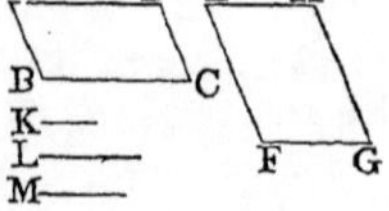

From this it is plain how the ratio of two equiangular parallelograms may be found, when the ratios of their sides are given.

(70.) PROPOSITION LXVIII.

See N. *If the sides of two parallelograms which have unequal, but given angles, have given ratios to one another, the parallelograms shall have a given ratio to one another.*

Let two parallelograms ABCD, EFGH which have the given unequal angles ABC, EFG, have the ratios of their sides, viz. of AB to EF, and of BC to FG, given: the ratio of the parallelogram AC to EG is given.

At the point B of the straight line BC, make the angle CBK equal to the given angle EFG, and complete the parallelogram KBCL: and because each of the angles BAK, BKA, is given,
* 43 Dat. the triangle ABK is given * in species: therefore the ratio of AB to BK is given; and the ratio of AB to EF is
* 9 Dat. given, wherefore * the ratio of BK to EF is given. And the ratio of BC to FG is given; and the angle KBC is

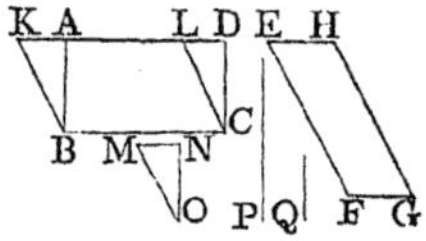

* 67 Dat. equal to the angle EFG; therefore * the ratio of the parallelo-
* 35. 1. gram KC to EG is given: but KC is equal * to AC; therefore the ratio of AC to EG is given.

The ratio of the parallelogram AC to EG may be found thus: take the straight line MN given in position and magnitude, and make the angle MNO equal to the given angle KAB, and the angle NMO equal to the given angle AKB or FEH: and because the ratio of AB to EF is given, make the ratio of NO to P the same; also make the ratio of P to Q the same with the given ratio of BC to FG, the parallelogram AC is to EG, as MO to Q.

Because the angle KAB is equal to the angle MNO, and the angle AKB equal to the angle NMO; the triangle AKB is equiangular to NMO: therefore as KB to BA, so is MO to ON; and as BA to EF, so is NO to P; wherefore, ex æquali, as KB to EF, so is MO to P: and BC is to FG, as P to Q, and the parallelograms KC, EG are equiangular; therefore, as was shewn in Prop. 67., the parallelogram KC, that is AC, is to EG, as MO to Q.

(71.) Cor. 1. If two triangles ABC, DEF have two equal angles, or two unequal but given angles ABC, DEF, and if the ratios of the sides about these angles, viz. the ratios of AB to DE, and of BC to EF, be given; the triangles shall have a given ratio to one another.

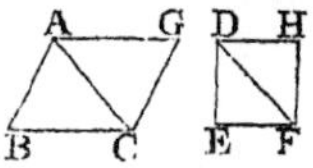

Complete the parallelograms BG, EH; the ratio of BG to EH is given *; and therefore the triangles which are the halves * of them have a given * ratio to one another. * 67 or 68 Dat. * 34. 1. * 15. 5.

(72.)

Cor. 2. If the bases BC, EF, of two triangles ABC, DEF, have a given ratio to one another, and if also the straight lines AG, DH, which are drawn to the bases from the opposite angles, either in equal angles, or unequal but given angles, AGC, DHF have a given ratio to one another; the triangles shall have a given ratio to one another.

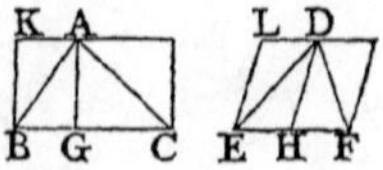

Draw BK, EL parallel to AG, DH, and complete the parallelograms KC, LF: and because the angles AGC, DHF, or their equals, the angles KBC, LEF, are either equal, or unequal but given; and that the ratio of AG to DH, that is, of KB to LE, is given; as also the ratio of BC to EF; therefore * the ratio of the parallelogram KC to LF is given; wherefore also the ratio of the triangle ABC to DEF is given *. * 67 or 68 Dat. * 41. 1. and 15. 5.

PROPOSITION LXIX.

(61.)

If a parallelogram which has a given angle be applied to one side of a rectilineal figure given in species; if the figure have a given ratio to the parallelogram, the parallelogram is given in species.

Let ABCD be a rectilineal figure given in species, and to one side of it AB, let the parallelogram ABEF having the given angle ABE, be applied: if the figure ABCD have a given ratio to the parallelogram BF, the parallelogram BF is given in species.

Through the point A, draw AG parallel to BC, and through the point C, draw CG parallel to AB, and produce GA, CB to the points H, K; because the angle ABC is given *, and the ratio of AB to BC is given, the figure ABCD being given in species, therefore the parallelogram BG is given * in species. And because upon the same straight line AB the two rectilineal figures BD, BG, given in species, are described, the ratio of BD to BG is given *; and, by hypothesis, the ratio of BD to the parallelogram BF is given; wherefore * the ratio of BF, that is *, of the parallelogram BH, to BG, is given, and therefore * the ratio of the straight line KB to BC is given; and the ratio of BC to BA is given, wherefore the ratio of KB to * 3 Def. * 3 Def. * 53 Dat. * 9 Dat. * 35. 1. * 1. 6.

* 9 Dat. BA is given *: and because the angle ABC is given, the adjacent angle ABK is given; and the angle ABE is given, therefore the remaining angle KBE is given. The angle EKB is also given, because it is equal to the angle ABK; therefore the triangle BKE is given in species, and consequently the ratio of EB to BK is given; and the ratio of KB to BA is given, where-

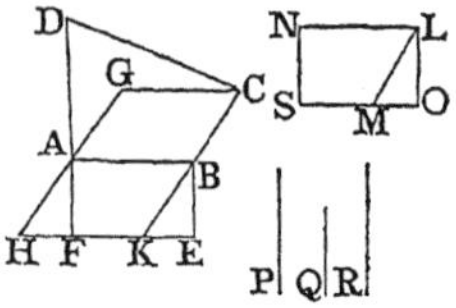

* 9 Dat. fore * the ratio of EB to BA is given; and the angle ABE is
* 3 Def. given, therefore the parallelogram BF is given * in species.

A parallelogram similar to BF may be found thus: take a straight line LM given in position and magnitude; and because the angles ABK, ABE, are given, make the angle NLM equal to ABK, and the angle NLO equal to ABE. And because the ratio of BF to BD is given, make the ratio of LM to P the same with it; and because the ratio of the figure BD
* 53 Dat. to BG is given, find * this ratio, and make the ratio of P to Q the same. Also, because the ratio of CB to BA is given, make the ratio of Q to R the same; and take LN equal to R; through the point M draw OM parallel to LN, and complete the parallelogram NLOS; then this is similar to the parallelogram BF.

Because the angle ABK is equal to NLM, and the angle ABE to NLO, the angle KBE is equal to MLO; and the angles BKE, LMO are equal, because the angle ABK is equal to NLM; therefore the triangles BKE, LMO are equiangular to one another; wherefore, as BE to BK, so is LO to LM; and because as the figure BF to BD, so is the straight line LM to
* 35. 1. P; and as BD to BG, so is P to Q; ex æquali, as BF, that is *,
* 1. 6. BH, to BG, so is LM to Q: but BH is to * BG, as KB to BC; as therefore KB to BC, so is LM to Q; and because BE is to BK, as LO to LM; and as BK to BC, so is LM to Q; and as BC to BA, so Q was made to R; therefore, ex æquali, as BE to BA, so is LO to R, that is, to LN; and the angles ABE, NLO, are equal; therefore the parallelogram BF is similar to LS.

(62. 78.)

PROPOSITION LXX.

See N. *If two straight lines have a given ratio to one another, and upon one of them be described a rectilineal figure given in species, and upon the other a parallelogram having a given angle; if the figure have a given ratio to the parallelogram, the parallelogram is given in species.*

Let the two straight lines AB, CD have a given ratio to one another, and upon AB let the figure AEB given in species be described, and upon CD the parallelogram DF having the given angle FCD: if the ratio of AEB to DF be given, the parallelogram DF is given in species.

Upon the straight line AB, conceive the parallelogram AG to be described, similar and similarly placed to FD; and because the ratio of AB to CD is given, and upon them are described the similar rectilineal figures AG, FD; the ratio of AG to FD is given *; and the ratio of FD to AEB is given; therefore * the ratio of AEB to AG is given; and the angle ABG is given, because it is equal to the angle FCD; because therefore the parallelogram AG, which has a given angle ABG, is applied to a side AB of the figure AEB given in species, and the ratio of AEB to AG is given, the parallelogram AG is given * in species: but FD is similar to AG; therefore FD is given in species.

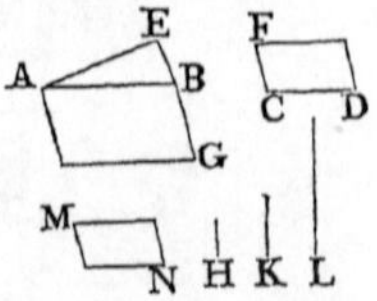

* 54 Dat.
* 9 Dat.
* 69 Dat.

A parallelogram similar to FD may be found thus: take a straight line H given in magnitude; and because the ratio of the figure AEB to FD is given, make the ratio of H to K the same with it: also because the ratio of the straight line CD to AB is given, find by the 54th Dat. the ratio which the figure FD described upon CD has to the figure AG described upon AB similar to FD; and make the ratio of K to L the same with that ratio: and because the ratios of H to K, and of K to L, are given, the ratio of H to L is given *; because therefore, as AEB to FD, so is H to K; and as FD to AG, so is K to L: ex æquali, as AEB to AG, so is H to L; therefore the ratio of AEB to AG is given; and the figure AEB is given in species, and to its side AB the parallelogram AG is applied in the given angle ABG; therefore by the 69th Dat. a parallelogram may be found similar to AG: let this be the parallelogram MN; MN also is similar to FD; for, by the construction, MN is similar to AG, and AG is similar to FD; therefore the parallelogram FD is similar to MN.

* 9 Dat.

(81.) PROPOSITION LXXI.

If the extremes of three proportional straight lines have given ratios to the extremes of other three proportional straight lines, the means shall also have a given ratio to one another. And if one extreme have a given ratio to one extreme, and the mean to the mean, likewise the other extreme shall have to the other a given ratio.

Let A, B, C be three proportional straight lines, and D, E, F, three other; and let the ratios of A to D, and of C to F, be given: then the ratio of B to E is also given.

Because the ratio of A to D, as also of C to F, is given, the
* 67 Dat. ratio of the rectangle A, C, to the rectangle D, F, is given *;
* 17. 6. but the square of B is equal * to the rectangle A, C;
* 17. 6. and the square of E to the rectangle * D, F; therefore the ratio of the square of B to the square of E is
* 58 Dat. given; wherefore * also the ratio of the straight line
* 67 Dat. B to E is given *.

Next, let the ratio of A to D, and of B to E, be given; then the ratio of C to F is also given.

Because the ratio of B to E is given, the ratio of the square
* 54 Dat. of B to the square of E is given *; therefore * the ratio of the
* 17. 6. rectangle A, C, to the rectangle D, F, is given; and the ratio of the side A to the side D is given; therefore the ratio of the
* 65 Dat. other side C to the other F is given *.

Cor. And if the extremes of four proportionals have to the extremes of four other proportionals given ratios, and one of the means a given ratio to one of the means; the other mean shall have a given ratio to the other mean, as may be shewn in the same manner as in the foregoing proposition.

(82.) PROPOSITION LXXII.

If four straight lines be proportionals; as the first is to the straight line to which the second has a given ratio, so is the third to a straight line to which the fourth has a given ratio.

Let A, B, C, D be four proportional straight lines, viz. as A to B, so C to D: as A is to the straight line to which B has a given ratio, so is C to a straight line to which D has a given ratio.

Let E be the straight line to which B has a given ratio, and as B to E, so make D to F: the ratio of B to E is given *, and therefore the ratio of D to F; * Hyp. and because as A to B, so is C to D; and as B to E so D to F: therefore, ex æquali, as A to E, so is C to F; and E is the straight line to which B has a given ratio, and F that to which D has a given ratio; therefore as A is to the straight line to which B has a given ratio, so is C to a line to which D has a given ratio.

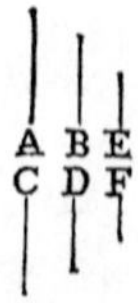

PROPOSITION LXXIII. (83.)

If four straight lines be proportionals; as the first is to the straight line to which the second has a given ratio, so is a straight line to which the third has a given ratio to the fourth. See N.

Let the straight line A be to B, as C to D: as A to the straight line to which B has a given ratio, so is a straight line to which C has a given ratio to D.

Let E be the straight line to which B has a given ratio, and as B to E, so make F to C; because the ratio of B to E is given, the ratio of C to F is given: and because A is to B, as C to D; and as B to E, so F to C; therefore, ex æquali in proportione perturbatâ *, A is to E, as F to D; that is, A is to E, to which B has a given ratio, as F, to which C has a given ratio, is to D. * 23. 5.

PROPOSITION LXXIV. (64.)

If a triangle have a given obtuse angle, the excess of the square of the side which subtends the obtuse angle, above the squares of the sides which contain it, shall have a given ratio to the triangle.

Let the triangle ABC have a given obtuse angle ABC; and produce the straight line CB, and from the point A, draw AD perpendicular to BC: the excess of the square of AC above the squares of AB, BC, that is *, the double of the rectangle contained by DB, BC, has a given ratio to the triangle ABC. * 12. 2.

Because the angle ABC is given, the angle ABD is also given; and the angle ADB is given; wherefore the triangle
* 43 Dat. ABD is given * in species; and therefore the ratio of AD to
* 1. 6. DB is given: and as AD to DB, so is * the rectangle AD, BC, to the rectangle DB, BC; wherefore the ratio of the rectangle AD, BC, to the rectangle DB, BC, is given, as also the ratio of twice the rectangle DB, BC, to the rectangle AD, BC: but the ratio of the rectangle AD, BC, to the triangle ABC, is
* 41. 1. given, because it is double * of the triangle; therefore the ratio of twice the rectangle DB,
* 9 Dat. BC, to the triangle ABC is given *: and twice the rectangle
* 12. 2. DB, BC, is the excess * of the square of AC above the squares of AB, BC; therefore this excess has a given ratio to the triangle ABC.

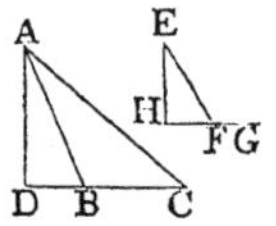

And the ratio of this excess to the triangle ABC may be found thus: take a straight line EF given in position and magnitude; and because the angle ABC is given, at the point F of the straight line EF make the angle EFG equal to the angle ABC; produce GF, and draw EH perpendicular to FG; then the ratio of the excess of the square of AC above the squares of AB, BC, to the triangle ABC, is the same with the ratio of quadruple the straight line HF to HE.

Because the angle ABD is equal to the angle EFH and the angle ADB to EHF, each being a right angle; the triangle
* 4. 6. ADB is equiangular to EHF; therefore * as BD to DA, so
* Cor. 4. 5. FH to HE; and as quadruple of BD to DA, so is * quadru-
* 9 Dat. ple of FH to HE: but as twice BD is to DA, so * is twice the rectangle DB, BC, to the rectangle AD, BC; and as DA to
* C. 5. the half of it, so is * the rectangle AD, BC, to its half the triangle ABC; therefore, ex æquali, as twice BD is to the half of DA, that is, as quadruple of BD is to DA, that is, as quadruple of FH to HE, so is twice the rectangle DB, BC to the triangle ABC.

(65.) PROPOSITION LXXV.

If a triangle have a given acute angle, the space by which the square of the side subtending the acute angle is less than the squares of the sides which contain it, shall have a given ratio to the triangle.

Let the triangle ABC have a given acute angle ABC, and

draw AD perpendicular to BC: the space by which the square of AC is less than the squares of AB, BC, that is * the double (* 13. 2.) of the rectangle contained by CB, BD, has a given ratio to the triangle ABC.

Because the angles ABD, ADB, are each of them given, the triangle ABD is given in species; and therefore the ratio of BD to DA is given: and as BD to DA, so is the rectangle CB, BD to the rectangle CB, AD: therefore the ratio of these rectangles is given, as also the ratio of twice the rectangle CB, BD to the rectangle CB, AD: but the rectangle CB, AD, has a given ratio to its half the triangle ABC: therefore * the ratio of twice the rectangle CB, BD, to the triangle (* 9 Dat.) ABC is given; and twice the rectangle CB, BD, is * the space by (* 13. 2.) which the square of AC is less than the squares of AB, BC; therefore the ratio of this space to the triangle ABC is given: and the ratio may be found as in the preceding proposition.

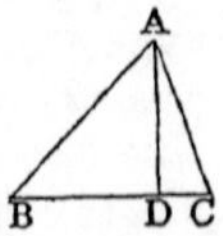

LEMMA.

If from the vertex A *of an isosceles triangle* ABC, *any straight line* AD *be drawn to the base* BC, *the square of the side* AB *is equal to the rectangle* BD, DC, *of the segments of the base together with the square of* AD; *but if* AD *be drawn to the base produced, the square of* AD *is equal to the rectangle* BD, DC, *together with the square of* AB.

CASE 1. Bisect the base BC in E, and join AE, which will be perpendicular * to BC; wherefore the square (* 8. 1.) of AB is equal * to the squares of AE, EB; (* 47. 1.) but the square of EB is equal * to the rectangle BD, DC, together with the square of (* 5. 2.) DE; therefore the square of AB is equal to the squares of AE, ED, that is, to * the square of AD, together with the rectangle BD, DC. (* 47. 1.)

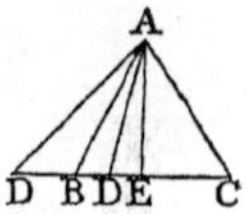

The other case is shewn in the same way by 6. 2. Elem.

PROPOSITION LXXVI. (67.)

If a triangle have a given angle, the excess of the square of the straight line which is equal to the two sides that contain the given angle, above the square of the third side, shall have a given ratio to the triangle.

Let the triangle ABC have the given angle BAC: the excess of the square of the straight line which is equal to BA, AC, together above the square of BC, shall have a given ratio to the triangle ABC.

Produce BA, and take AD equal to AC, join DC, and produce it to E, and through the point B, draw BE parallel to AC: join AE, and draw AF perpendicular to DC; and because AD is equal to AC, BD is equal to BE; and BC is drawn from the vertex B of the isosceles triangle DBE: therefore by the lemma, the square of BD, that is, of BA and AC together, is equal to the rectangle DC, CE, together with the square of BC: and therefore the square of BA, AC, together, that is, of BD, is greater than the square of BC by the rectangle DC, CE; and this rectangle has a given ratio to the triangle ABC: because the angle BAC is given, the adjacent angle CAD is given; and each of the angles ADC, DCA is given, for each of

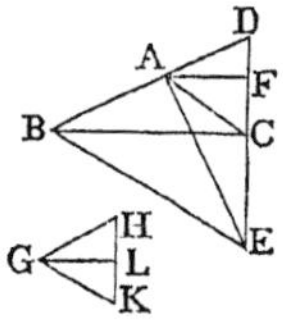

* 5 & 32. 1. them is the half * of the given angle BAC; therefore the tri-
* 43 Dat. angle ADC is given in species *; and AF is drawn from its vertex to the base in a given angle; wherefore the ratio of
* 50 Dat. AF to the base CD is given *; and as CD to AF, so is * the
* 1. 6. rectangle DC, CE, to the rectangle AF, CE; and the ratio of
* 41. 1. the rectangle AF, CE, to its half *, the triangle ACE is given; therefore the ratio of the rectangle DC, CE, to the triangle
* 37. 1. ACE, that is *, to the triangle ABC, is given *; and the rect-
* 9 Dat. angle DC, CE is the excess of the square of BA, AC together, above the square of BC: therefore the ratio of this excess to the triangle ABC is given.

The ratio which the rectangle DC, CE has to the triangle ABC, is found thus: take the straight line GH given in position and magnitude, and at the point G in GH, make the angle HGK equal to the given angle CAD, and take GK equal to GH, join KH, and draw GL perpendicular to it: then the ratio of HK to the half of GL is the same with the ratio of the rectangle DC, CE, to the triangle ABC: because the angles HGK, DAC, at the vertices of the isosceles triangles GHK, ADC, are equal to one another, these triangles are similar; and because GL, AF are perpendicular to the bases HK, DC, as HK to GL,
* 4. 6. & 22. 5. so is * (DC to AF, and so is) the rectangle DC, CE, to the rectangle AF, CE; but as GE to its half, so is the rectangle AF, CE, to its half, which is the triangle ACE, or the triangle

ABC; therefore ex æquali, HK is to the half of the straight line GL, as the rectangle DC, CE is to the triangle ABC.

Cor. And if a triangle have a given angle, the space by which the square of the straight line, which is the difference of the sides which contain the given angle, is less than the square of the third side, shall have a given ratio to the triangle. This is demonstrated the same way as in the preceding proposition, by help of the second case of the Lemma.

PROPOSITION LXXVII. (I.)

If the perpendicular drawn from a given angle of a triangle to the opposite side, or base, have a given ratio to the base, the triangle is given in species. See N.

Let the triangle ABC have the given angle BAC, and let the perpendicular AD drawn to the base BC, have a given ratio to it: the triangle ABC is given in species.

If ABC be an isosceles triangle, it is evident *, that if any one of its angles be given, the rest are also given; and therefore the triangle is given in species, without the consideration of the ratio of the perpendicular to the base which in this case is given by Prop. 50. * 5. & 32. 1.

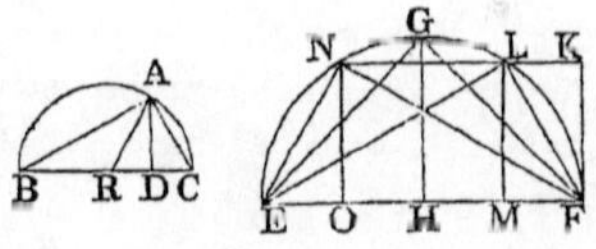

But when ABC is not an isosceles triangle, take any straight line EF given in position and magnitude, and upon it describe the segment of a circle EGF, containing an angle equal to the given angle BAC, draw GH bisecting EF at right angles, and join EG, GF: then, since the angle EGF is equal to the angle BAC, and that EGF is an isosceles triangle, and ABC is not, the angle FEG is not equal to the angle CBA: draw EL, making the angle FEL equal to the angle CBA; join FL, and draw LM perpendicular to EF; then because the triangles ELF, BAC are equiangular, as also are the triangles MLE, DAB, as ML to LE, so is DA to AB: and as LE to EF, so is AB to BC; wherefore, ex æquali, as LM to EF, so is AD to BC; and because the ratio of AD to BC is given, therefore the ratio of LM to EF is given; and EF is given, wherefore * LM also is given. Complete the parallelogram LMFK; and because LM is given, FK is given in magnitude; it is also given in position; and the point F is given, and consequently * * 2 Dat. * 30 Dat.

the point K; and because through K the straight line KL is
* 31 Dat. drawn parallel to EF, which is given in position, therefore* KL is given in position: and the circumference ELF is given
* 28 Dat. in position; therefore the point L is given*: and because the points L, E, F are given, the straight lines LE, EF, FL are
* 29 Dat. given* in magnitude; therefore the triangle LEF is given in
* 42 Dat. species*; and the triangle ABC is similar to LEF, wherefore also ABC is given in species.

Because LM is less than GH, the ratio of LM to EF, that is, the given ratio of AD to BC, must be less than the ratio of GH to EF, which the straight line in a segment of a circle containing an angle equal to the given angle, that bisects the base of the segment at right angles, has unto the base.

Cor. 1. If two triangles ABC, LEF have one angle BAC equal to one angle ELF, and if the perpendicular AD be to the base BC, as the perpendicular LM to the base EF, the triangles ABC, LEF are similar.

Describe the circle EGF about the triangle ELF, and draw LN parallel to EF, join EN, NF, and draw NO perpendicular to EF: because the angles ENF, ELF, are equal, and that the angle EFN is equal to the alternate angle FNL, that is, to the angle FEL, in the same segment, therefore the triangle NEF is similar to LEF; and in the segment EGF there can be no other triangle upon the base EF, which has the ratio of its perpendicular to that base the same with the ratio of LM or NO to EF, because the perpendicular must be greater or less than LM or NO; but, as has been shewn in the preceding demonstration, a triangle, similar to ABC, can be described in the segment EGF, upon the base EF, and the ratio of its perpendicular to the base is the same, as was there shewn, with the ratio of AD to BC, that is, of LM to EF; therefore that triangle must be either LEF, or NEF, which therefore are similar to the triangle ABC.

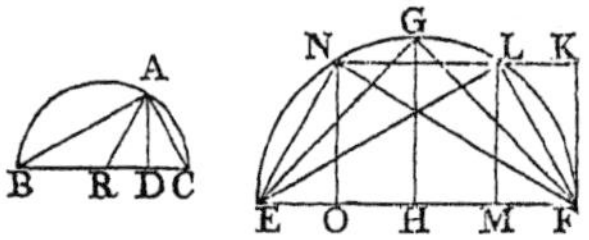

Cor. 2. If a triangle ABC have a given angle BAC, and if the straight line AR drawn from the given angle to the opposite side BC, in a given angle ARC, have a given ratio to BC, the triangle ABC is given in species.

Draw AD perpendicular to BC; therefore the triangle ARD is given in species; wherefore the ratio of AD to AR is given:

and the ratio of AR to BC is given, and consequently* the ratio of AD to BC is given; and the triangle ABC is therefore given in species *. * 9 Dat. * 77 Dat.

Cor. 3. If two triangles ABC, LEF have one angle BAC equal to one angle ELF, and if straight lines drawn from these angles to the bases, making with them given and equal angles, have the same ratio to the bases, each to each; then the triangles are similar: for having drawn perpendiculars to the bases from the equal angles, as one perpendicular is to its base, so is the other to its base *; wherefore, by Cor. 1. the triangles are similar. * 4. 6. and 22. 5.

A triangle similar to ABC may be found thus: having described the segment EGF, and drawn the straight line GH as was directed in the proposition, find FK, which has to EF the given ratio of AD to BC; and place FK at right angles to EF from the point F; then because, as has been shewn, the ratio of AD to BC, that is, of FK to EF, must be less than the ratio of GH to EF, therefore FK is less than GH; and consequently, the parallel to EF drawn through the point K, must meet the circumference of the segment in two points: let L be either of them, and join EL, LF, and draw LM perpendicular to EF; then, because the angle BAC is equal to the angle ELF, and that AD is to BC, as KF, that is, LM, to EF, the triangle ABC is similar to the triangle LEF, by Cor. 1.

PROPOSITION LXXVIII. (80.)

If a triangle have one angle given, and if the ratio of the rectangle of the sides which contain the given angle to the square of the third side be given, the triangle is given in species.

Let the triangle ABC have the given angle BAC, and let the ratio of the rectangle BA, AC, to the square of BC be given: the triangle ABC is given in species.

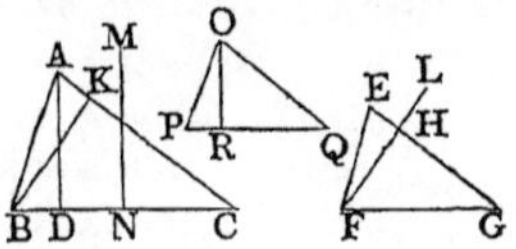

From the point A, draw AD perpendicular to BC, the rectangle AD, BC, has a given ratio to its half * the triangle ABC; * 41. 1.
and because the angle BAC is given, the ratio of the triangle ABC to the rectangle BA, AC is given *; and by the hypothesis, the ratio of the rectangle BA, AC, to the square of BC * Cor. 62 Dat.

* 9 Dat. * 1. 6. * 77 Dat. is given; therefore * the ratio of the rectangle AD, BC, to the square of BC, that is *, the ratio of the straight line AD to BC, is given; wherefore the triangle ABC is given in species*.

A triangle similar to ABC may be found thus: take a straight line EF given in position and magnitude, and make the angle FEG equal to the given angle BAC, and draw FH perpendicular to EG, and BK perpendicular to AC; therefore the triangles ABK, EFH are similar, and the rectangle AD, BC, or the rectangle BK, AC which is equal to it, is to the rectangle BA, AC, as the straight line BK to BA, that is, as FH to FE. Let the given ratio of the rectangle BA, AC, to the square of BC, be the same with the ratio of the straight line EF to FL; therefore, ex æquali, the ratio of the rectangle AD, BC, to the square of BC, that is, the ratio of the straight line AD to BC, is the same with the ratio of HF to FL; and because AD is not greater than the straight line MN in the segment of the circle described about the triangle ABC, which bisects BC at right angles; the ratio of AD to BC, that is, of HF to FL, must not be greater than the ratio of MN to BC: * 77 Dat. let it be so; and find * a triangle OPQ which has one of its angles POQ equal to the given angle BAC, and the ratio of the perpendicular OR, drawn from that angle to the base PQ, the same with the ratio of HF to FL; then the triangle ABC is similar to OPQ: because, as has been shewn, the ratio of AD to BC is the same with the ratio of (HF to FL, that is, by the construction, with the ratio of) OR to PQ; and the angle BAC is equal to the angle POQ. Therefore the triangle ABC * 1 Cor. 77 Dat. is similar * to the triangle POQ.

Otherwise:

Let the triangle ABC have the given angle BAC, and let the ratio of the rectangle BA, AC, to the square of BC, be given: the triangle ABC is given in species.

Because the angle BAC is given, the excess of the square of both the sides BA, AC, together, above the square of the * 76 Dat. third side BC has a given * ratio to the triangle ABC. Let the figure D be equal to this excess; therefore the ratio of D to the triangle ABC is given: and the ratio of the triangle ABC to the rectangle BA, AC, is * Cor. 62 Dat. given*, because BAC is a given angle; and the rectangle BA, * 10 Dat. AC has a given ratio to the square of BC; wherefore * the

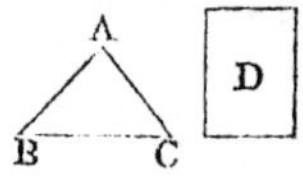

ratio of D to the square of BC is given; and by composition * * 7 Dat.
the ratio of the space D, together with the square of BC, to the square of BC is given; but D, together with the square of BC, is equal to the square of both BA and AC together; therefore the ratio of the square of BA, AC, together, to the square of BC, is given; and the ratio of BA, AC, together, to BC is therefore given*; and the angle BAC is given, wherefore * * 59 Dat. * 48 Dat.
the triangle ABC is given in species.

The composition of this, which depends upon those of the 76th and 48th Propositions, is more complex than the preceding composition, which depends upon that of Prop. 77, which is easy.

PROPOSITION LXXIX. (K.)

If a triangle have a given angle, and if the straight line drawn from that angle to the base, making a given angle with it, divide the base into segments which have a given ratio to one another, the triangle is given in species. See N.

Let the triangle ABC have the given angle BAC, and let the straight line AD, drawn to the base DC, making the given angle ADB, divide BC into the segments BD, DC, which have a given ratio to one another: the triangle ABC is given in species.

Describe * the circle BAC about the triangle, and from its * 5. 4.
centre E, draw EA, EB, EC, ED; because the angle BAC is given, the angle BEC at the centre, which is the double * of * 20. 3.
it, is given. And the ratio of BE to EC is given, because they are equal to one another; therefore * the * 44 Dat.
triangle BEC is given in species, and the ratio of EB to BC is given; also the ratio of CB to BD is given *; because the ratio of BD to DC * 7 Dat.
is given, therefore the ratio of EB to BD is given *, and the angle EBC is given, where- * 9 Dat.

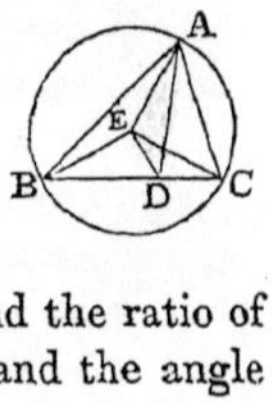

fore the triangle EBD is given * in species, and the ratio of * 44 Dat.
EB, that is, of EA, to ED, is therefore given; and the angle EDA is given, because each of the angles BDE, BDA, is given; therefore the triangle AED is given * in species, and the angle * 47 Dat.
AED given: also the angle DEC is given, because each of the angles BED, BEC, is given; therefore the angle AEC is given, and the ratio of EA to EC, which are equal, is given; and the triangle AEC is therefore given* in species, and the angle * 44 Dat.

ECA is given; and the angle ECB is given, wherefore the an-
* 43 Dat. gle ACB is given; and the angle BAC is also given; therefore* the triangle ABC is given in species.

A triangle similar to ABC may be found, by taking a straight line given in position and magnitude, and dividing it in the given ratio which the segments BD, DC, are required to have to one another; then, if upon that straight line a segment of a circle be described containing an angle equal to the given angle BAC, and a straight line be drawn from the point of division in an angle equal to the given angle ADB, and from the point where it meets the circumference, straight lines be drawn to the extremity of the first line, these together with the first line, shall contain a triangle similar to ABC, as may easily be shewn.

The demonstration may be also made in the manner of that of the 77th Prop., and that of the 77th may be made in the manner of this.

(L.)

PROPOSITION LXXX.

If the sides about an angle of a triangle have a given ratio to one another, and if the perpendicular drawn from that angle to the base have a given ratio to the base, the triangle is given in species.

Let the sides BA, AC, about the angle BAC of the triangle ABC have a given ratio to one another, and let the perpendicular AD have a given ratio to the base BC: the triangle ABC is given in species.

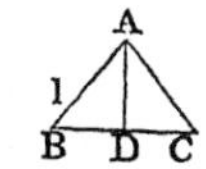

First, let the sides AB, AC be equal to one another,
* 26. 1. therefore the perpendicular AD bisects* the base BC; and the ratio of AD to BC, and therefore to its half DB, is given: and the angle ADB is
* 43 Dat. given; wherefore the triangle* ABD, and consequently the
* 44 Dat. triangle ABC, is given* in species.

But let the sides be unequal, and BA be greater than AC; and make the angle CAE equal to the angle ABC; because the angle AEB is common to the triangles AEB, CEA, they are similar; therefore as AB to BE, so is CA to AE, and, by permutation, as BA to AC, so is BE to EA, and so is EA to EC; and the ratio of BA to AC is given, therefore the ratio of BE to EA, and the ratio of EA to EC, as

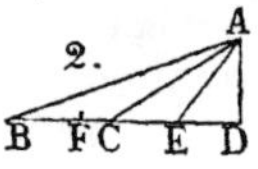

also the ratio of BE to EC, is given *; wherefore the ratio of EB to BC is given *; and the ratio of AD to BC is given by the hypothesis, therefore * the ratio of AD to BE is given; and the ratio of BE to EA was shewn to be given; wherefore the ratio of AD to EA is given; and ADE is a right angle, therefore the triangle ADE is given * in species, and the angle AEB given; the ratio of BE to EA is likewise given, therefore * the triangle ABE is given in species, and consequently the angle EAB, as also the angle ABE, that is, the angle CAE, is given: therefore the angle BAC is given, and the angle ABC being also given, the triangle ABC is given * in species.

* 9 Dat.
* 6 Dat.
* 9 Dat.
* 46 Dat.
* 44 Dat.
* 43 Dat.

How to find a triangle which shall have the things which are mentioned to be given in the proposition, is evident in the first case; and to find it the more easily in the other case, it is to be observed that, if the straight line EF equal to EA be placed in EB towards B, the point F divides the base BC into the segments BF, FC, which have to one another the ratio of the sides BA, CA, because BE, EA, or EF, and EC, were shewn to be proportionals, therefore * BF is to FC, as BE to EF or EA, that is, as BA to AC; and AE cannot be less than the altitude of the triangle ABC, but it may be equal to it, which if it be, the triangle, in this case, as also the ratio of the sides, may be thus found: having given the ratio of the perpendicular to the base, take the straight line GH, given in position and magnitude, for the base of the triangle to be found; and let the given ratio of the perpendicular to the base be that of the straight line K to GH, that is, let K be equal to the perpendicular; and suppose GLH to be the triangle which is to be found, therefore having made the angle HLM equal to LGH, it is required that LM be perpendicular to GM, and equal to K; and because GM, ML, MH, are proportionals, as was shewn of BE, EA, EC, the rectangle GMH is equal to the square of ML. Add the common square of NH (having bisected GH in N), and the square of NM is equal * to the squares of the given straight lines NH and ML, or K: therefore the square of NM, and its side NM, is given, as also the point M, viz. by taking the straight line NM, the square of which is equal to the squares of NH, ML. Draw

* 19. 5.

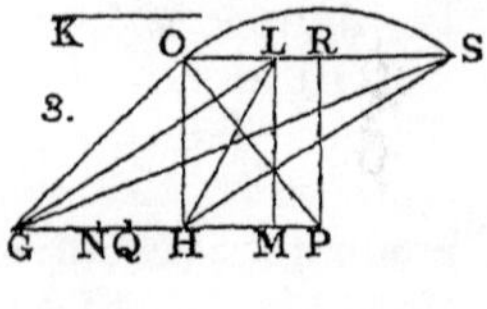

* 6. 2.

ML equal to K, at right angles to GM; and because ML is given in position and magnitude, therefore the point L is given: join LG, LH; then the triangle LGH is that which was to be found; for the square of NM is equal to the squares of NH and ML, and taking away the common square of NH,
* 6. 2. the rectangle GMH is equal* to the square of ML; therefore
* 6. 6. as GM to ML, so is ML to MH; and the triangle LGM is* therefore equiangular to HLM, and the angle HLM equal to the angle LGM, and the straight line LM, drawn from the vertex of the triangle, making the angle HLM equal to LGH, is perpendicular to the base, and equal to the given straight line K, as was required; and the ratio of the sides GL, LH, is the same with the ratio of GM to ML, that is, with the ratio of the straight line which is made up of GN, the half of the given base, and of NM, the square of which is equal to the squares of GN and K, to the straight line K.

And whether this ratio of GM to ML be greater or less than the ratio of the sides of any other triangle upon the base GH, and of which the altitude is equal to the straight like K, that is, the vertex of which is in the parallel to GH drawn through the point L, may be thus found. Let OGH be any such triangle, and draw OP, making the angle HOP equal to the angle OGH; therefore, as before, GP, PO, PH are proportionals, and PO cannot be equal to LM, because the rectangle GPH would be equal to the rectangle GMH, which is impossible; for the point P cannot fall upon M, because O would then fall on L; nor can PO be less than LM, therefore it is greater; and consequently the rectangle GPH is greater than the rectangle GMH, and the straight line GP greater than GM: therefore the ratio of GM to MH is greater than the ratio of GP to PH, and the ratio of the square of GM to the square of ML is
* 2 Cor. 20. 6. therefore* greater than the ratio of the square of GP to the square of PO, and the ratio of the straight line GM to ML greater than the ratio of GP to PO. But as GM to ML, so is GL to LH; and as GP to PO, so is GO to OH; therefore the ratio of GL to LH is greater than the ratio of GO to OH; wherefore the ratio of GL to LH is the greatest of all others; and consequently the given ratio of the greater side to the less must not be greater than this ratio.

But if the ratio of the sides be not the same with this greatest ratio of GM to ML, it must necessarily be less than it: let any less ratio be given, and the same things being supposed, viz. that GH is the base, and K equal to the altitude of the triangle, it may be found as follows: divide GH in the point Q, so that the ratio of GQ to QH may be the same with the given ratio of the sides; and as GQ to QH, so make GP to PQ, and so will* PQ be to PH; wherefore the square of GP * 19. 5. is to the square of PQ, as* the straight line GP to PH: and * 2 Cor. 20. 6. because GM, ML, MH are proportionals, the square of GM is to the square of ML, as * the straight line GM to MH: but * 2 Cor. 20. 6. the ratio of GQ to QH, that is, the ratio of GP to PQ, is less than the ratio of GM to ML; and therefore the ratio of the square of GP to the square of PQ is less than the ratio of the square of GM to that of ML; and consequently the ratio of the straight line GP to PH is less than the ratio of GM to MH; and, by division, the ratio of GH to HP is less than that of GH to HM; wherefore* the straight line HP is greater * 10. 5. than HM, and the rectangle GPH, that is, the square of PQ, greater than the rectangle GMH, that is, than the square of ML, and the straight line PQ is therefore greater than ML. Draw LR parallel to GP, and from P draw PR at right angles to GP. Because PQ is greater than ML, or PR, the circle described from the centre P, at the distance PQ, must necessarily cut LR in two points; let these be O, S, and join OG, OH; SG, SH: each of the triangles OGH, SGH, have the things mentioned to be given in the proposition: join OP, SP; and because as GP to PQ, or PO, so is PO to PH, the triangle OGP is equiangular to HOP; as therefore OG to GP, so is HO to OP; and, by permutation, as GO to OH, so is GP to PO, or PQ; and so is GQ to QH; therefore the triangle OGH has the ratio of its sides GO, OH, the same with the given ratio of GQ to QH; and the perpendicular has to the base the given ratio of K to GH, because the perpendicular is equal to LM, or K: the like may be shewn in the same way of the triangle SGH.

This construction, by which the triangle OGH is found, is shorter than that which would be deduced from the demonstration of the datum, by reason that the base GH is given in position and magnitude, which was not supposed in the demonstration: the same thing is to be observed in the next proposition.

PROPOSITION LXXXI. (M.)

If the sides about an angle of a triangle be unequal, and have a given ratio to one another, and if the perpendicular from that angle to the base divide it into segments that have a given ratio to one another, the triangle is given in species.

Let ABC be a triangle, the sides of which about the angle BAC are unequal, and have a given ratio to one another, and let the perpendicular AD to the base BC divide it into the segments BD, DC, which have a given ratio to one another: the triangle ABC is given in species.

Let AB be greater than AC, and make the angle CAE equal to the angle ABC; and because the angle AEB is common to
* 4. 6. the triangles ABE, CAE, they are * equiangular to one another: therefore as AB to BE, so is CA to AE, and, by permutation, as AB to AC, so is BE to EA, and so is EA to EC: but the ratio of BA to AC is given, therefore the ratio of BE to EA, as also the ratio of EA to EC, is given;

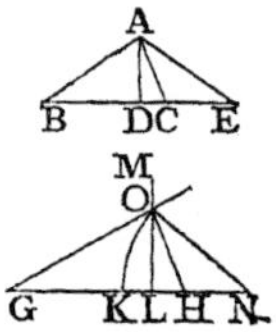

* 9 Dat. wherefore * the ratio of BE to EC, as also *
* Cor. 6. Dat. the ratio of EC to CB, is given: and the ratio
* 7 Dat. of BC to CD is given *, because the ratio of
* 9 Dat. BD to DC is given; therefore * the ratio of EC to CD is given,
* 7 Dat. and consequently * the ratio of DE to EC: and the ratio of
* 9 Dat. EC to EA was shewn to be given, therefore * the ratio of DE
* 46 Dat. to EA is given: and ADE is a right angle, wherefore * the triangle ADE is given in species, and the angle AED given;
* 44 Dat. and the ratio of CE to EA is given, therefore* the triangle AEC is given in species, and consequently the angle ACE is given, as also the adjacent angle ACB. In the same manner, because the ratio of BE to EA is given, the triangle BEA is given in species, and the angle ABE is therefore given: and
* 43 Dat. the angle ACB is given; wherefore the triangle ABC is given* in species.

But the ratio of the greater side BA to the other AC must be less than the ratio of the greater segment BD to DC; because the square of BA is to the square of AC, as the squares of BD, DA to the squares of DC, DA; and the squares of BD, DA have to the squares of DC, DA, a less ratio than the square of BD has to the square of DC †, because the square of BD is

† If A be greater than B, and C any third magnitude; then A and C together have to B and C together a less ratio than A has to B.

Let A be to B as C to D, and because A is greater than B, C is greater than D: but as A is to B, so A and C to B and D, and A and C have to B greater than the square of DC, therefore the square of BA has to the square of AC a less ratio than the square of BD has to that of DC: and consequently the ratio of BA to AC is less than the ratio of BD to DC.

This being premised, a triangle which shall have the things mentioned to be given in the proposition, and to which the triangle ABC is similar, may be found thus: take a straight line GH given in position and magnitude, and divide it in K, so that the ratio of GK to KH may be the same with the given ratio of BA to AC: divide also GH in L, so that the ratio of GL to LH may be the same with the given ratio of BD to DC, and draw LM at right angles to GH: and because the ratio of the sides of a triangle is less than the ratio of the segments of the base, as has been shewn, the ratio of GK to KH, is less than the ratio of GL to LH; wherefore the point L must fall betwixt K and H: also make as GK to KH, so GN to NK, and so shall* NK be to NH. And from the centre N, at the distance NK, describe a circle, and let its circumference meet LM in O, and join OG, OH; then OGH is the triangle which was to be described: because GN is to NK, or NO, as NO to NH, the triangle OGN, is equiangular to HON; therefore as OG to GN, so is HO to ON, and, by permutation, as GO to OH, so is GN to NO, or NK, that is, as GK to KH, that is, in the given ratio of the sides, and by the construction, GL, LH have to one another the given ratio of the segments of the base.

* 19. 5.

PROPOSITION LXXXII. (60.)

If a parallelogram given in species and magnitude be increased or diminished by a gnomon given in magnitude, the sides of the gnomon are given in magnitude.

First, let the parallelogram AB given in species and magnitude, be increased by the given gnomon ECBDFG: each of the straight lines CE, DF, is given.

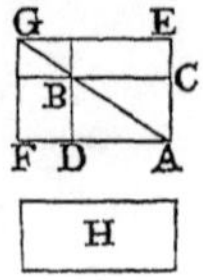

Because AB is given in species and magnitude, and that the gnomon ECBDFG is given, therefore the whole space AG is given in magnitude: but AG is also given in species, because it is similar* to AB; therefore the sides of AG are given*: each of the straight lines AE, AF, is therefore

* 2 Def. 2. and 24. 6.
* 60 Dat.

and C a less ratio than A and C have to B and D, because C is greater than D, therefore A and C have to B and C a less ratio than A to B.

* 60 Dat. given; and each of the straight lines CA, AD, is given*,
* 4 Dat. therefore each of the remainders EC, DF, is given*.

Next, let the parallelogram AG, given in species and magnitude, be diminished by the given gnomon ECBDFG, each of the straight lines CE, DF, is given.

Because the parallelogram AG is given, as also its gnomon ECBDFG, the remaining space AB is given in magnitude: but
* 2 Def. 2. and 24. 6. it is also given in species; because it is similar* to AG; there-
* 60 Dat. fore* its sides CA, AD are given, and each of the straight lines EA, AF is given; therefore EC, DF are each of them given.

The gnomon and its sides CE, DF, may be found thus in the first case. Let H be the given space to which the gnomon
* 25. 6. must be made equal, and find* a parallelogram similar to AB and equal to the figures AB and H together, and place its sides AE, AF, from the point A, upon the straight lines AC, AD, and complete the parallelogram AG which is about the same
* 26. 6. diameter* with AB; because therefore AG is equal to both AB and H, take away the common part AB, the remaining gnomon ECBDFG is equal to the remaining figure H; therefore a gnomon equal to H, and its sides CE, DF, are found: and in like manner they may be found in the other case, in which the given figure H must be less than the figure FE from which it is to be taken.

(58.)

PROPOSITION LXXXIII.

If a parallelogram equal to a given space be applied to a given straight line, deficient by a parallelogram given in species, the sides of the defect are given.

Let the parallelogram AC equal to a given space be applied to the given straight line AB, deficient by the parallelogram BDCL given in species: each of the straight lines CD, DB, is given.

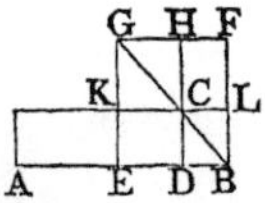

Bisect AB in E; therefore EB is given in
* 18. 6. magnitude; upon EB describe* the parallelogram EF similar to DL, and similarly placed; therefore EF is given in species, and is about
* 26. 6. the same diameter* with DL; let BCG be the diameter, and construct the figure; therefore because the figure EF given in species, is described upon the given straight
* 56 Dat. line EB, EF is given* in magnitude, and the gnomon ELH is
* 36 & 43. 1. * 82 Dat. equal* to the given figure AC; therefore* since EF is dimi-

nished by the given gnomon ELH, the sides EK, FH of the gnomon are given; but EK is equal to DC, and FH to BD; wherefore CD, DB are each of them given.

This demonstration is the analysis of the problem in the 28th Prop. of Book 6. the construction and demonstration of which proposition is the composition of the analysis; and because the given space AC, or its equal the gnomon ELH is to be taken from the figure EF, described upon the half of AB similar to BC, therefore AC must not be greater than EF, as is shewn in the 27th Prop. Book 6.

PROPOSITION LXXXIV. (59.)

If a parallelogram equal to a given space, be applied to a given straight line exceeding by a parallelogram given in species, the sides of the excess are given.

Let the parallelogram AC equal to a given space, be applied to the given straight line AB, exceeding by the parallelogram BDCL given in species: each of the straight lines CD, DB, is given.

Bisect AB in E; therefore EB is given in magnitude: upon BE describe * the parallelogram EF similar to LD, and similarly placed; therefore EF is given in species, * 18. 6.
and is about the same diameter * with LD. Let * 26. 6.
CBG be the diameter, and construct the figure: therefore, because the figure EF given in species is described upon the given straight line

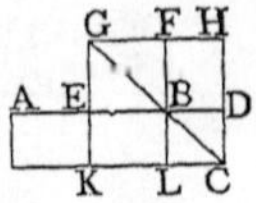

EB, EF is given in magnitude *, and the gnomon ELH is equal * 56 Dat.
to the given figure * AC; wherefore, since EF is increased by * 36 & 43. 1.
the given gnomon ELH, its sides EK, FH are given *; but * 82 Dat.
EK is equal to CD, and FH to BD, therefore CD, DB are each of them given.

This demonstration is the analysis of the problem in the 29th Prop. Book 6., the construction and demonstration of which is the composition of the analysis.

Cor. If a parallelogram given in species be applied to a given straight line, exceeding by a parallelogram equal to a given space; the sides of the parallelogram are given.

Let the parallelogram ADCE given in species, be applied to the given straight line AB, exceeding by the parallelogram BDCG equal to a given space: the sides AD, DC, of the parallelogram are given.

Draw the diameter DE of the parallelogram AC, and con-
* 43. 1. struct the figure. Because the parallelogram AK is equal* to BC which is given, therefore AK is given;
* 24. 6. and BK is similar* to AC, therefore BK is given in species. And since the parallelogram AK given in magnitude is applied to the given straight line AB, exceeding by the parallelogram BK given in species, therefore, by this proposition, BD, DK, the sides of the excess, are given, and the straight line AB is given; therefore the whole AD, as also DC, to which it has a given ratio, is given.

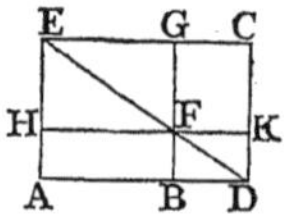

PROB.—*To apply a parallelogram similar to a given one, to a given straight line* AB, *exceeding by a parallelogram equal to a given space.*

* 29. 6. To the given straight line AB apply* the parallelogram AK equal to the given space, exceeding by the parallelogram BK similar to the one given. Draw DF the diameter of BK, and through the point A, draw AE parallel to BF, meeting DF produced in E, and complete the parallelogram AC.

* 43. 1. The parallelogram BC is equal* to AK, that is, to the given
* 24. 6. space; and the parallelogram AC is similar* to BK; therefore the parallelogram AC is applied to the straight line AB similar to the one given, and exceeding by the parallelogram BC which is equal to the given space.

(84.) PROPOSITION LXXXV.

If two straight lines contain a parallelogram given in magnitude in a given angle; if the difference of the straight lines be given, they shall each of them be given.

Let AB, BC contain the parallelogram AC given in magnitude, in the given angle ABC, and let the excess of BC above AB be given: each of the straight lines AB, BC is given.

Let DC be the given excess of BC above BA, therefore the remainder BD is equal to BA. Complete the parallelogram AD; and because AB is equal to BD, the ratio of AB to BD is given; and the angle ABD is given, therefore the parallelogram AD is given in species; and because the given parallelogram AC is applied to the given straight line DC, exceeding by the parallelogram AD given in species, the sides

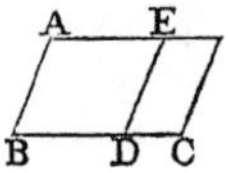

of the excess are given*: therefore BD is given; and DC is given, wherefore the whole BC is given: and AB is given, therefore AB, BC are each of them given. * 84 Dat.

PROPOSITION LXXXVI. (85.)

If two straight lines contain a parallelogram given in magnitude, in a given angle; if both of them together be given, they shall each of them be given.

Let the two straight lines AB, BC contain the parallelogram AC given in magnitude, in the given angle ABC, and let AB, BC, together be given: each of the straight lines AB, BC, is given.

Produce CB, and make DB equal to BA, and complete the parallelogram ABDE. Because DB is equal to BA, and the angle ABD given, because the adjacent angle ABC is given, the parallelogram AD is given in species: and because AB, BC together are given, and AB is equal to BD; therefore DC is given: and because the given parallelogram AC is applied to the given straight line DC, deficient by the parallelogram AD given in species, the sides AB, BD, of the defect, are given*; and DC is given, wherefore the remainder BC is given; and each of the straight lines AB, BC, is therefore given. * 83 Dat.

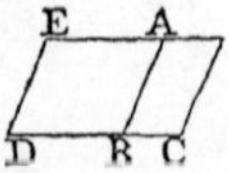

PROPOSITION LXXXVII. (87.)

If two straight lines contain a parallelogram given in magnitude, in a given angle; if the excess of the square of the greater above the square of the lesser be given, each of the straight lines shall be given.

Let the two straight lines AB, BC contain the given parallelogram AC in the given angle ABC: if the excess of the square of BC above the square of BA be given, AB and BC are each of them given.

Let the given excess of the square of BC above the square of BA be the rectangle CB, BD; take this from the square of BC; the remainder, which is* the rectangle BC, CD, is equal to the square of AB: and because the angle ABC of the parallelogram AC is given, the ratio of the sides of the rectangle AB, BC, to the parallelogram AC is given*; and AC is given, * 2. 2. * 62 Dat.

therefore the rectangle AB, BC is given; and the rectangle CB, BD is given; therefore the ratio of the rectangle CB, BD,
* 1. 6. to the rectangle AB, BC, that is *, the ratio of
* 54 Dat. the straight line DB to BA is given; therefore* the ratio of the square of DB to the square of BA is given: and the square of BA is equal to the rectangle BC, CD: wherefore the ratio of

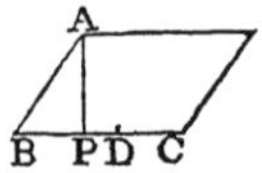

the rectangle BC, CD, to the square of BD is given, as also the ratio of four times the rectangle BC, CD, to the square of
* 7 Dat. BD; and, by composition *, the ratio of four times the rectangle BC, CD, together with the square of BD, to the square of BD is given: but four times the rectangle BC, CD, together
* 8. 2. with the square of BD, is equal * to the square of the straight lines BC, CD, taken together: therefore the ratio of the square
* 58 Dat. of BC, CD, together to the square of BD is given; wherefore * the ratio of the straight line BC, together with CD to BD is given: and, by composition, the ratio of BC together with CD and DB, that is, the ratio of twice BC to BD is given: there-
* 1. 6. fore the ratio of BC to BD is given, as also * the ratio of the square of BC to the rectangle CB, BD: but the rectangle CB, BD, is given, being the given excess of the squares of BC, BA; therefore the square of BC, and the straight line BC is given: and the ratio of BC to BD, as also of BD to BA, has been shewn
* 9 Dat. to be given: therefore * the ratio of BC to BA is given; and BC is given, wherefore BA is given.

The preceding demonstration is the analysis of this problem, viz.

A parallelogram AC, which has a given angle ABC, being given in magnitude, and the excess of the square of BC, one of its sides above the square of the other BA, being given; to find the sides: and the composition is as follows:

Let EFG be the given angle to which the angle ABC is required to be equal, and from any point E in FE, draw EG perpendicular to FG; let the rectangle EG, GH be the given space to which the parallelogram AC is to be made equal; and the rectangle HG, GL, be the given excess of the squares of BC, BA.

Take, in the straight line GE, GK equal to FE, and make GM double of GK; join ML, and in GL produced, take LN equal to LM: bisect GN in O, and between GH, GO, find a mean proportional BC: as OG to GL, so make CB to BD; and make the angle CBA equal to GFE, and as LG to GK, so make

DB to BA, and complete the parallelogram AC: AC is equal to the rectangle EG, GH, and the excess of the squares of CB, BA is equal to the rectangle HG, GL.

Because as CB to BD, so is OG to GL, the square of CB is to the rectangle CB, BD, as* the rectangle HG, GO, to the rectangle HG, GL: and the square of CB is equal to the rectangle HG, GO, because GO, BC, GH are proportionals; therefore the rectangle CB, BD is equal* to HG, GL: and because as CB to BD, so is OG to GL; twice CB is to BD, as twice OG, that is GN to GL; and, by division, as BC together with CD is to BD, so is NL, that is LM, to LG: therefore* the square of BC together with CD is to the square of BD, as the square of ML to the square of LG: but the square of BC and CD together is equal* to four times the rectangle BC, CD, together with the square of BD; therefore four times the rectangle BC, CD, together with the square of BD is to the square of BD, as the square of ML to the square of LG: and, by division, four times the rectangle BC, CD is to the square of BD, as the square of MG to the square of GL; wherefore the rectangle BC, CD, is to the square of BD, as (the square of KG the half of MG to the square of GL, that is, as) the square of AB to the square of BD, because as LG to GK, so DB was made to BA: therefore* the rectangle BC, CD is equal to the square of AB. To each of these add the rectangle CB, BD, and the square of BC becomes equal to the square of AB, together with the rectangle CB, BD; therefore this rectangle, that is, the given rectangle HG, GL, is the excess of the squares of BC, AB. From the point A draw AP perpendicular to BC, and because the angle ABP is equal to the angle EFG, the triangle ABP is equiangular to EFG: and DB was made to BA, as LG to GK; therefore as the rectangle CB, BD to CB, BA, so is the rectangle HG, GL to HG, GK; and as the rectangle CB, BA to AP, BC, so is (the straight line BA to AP, and so is FE or GK to EG, and so is) the rectangle HG, GK to HG, GE; therefore, ex æquali, as the rectangle CB, BD to AP, BC, so is the rectangle HG, GL to EG, GH: and the rectangle CB, BD is equal to HG, GL; therfore the rectangle AP, BC, that is, the parallelogram AC, is equal to the given rectangle EG, GH.

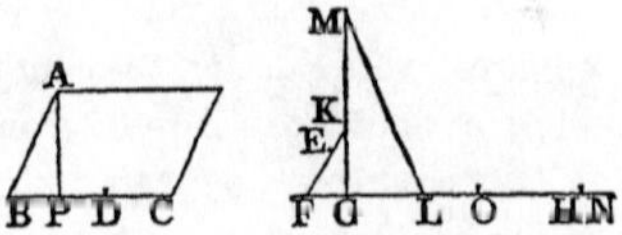

* 1. 6.
* 14. 5.
* 22. 6.
* 8. 2.
* 14. 5.

(N.)

PROPOSITION LXXXVIII.

If two straight lines contain a parallelogram given in magnitude in a given angle; if the sum of the squares of its sides be given, the sides shall each of them be given.

Let the two straight lines AB, BC contain the parallelogram ABCD given in magnitude in the given angle ABC, and let the sum of the squares of AB, BC be given: AB, BC are each of them given.

First, let ABC be a right angle: and because twice the rectangle contained by two equal straight lines is equal to both their squares; but if two straight lines are unequal, twice the rectangle contained by them is less than the sum of their squares, as is evident from the 7th Prop. Book 2. Elem.; therefore twice the given space, to which space the rectangle of which the sides are to be found is equal, must not be greater than the given sum of the squares of the sides: and if twice that space be equal to the given sum of the squares, the sides of the rectangle must necessarily be equal to one another: therefore in this case describe a square ABCD equal to the given rectangle, and its sides AB, BC are those which were to be found: for the rectangle AC is equal to the given space, and the sum of the squares of its sides AB, BC, is equal to twice the rectangle AC, that is, by the hypothesis, to the given space to which the sum of the squares was required to be equal.

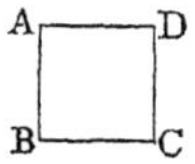

But if twice the given rectangle be not equal to the given sum of the squares of the sides, it must be less than it, as has been shewn. Let ABCD be the rectangle, join AC, and draw BE perpendicular to it, and complete the rectangle AEBF, and describe the circle ABC about the triangle ABC; AC is its diameter*: and because the triangle ABC is similar* to AEB, as AC to CB, so is AB to BE; therefore the rectangle AC, BE is equal to AB, BC; and the rectangle AB, BC is given, wherefore AC, BE is given: and because the sum of the squares of AB, BC, is given, the square of AC which is equal* to that sum is given; and AC itself

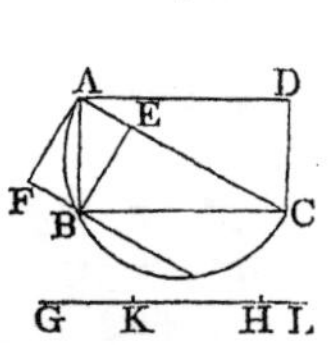

* Cor. 5. 4.

* 8. 6.

* 47. 1.

is therefore given in magnitude; let AC be likewise given in position, and the point A; therefore AF is given * in position: * 32 Dat.
and the rectangle AC, BE is given, as has been shewn, and AC is given, wherefore * BE is given in magnitude, as also * 61 Dat.
AF which is equal to it; and AF is also given in position, and the point A is given, wherefore * the point F is given, * 30 Dat.
and the straight line FB in position *: and the circumference * 31 Dat.
ABC is given in position, wherefore * the point B is given. * 28 Dat.
And the points A, C are given; therefore the straight lines AB, BC are given * in position and magnitude. * 29 Dat.

The sides AB, BC of the rectangle, may be found thus: let the rectangle GH, GK be the given space to which the rectangle AB, BC is equal; and let GH, GL be the given rectangle to which the sum of the squares of AB, BC, is equal: find * a square equal to the rectangle GH, GL: and let its * 14. 2.
side AC be given in position; upon AC as a diameter describe the semicircle ABC, and as AC to GH, so make GK to AF, and from the point A, place AF at right angles to AC: therefore the rectangle CA, AF is equal * to GH, GK; and, by the * 16. 6.
hypothesis, twice the rectangle GH, GK, is less than GH, GL, that is, than the square of AC; wherefore twice the rectangle CA, AF, is less than the square of AC, and the rectangle CA, AF itself, less than half the square of AC, that is, than the rectangle contained by the diameter AC and its half; wherefore AF is less than the semidiameter of the circle, and consequently the straight line drawn through the point F parallel to AC, must meet the circumference in two points: let B be either of them, and join AB, BC, and complete the rectangle ABCD; ABCD is the rectangle which was to be found: draw BE perpendicular to AC; therefore BE is equal* * 34. 1.
to AF, and because the angle ABC in a semicircle is a right angle, the rectangle AB, BC is equal * to AC, BE, that is, to * 8. 6.
the rectangle CA, AF, which is equal to the given rectangle GH, GK: and the squares of AB, BC are together equal * to * 47. 1.
the square of AC, that is, to the given rectangle GH, GL.

But if the given angle ABC of the parallelogram AC be not a right angle; in this case, because ABC is a given angle, the ratio of the rectangle contained by the sides AB, BC, to the parallelogram AC is given *; and AC is given, therefore the * 62 Dat.
rectangle AB, BC is given: and the sum of the squares of AB, BC is given: therefore the sides AB, BC are given by the preceding case.

The sides AB, BC, and the parallelogram AC, may be found thus: let EFG be the given angle of the parallelogram, and from any point E in FE, draw EG perpendicular to FG; and let the rectangle EG, FH be the given space to which the parallelogram is to be made equal, and let EF, FK be the given rectangle to which the sum of the squares of the sides is to be equal. And, by the preceding case, find the sides of a rectangle which is equal to the given rectangle EF, FH, and the squares of the sides of which are together equal to the given rectangle EF, FK; therefore as was shewn in that case, twice the rectangle EF, FH must not be greater than the rectangle EF, FK: let it be so, and let AB, BC be the sides of the rectangle joined in the angle ABC equal to the given angle EFG, and complete the parallelogram ABCD, which will be that which was to be found: draw AL perpendicular to BC, and because the angle ABL is equal to EFG, the triangle ABL is equiangular to EFG; and the parallelogram AC, that is, the rectangle AL, BC, is to the rectangle AB, BC, as (the straight line AL to AB, that is, as EG to EF, that is, as) the rectangle EG, FH to EF, FH; and, by the construction, the rectangle AB, BC is equal to EF, FH, therefore the rectangle AL, BC, or its equal, the parallelogram AC, is equal to the given rectangle EG, FH: and the squares of AB, BC are together equal, by construction, to the given rectangle EF, FK.

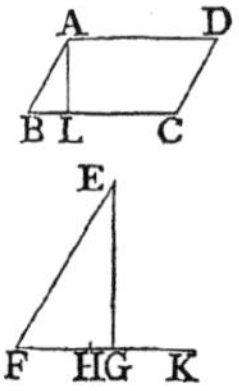

(86.) PROPOSITION LXXXIX.

If two straight lines contain a given parallelogram in a given angle, and if the excess of the square of one of them above a given space have a given ratio to the square of the other, each of the straight lines shall be given.

Let the two straight lines AB, BC contain the given parallelogram AC in the given angle ABC, and let the excess of the square of BC above a given space have a given ratio to the square of AB: each of the straight lines AB, BC is given.

Because the excess of the square of BC above a given space has a given ratio to the square of BA, let the rectangle CB, BD be the given space; take this from the square of BC; the remainder, to wit, the rectangle* BC, CD has a given ratio

* 2. 2.

to the square of BA: draw AE perpendicular to BC, and let the square of BF be equal to the rectangle BC, CD; then because the angle ABC, as also BEA, is given, the triangle ABE is given * in species, and the ratio of AE to AB * 43 Dat.
is given: and because the ratio of the rectangle BC, CD, that is, of the square of BF to the square of BA, is given, the ratio of the straight line BF to BA is given *; and the ratio of AE * 58 Dat.
to AB is given, wherefore * the ratio of AE to BF is given; * 9 Dat.
as also the ratio of the rectangle AE, BC, that is *, of the pa- * 35. 1.
rallelogram AC to the rectangle FB, BC; and AC is given, wherefore the rectangle FB, BC, is given. The excess of the square of BC above the square of BF, that is, above the rectangle BC, CD, is also given, for it is equal * to the given rect- * 2. 2.
angle CB, BD: therefore, because the rectangle contained by the straight lines FB, BC, is given, and also the excess of the square of BC above the square of BF; FB, BC are each of them given *: and the ratio of FB to BA is given; therefore * 87 Dat.
AB, BC are given.

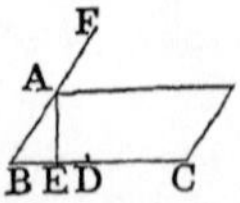

The Composition is as follows:

Let GHK be the given angle to which the angle of the parallelogram is to be made equal, and from any point G in HG, draw GK perpendicular to HK; let GK, HL be the rectangle to which the parallelogram is to be made equal, and let LH, HM be the rectangle equal to the given space which is to be taken from the square of one of the sides; and let the ratio of the remainder to the square of the other side be the same with the ratio of the square of the given straight line NH to the square of the given straight line HG.

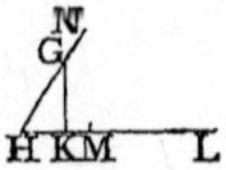

By help of the 87th dat. find two straight lines BC, BF, which contain a rectangle equal to the given rectangle NH, HL, and such that the excess of the square of BC above the square of BF be equal to the given rectangle LH, HM; and join CB, BF in the angle FBC equal to the given angle GHK: and as NH to HG, so make FB to BA, and complete the parallelogram AC, and draw AE perpendicular to BC; then AC is equal to the rectangle GK, HL: and if from the square of BC, the given rectangle LH, HM, be taken, the

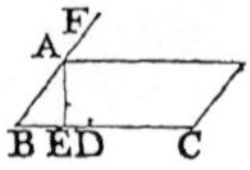

remainder shall have to the square of BA the same ratio which the square of NH has to the square of HG.

Because, by the construction, the square of BC is equal to the square of BF together with the rectangle LH, HM; if from the square of BC there be taken the rectangle LH, HM,
* 22. 6. there remains the square of BF, which has * to the square of BA the same ratio which the square of NH has to the square of HG, because, as NH to HG, so FB was made to BA; but as HG to GK, so is BA to AE, because the triangle GHK is equiangular to ABE; therefore, ex æquali, as NH to GK, so
* 1. 6. is FB to AE; wherefore * the rectangle NH, HL is to the rectangle GK, HL, as the rectangle FB, BC to AE, BC; but, by the construction, the rectangle NH, HL is equal to FB,
* 14. 5. BC; therefore * the rectangle GK, HL is equal to the rectangle AE, BC, that is, to the parallelogram AC.

The analysis of this problem might have been made as in the 86th Prop. in the Greek, and the composition of it may be made as that which is in Prop. 87th of this edition.

(O.)

PROPOSITION XC.

If two straight lines contain a given parallelogram in a given angle, and if the square of one of them, together with the space which has a given ratio to the square of the other, be given, each of the straight lines shall be given.

Let the two straight lines AB, BC contain the given parallelogram AC in the given angle ABC, and let the square of BC together with the space which has a given ratio to the square of AB be given: AB, BC are each of them given.

Let the square of BD be the space which has the given ratio to the square of AB: therefore, by the hypothesis, the square of BC together with the square of BD is given. From the point A, draw AE perpendicular to BC; and because the an-
* 43 Dat. gles ABE, BEA are given, the triangle ABE is given * in species: therefore the ratio of BA to AE is given: and because the ratio of the square of BD to the square of BA is given, the ratio of the straight
* 58 Dat. line BD to BA is given *; and the ratio of BA
* 9 Dat. to AE is given; therefore * the ratio of AE to BD is given, as also the ratio of the rectangle AE,

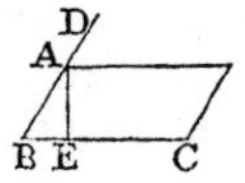

BC, that is, of the parallelogram AC, to the rectangle DB, BC; and AC is given, therefore the rectangle DB, BC is given;

and the square of BC together with the square of BD is given: therefore * because the rectangle contained by the two straight lines DB, BC is given, and the sum of their squares is given, the straight lines DB, BC are each of them given; and the ratio of DB to BA is given; therefore AB, BC are given. * 88 Dat.

The Composition is as follows:

Let FGH be the given angle to which the angle of the parallelogram is to be made equal, and from any point F in GF, draw FH perpendicular to GH; and let the rectangle FH, GK be that to which the parallelogram is to be made equal; and let the rectangle KG, GL be the space to which the square of one of the sides of the parallelogram, together with the space which has a given ratio to the square of the other side is to be made equal; and let this given ratio be the same which the square of the given straight line MG has to the square of GF.

By the 88th dat. find two straight lines DB, BC, which contain a rectangle equal to the given rectangle MG, GK, and such that the sum of their squares is equal to the given rectangle KG, GL; therefore, by the determination of the problem in that proposition, twice the rectangle MG, GK must not be greater than the rectangle KG, GL. Let it be so, and join the straight lines DB, BC, in the angle DBC, equal to the given angle FGH; and, as MG to GF, so make DB to BA, and complete the parallelogram AC; AC is equal to the rectangle FH, GK: and the square of BC together with the square of BD, which, by the construction, has to the square of BA the given ratio which the square of MG has to the square of GF, is equal, by the construction, to the given rectangle KG, GL. Draw AE perpendicular to BC.

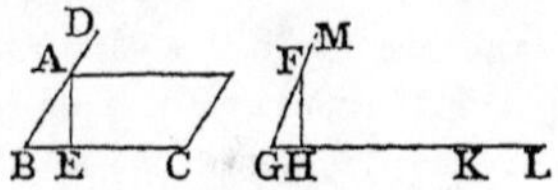

Because, as DB to BA, so is MG to GF; and as BA to AE, so GF to FH; ex æquali, as DB to AE, so is MG to FH; therefore as the rectangle DB, BC to AE, BC, so is the rectangle MG, GK to FH, GK; and the rectangle DB, BC is equal to the rectangle MG, GK; therefore the rectangle AE, BC, that is, the parallelogram AC, is equal to the rectangle FH, GK.

D D

(88.)

PROPOSITION XCI.

If a straight line drawn within a circle given in magnitude, cut off a segment which contains a given angle, the straight line is given in magnitude.

In the circle ABC given in magnitude, let the straight line AC be drawn, cutting off the segment AEC which contains the given angle AEC: the straight line AC is given in magnitude.

* 1. 3. Take D the centre of the circle*, join AD, and produce it to E, and join EC: the angle ACE being a
* 31. 3. right * angle, is given; and the angle AEC is
* 43 Dat. given; therefore* the triangle ACE is given in species, and the ratio of EA to AC is therefore given, and EA is given in magnitude, be-
* 5 Def. cause the circle is given * in magnitude: AC
* 2 Dat. is therefore given * in magnitude.

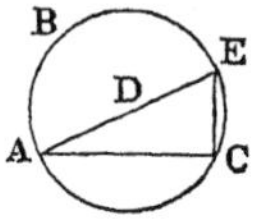

(89.)

PROPOSITION XCII.

If a straight line given in magnitude be drawn within a circle given in magnitude, it shall cut off a segment containing a given angle.

Let the straight line AC given in magnitude be drawn within the circle ABC given in magnitude: it shall cut off a segment containing a given angle.

Take D the centre of the circle, join AD and produce it to E, and join EC: and because each of the straight lines EA and AC is given,
* 1 Dat. their ratio is given *: and the angle ACE is a
* 46 Dat. right angle, therefore the triangle ACE is given * in species, and consequently the angle AEC is given.

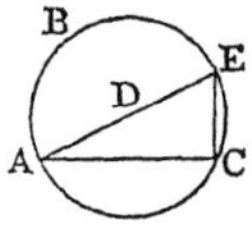

(90.)

PROPOSITION XCIII.

If from any point in the circumference of a circle given in position, two straight lines be drawn, meeting the circumference and containing a given angle; if the point in which one of them meets the circumference again be given, the point in which the other meets it is also given.

From any point A, in the circumference of a circle ABC given in position, let AB, AC be drawn to the circumference, making the given angle BAC: if the point B be given, the point C is also given.

Take D the centre of the circle, and join BD, DC, and because each of the points B, D is given, BD is given * in position; and because the angle * 29 Dat. BAC is given, the angle BDC is given *, there- * 20. 3. fore because the straight line DC is drawn to the given point D, in the straight line BD given in position, in the given angle BDC, DC is given * in position. And the circumference ABC is given * 32 Dat. in position, therefore * the point C is given. * 28 Dat.

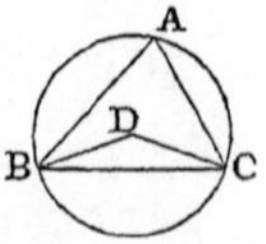

PROPOSITION XCIV. (91.)

If from a given point a straight line be drawn touching a circle given in position, the straight line is given in position and magnitude.

Let the straight line AB be drawn from the given point A, touching the circle BC given in position: AB is given in position and magnitude.

Take D the centre of the circle, and join DA, DB: Because each of the points D, A is given, the straight line AD is given* * 29 Dat. in position and magnitude: and BDA is a right * angle, wherefore DA is a diameter * * 18. 3. * Cor. 5. 4. of the circle DBA described about the triangle DBA: and that circle is therefore given in position: and the circle BC is given* * 6 Def. in position, therefore the point B is given *. * 28 Dat. The point A is also given: therefore the straight line AB is given * in position and magnitude. * 29 Dat.

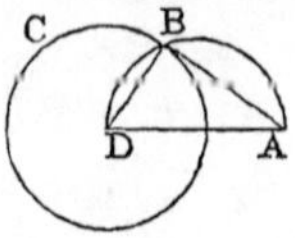

PROPOSITION XCV. (92.)

If a straight line be drawn from a given point without a circle given in position, the rectangle contained by the segments betwixt the point and the circumference of the circle is given.

Let the straight line ABC be drawn from the given point A without the circle BCD given in position, cutting it in B, C: the rectangle BA, AC is given.

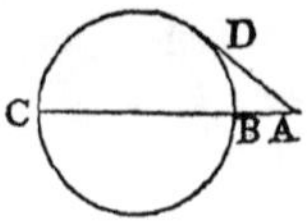

From the point A, draw * AD touching * 17. 3. the circle; wherefore AD is given * in position and magni- * 94 Dat.

* 56 Dat. tude: and because AD is given, the square of AD is given *,
* 36. 3. which is equal * to the rectangle BA, AC: therefore the rectangle BA, AC is given.

(93.)

PROPOSITION XCVI.

If a straight line be drawn through a given point within a circle given in position, the rectangle contained by the segments betwixt the point and the circumference of the circle is given.

Let the straight line BAC be drawn through the given point A, within the circle BCE, given in position: the rectangle BA, AC is given.

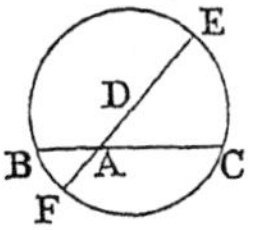

Take D, the centre of the circle, join AD, and produce it to the points E, F: because the points A, D are given, the straight line AD is
* 29 Dat. given * in position; and the circle BEC is given in position; therefore the points E, F are
* 28 Dat. given *; and the point A is given, therefore EA, AF are each
* 29 Dat. of them given *, and the rectangle EA, AF is therefore given;
* 35. 3. and it is equal * to the rectangle BA, AC; which consequently is given.

(94.)

PROPOSITION XCVII.

If a straight line be drawn within a circle given in magnitude, cutting off a segment containing a given angle; if the angle in the segment be bisected by a straight line produced till it meets the circumference, the straight lines which contain the given angle shall both of them together have a given ratio to the straight line which bisects the angle: and the rectangle contained by both these lines together which contain the given angle, and the part of the bisecting line cut off below the base of the segment, shall be given.

Let the straight line BC be drawn within the circle ABC given in magnitude, cutting off a segment containing the given angle BAC, and let the angle BAC be bisected by the straight line AD: BA together with AC has a given ratio to AD; and the rectangle contained by BA and AC together, and the straight line ED cut off from AB below BC the base of the segment, is given.

Join BD; and because BC is drawn within the circle ABC given in magnitude, cutting off the segment BAC, containing the given angle BAC; BC is given * in magnitude: by the same reason BD is given; therefore * the ratio of BC to BD is given: and because the angle BAC is bisected by AD, as BA to AC, so is * BE to EC; and, by permutation, as AB to BE, so is AC to CE; wherefore *, as BA and AC together to BC, so is AC to CE; and because the angle BAE is equal to EAC, and the angle ACE to ADB *, the triangle ACE is equiangular to the triangle ADB; therefore as AC to CE, so is AD to DB: but as AC to CE, so is BA together with AC to BC; as therefore BA and AC to BC, so is AD to DB; and, by permutation, as BA and AC to AD, so is BC to BD: and the ratio of BC to BD is given, therefore the ratio of BA together with AC to AD is given.

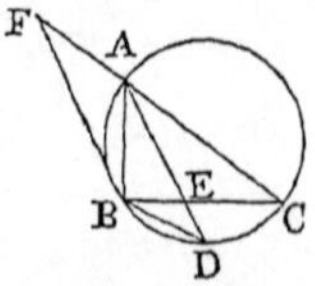

* 91 Dat.
* 1 Dat.
* 3. 6.
* 12. 5.
* 21. 3.

Also the rectangle contained by BA and AC together, and DE, is given.

Because the triangle BDE is equiangular to the triangle ACE, as BD to DE, so is AC to CE; and as AC to CE, so is BA and AC to BC; therefore as BA and AC to BC, so is BD to DE; wherefore the rectangle contained by BA and AC together, and DE, is equal to the rectangle CB, BD: but CB, BD is given; therefore the rectangle contained by BA and AC together, and DE, is given.

Otherwise,

Produce CA, and make AF equal to AB, and join BF; and because the angle BAC is double * of each of the angles BFA, BAD; the angle BFA is equal to BAD; and the angle BCA is equal to BDA, therefore the triangle FCB is equiangular to ABD: as therefore FC to CB, so is AD to DB; and, by permutation, as FC, that is BA and AC together, to AD, so is CB to BD: and the ratio of CB to BD is given, therefore the ratio of BA and AC to AD is given.

* 5. & 32. 1.

And because the angle BFC is equal to the angle DAC, that is, to the angle DBC, and the angle ACB equal to the angle ADB; the triangle FCB is equiangular to BDE: as therefore FC to CB, so is BD to DE: therefore the rectangle contained by FC, that is, BA and AC together, and DE, is equal to the rectangle CB, BD, which is given, and therefore the rectangle contained by BA, AC together, and DE, is given.

(P.)

PROPOSITION XCVIII.

If a straight line be drawn within a circle given in magnitude, cutting off a segment containing a given angle; if the angle adjacent to the angle in the segment be bisected by a straight line produced till it meet the circumference again, and the base of the segment; the excess of the straight lines which contain the given angle shall have a given ratio to the segment of the bisecting line which is within the circle; and the rectangle contained by the same excess, and the segment of the bisecting line betwixt the base produced and the point where it again meets the circumference, shall be given.

Let the straight line BC be drawn within the circle ABC given in magnitude, cutting off a segment containing the given angle BAC, and let the angle CAF adjacent to BAC be bisected by the straight line DAE, meeting the circumference again in D, and BC the base of the segment produced in E; the excess of BA, AC, has a given ratio to AD; and the rectangle which is contained by the same excess and the straight line ED is given.

Join BD, and through B, draw BG parallel to DE meeting AC produced in G: and because BC cuts off from the circle ABC given in magnitude, the segment BAC containing a given angle, BC is therefore given * in magnitude: by the same reason BD is given, because the angle BAD is equal to the given angle EAF; therefore the ratio of BC to BD is given: and because the angle CAE is equal to

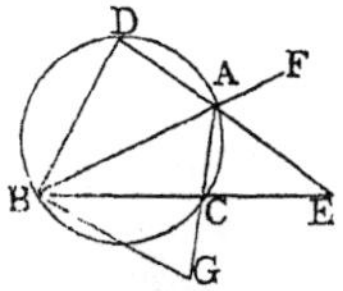

* 91 Dat.

EAF, of which CAE is equal to the alternate angle AGB, and EAF to the interior and opposite angle ABG, therefore the angle AGB is equal to ABG, and the straight line AB equal to AG; so that GC is the excess of BA, AC: and because the angle BGC is equal to GAE, that is, to EAF, or the angle BAD; and that the angle BCG is equal to the opposite interior angle BDA of the quadrilateral BCAD in the circle; therefore the triangle BGC is equiangular to BDA. Therefore as GC to CB, so is AD to DB; and, by permutation, as GC which is the excess of BA, AC, to AD, so is BC to BD: and the ratio of CB to BD is given; therefore the ratio of the excess of BA, AC, to AD is given.

And because the angle GBC is equal to the alternate angle DEB, and the angle BCG equal to BDE; the triangle BCG is equiangular to BDE: therefore as GC to CB, so is BD to DE; and consequently the rectangle GC, DE is equal to the rectangle CB, BD which is given, because its sides CB, BD are given: therefore the rectangle contained by the excess of BA, AC, and the straight line DE is given.

PROPOSITION XCIX. (95.)

If from a given point in the diameter of a circle given in position, or in the diameter produced, a straight line be drawn to any point in the circumference, and from that point a straight line be drawn at right angles to the first, and from the point in which this meets the circumference again, a straight line be drawn parallel to the first, the point in which this parallel meets the diameter is given; and the rectangle contained by the two parallels is given.

In BC the diameter of the circle ABC given in position, or in BC produced, let the given point D be taken; and from D, let a straight line DA be drawn to any point A in the circumference, and let AE be drawn at right angles to DA; and from the point E, where it meets the circumference again, let EF be drawn parallel to DA meeting BC in F: the point F is given, as also the rectangle AD, EF.

Produce EF to the circumference in G, and join AG: because GEA is a right angle, the straight line AG is * the diameter of the circle ABC; and BC is also a diameter of it; therefore the point H, where they meet, is the centre of the circle, and consequently H is given: and the point D is given, wherefore DH is given in magnitude. And because AD is parallel to FG, and GH equal to HA; DH is equal * to HF, and AD equal to GF: and DH is given, therefore HF is given in magnitude; and it is also given in position, and the point H is given, therefore * the point F is given.

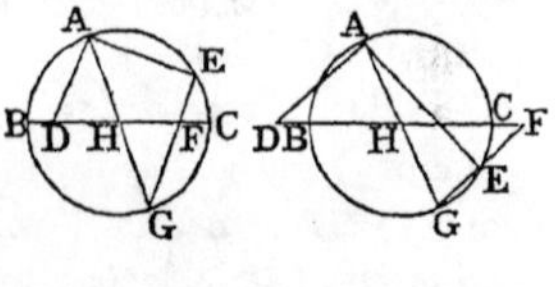

* Cor. 5. 4.

* 4. 6.

* 30 Dat.

And because the straight line EFG is drawn from a given point F, without or within the circle ABC given in position,

* 95 or 96 Dat. therefore * the rectangle EF, FG is given: and GF is equal to AD, wherefore the rectangle AD, EF is given.

(Q.) PROPOSITION C.

If from a given point in a straight line given in position, a straight line be drawn to any point in the circumference of a circle given in position; and from this point a straight line be drawn, making with the first an angle equal to the difference of a right angle, and the angle contained by the straight line given in position, and the straight line which joins the given point and the centre of the circle; and from the point in which the second line meets the circumference again, a third straight line be drawn, making with the second an angle equal to that which the first makes with the second; the point in which this third line meets the straight line given in position is given, as also the rectangle contained by the first straight line, and the segment of the third betwixt the circumference and the straight line given in position, is given.

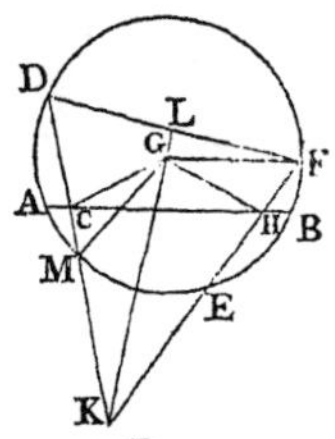

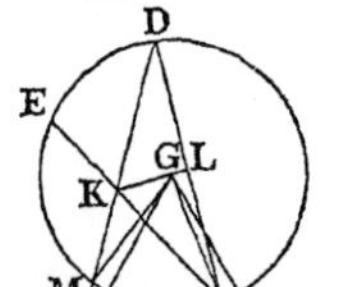

Let the straight line CD be drawn from the given point C, in the straight line AB given in position, to the circumference of the circle DEF given in position, of which G is the centre; join CG, and from the point D, let DF be drawn, making the angle CDF equal to the difference of a right angle, and the angle BCG; and from the point F let FE be drawn, making the angle DFE equal to CDF, meeting AB in H: the point H is given; as also the rectangle CD, FH.

Let CD, FH meet one another in the point K, from which draw KL perpendicular to DF; and let DC meet the circumference again in M, and let FH meet the same in E, and join MG, GF, GH.

Because the angles MDF, DFE are equal to one another, the circumferences
* 26. 3. MF, DE are equal *; and adding or taking away the common part ME, the circumference DM is equal to EF; therefore the straight line DM is equal to the straight line EF, and the
* 8. 1. angle GMD to the angle * GFE; and the angles GMC, GFH

are equal to one another, because they are either the same with the angles GMD, GFE, or adjacent to them: and because the angles KDL, LKD are together equal * to a right angle, that is, by the hypothesis, to the angles KDL, GCB; the angle GCB or GCH is equal to the angle (LKD, that is to the angle) LKF or GKH: therefore the points C, K, H, G are in the circumference of a circle: and the angle GCK is therefore equal to the angle GFH: and the angle GMC is equal to GFH, and the straight line GM to GF; therefore * CG is equal to GH, and CM to HF: and because CG is equal to GH, the angle GCH is equal to GHC; but the angle GCH is given: therefore GHC is given, and consequently the angle CGH is given; and CG is given in position, and the point G; therefore * GH is given in position; and CB is also given in position, wherefore the point H is given.

* 32. 1.

* 26. 1.

* 32 Dat.

And because HF is equal to CM, the rectangle DC, FH is equal to DC, CM: but DC, CM is given *, because the point C is given, therefore the rectangle DC, FH is given.

* 95 or 96 Dat.

END OF THE DATA.

NOTES

ON

EUCLID'S DATA.

DEFINITIONS.

II. This is made more explicit than in the Greek text, to prevent a mistake which the author of the second demonstration of the 24th Prop. in the Greek edition has fallen into, of thinking that a ratio is given to which another ratio is shewn to be equal, though this other be not exhibited in given magnitudes. See the Notes on the Prop. which is the 13th in this edition. Besides, by this definition, as it is now given, some propositions are demonstrated, which in the Greek are not so well done by help of Prop. 2.

IV. In the Greek text, Def. 4. is thus: "Points, lines, "spaces, and angles, are said to be given in position which "have always the same situation;" but this is imperfect and useless, because there are innumerable cases in which things may be given according to this definition, and yet their position cannot be found: for instance, let the triangle ABC be given in position, and let it be proposed to draw a straight line BD from the angle at B to the opposite side AC, which shall cut off the angle DBC, which shall be the seventh part of the angle ABC; suppose this is done, therefore the straight line BD is invariable in its position, that is, has always the same situation; for any other straight line drawn from the point B on either side of BD cuts off an angle greater

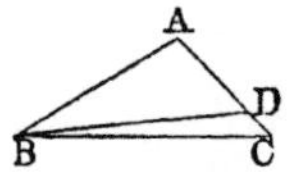

or less than the seventh part of the angle ABC; therefore, according to this definition, the straight line BD is given in position, as also * the point D in which it meets the straight line AC, which is given in position. But from the things here given, neither the straight line BD, nor the point D can be found by the help of Euclid's Elements only, by which every thing in his Data is supposed may be found. This definition is therefore of no use. We have amended it by adding, " and which are either actually exhibited, or can be found," for nothing is to be reckoned given, which cannot be found, or is not actually exhibited. * 28 Dat.

The definition of an angle given by position is taken out of the 4th, and given more distinctly by itself in the definition marked A.

XI to XV. The 11th and 12th are omitted, because they cannot be given in English so as to have any tolerable sense: and therefore, wherever the terms defined occur, the words which express their meaning are made use of in their place.

The 13th, 14th, 15th, are omitted, as being of no use.

It is to be observed in general of the data in this book that they are to be understood to be given geometrically, not always arithmetically, that is, they cannot always be exhibited in numbers; for instance, if the side of a square be given, the ratio of it to its diameter is given geometrically *, but not in numbers; and the diameter is given*; but though the number of any equal parts in the side be given, for example 10, the number of them in the diameter cannot be given; and the like holds in many other cases. * 44 Dat. * 2 Dat.

PROPOSITION I.

In this it is shown that A is to B as C to D, from this, that A is to C as B to D, and then by permutation; but it follows directly without these two steps, from 7. 5.

PROPOSITION II.

The limitation added to the end of this proposition between the inverted commas, is quite necessary, because without it the proposition cannot always be demonstrated: For

the author having said‡, "because A is given, a magnitude
* 1 Def. "equal to it can be found*; let this be C; and because the "ratio of A to B is given, a ratio which is the same to it can be
* 2 Def. "found*," adds, "let it be found, and let it be the ratio of "C to Δ." Now, from the second definition, nothing more follows than that some ratio, suppose the ratio of E to Z, can be found, which is the same with the ratio of A to B, and when the author supposes that the ratio of C to Δ, which is also the same with the ratio of A to B, can be found, he necessarily supposes that to the three magnitudes E, Z, C, a fourth proportional Δ may be found; but this cannot always be done by the Elements of Euclid; from which it is plain Euclid must have understood the proposition under the limitation which is now added to his text. An example will make this clear: Let A be a given angle, and B another angle to which A has a given ratio; for instance, the ratio of the given straight line E to the given one Z; then, having found an angle C equal to A, how can the angle Δ be found to which C has the same ratio that E has to Z? Certainly no way, until it be shewn how to find an angle to which a given angle has a given ratio, which cannot be done by Euclid's Elements, nor probably by any Geometry known in his time. Therefore in all the propositions of this book which depend upon this second, the above-mentioned limitation must be understood, though it be not explicitly mentioned.

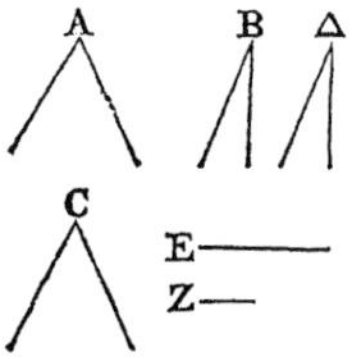

PROPOSITION V.

The order of the Propositions in the Greek text between Prop. 4. and Prop. 25. is now changed into another which is more natural, by placing those which are more simple before those which are more complex; and by placing together those which are of the same kind, some of which were mixed among others of a different kind. Thus, Prop. 12. in the Greek is now made the 5th, and those which were the 22d and 23d are made the 11th and 12th, as they are more simple than the propositions concerning magnitudes, the excess of

‡ See Dr. Gregory's edition of the Data.

one of which above a given magnitude has a given ratio to the other, after which these two were placed; and the 24th in the Greek text is, for the same reason, made the 13th.

PROPOSITIONS VI. VII.

These are universally true, though, in the Greek text, they are demonstrated by Prop. 2. which has a limitation; they are therefore now shewn without it.

PROPOSITION XII.

In the 23d Prop. in the Greek text, which here is the 12th, the words "μὴ τοὺς αὐτοὺς δὲ" are wrong translated by Claud. Hardy, in his edition of Euclid's Data, printed at Paris, anno 1625, which was the first edition of the Greek text; and Dr. Gregory follows him in translating them by the words, "etsi non easdem", as if the Greek had been εἰ καὶ μὴ τοὺς αὐτοὺς, as in Prop. 9. of the Greek text. Euclid's meaning is, that the ratios mentioned in the proposition must not be the same: for if they were, the proposition would not be true. Whatever ratio the whole has to the whole, if the ratios of the parts of the first to the parts of the other be the same with this ratio, one part of the first may be double, triple, &c. of the other part of it, or have any other ratio to it, and consequently cannot have a given ratio to it, wherefore, these words must be rendered by "non autem, "easdem", but not the same ratios, as Zambertus has translated them in his edition.

PROPOSITION XIII.

Some very ignorant editor has given a second demonstration of this proposition in the Greek text, which has been as ignorantly kept in by Claud. Hardy and Dr. Gregory, and has been retained in the translations of Zambertus and others. Carolus Renaldinus gives it only. The author of it has thought that a ratio was given, if another ratio could be shewn to be the same to it, though this last ratio be not found. But this is altogether absurd, because from it would be deduced that the ratio of the sides of any two squares is given, and the ratio of the diameters of any two circles, &c. And it is to be observed that the moderns frequently take given ratios, and ratios that are always the same, for one and the same thing; and Sir Isaac Newton has fallen into this mistake in the 17th Lemma of his Principia, edit. 1713, and in other places; but this should be carefully avoided, as it may lead into other errors.

PROPOSITIONS XIV. XV.

Euclid in this book has several propositions concerning magnitudes, the excess of one of which above a given magnitude has a given ratio to the other; but he has given none concerning magnitudes whereof one together with a given magnitude has a given ratio to the other; though these last occur as frequently in the solution of problems as the first; the reason of which is, that the last may be all demonstrated by help of the first; for if a magnitude, together with a given magnitude, has a given ratio to another magnitude, the excess of this other above a given magnitude shall have a given ratio to the first, and on the contrary; as we have demonstrated in Prop. 14. And for a like reason, Prop. 15. has been added to the Data. One example will make the thing clear: Suppose it were to be demonstrated, that if a magnitude A together with a given magnitude has a given ratio to another magnitude B, that the two magnitudes A and B, together with a given magnitude, have a given ratio to that other magnitude B; which is the same proposition with respect to the last kind of magnitudes above-mentioned, that the first part of Prop. 16. in this edition, is in respect of the first kind: this is shewn thus, from the hypothesis, and by the first part of Prop. 14. the excess of B above a given magnitude has unto A a given ratio; and, therefore, by the first part of Prop. 17. the excess of B above a given magnitude has unto B and A together a given ratio; and by the second part of Prop. 14. A and B together with a given magnitude have unto B a given ratio; which is the thing that was to be demonstrated. In like manner, the other propositions concerning the last kind of magnitudes may be shewn.

PROPOSITIONS XVI. XVII.

In the third part of Prop. 10. in the Greek text, which is the 16th in this edition, after the ratio of EC to CB has been shewn to be given; from this, by inversion and conversion, the ratio of BC to BE is demonstrated to be given; but without these two steps, the conclusion should have been made only by citing the 6th Proposition. And in like manner, in the first part of Prop. 11. in the Greek, which in this edition is the 17th, from the ratio of DB to DC being given, the ratio of DC to DB is shown to be given, by inversion and composition, instead of citing Prop. 7. and the same fault occurs in the second part of the same Prop. 11.

PROPOSITIONS XXI. XXII.

These now are added, as being wanting to complete the subject treated of in the four preceding propositions.

PROPOSITION XXIII.

This, which is Prop. 20. in the Greek text, was separated from Prop. 14. 15. 16, in that text, after which it should have been immediately placed, as being of the same kind; it is now put into its proper place; but Prop. 21. in the Greek is left out, as being the same with Prop. 14. in that text, which is here Prop. 18.

PROPOSITION XXIV.

This, which is Prop. 13. in the Greek, is now put into its proper place, having been disjoined from the three following it in this edition, which are of the same kind.

PROPOSITION XXVIII.

This, which in the Greek text is Prop. 25. and several of the following propositions, are there deduced from Def. 4. which is not sufficient, as has been mentioned in the note on that definition. They are therefore now shewn more explicitly.

PROPOSITIONS XXXIV. XXXVI.

Each of these has a determination, which is now added, which occasions a change in their demonstrations.

PROPOSITIONS XXXVII. XXXIX. XL. XLI.

The 35th and 36th Propositions in the Greek text are joined into one, which makes the 39th in this edition, because the same enunciation and demonstration serves both: and for the same reason Prop. 37. 38. in the Greek are joined into one, which is here the 40th.

Proposition 37. is added to the Data, as it frequently occurs in the solution of problems; and Prop. 41. is added, to complete the rest.

PROPOSITION XLII.

This is Prop. 39. in the Greek text, where the whole construction of Prop. 22. of Book 1. of the Elements is put, without need, into the demonstration, but is now only cited.

PROPOSITION XLV.

This is Prop. 42. in the Greek, where the three straight lines made use of in the construction are said, but not shewn, to be such that any two of them are greater than the third, which is now done.

PROPOSITION XLVII.

This is Prop. 44. in the Greek text; but the demonstration of it is changed into another, wherein the several cases of it are shewn, which, though necessary, is not done in the Greek.

PROPOSITION XLVIII.

There are two cases in this Proposition, arising from the two cases of the third part of Prop. 47. on which the 48th depends: and in the composition these two cases are explicitly given.

PROPOSITION LII.

The construction, and demonstration of this, which is Prop. 48. in the Greek, are made something shorter than in that text.

PROPOSITION LIII.

Prop. 63. in the Greek text is omitted, being only a case of Prop. 49. in that text, which is Prop. 53. in this edition.

PROPOSITION LVIII.

This is not in the Greek text, but its demonstration is contained in that of the first part of Prop. 54. in that text; which proposition is concerning figures that are given in species; this 58th is true of similar figures, though they be not given in species, and as it frequently occurs, it was necessary to add it.

PROPOSITIONS LIX. LXI.

This is the 54th in the Greek; and the 77th in the Greek, being the very same with it, is left out, and a shorter demonstration is given of Prop. 61.

PROPOSITION LXII.

This, which is most frequently useful, is not in the Greek, and is necessary to Prop. 87. 88. in this edition, as also though not mentioned, to Prop. 86. 87. in the former editions. Prop. 66. in the Greek text is made a corollary to it.

PROPOSITION LXIV.

This contains both Prop. 74. and 73. in the Greek text; the first case of the 74th is a repetition of Prop. 56. from which it is separated in that text by many propositions; and as there is no order in these propositions, as they stand in the Greek, they are now put into the order which seemed most convenient and natural.

The demonstration of the first part of Prop. 73. in the Greek is grossly vitiated. Dr. Gregory says, that the sentences he has inclosed betwixt two stars are superfluous, and ought to be cancelled; but he has not observed that what follows them is absurd, being to prove that the ratio [see his figure] of ΑΓ to ΓΚ is given, which, by the hypothesis at the beginning of the proposition, is expressly given; so that the whole of this part was to be altered, which is done in this Prop. 64.

PROPOSITIONS LXVII. LXVIII.

Prop. 70. in the Greek text, is divided into these two, for the sake of distinctness; and the demonstration of the 67th is rendered shorter than that of the first part of Prop. 70. in the Greek, by means of Prop. 23. of Book 6. of the Elements.

PROPOSITION LXX.

This is Prop. 62. in the Greek text; Prop. 78. in that text is only a particular case of it, and is therefore omitted.

Dr. Gregory, in the demonstration of Prop. 62. cites the 49th Prop. Dat. to prove that the ratio of the figure AEB to the parallelogram AH is given; whereas this was shewn a few lines before: and besides, the 49th Prop. is not applicable to these two figures; because AH is not given in species, but is, by the step for which the citation is brought, proved to be given in species.

PROPOSITION LXXIII.

Prop. 83. in the Greek text is neither well enunciated nor demonstrated. The 73d, which in this edition is put in place of it, is really the same, as will appear by considering [see Dr. Gregory's edition], that A, B, Γ, E, in the Greek text, are four proportionals, and that the proposition is to shew that Δ, which has a given ratio to E, is to Γ, as B is to a straight line to which A has a given ratio; or, by inversion, that Γ is to Δ, as a straight line to which A has a given ratio is to B; that is, if the proportionals be placed in this order, viz. Γ, E, A, B, that the first Γ is to Δ, to which the second E has a given ratio, as a straight line to which the third A has a given ratio is to the fourth B; which is the enunciation of this 73d, and was thus changed that it might be made like to that of Prop. 72. in this edition, which is the 82d in the Greek text: and the demonstration of Prop. 73. is the same with that of Prop. 72. only making use of Prop. 23. instead of Prop. 22. of Book 5. of the Elements.

PROPOSITION LXXVII.

This is put in place of Prop. 79. in the Greek text, which is not a datum, but a theorem premised as a Lemma to Prop. 80. in that text: and Prop. 79. is made Cor. 1. to Prop. 77. in this edition. Cl. Hardy, in his edition of the Data, takes notice, that in Prop. 80. of the Greek text, the parallel KL in the figure of Prop. 77. in this edition, must meet the circumference, but does not demonstrate it, which is done here at the end of Cor. 3. Prop. 77 in the construction for finding a triangle similar to ABC.

PROPOSITION LXXVIII.

The demonstration of this, which is Prop. 80. in the Greek, is rendered a good deal shorter by help of Prop. 77.

PROPOSITIONS LXXIX. LXXX. LXXXI.

These are added to Euclid's Data, as propositions which are often useful in the solution of Problems.

PROPOSITION LXXXII.

This, which is Prop. 60. in the Greek text, is placed before the 83d and 84th, which in the Greek are the 58th and 59th, because the demonstrations of these two in this edition are deduced from that of Prop. 82. from which they naturally follow.

PROPOSITIONS LXXXVIII. XC.

Dr. Gregory, in his preface to Euclid's works, which he published at Oxford in 1703, after having told that he had supplied the defects of the Greek text of the Data in innumerable places from several manuscripts, and corrected Cl. Hardy's translation by Mr. Bernard's, adds, that the 86th theorem, "or proposition", seemed to be remarkably vitiated, but which could not be restored by help of the manuscripts; then he gives three different translations of it in Latin, according to which he thinks it may be read; the two first have no distinct meaning, and the third, which he says is the best, though it contains a true proposition, which is the 90th in this edition, has no connexion in the least with the Greek text. And it is strange that Dr. Gregory did not observe, that, if Prop. 86. were changed into this, the demonstration of the 86th must be cancelled, and another put into its place: but the truth is, both the enunciation and the demonstration of Prop. 86. are quite entire and right, only Prop. 87. which is more simple, ought to have been placed before it; and the deficiency which the Doctor justly observes to be in this part of Euclid's Data, and which, no doubt, is owing to the carelessness and ignorance of the Greek editors, should have been supplied, not by changing Prop. 86. which is both entire and necessary, but by adding the two propositions, which are the 88th and 90th in this edition.

PROPOSITIONS XCVIII. C.

These were communicated to me by two excellent geometers, the first of them by the Right Honourable the Earl Stanhope, and the other by Dr. Matthew Stewart; to which I have added the demonstrations.

Though the order of the propositions has been in many places changed from that in former editions, yet this will be of little disadvantage, as the ancient geometers never cite the Data, and the moderns very rarely.

As that part of the composition of a problem which is its construction may not be so readily deduced from the analysis by beginners, for their sake the following example is given: in which the derivation of the several parts of the construction from the analysis is particularly shewn, that they may be assisted to do the like in other problems.

PROBLEM.

Having given the magnitude of a parallelogram, the angle of which ABC *is given, and also the excess of the square of its side* BC *above the square of the side* AB; *to find its sides and describe it.*

The analysis of this is the same with the demonstration of the 87th Prop. of the Data, and the construction that is given of the problem at the end of that proposition is thus derived from the analysis.

Let EFG be equal to the given angle ABC, and because in the analysis it is said that the ratio of the rectangle AB, BC, to the parallelogram AC, is given by the 62d Prop. Dat. therefore, from a point in FE, the perpendicular EG is drawn to FG, as the ratio of FE to EG is the ratio of the rectangle AB, BC, to the parallelogram AC, by what is shewn at the end of Prop. 62. Next

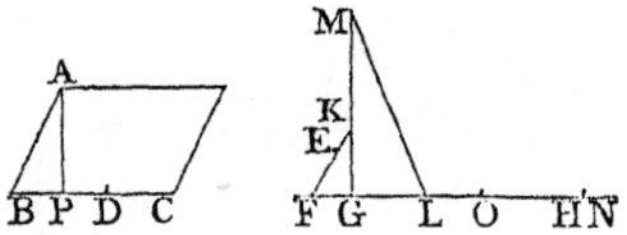

the magnitude of AC is exhibited by making the rectangle EG, GH equal to it; and the given excess of the square of BC above the square of BA, to which excess the rectangle CB, BD is equal, is exhibited by the rectangle HG, GL: then, in the analysis, the rectangle AB, BC is said to be given, and this is equal to the rectangle FE, GH, because the rectangle AB, BC is to the parallelogram AC, as (FE to EG, that is, as) the rectangle FE, GH to EG, GH; and the parallelogram AC is equal to the rectangle EG, GH; therefore the rectangle AB, BC is equal to FE, GH; and consequently the ratio of the rectangle CB, BD, that is, of the rectangle HG, GL, to AB, BC, that is, of the straight line DB to BA, is the same with the ratio (of the rectangle GL, GH to FE, GH, that is) of the straight line GL to FE, which ratio of DB to BA is the next thing said to be given in the analysis: from this it is plain that the square of FE is to the square of GL, as the square of BA, which is equal to the rectangle BC, CD, is to the square of BD: the ratio of which spaces is the next thing said to be given: and from this it follows, that four times the square of FE is to the square of GL, as four times the rectangle BC, CD is to the square of BD; and by composition, four times the square of FE, together with the square of GL, is to the square of GL, as four times the rectangle BC, CD, together with the square of BD, is to the square of BD, that is* as the square of the straight lines BC, CD, taken together is to the square of BD, which ratio is the next thing said to be given in the analysis: and because four times the square of FE and the square of GL are to be added together; therefore, in the perpendicular EG there is taken KG equal to FE, and MG equal to the double of it, because thereby the squares of MG, GL, that is, joining ML, the square of ML is equal to four times the square of FE, and to the square of GL: and because the square of ML is to the square of GL, as the square of the straight line made up of BC and CD is to the square of BD, therefore* ML is to LG, as BC together with CD is to BD; and, by composition, ML and LG together, that is, producing GL to N, so that ML be equal to LN, the straight line NG, is to GL, as twice BC is to BD; and by taking GO equal to the half of NG, GO is to GL, as BC to BD, the ratio of which is said to be given in the analysis: and from this it follows, that the rectangle HG, GO is to HG, GL, as the square of BC is to the rectangle CB, BD, which is equal

* 8. 2.

* 22. 6.

to the rectangle HG, GL; and therefore the square of BC is equal to the rectangle HG, GO; and BC is consequently found by taking a mean proportional betwixt HG and GO, as is said in the construction: and because it was shewn that GO is to GL, as BC to BD, and that now the three first are found, the fourth BD is found by 12. 6. It was likewise shewn that LG is to FE, or GK, as DB to BA, and the three first are now found, and thereby the fourth BA. Make the angle ABC equal to EFG, and complete the parallelogram of which the sides are AB, BC, and the construction is finished; the rest of the composition contains the demonstration.

As the propositions from the 13th to the 28th may be thought by beginners to be less useful than the rest, because they cannot so readily see how they are to be made use of in the solution of problems; on this account the two following problems are added, to shew that they are equally useful with the other propositions, and from which it may be easily judged that many other problems depend upon these propositions.

PROBLEM I.

To find three straight lines such, that the ratio of the first to the second is given; and if a given straight line be taken from the second, the ratio of the remainder to the third is given; also the rectangle contained by the first and third is given.

Let AB be the first straight line, CD the second, and EF the third: and because the ratio of AB to CD is given, and that if a given straight line be taken from CD, the ratio of the
* 24 Dat. remainder to EF is given; therefore * the excess of the first AB above a given straight line has a given ratio to the third EF: let BH be that given straight line; therefore AH the excess of AB above it, has a given ratio to EF; and conse-

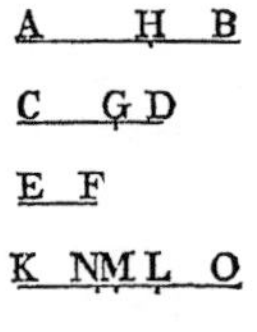

* 1. 6. quently * the rectangle BA, AH has a given ratio to the rectangle AB, EF, which last rect-
* 2 Dat. angle is given by the hypothesis: therefore * the rectangle BA, AH is given, and BH the excess of its sides
* 85 Dat. is given: wherefore the sides AB, AH are given *: And be-

cause the ratios of AB to CD, and of AH to EF, are given, CD and EF are * given. * 2 Dat.

The Composition.—Let the given ratio of KL to KM be that which AB is required to have to CD: and let DG be the given straight line which is to be taken from CD, and let the given ratio of KM to KN be that which the remainder must have to EF; also let the given rectangle NK, KO be that to which the rectangle AB, EF is required to be equal: find the given straight line BH which is to be taken from AB, which is done, as plainly appears from Prop. 24. Dat. by making as KM to KL, so GD to HB. To the given straight line BH, apply * a rectangle equal to LK, KO, exceeding by a square, * 29. 6. and let BA, AH be its sides: then is AB the first of the straight lines required to be found, and by making as LK to KM, so AB to DC, DC will be the second: and lastly, make as KM to KN, so CG to EF, and EF is the third.

For as AB to CD, so is HB to GD, each of these ratios being the same with the ratio of LK to KM; therefore * AH is to * 19. 5. CG, as (AB to CD, that is, as) LK to KM; and as CG to EF, so is KM to KN; wherefore, ex æquali, as AH to EF, so is LK to KN: and as the rectangle BA, AH to the rectangle BA, EF, so is * the rectangle LK, KO to the rectangle KN, KO: and, * 1. 6. by the construction, the rectangle BA, AH is equal to LK, KO: therefore * the rectangle AB, EF is equal to the given * 14. 5. rectangle NK, KO: and AB has to CD the given ratio of KL to KM; and from CD the given straight line GD being taken, the remainder CG has to EF the given ratio of KM to KN. Q. E. D.

PROBLEM II.

To find three straight lines such, that the ratio of the first to the second is given; and if a given straight line be taken from the second, the ratio of the remainder to the third is given; also the sum of the squares of the first and third is given.

Let AB be the first straight line, BC the second, and BD the third: and because the ratio of AB to BC is given, and that if a given straight line be taken from BC, the ratio of the remainder to BD is given; therefore * the excess of the first * 24 Dat. AB above a given straight line, has a given ratio to the third BD: let AE be that given straight line, therefore the remainder EB has a given ratio to BD: let BD be placed at right

* 44 Dat. angles to EB, and join DE; then the triangle EBD is * given in species; wherefore the angle BED is given: let AE, which is given in magnitude, be given also in position, as also the
* 32 Dat. point E, and the straight line ED will be given * in position: join AD, and because the sum of the squares of AB, BD, that
* 47. 1. is *, the square of AD is given, therefore the straight line AD
* 34 Dat. is given in magnitude; and it is also given * in position, because from the given point A it is drawn to the straight line ED given in position: therefore the point D, in which the two straight lines AD, ED given in position, cut one another, is
* 28 Dat. given *: and the straight line DB, which is at right angles to
* 33 Dat. AB, is given * in position, and AB is given in position, there-
* 28 Dat. fore * the point B is given: and the points A, D are given,
* 29 Dat. wherefore * the straight lines AB, BD are given: and the ra-
* 2 Dat. tio of AB to BC is given, and therefore * BC is given.

THE COMPOSITION.—Let the given ratio of FG to GH be that which AB is required to have to BC, and let HK be the given straight line which is to be taken from BC, and let the ratio which the remainder is required to have to BD be the given ratio of HG to GL, and place GL at right angles to FH, and join LF, LH: next, as HG is to GF, so make HK to AE;

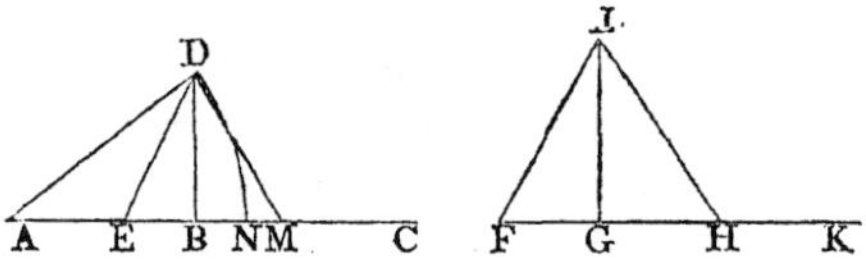

produce AE to N, so that AN be the straight line to the square of which the sum of the squares of AB, BD, is required to be equal; and make the angle NED equal to the angle GFL; and from the centre A, at the distance AN, describe a circle, and let its circumference meet ED in D, and draw DB perpendicular to AN and DM, making the angle BDM equal to the angle GLH. Lastly, produce BM to C, so that MC be equal to HK; then is AB the first, BC the second, and BD the third of the straight lines that were to be found.

For the triangles EBD, FGL, as also DBM, LGH, being equiangular, as EB to BD, so is FG to GL; and as DB to BM, so is LG to GH; therefore, ex æquali, as EB to BM, so is (FG
* 12. 5. to GH, and so is) AE to HK or MC; wherefore *, AB is to BC, as AE to HK, that is, as FG to GH, that is, in the given ratio:

and from the straight line BC taking MC, which is equal to the given straight line HK, the remainder BM has to BD the given ratio of HG to GL: and the sum of the squares of AB, BD, is equal* to the square of AD or AN, which is the given space. *Q. E. D.* * 47. 1.

I believe it would be in vain to try to deduce the preceding construction from an algebraical solution of the problem.

END OF THE NOTES TO THE DATA.

THE

ELEMENTS

OF

PLANE AND SPHERICAL TRIGONOMETRY.

PLANE TRIGONOMETRY.

LEMMA I.

LET ABC be a rectilineal angle: if about the point B as a centre, and with any distance BA, a circle be described, meeting BA, BC, the straight lines including the angle ABC in A, C; the angle ABC will be to four right angles, as the arch AC to the whole circumference.

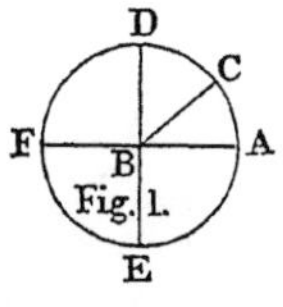

Fig. 1.

Produce AB till it meet the circle again in F, and through B, draw DE perpendicular to AB, meeting the circle in D, E.

By 33. 6. Elem. the angle ABC is to a right angle ABD, as the arch AC to the arch AD; and quadrupling the consequents, the angle ABC will be to four right angles, as the arch AC to four times the arch AD, or to the whole circumference.

LEMMA II.

Let ABC be a plane rectilineal angle as before: about B as a centre with any two distances BD, BA, let two circles be described meeting BA, BC, in D, E; A, C: the arch AC will be to the whole circumference of which it is an arch, as the arch DE is to the whole circumference of which it is an arch.

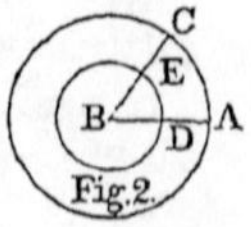

By Lemma 1. the arch AC is to the whole circumference of which it is an arch, as the angle ABC is to four right angles; and by the same Lemma 1. the arch DE is to the whole circumference of which it is an arch, as the angle ABC is to four right angles; therefore the arch AC is to the whole circumference of which it is an arch, as the arch DE to the whole circumference of which it is an arch.

DEFINITIONS.

I. Let ABC be a plane rectilineal angle; if about B as a centre, with BA any distance, a circle ACF be described, meeting BA, BC, in A, C; the arch AC is called the measure of the angle ABC.

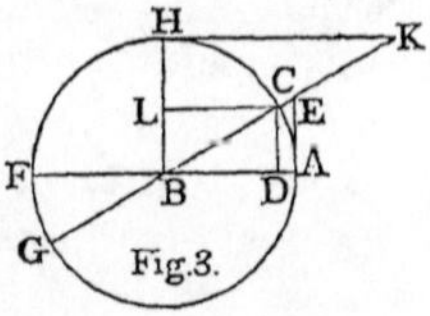

II. The circumference of a circle is supposed to be divided into 360 equal parts called degrees, and each degree into 60 equal parts called minutes, and each minute into 60 equal parts called seconds, &c. And as many degrees, minutes, seconds, &c. as are contained in any arch, of so many degrees, minutes, seconds, &c. is the angle, of which that arch is the measure, said to be.

Corollary. Whatever be the radius of the circle of which the measure of a given angle is an arch, that arch will contain the same number of degrees, minutes, seconds, &c. as is manifest from Lemma 2.

III. Let AB be produced till it meet the circle again in F; the angle CBF, which together with ABC is equal to two right angles, is called the *Supplement* of the angle ABC. See fig. 3.

IV. A straight line CD drawn through C, one of the extremities of the arch AC, perpendicular upon the diameter passing See fig. 3.

through the other extremity A, is called the *Sine* of the arch AC, or of the angle ABC, of which it is the measure.

Cor. The *Sine* of a quadrant, or of a right angle, is equal to the radius.

See fig. 3. V. The segment DA of the diameter passing through A, one extremity of the arch AC, between the sine CD, and that extremity is called the *Versed Sine* of the arch AC, or angle ABC.

See fig. 3. VI. A straight line AE touching the circle at A, one extremity of the arch AC, and meeting the diameter BC passing through the other extremity C in E, is called the *Tangent* of the arch AC, or of the angle ABC.

See fig. 3. VII. The straight line BE between the centre and the extremity of the tangent AE, is called the *Secant* of the arch AC or angle ABC.

Cor. to Def. 4. 6. 7. The sine, tangent, and secant of any angle ABC, are likewise the sine, tangent, and secant of its supplement CBF.

It is manifest from Def. 4. that CD is the sine of the angle CBF. Let CB be produced till it meet the circle again in G; and it is manifest that AE is the tangent, and BE the secant of the angle ABG or EBF, from Def. 6. 7.

Cor. to Def. 4. 5. 6. 7. The sine, versed sine, tangent, and secant, of any arch, which is the measure of any given angle ABC, is to the sine, versed sine, tangent, and secant, of any other arch which is the measure of the same angle, as the radius of the first is to the radius of the second.

Fig. 4.

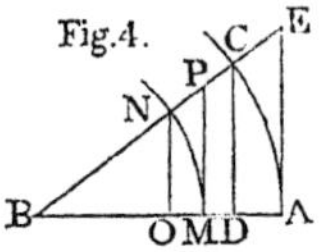

Let AC, MN be measures of the angle ABC, according to Def. 1.; CD the sine, DA the versed sine, AE the tangent, and BE the secant, of the arch AC, according to Def. 4. 5. 6. 7.; and NO the sine, OM the versed sine, MP the tangent, and BP the secant of the arch MN, according to the same definitions. Since CD, NO, AE, MP are parallel, CD is to NO as the radius CB to the radius NB, and AE to MP as AB to BM, and BC or BA to BD, as BN or BM to BO; and by conversion, DA to MO as AB to MB. Hence the corollary is manifest: therefore, if the radius be supposed to be divided into any given number of equal parts, the sine, versed sine, tangent, and secant of any given angle will each contain a given number of these parts; and, by trigonometrical tables, the length of the sine, versed sine, tangent, and secant of any angle may be found in parts of which the radius contains a given number; and, vice versâ, a number expressing the length of the sine, versed sine, tangent, and secant, being given, the angle of which it is the sine, versed sine, tangent, and secant, may be found.

VIII. The difference of an angle from a right angle, is called the *complement* of that angle. Thus, if BH be drawn perpendicular to AB, the angle CBH will be the complement of the angle ABC, or of CBF. See fig. 3.

IX. Let HK be the tangent; CL or DB, which is equal to it, the sine; and BK the secant of CBH, the complement of ABC, according to Def. 4. 6. 7. HK is called the *cotangent*, BD the *cosine*, and BK the *cosecant*, of the angle ABC. See fig. 3.

Cor. 1. The radius is a mean proportional between the tangent and cotangent.

For, since HK, BA are parallel, the angles HKB, ABC will be equal, and the angles KHB, BAE are right; therefore the triangles BAE, KHB are similar, and therefore AE is to AB, as BH or BA to HK.

Cor. 2. The radius is a mean proportional between the cosine and secant of any angle ABC.

Since CD, AE are parallel, BD is to BC or BA as BA to BE.

PROPOSITION I.

In a right-angled plane triangle, if the hypothenuse be made radius, the sides become the sines of the angles opposite to them: and if either side be made radius, the remaining side is the tangent of the angle opposite to it, and the hypothenuse the secant of the same angle.

Fig. 5.

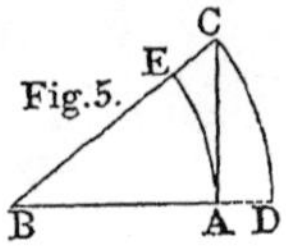

Let ABC be a right-angled triangle: if the hypothenuse BC be made radius, either of the sides AC will be the sine of the angle ABC opposite to it; and if either side BA be made radius, the other side AC will be the tangent of the angle ABC opposite to it, and the hypothenuse BC the secant of the same angle.

About B, as a centre, with BC, BA for distances, let two circles CD, EA be described, meeting BA, BC, in D, E: since CAB is a right angle, BC being radius, AC is the sine of the angle ABC, by def. 4. and BA being radius, AC is the tangent, and BC the secant, of the angle ABC, by def. 6. 7.

COR. 1. Of the hypothenuse, a side, and an angle of a right-angled triangle, any two being given, the third is also given.

COR. 2. Of the two sides and an angle of a right-angled triangle, any two being given, the third is also given.

PROPOSITION II.

The sides of a plane triangle are to one another, as the sines of the angles opposite to them.

In right-angled triangles, this Proposition is manifest from Prop. 1. for if the hypothenuse be made radius, the sides are the sines of the angles opposite to them, and the radius is the sine of a right angle * which is opposite to the hypothenuse.

* Cor. Def. 4. of this.

In any oblique-angled triangle ABC, any two sides AB, AC will be to one another as the sines of the angles ACB, ABC, which are opposite to them.

From C, B, draw CE, BD, perpendicular upon the opposite sides AB, AC, produced if need be. Since CEB, CDB are right angles, BC being radius, CE is the sine of the angle CBA, and BD the sine of the angle ACB; but the two triangles CAE, DAB have each a right angle at D and E; and likewise the common angle CAB; therefore they are similar, and consequently, CA is to AB, as CE to DB; that is, the sides are as the sines of the angles opposite them.

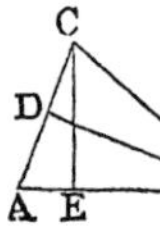

Fig.6.

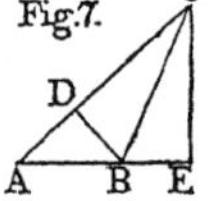

Fig.7.

COR. Hence of two sides, and two angles opposite to them, in a plane triangle, any three being given, the fourth is also given.

PROPOSITION III.

In a plane triangle, the sum of any two sides is to their difference, as the tangent of half the sum of the angles at the base, to the tangent of half their difference.

Let ABC be a plane triangle, the sum of any two sides AB, AC will be to their difference as the tangent of half the sum of the angles at the base ABC, ACB, to the tangent of half their difference.

About A as a centre, with AB the greater side for a distance, let a circle be described, meeting AC, produced in E, F, and BC in D; join DA, EB, FB; and draw FG parallel to BC, meeting EB in G.

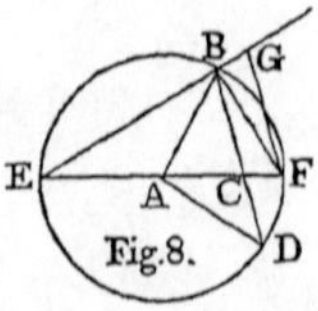

Fig. 8.

The angle EAB* is equal to the sum of the angles at the base, and the angle EFB at the circumference is equal to the half of EAB at the centre*; therefore EFB is half the sum of the angles at the base; but the angle ACB * is equal to the angles CAD and ADC or ABC together: therefore FAD is the difference of the angles at the base, and FBD at the circumference, or BFG, on account of the parallels FG, BD, is the half of that difference: but since the angle EBF in a semicircle is a right angle*, FB being radius, BE, BG are the tangents of the angles EFB, BFG; but it is manifest that EC is the sum of the sides BA, AC, and CF their difference; and since BC, FG are parallel*, EC is to CF, as EB to BG; that is, the sum of the sides is to their difference, as the tangent of half the sum of the angles at the base, to the tangent of half their difference.

* 32. 1.‡ * 20. 3. * 32. 1. * 1. of this. * 2. 6.

PROPOSITION IV.

In any plane triangle BAC, *whose two sides are* BA, AC, *and base* BC, *the less of the two sides, which let be* BA, *is to the greater* AC, *as the radius is to the tangent of an angle; and the radius is to the tangent of the excess of this angle above half a right angle, as the tangent of half the sum of the angles* B *and* C *at the base, is to the tangent of half their difference.*

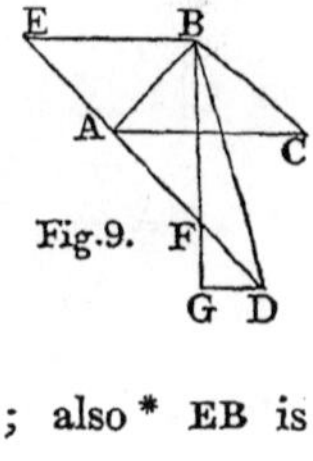

Fig. 9.

At the point A, draw the straight line EAD perpendicular to BA; make AE, AF, each equal to AB, and AD to AC; join BE, BF, BD, and from D draw DG perpendicular upon BF. And because BA is at right angles to EF, and EA, AB, AF are equal, each of the angles EBA, ABF is half a right angle, and the whole EBF is a right angle; also* EB is

* 4. 1.

‡ The figures in the margin refer to the Elements of Euclid, unless otherwise expressed.

equal to BF. And since EBF, FGD are right angles, EB is parallel to GD, and the triangles EBF, FGD are similar; therefore EB is to BF, as DG to GF, and EB being equal to BF, FG must be equal to GD. And because BAD is a right angle, BA the less side is to AD or AC the greater, as the radius is to the tangent of the angle ABD; and because BGD is a right angle, BG is to GD or GF, as the radius is to the tangent of GBD, which is the excess of the angle ABD above ABF half a right angle. But because EB is parallel to GD, BG is to GF, as ED is to DF; that is, since ED is the sum of
* 3. of this. the sides BA, AC, and FD their difference *, as the tangent of half the sum of the angles B, C, at the base, to the tangent of half their difference. Therefore, in any plane triangle, &c. Q. E. D.

PROPOSITION V.

In any triangle, twice the rectangle contained by any two sides, is to the difference of the sum of the squares of these two sides, and the square of the base, as the radius is to the cosine of the angle included by the two sides.

Let ABC be a plane triangle, twice the rectangle ABC contained by any two sides BA, BC, is to the difference of the sum of the squares of BA, BC, and the square of the base AC, as the radius to the cosine of the angle ABC.

Fig.10. Fig.11.

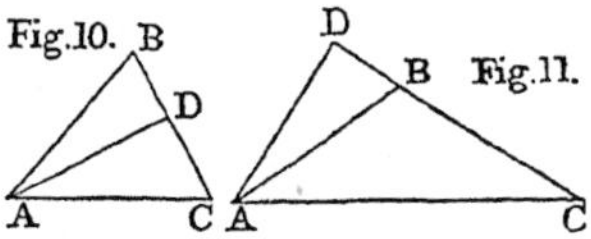

From A, draw AD perpendicular upon the opposite side
* 12. & 13. 2. BC, then * the difference of the sum of the squares of AB, BC, and the square of the base AC, is equal to twice the rectangle CBD; but twice the rectangle CBA is to twice the rectangle CBD, that is to the difference of the sum of the squares of
* 1. 6. AB, BC, and the square of AC * as AB to BD; that is *, as
* 1. of this. radius to the sine of BAD, which is the complement of the angle ABC; that is, as radius to the cosine of ABC.

PROPOSITION VI.

In any triangle ABC, *whose two sides are* AB, AC, *and base* BC, *the rectangle contained by half the perimeter, and the excess of it above the base* BC, *is to the rectangle contained by the straight lines by which the half of the perimeter ex-*

ceeds the other two sides AB, AC, *as the square of the radius is to the square of the tangent of half the angle* BAC *opposite to the base.*

Let the angles BAC, ABC be bisected by the straight lines AG, BG; and producing the side AB, let the exterior angle CBH be bisected by the straight line BK, meeting AG in K; and from the points G, K, let there be drawn perpendicular upon the sides, the straight lines GD, GE, GF, KH, KL, KM. Since therefore * G is the centre of the circle inscribed in the triangle ABC; GD, GF, GE will be equal, and AD will be equal to AE, BD to BF, and CE to CF. In like manner KH, KL, KM will be equal, and BH will be equal to BM, and AH to AL, because the angles HBM, HAL are bisected by the straight lines BK, KA: and because in the triangles KCL, KCM, the sides LK, KM are equal, KC is common, and KLC, KMC are right angles, CL will be equal to CM: since therefore BM is equal to BH, and CM to CL; BC will be equal to BH and CL together; and, adding AB and AC together, AB, AC, and BC will together be equal to AH and AL together: but AH, AL are equal: wherefore each of them is equal to half the perimeter of the triangle ABC: but since AD, AE are equal, and BD, BF, and also CE, CF; AB, together with FC, will be equal to half the perimeter of the triangle to which AH or AL was shewn to be equal; taking away therefore the common AB, the remainder FC will be equal to the remainder BH: in the same manner it is demonstrated, that BF is equal to CL: and since the points B, D, G, F are in a circle, the angle DGF will be equal to the exterior and opposite angle FBH *; wherefore their halves BGD, HBK will be equal to one another: the right-angled triangles BGD, HBK will therefore be equiangular, and GD will be to BD, as BH to HK: and the rectangle contained by GD, HK, will be equal to the rectangle DBH or BFC: but since AH is to HK, as AD to DG, the rectangle HAD * will be to the rectangle contained by HK, DG, or the rectangle BFC (as the square of AD is to the square of DG, that is), as the square of the radius is to the square of the tangent of the angle DAG,

* 4. 4.

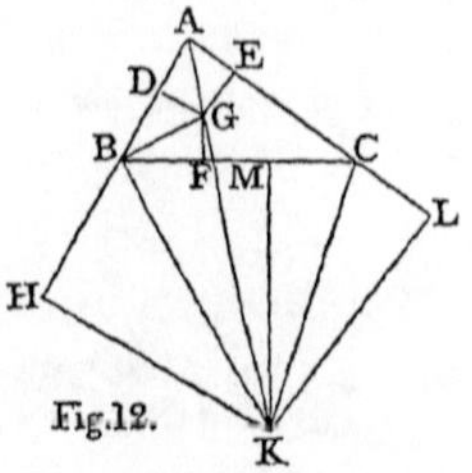

Fig. 12.

* 22. 3.

* 22. 6.

that is, the half of BAC: but HA is half the perimeter of the triangle ABC, and AD is the excess of the same above HD, that is, above the base BC; but BF or CL is the excess of HA or AL above the side AC; and FC, or HB, is the excess of the same HA above the side AB; therefore the rectangle contained by half the perimeter, and the excess of the same above the base, viz. the rectangle HAD, is to the rectangle contained by the straight lines by which the half of the perimeter exceeds the other two sides, that is, the rectangle BFC, as the square of the radius is to the square of the tangent of half the angle BAC opposite to the base. Q. E. D.

PROPOSITION VII.

In a plane triangle, the base is to the sum of the sides as the difference of the sides is to the sum or difference of the segments of the base made by the perpendicular upon it from the vertex, according as the square of the greater side is greater or less than the sum of the squares of the lesser side and the base.

Let ABC be a plane triangle; if from A the vertex be drawn a straight line AD perpendicular upon the base BC, the base BC will be to the sum of the sides BA, AC, as the difference of the same sides is to the sum or difference of the segments CD, BD, according as the square of AC the greater side is greater or less than the sum of the squares of the lesser side AB, and the base BC.

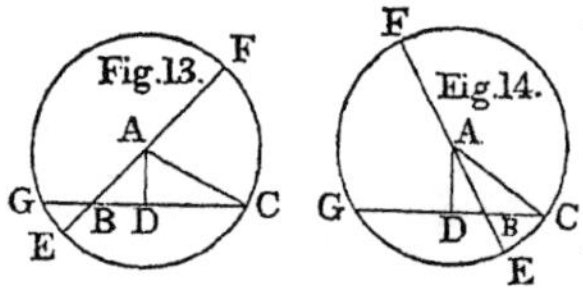

About A, as a centre, with AC the greater side for a distance, let a circle be described meeting AB produced in E, F, and CB in G: it is manifest, that FB is the sum, and BE the difference of the sides; and since AD is perpendicular to GC; GD, CD will be equal; consequently GB will be equal to the sum or difference of the segments CD, BD, according as the perpendicular AD meets the base produced, or the base; that is (by Conv. 12. 13. 2.), according as the square of AC is greater or less than the sum of

* 35. 3. the squares of AB, BC: but * the rectangle CBG is equal to

* 16. 6. the rectangle EBF; that is * BC is to BF, as BE is to BG; that is, the base is to the sum of the sides, as the difference of

the sides is to the sum or difference of the segments of the base made by the perpendicular from the vertex, according as the square of the greater side is greater or less than the sum of the squares of the lesser side and the base. Q. E. D.

PROPOSITION VIII.

PROB.—*The sum and difference of two magnitudes being given, to find them.*

Half the given sum added to half the given difference, will be the greater, and half the difference subtracted from half the sum, will be the less.

For let AB be the given sum, AC the greater, and BC the less. Let AD be half the given sum; and to AD, DB, which are equal, let DC be added; then AC will be equal to BD and DC together: that is, to BC, and twice DC; consequently, twice DC is the difference, and DC half that difference; but AC the greater is equal to AD, DC; that is, to half the sum added to half the difference, and BC the less is equal to the excess of BD, half the sum, above DC half the difference. Q. E. D.

Fig.15.

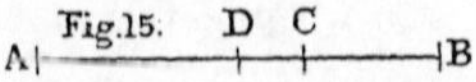

SCHOLIUM.

Of the six parts of a plane triangle (the three sides and three angles) any three being given, to find the other three is the business of plane trigonometry; and the several cases of that problem may be resolved by means of the preceding propositions, as in the two following, with the tables annexed. In these, the solution is expressed by a fourth proportional to three given lines; but if the given parts be expressed by numbers from trigonometrical tables, it may be obtained arithmetically by the common Rule of Three.

NOTE. In the tables the following abbreviations are used: R. is put for the Radius; T. for Tangent; and S. for Sine. Degrees, minutes, seconds, &c. are written in this manner; 30° 25′ 13″, &c. which signify 30 degrees, 25 minutes, 13 seconds, &c.

SOLUTION OF THE CASES OF RIGHT-ANGLED TRIANGLES.

GENERAL PROPOSITION.

In a right-angled triangle, of the three sides and three angles, any two being given besides the right angle, the other three may be found, except when the two acute angles are given; in which case the ratios of the sides are only given, being the same with the ratios of the sines of the angles opposite to them.

It is manifest from 47. 1. that of the two sides and hypothenuse, if any two be given, the third may also be found. It is also manifest from 32. 1. that if one of the acute angles of a right-angled triangle be given, the other is also given, for it is the complement of the former to a right angle.

If two angles of any triangle be given, the third is also given, being the supplement of the two given angles to two right angles.

The other cases may be resolved by help of the preceding propositions, as in the following table:

Fig.16.

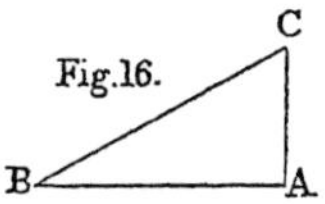

Case.	Given.	Sought.	
1	Two sides, AB, AC.	The angles B, C.	AB : AC :: R : T, B, of which C is the complement.
2	AB, BC, a side and the hypothenuse.	The angles B, C.	BC : BA :: R : S, C, of which B is the complement.
3	AB, B, a side and an angle.	The other side AC.	R : T, B :: BA : AC.
4	AB and B, a side and an angle.	The hypothenuse BC.	S, C : R :: BA : BC.
5	BC and B, the hypothenuse and an angle.	The side AC.	R : S, B :: BC : CA.

These five cases are resolved by Prop. 1.

SOLUTION OF THE CASES OF OBLIQUE-ANGLED TRIANGLES.

GENERAL PROPOSITION.

In an oblique-angled triangle, of the three sides and three angles, any three being given, the other three may be found, except when the three angles are given; in which case the ratios of the sides are only given, being the same with the ratios of the sines of the angles opposite to them.

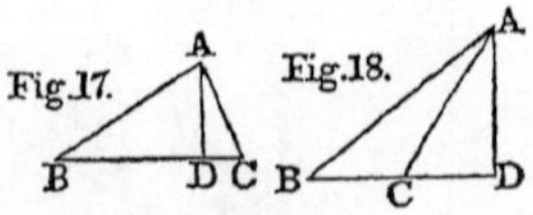

Case.	Given.	Sought.	
1	A, B, and therefore C, and the side AB.	BC, AC.	S, C : S, A : : AB : BC, and also S, C : S, B : : AB : AC. (2.)
2	AB, AC, and B, two sides and an angle opposite to one of them.	The angles A and C.	AC : AB : : S, B : S, C. (2.) This case admits of two solutions; for C may be greater or less than a quadrant. (Cor. to def. 4.)
3	AB, AC, and A, two sides, and the included angle.	The angles B and C.	AB + AC : AB — AC : : T, $\frac{C+B}{2}$: T, $\frac{C-B}{2}$. (3.) the sum and difference of the angles C, B, being given, each of them is given. (7.) *Otherwise;* BA : AC : : R : T, ABD, and also R : T, ABD — 45° : : T, $\frac{B+C}{2}$: T, $\frac{B-C}{2}$ (4.) therefore B and C are given as before. (7.)

See fig. 9. Prop. 4. page 431.

	Case.	Given.	Sought.	
See fig. 17. See fig. 18. See fig. 17. See fig. 18.	4	AB, BC, CA the three sides.	A, B, C the three angles.	2 AC × CB : ACq + CBq — ABq :: R : CoS, C. If ACq + CBq be greater than ABq. 2 AC × CB : ABq — ACq — CBq :: R : CoS, C. If ABq be greater than ACq + CBq. (4.) *Otherwise;* Let AB + BC + AC = 2 P. P × (P — AB) : (P — AC) × (P — BC) :: Rq : Tq, ½ C, and hence C is known. (5.) *Otherwise;* Let AD be perpendicular to BC. 1. If ABq be less than ACq + CBq. BC : BA + AC :: BA — AC : BD — DC, and BC the sum of BD, DC is given; therefore each of them is given. (7.) 2. If ABq be greater than ACq + CBq. BC : BA + AC :: BA — AC : BD + DC; and BC the difference of BD, DC, is given, therefore each of them is given. (7.) And CA : CD :: R : CoS, C. (1.) and C being found, A and B are found by case 2 or 3.

CONSTRUCTIONS

OF THE

TRIGONOMETRICAL CANON.

A Trigonometrical Canon is a Table, which, beginning from one second or one minute, orderly expresses the lengths that every sine, tangent, and secant have, in respect of the radius, which is supposed unity; and is conceived to be divided into 10000000 or more decimal parts. And so the sine, tangent, or secant of an arc, may be had by the help of this table; and, contrariwise, a sine, tangent, or secant being given, we may find the arc it expresses. Take notice that in the following tract, R signifies the radius, S a sine, Cos. a cosine, T a tangent, and Cot. a cotangent; also ACq signifies the square of the right line AC; and the marks or characters, +, —, =, :, : :, and √, are severally used to signify addition, subtraction, equality, proportionality, and the extraction of the square root. Again, when the sum or difference of two quantities is between parentheses, then that sum or difference is to be considered as one quantity.

PROPOSITION I.

THEOREM.—*The two sides of any right-angled triangle being given, the other side is also given.*

For * ACq=ABq+BCq, and ACq—BCq=ABq, and interchangeably ACq—ABq=BCq. Whence, by the extraction of the square root, there is given AC=√(ABq+BCq); and AB=√(ACq—BCq); and BC=√(ACq—ABq). * 47. 1.

Fig. 19.

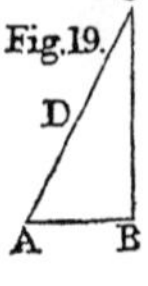

PROPOSITION II.

PROBLEM.—*The sine* DE *of the arc* BD, *and the radius* CD, *being given, to find the cosine* DF.

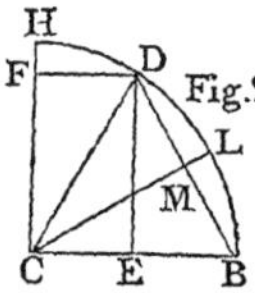
Fig. 20.

The radius CD, and the sine DE, being given in the right-angled triangle CDE, there will be given (by the last Prop.) $\sqrt{(\text{CDq}-\text{DEq})}=(\text{CE}=)$ DF.

PROPOSITION III.

See fig. 20. PROB.—*The sine* DE *of any arc* DB *being given, to find* DM *or* BM, *the sine of half the arc.*

DE being given, CE (by the last Prop.) will be given, and accordingly EB, which is the difference between the cosine and radius. Therefore DE, EB, being given, in the right-angled triangle DBE, there will be given DB, whose half DM is the sine of the arc $\text{DL}=\frac{1}{2}$ the arc BD.

PROPOSITION IV.

See fig. 20. PROB.—*The sine* BM *of the arc* BL *being given, to find the sine of double that arc.*

The sine BM being given, there will be given (by Prop. 2.) the cosine CM. But the triangles CBM, DBE are equiangular, because the angles at E and M are right angles, and the angle at B common: wherefore* we have CB : CM : : (BD, or) 2 BM : DE. Whence, since the three first terms of this analogy are given, the fourth also, which is the sine of the arc DB, will be known.

* 4. 6.

COROLLARY. Hence CB : 2 CM : : BD : 2 DE; that is, the radius is to double the cosine of one half of the arc DB, as the subtense of the arc DB is to the subtense of double that arc. Also CB : 2 CM : : (2 BM : 2 DE : :) BM : DE : : $\frac{1}{2}$ CB : CM. Wherefore the sine of an arc, and the sine of its double, being given, the cosine of the arc itself is given.

PROPOSITION V.

PROB.—*The sines of two arcs* BD, FD, *being given, to find*

FI, *the sine of the sum, as likewise* EL, *the sine of their difference.*

Let the radius CD be drawn, and then CO is the cosine of the arc FD, which accordingly is given, and draw OP through O parallel to DK; also let OM, GE be drawn parallel to CB. Then because the triangles CDK, COP, CHI, FOH, FOM are equiangular; in the first place, CD : DK : : CO : OP, which, consequently, is known. Also we have CD : CK : : FO : FM; and so, likewise, this will be known. But because FO = EO, then will FM = MG = ON; and so OP + FM = FI = sine of the sum of the arcs; and OP — FM: that is, OP — ON = EL = sine of the difference of the arcs: which were to be found.

Fig. 21.

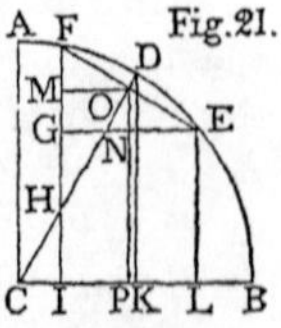

Cor. Because the differences of the arcs BE, BD, BF are equal, the arc BD is an arithmetical mean between the arcs BE, BF.

PROPOSITION VI.

Theor.—*The same things being supposed, the radius is to double the cosine of the mean arc, as the sine of the difference is to the difference of the sines of the extremes.* See fig. 21.

For we have CD : CK : : FO : FM; whence by doubling the consequents, CD : 2 CK : : FO : (2 FM, or) to FG, which is the difference of the sines EL, FI. Q. E. D.

Cor. 1. If the arc BD be 60 degrees, the difference of the sines FI, EL, will be equal to the sine FO of the difference. For, in this case, CK is the sine of 30 degrees; the double whereof is equal* to the radius; and so, since CD = 2 CK, we shall have FO = FG. And, consequently, if the two arcs BE, BF be equidistant from the arc of 60 degrees, the difference of the sines will be equal to the sine of the difference FD. * 15. 4.

Cor. 2. Hence, if the sines of all arcs distant from one another by a given interval, be given, from the beginning of a quadrant to 60 degrees, the other sines may be found by one addition only. For the sine of 61 degrees = the sine of 59 degrees + the sine of one degree; and the sine of 62 degrees = the sine of 58 degrees + the sine of two degrees. Also, the sine of 63 degrees = the sine of 57 degrees + the sine of 3 degrees, and so on.

Cor. 3. If the sines of all arcs, from the beginning of a quadrant to any part of a quadrant, distant from each other by a given interval, be given, thence we may find the sines of all arcs to the double of that part. For example: let all the sines to 15 degrees be given; then, by the preceding analogy, all the sines to 30 degrees may be found. For the radius is to double the cosine of 15 degrees, as the sine of 1 degree is to the difference of the sines of 14 degrees, and 16 degrees; so, also, is the sine of 3 degrees to the difference between the sines of 12 and 18 degrees; and so on continually, until you come to the sine of 30 degrees.

After the same manner, as the radius is to double the cosine of 30 degrees, or to double the sine of 60 degrees, so is the sine of 1 degree to the difference of the sines of 29 and 31 degrees : : sine 2 degrees to the difference of the sines of 28 and 32 degrees : : sine 3 degrees to the difference of the sines of 27 and 33 degrees. But, in this case, the radius is to double the cosine of 30 degrees, as 1 to $\surd 3$.

For * the angle BGC$=$60 degrees, as the arc BC, its measure, is a sixth part of the whole circumference; and the straight line BC$=$R. Hence it is evident that the sine of 30 degrees is equal to half the radius; and therefore, by Prop. 2. the cosine of 30 degrees $=\surd\left(R^2-\frac{R^2}{2}\right)=\frac{\surd 3R^2}{2}$, and its double $=\surd 3R^2=R\times\surd 3$. Consequently, radius is to double the cosine of 30^0 : : $R : R\times\surd 3 :: 1 : \surd 3$.

* See Fig. Prop. 15. 4. of the Elements.

And, accordingly, if the sines of the distances from the arc of 30 degrees, be multiplied by $\surd 3$, the differences of the sines will be had.

So, likewise, may the sines of the minutes in the beginning of the quadrant be found, by having the sines and cosines of one and two minutes given. For, as the radius is to double the cosine of $2'$: : sine $1'$: difference of the sines of $1'$ and $3'$: : sine $2'$: difference of the sines of $0'$ and $4'$; that is, to the sine of $4'$. And so, the sines of the first four minutes being given, we may thereby find the sines of the others to $8'$, and from thence to $16'$, and so on.

PROPOSITION VII.

THEOR.—*In small arcs, the sines and tangents of the same arcs are nearly to one another, in a ratio of equality.*

For, because the triangles CED, CBG, are equiangular, CE : CB : : ED : BG. But as the point E approaches B, EB will vanish in respect of the arc BD; whence CE will become nearly equal to CB, and so ED will be also nearly equal to BG. If EB be less than the $\frac{1}{10000000}$ part of the radius, then the difference between the sine and the tangent will be also less than the $\frac{1}{10000000}$ part of the tangent.

Fig. 22.

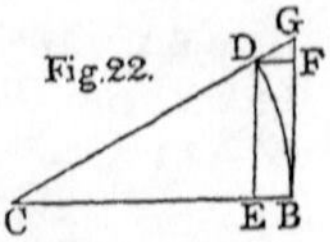

Cor. Since any arc is less than the tangent, and greater than its sine, and the sine and tangent of a very small arc are nearly equal; it follows, that the arc will be nearly equal to its sine: and so in very small arcs, it will be, as arc is to arc, so is sine to sine.

PROPOSITION VIII.

Prob.—*To find the sine of the arc of one minute.*

The side of a hexagon inscribed in a circle, that is, the subtense of 60 degrees, is equal to the radius (by Cor. 15th of the 4th); and so the half of the radius will be the sine of the arc of 30 degrees. Wherefore the sine of the arc of 30 degrees being given, the sine of the arc of 15 degrees may be found (by Prop. 3.). Also the sine of the arc of 15 degrees being given, (by the same Prop.) we may have the sine of 7 degrees 30 minutes. So, likewise, can we find the sine of the half of this, viz. 3 degrees 45 minutes; and so on, until 12 bisections being made, we come to an arc of 52^{2}, 44^{3}, 03^{4}, 45^{5}, whose cosine is nearly equal to the radius; in which case (as is manifest from Prop. 7.) arcs are proportional to their sines: and so, as the arc of 52^{2}, 44^{3}, 03^{4}, 45^{5}, is to an arc of one minute, so will the sine before found be to the sine of an arc of one minute, which therefore will be given. And when the sine of one minute is found, then (by Prop. 2. and 4.) the sine and cosine of two minutes will be had.

PROPOSITION IX.

Theor.—*If the angle* BAC, *being in the periphery of a circle, be bisected by the right line* AD, *and if* AB *be produced until* DE=AD *meets it in* E; *then will* CE=AB.

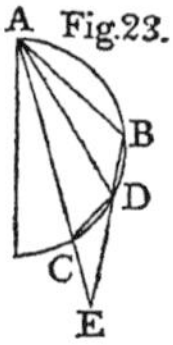

Fig. 23.

In the quadrilateral figure ABDC the angles B
* 22. 3. and DCA * are equal to two right angles=DCE+
* 13. 1. DCA* whence the angle B=DCE. But, likewise,
* 5. 1. angle E=DAC*=DAB, and DC=DB: wherefore the triangles BAD and CED are congruous, and so CE is equal to AB. Q. E. D.

PROPOSITION X.

THEOR.—*Let the arcs* AB, BC, CD, DE, EF, *&c. be equal: and let the subtenses of the arcs,* AB, AC, AD, AE, *&c. be drawn; then will* AB : AC :: AC : AB+AD :: AD : AC+AE :: AE : AD+AF :: AF : AE+AG.

Let AD be produced to H, AE to I, AF to K, and AG to L, so that the triangles ACH, ADI, AEK, AFL, be isosceles ones: then because the angle BAD is bisected, we shall have DH=AB (by the last Prop.): so likewise EI=AC, FK=AD, also GL=AE.

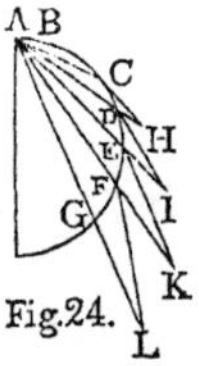

Fig. 24.

But the isosceles triangles ABC, ACH, ADI, AEK, AFL, because of the equal angles at the bases, are equiangular: wherefore it will be, as AB : AC :: AC : (AH=)AB+AD :: AD : (AI=) AC+AE :: AE : (AK=)AD+AF :: AF : (AL=) AE+AG. Q. E. D.

COR. 1. Because AB is to AC, as radius is to double the cosine of $\frac{1}{2}$ the arc AB (by Coroll. Prop. 4.), it will also be, as radius is to double the cosine of $\frac{1}{2}$ the arc AB, so is $\frac{1}{2}$ AB : $\frac{1}{2}$ AC :: $\frac{1}{2}$ AC : $\frac{1}{2}$ AB + $\frac{1}{2}$ AD :: $\frac{1}{2}$ AD : $\frac{1}{2}$ AC + $\frac{1}{2}$ AE :: $\frac{1}{2}$ AE : $\frac{1}{2}$ AD + $\frac{1}{2}$ AF, &c. Now let each of the arcs AB, BC, CD, &c. be 2′; then will $\frac{1}{2}$ AB be the sine of one minute, $\frac{1}{2}$ AC the sine of 2 minutes, $\frac{1}{2}$ AD the sine of 3 minutes, $\frac{1}{2}$ AE the sine of 4 minutes, &c. Whence if the sines of one and two minutes be given, we may easily find all the other sines in the following manner.

Let the cosine of the arc of one minute, that is, the sine of the arc of 89 deg. 59′, be called Q; and make the following analogies; R. : 2 Q :: Sin. 2′ : S. 1′+S. 3′. Wherefore the sine of 3 minutes will be given. Also, R. : 2 Q :: S. 3′ : S. 2′+S. 4′. Wherefore the S. 4′ is given. And R. : 2 Q :: S. 4′ : S. 3′+S. 5′; and so the sine of 5′ will be had. Likewise, R. : 2 Q :: S. 5′ : S. 4′+S. 6′; and so we shall have the sine of 6′. And in like manner, the sines of every minute of the quadrant will be given. And because the radius, or the first term of the analogy, is unity, the operations will be with great ease and expedition calculated by multiplication, and contracted by addition. When the sines are found to 60 degrees, all the other sines may be had by addition only, by Cor. 1. Prop. 6.

The sines being given, the tangents and secants may be found from the following analogies; because the triangles BDC, BAE, BHK, are equiangular, we have See fig. 3. p. 427.

BD : DC :: BA : AE ; that is, Cos. : S. :: R. : T.
AE : BA :: BH : HK ; that is, T. : R. :: R. : Cot.
BD : BC :: BA : BE ; that is, Cos. : R. :: R. : Secant.
CD : BC :: BH : BK ; that is, S. : R. :: R. : Cosec.

OF LOGARITHMS.

ARTICLE 1. The indices or exponents of a series of numbers in geometrical progression, proceeding from 1, are also called the logarithms of the numbers in that series*. Thus if a denote any number, and the geometrical series, 1, a^1, a^2, a^3, a^4, &c. be produced by actual multiplication, then 1, 2, 3, 4, &c. are called the logarithms of the first, second, third, and fourth powers of a respectively. Consequently, if, in the above, a be equal to the number 2, then 1 is the logarithm of 2, 2 is the logarithm of 4, 3 is the logarithm of 8, 4 is the logarithm of 16, &c. But if a be equal to 10, then 1 is the logarithm of 10, 2 is the logarithm of 100, 3 is the logarithm of 1000, 4 is the logarithm of 10000, &c. The series may be continued both ways from 1. Thus $\frac{1}{a^4}, \frac{1}{a^3}, \frac{1}{a^2}, \frac{1}{a}$, 1, a^1, a^2, a^3, a^4, &c. constitute a series in geometrical progression, and, agreeable to the established notation in algebra, the indices, or logarithms are, —4, —3, —2, —1, 0, 1, 2, 3, 4, &c. If

* The reader ought to be acquainted with arithmetical and geometrical progression and the binomial theorem, before he enters on a perusal of any account of logarithms.

a be equal to the number 2, then -4 is the logarithm of $\frac{1}{16}$, -3 is the logarithm of $\frac{1}{8}$, -2 is the logarithm of $\frac{1}{4}$, -1 is the logarithm of $\frac{1}{2}$, 0 is the logarithm of 1, 1 is the logarithm of 2, &c. If a be equal to 10, then -4 is the logarithm of $\frac{1}{10000}$, -3 is the logarithm of $\frac{1}{1000}$, -2 is the logarithm of $\frac{1}{100}$, -1 is the logarithm of $\frac{1}{10}$, 0 is the logarithm of 1, and 1 is the logarithm of 10, &c.

Art. 2. From the above it is evident that the logarithms of a series of numbers in geometrical progression, constitute a series of numbers in arithmetical progression. Beginning with 1, and proceeding towards the right hand, the terms in the geometrical series are produced by multiplication, but their corresponding logarithms are produced by addition. On the contrary, beginning with 1, and proceeding towards the left hand, the terms in the geometrical progression are produced by division, but their corresponding logarithms are produced by subtraction.

Art. 3. The same observations apply to logarithms when they are fractions. Thus if $a^{\frac{1}{n}}$ denote any number, $\frac{1}{a^{\frac{4}{n}}}$, $\frac{1}{a^{\frac{3}{n}}}$, $\frac{1}{a^{\frac{2}{n}}}$, $\frac{1}{a^{\frac{1}{n}}}$, 1, $a^{\frac{1}{n}}$, $a^{\frac{2}{n}}$, $a^{\frac{3}{n}}$, $a^{\frac{4}{n}}$, &c. constitute a series of numbers in geometrical progression, of which $-\frac{4}{n}$, $-\frac{3}{n}$, $-\frac{2}{n}$, $-\frac{1}{n}$, 0, $\frac{1}{n}$, $\frac{2}{n}$, $\frac{3}{n}$, $\frac{4}{n}$, &c. are the logarithms; and it is evident that the assertions in the last article hold true, both with respect to the numbers in geometrical progression and their corresponding logarithms. As a and n may be taken at pleasure, it follows that numbers in very different geometrical progressions may have the same logarithms; and that the same series of numbers in geometrical progression may have different series of logarithms corresponding to them.

Art. 4. If a be an indefinitely small decimal fraction, and successive powers of $1+a$ be raised, then the excess of any power of $1+a$ above that immediately preceding it will be indefinitely small. Thus let $a = \cdot 0000000001$, and then $(1+a)^2 = 1\cdot 00000000002000000000001$; and $(1+a)^3 = 1\cdot 000000000030000000000300000000001$; and proceeding by actual multiplication to obtain higher powers of $1\cdot 0000000001$, it will be found that the difference between two successive powers is very small. If, instead of supposing, as above, that $a = \cdot 0000000001$, we suppose it only one millionth part of this value, then the successive powers of $1+a$ will differ from one another by much smaller decimal fractions.

ART. 5. If, therefore, a be indefinitely small, and successive powers of $1+a$ be raised, a series of numbers in geometrical progression will be produced, of which the common numbers 2, 3, 4, 5, &c. will become terms. For on every multiplication by $1+a$, an indefinitely small addition is made to the power multiplied, and by this indefinitely small addition, the next higher power is produced. Some power of $1+a$ will, therefore, be equal to the number 2, or so nearly equal to it that they may be considered as equal. Continuing the advancement of the powers of $1+a$, the numbers 3, 4, 5, &c. for the same reasons, will fall into the series.

ART. 6. The sum of the logarithms of any two numbers is equal to the logarithm of the product of the same two numbers. Thus if $1+a$ raised to the nth power be equal to the number N, and if $1+a$ raised to the mth power be equal to number M, then, by the preceding articles, n is the logarithm of $(1+a)^n$ or of its equal N, and for the same reason, m is the logarithm of M. Hence it follows that $n+m =$ the logarithm of $N \times M$, for $N \times M = (1+a)^n\,(1+a)^m = (1+a)^{n+m}$ by the nature of indices. If the logarithm of N be subtracted from the logarithm of M, the difference is equal to the logarithm of the quotient which arises from the division of M by N. For $\frac{M}{N} = \frac{(1+a)^m}{(1+a)^n} = (1+a)^{m-n}$, by the nature of indices. The addition of logarithms, therefore, answers to the multiplication of the natural numbers to which they belong; and the subtraction of logarithms answers to the division by the natural numbers to which they belong.

ART. 7. If the logarithms of a series of natural numbers be all multiplied by the same number, the several products will have the last-mentioned properties of logarithms. Thus, if the indices of all the powers of $1+a$ be multiplied by l, then, using the notation stated in the last article, the logarithm of N is nl, and the logarithm of M is ml, and the logarithm of $N \times M$ is $nl+ml$; for $N \times M = (1+a)^{nl}\,(1+a)^{ml} = (1+a)^{nl+ml}$, by the nature of indices. Also $ml - nl =$ the logarithm of $\frac{M}{N}$, for $\frac{M}{N} = \frac{(1+a)^{ml}}{(1+a)^{nl}} = (1+a)^{ml-nl}$. Hence the products arising from the multiplication of l into the indices of the powers of $1+a$, are termed logarithms, as are also all numbers which have the properties stated at the end of article 6. It is on account of these properties that logarithms are so very useful in calculations of the highest importance.

ART. 8. If the indices of the powers of $1+a$, be multiplied by a, the products are called the hyperbolic logarithms of the numbers equal to the powers of $1+a$. Thus, if the number N be equal to $(1+a)^n$, then na is the hyperbolic logarithm of N; and if the number M be equal to $(1+a)^m$, then ma is the hyperbolic logarithm of M. Hyperbolic logarithms are not those in common use, but they can be calculated with less labour than any other kind, and common logarithms are obtained from them.

ART. 9. If successive powers of a very small fraction be raised, they will successively be less and less in value. This truth appears most evident by putting the value in the form of a vulgar fraction. Thus $\left(\frac{1}{100000}\right)^2 = \frac{1}{10000000000}$; $\left(\frac{1}{100000}\right)^3 = \frac{1}{1000000000000000}$, &c.

ART. 10. Let it be required to determine the hyperbolic logarithm L, of any number N. Using the same notation as in the preceding articles, $(1+a)^n = N$, and by extracting the nth root of each side of the equation, $1+a = N^{\frac{1}{n}}$. Put $m = \frac{1}{n}$, and $1+x = N$, and then $N^{\frac{1}{n}} = (1+x)^m =$ (by the binomial theorem) $1+mx + m \times \frac{m-1}{2} \times x^2 + m \times \frac{m-1}{2} \times \frac{m-2}{3} \times x^3 + \&c. = 1+a$. Now, as a is indefinitely small, the power of $1+a$, which is equal to the number N, must be indefinitely high; or, which is the same thing, n must be indefinitely great. Consequently m must be indefinitely small, and therefore may be rejected from the expressions $m-1$, $m-2$, $m-3$, &c. Hence 1 being taken from each side of the above equation, we have $a = mx - \frac{mx^2}{2} + \frac{mx^3}{3} - \frac{mx^4}{4} + \frac{mx^5}{5}$ &c. Each side of this equation being divided by m, we have $\frac{a}{m} = x - \frac{x^2}{2} + \frac{x^3}{3} - \frac{x^4}{4} + \frac{x^5}{5} -$ &c. But $m = \frac{1}{n}$, and therefore $\frac{a}{m} = an = x - \frac{x^2}{2} + \frac{x^3}{3} - \frac{x^4}{4} + \frac{x^5}{5} -$ &c. $=$ L, the hyperbolic logarithm of

N, by article 8. This series, however, if x be a whole number, does not converge. Let M be a whole number, and $M = \frac{1}{1-x}$, and then x is less than 1. For, multiplying both sides of the equation by $1-x$, we have $M - Mx = 1$, and therefore $1 - \frac{1}{M} = x$. Now, let $M = \frac{1}{1-x} = (1+x)^r$. Then we have $1 + a = \frac{1}{(1-x)^{\frac{1}{p}}} = (1-x)^{-\frac{1}{p}} = (1-x)^r$ (by putting $r = -\frac{1}{p}$) $= 1 - rx + r \times \frac{r-1}{2} \times x^2 - r \times \frac{r-1}{2} \times \frac{r-2}{3} \times x^3 +$ &c. But for the same reasons as above, r must be indefinitely small, and therefore may be rejected from the factors $r-1$, $r-2$, $r-3$, &c. Consequently, taking 1 from each side of the above equation, $a = -rx - \frac{rx^2}{2} - \frac{rx^3}{3} - \frac{rx^4}{4} - \frac{rx^5}{5} -$&c. But $-r = \frac{1}{p}$, and therefore, dividing the left hand side of the equation by $\frac{1}{p}$, and the other by $-r$, we have $ap = x + \frac{x^2}{2} + \frac{x^3}{3} + \frac{x^4}{4} + \frac{x^5}{5} +$ &c. = the hyperbolic logarithm of M.

Art. 11. As, by the last article, the hyperbolic logarithm of N or $1+x$ is $x - \frac{x^2}{2} + \frac{x^3}{3} - \frac{x^4}{4} + \frac{x^5}{5} - \frac{x^6}{6} + \frac{x^7}{7} -$ &c. and as the hyperbolic logarithm of M or $\frac{1}{1-x}$ is $x + \frac{x^2}{2} + \frac{x^3}{3} + \frac{x^4}{4} + \frac{x^5}{5} + \frac{x^6}{6} + \frac{x^7}{7} +$ &c. the hyperbolic logarithm of $N \times M$, or $\frac{1+x}{1-x}$ is equal to the sum of these two series, that is, equal to $2x + \frac{2x^3}{3} + \frac{2x^5}{5} + \frac{2x^7}{7} +$ &c. This series converges faster than either of the preceding, and its value may be expressed thus: $2 \times (x + \frac{x^3}{3} + \frac{x^5}{5} + \frac{x^7}{7} +$ &c.).

Art. 12. The logarithm of $\frac{n+1}{n} = 2 \times$ logarithm of $\frac{2n+2}{2n+1} +$ the logarithm of $\frac{(2n+1)^2}{(2n+1)^2-1}$. For, as the addition of logarithms answers to the multiplication of the numbers to which they belong, the logarithm of the square of any number, is the logarithm of the number multiplied by 2. Hence the logarithm of $\frac{(2n+2)^2}{(2n+1)^2}$ is $2 \times$ logarithm of $\frac{2n+2}{2n+1}$.

But $\frac{(2n+2)^2}{(2n+1)^2} \times \frac{(2n+1)^2}{(2n+1)^2-1} = \frac{(2n+2)^2}{(2n+1)^2-1} = \frac{4n^2+8n+4}{4n^2+4n} = \frac{n^2+2n+1}{n^2+n} = \frac{(n+1)\ (n+1)}{n\ (n+1)} = \frac{n+1}{n}$.

From the preceding articles, hyperbolic logarithms may be calculated, as in the following examples.

Example 1. Required the hyperbolic logarithm of 2.

Put $\frac{n+1}{n}=2$, and then $n=1$, $\frac{2n+2}{2n+1}=\frac{4}{3}$, and $\frac{(2n+1)^2}{(2n+1)^2-1} = \frac{9}{8}$. In order to proceed by the series in article 11, let $\frac{1+x}{1-x}=\frac{4}{3}$, and then $x=\frac{1}{7}$. Consequently,

$$x = 0.14285714286$$
$$\frac{x^3}{3} = 0.00097181730$$
$$\frac{x^5}{5} = 0.00001189980$$
$$\frac{x^7}{7} = 0.00000017347$$
$$\frac{x^9}{9} = 0.00000000275$$
$$\frac{x^{11}}{11} = 0.00000000004$$

Sum of the above terms - - - - 0.14384103622
2

Log. of $\frac{1+x}{1-x}$ or $\frac{2n+2}{2n+1}$ or $\frac{4}{3}$ - 0.28768207244

The double of which is 0.57536414488, and answers to the first part of the expression in article 12.

Secondly, let $\frac{1+x}{1-x}=\frac{9}{8}$, and then $8+8x=9-9x$, and x now is equal to $\frac{1}{17}$. Consequently,

$$x = 0.05882352941$$
$$\frac{x^3}{3} = 0.00006784721$$
$$\frac{x^5}{5} = 0.00000014086$$
$$\frac{x^7}{7} = 0.00000000035$$

Sum of the above terms - - - - 0.05889151783
2

Log. of $\frac{(2n+1)^2}{(2n+1)^2-1}$ or $\frac{9}{8}$ - - - - - - 0.11778303566

which answers to the second part of the expression in article 12. Consequently the hyperbolic logarithm of the number 2 is 0.57536414488 + 0.11778303566 = 0.69314718054.

The hyperbolic logarithm of 2 being thus found, that of 4, 8, 16, and all the other powers of 2 may be obtained by multiplying the logarithm of 2, by 2, 3, 4, &c. respectively, as is evident from the properties of logarithms stated in article 6. Thus by multiplication, the hyperbolic logarithm of

4 = 1.38629436108
of 8 = 2.07944154162
&c.

From the above, the logarithm of 3 may easily be obtained. For $4 \div \frac{4}{3} = 4 \times \frac{3}{4} = 3$; and therefore as the logarithm of $\frac{4}{3}$ was determined above, and also the logarithm of 4,

From the logarithm of 4, viz. - - 1.38629436108,
Subtract the logarithm of $\frac{4}{3}$, viz. - - 0.28768207244,

And the logarithm of 3 is - - - 1.09861228864.

Having found the logarithms of 2 and 3, we can find, by addition only, the logarithms of all the powers of 2 and 3, and also the logarithms of all the numbers which can be produced by multiplication from 2 and 3. Thus,

To the logarithm of 3, viz. - - - - 1.09861228864
Add the logarithm of 2, viz. - - - - 0.69314718054

And the sum is the logarithm of 6 - 1.79175946918

To this last found add the logarithm of 2, and the sum 2.48490664972 is the logarithm of 12.

The hyperbolic logarithms of other prime numbers may be more readily calculated by attending to the following article.

ART. 13. Let a, b, c, be three numbers in arithmetical progression, whose common difference is 1. Let b be the prime number, whose logarithm is sought, and a and c even numbers whose logarithms are known, or easily obtained from others already computed. Then, a being the least of the three, and the common difference being 1, $a=b-1$, and $c=b+1$. Consequently $a\times c=(b-1)\ (b+1)\ =b^2-1$, and $ac+1=b^2$; and therefore $\frac{b^2}{ac}=\frac{ac+1}{ac}$. This is a general expression for the fraction which it will be proper to put $=\frac{1+x}{1-x}$, that the series expressing the hyperbolic logarithm may converge quickly. For as $\frac{1+x}{1-x}=\frac{ac+1}{ac}$, $ac+acx=ac+1-acx-x$, and therefore $2acx+x=1$, and $x=\frac{1}{2ac+1}$.

Example 2. Required the hyperbolic logarithm of 5.

Here $a=4$, $c=6$, and $x=\frac{1}{2ac+1}=\frac{1}{49}$. Consequently,

	$x=0.0204081632$
	$\frac{x^3}{3}=0.0000028332$
	$\frac{x^5}{5}=0.0000000007$
Sum of the above terms - - -	0.0204109971
	2
Log. of $\frac{1+x}{1-x}$ or $\frac{25}{24}$ - - - -	0.0408219942

But $\frac{25}{24}\times 8\times 3=25$, and the addition of logarithms answers to the multiplication of the natural numbers to which they belong. Consequently,

To the log. of $\frac{25}{24}$ - - - - - - -	0.0408219942
Add the log. of 8 - - - - - - -	2.0794415422
And also the log. of 3 - - - - -	1.0986122890
And the sum is the log. of 25 -	3.2188758254

The half of this, viz. 1.6094379127, is the hyperbolic logarithm of 5; for $5\times5=25$.

Example 3. Required the hyperbolic logarithm of 7.

Here $a=6$, $c=8$, and $x=\frac{1}{2ac+1}=\frac{1}{97}$, and $\frac{ac+1}{ac}=\frac{1+x}{1-x}=\frac{49}{48}$. Consequently,

$$x=0.01030927835$$
$$\frac{x^3}{3}=0.00000036522$$
$$\frac{x^5}{5}=0.00000000002$$

Sum of the terms - - - - - - -	0.01030964359
	2
Log. of $\frac{49}{48}$ - - - - - - - - - - -	0.02061928718
To which add log. of 8 - - -	2.07944154162
And also log. of 6 - - - - - -	1.79175946918
The sum is the log. of 49 - -	3.89182029798

For $\frac{49}{48}\times6\times8=49$. Consequently the half of this, viz. 1.94591014899, is the hyperbolic logarithm of 7; for $7\times7=49$.

If the reader perfectly understand the investigations and examples already given, he will find no difficulty in calculating the hyperbolic logarithms of higher prime numbers. It will only be necessary for him, in order to guard against any embarrassment, to compute them as they advance in succession above those already mentioned. Thus, after what has been done, it would be proper, first of all, to calculate the hyperbolic logarithm of 11, then that of 13, &c.

Proceeding according to the method already explained, it will be found that

The hyperbolic logarithm of 11 is 2.397895273016
of 13 is 2.564999357538
of 17 is 2.833213344878
of 19 is 2.944438979941

Logarithms were invented by Lord Neper, Baron of Merchiston in Scotland. In the year 1614, he published at Edinburgh a small quarto, containing tables of them, of the hyperbolic kind, and an account of their construction and use. The discovery afforded the highest pleasure to mathematicians, as they were fully sensible of the very great utility of logarithms; but it was soon suggested by Mr. Briggs, afterwards Savilian Professor of Geometry in Oxford, that another kind of logarithms would be more convenient, for general purposes, than the hyperbolic. That one set of logarithms may be obtained from another will readily appear from the following article.

ART. 14. It appears from articles 1, 3, and 7, that if all the logarithms of the geometrical progression 1, $(1+a)^1$, $(1+a)^2$, $(1+a)^3$, $(1+a)^4$, $(1+a)^5$, &c. be multiplied or divided by any given number, the products and also the quotients will likewise be logarithms, for their addition or subtraction will answer to the multiplication or division of the terms in the geometrical progression to which they belong. The same terms in the geometrical progression may therefore be represented with different sets or kinds of logarithms in the following manner:

$$1,\ (1+a)^1,\ (1+a)^2,\ (1+a)^3,\ (1+a)^4,\ (1+a)^5,\ (1+a)^6,\ \&c.$$
$$1,\ (1+a)^l,\ (1+a)^{2l},\ (1+a)^{3l},\ (1+a)^{4l},\ 1+a)^{5l},\ (1+a)^{6l},\ \&c.$$
$$1,\ (1+a)^{\frac{1}{m}},\ (1+a)^{\frac{2}{m}},\ (1+a)^{\frac{3}{m}},\ (1+a)^{\frac{4}{m}},\ (1+a)^{\frac{5}{m}},\ (1+a)^{\frac{6}{m}},\ \&c.$$

In these expressions l and m denote any numbers, whole or fractional; and the positive value of the term in the geometrical progression, under the same number in the index, is understood to be the same in each of the three series. Thus if $(1+a)^4$ be equal to 7, then $(1+a)^{4l}$ is equal to 7, as is also $(1+a)^{\frac{4}{m}}$. If $(1+a)^6$ be equal to 10, then $(1+a)^{6l}$ is equal to 10, as is also $(1+a)^{\frac{6l}{m}}$, &c. If therefore l, $2l$, $3l$, &c. be hyperbolic logarithms, calculated by the methods already explained, the logarithms expressed by $\frac{1}{m}$, $\frac{2}{m}$, $\frac{3}{m}$, &c. may be derived from them; for the hyperbolic logarithm of any given number is to the logarithm in the last-mentioned set, of the same number, in a given ratio. Thus $4l : \frac{4}{m} :: 1 : \frac{4}{4lm} = \frac{1}{lm}$; also $6l : \frac{6}{m} :: 1 : \frac{6}{6lm} = \frac{1}{lm}$, &c.

ART. 15. Mr. Briggs's suggestion, above alluded to, was that 1 should be put for the logarithm of 10, and consequently 2 for the logarithm of 100, 3 for the logarithm of 1000, &c.

This proposed alteration appears to have met with the full approbation of Lord Neper; and Mr. Briggs afterwards, with incredible labour and perseverance, calculated extensive tables of logarithms of this new kind, which are now called common logarithms. If the expeditious methods for calculating hyperbolic logarithms explained in the foregoing articles *, had been known to Mr. Briggs, his trouble would have been comparatively trivial with that which he must have experienced in his operations.

Art. 16. It has been already determined that the hyperbolic logarithm of 5 is 1.6094379127, and that of 2 is 0.69314718054, and therefore the sum of these logarithms, viz. 2.30258509324 is the hyperbolic logarithm of 10. If, therefore, for the sake of illustration, as in article 14, we suppose $(1+a)^6 = 10$, and allow, in addition to the hypothesis there formed, that $\frac{1}{m}$, $\frac{2}{m}$, $\frac{3}{m}$, $\frac{4}{m}$, &c. denote common logarithms, then $6l = 2.30258509324$, and $\frac{6}{m} = 1$; and the ratio for reducing the hyperbolic logarithm of any number to the common logarithm of the same number, is that of 2.30258509324 to 1. Thus, in order to find the common logarithm of 2, 2.30258509324 : 1 : : 0.69314718054 : 0.3010299956, the common logarithm of 2. The common logarithms of 10 and 2 being known, we obtain the common logarithm of 5, by subtracting the common logarithm of 2 from 1, the common logarithm of 10; for 10 being divided by 2, the quotient is 5. Hence the common logarithm of 5 is 0.6989700044. Again, to find the common logarithm of 3, 2.30258509324 : 1 : : 1.09861228864 : 0.4771212546 the common logarithm of 3.

Art. 17. As the constant ratio, for the reduction of hyperbolic to common logarithms, is that of 2.30258509324 to 1, it is evident that the reduction may be made by multiplying the hyperbolic logarithm, of the number whose common logarithm is sought, by $\frac{1}{2.30258509324} = .4342944818$.

Thus 1.94591014899 the hyperbolic logarithm of 7, being multiplied by .4342944818, the product, viz. .8450980378, &c. is the common logarithm of 7.

* Some of the principal particulars of the foregoing methods were discovered by the celebrated Thomas Simpson. See also Mr. Hellin's Mathematical Essays published in 1788.

The common logarithms of prime numbers being derived from the hyperbolic, the common logarithms of other numbers may be obtained from those so derived, merely by addition or subtraction. For addition of logarithms, in any set or kind, answers to the multiplication of the natural numbers to which they belong, and consequently subtraction of logarithms to the division of the natural numbers. Hyperbolic logarithms are not only useful as a medium through which common logarithms may be obtained: they are absolutely necessary for finding the fluents of many fluxional expressions of the highest importance.

It is deemed unnecessary in this place to show the utility of logarithms by examples. Being once calculated and arranged in tables, not only for common numbers, but also for natural sines, tangents, and secants, it is manifest that a computer may save himself much time, and a great deal of labour, by means of their assistance; as otherwise multiplications and divisions of high numbers, or of decimals to a considerable number of places, would enter into his inquiries.

The writer of the foregoing articles now considers the design with which he set out as completed. He has endeavoured to explain, with perspicuity, the first principles of logarithms, and their relations to one another when of different sets or kinds; and he has laid before the young mathematical student the most improved and expeditious methods by which they may be calculated. If the reader should be desirous of farther information on the subject, he may meet with full gratification by the perusal of the history of discoveries and writings relating to logarithms, prefixed to Dr. Hutton's Mathematical Tables. He will also find the Tables of Logarithms, contained in that volume, the most useful for calculation, if in his computations he does not go beyond degrees and minutes: if he aims at a higher degree of accuracy, he will have recourse to Taylor's Tables, in which the Logarithmic Sines and Tangents are calculated to every second of the Quadrant.

A. ROBERTSON,
SAVILIAN PROFESSOR OF ASTRONOMY,
OXFORD.

SPHERICAL TRIGONOMETRY.

DEFINITIONS.

I. The pole of a circle of the sphere is a point in the superficies of the sphere, from which all straight lines drawn to the circumference of the circle are equal.

II. A great circle of the sphere is any whose plane passes through the centre of the sphere, and whose centre therefore is the same with that of the sphere.

III. A spherical triangle is a figure upon the superficies of a sphere comprehended by three arches of three great circles, each of which is less than a semicircle.

IV. A spherical angle is that which on the superficies of a sphere is contained by two arches of great circles, and is the same with the inclination of the planes of these great circles.

PROPOSITION I.

Great circles bisect one another.

As they have a common centre, their common section will be a diameter of each which will bisect them.

PROPOSITION II.

The arch of a great circle betwixt the pole and the circumference of another is a quadrant.

Let ABC be a great circle, and D its pole: if a great circle DC pass through D, and meet ABC in C, the arch DC will be a quadrant.

Fig. 25.

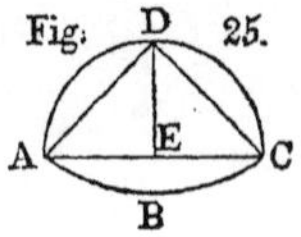

Let the great circle CD meet ABC again in A, and let AC be the common section of the great circles, which will pass through E, the centre of the sphere: join DE, DA, DC: by Def. 1. DA, DC are equal, and AE, EC are also equal, and DE is common; therefore* the angles DEA, DEC are * 8. 1.

equal; wherefore the arches DA, DC are equal, and consequently each of them is a quadrant. Q. E. D.

PROPOSITION III.

If a great circle be described meeting two great circles AB, AC *passing through its pole* A *in* B, C, *the angle at the centre of the sphere upon the circumference* BC, *is the same with the spherical angle* BAC, *and the arch* BC *is called the measure of the spherical angle* BAC.

Let the planes of the great circles AB, AC, intersect one another in the straight line AD, passing through D their common centre: join DB, DC.

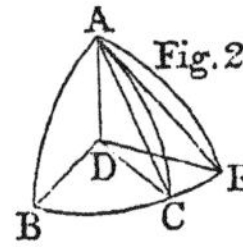

Fig. 26.

Since A is the pole of BC; AB, AC will be quadrants, and the angles ADB, ADC right angles; therefore * the angle CDB is the inclination of the planes of the circles AB, AC; that is (def. 4.), the spherical angle BAC. Q. E. D.

*6 Def. 11.

COROLLARY. If through the point A, two quadrants AB, AC be drawn, the point A will be the pole of the great circle BC, passing through their extremities B, C.

Join AC, and draw AE a straight line to any other point E, in BC; join DE: since AC, AB are quadrants, the angles ADB, ADC are right angles, and AD will be perpendicular to the plane of BC: therefore the angle ADE is a right angle, and AD, DC are equal to AD, DE, each to each; therefore AE, AC are equal, and A is the pole of BC, by Def. 1. Q. E. D.

PROPOSITION IV.

In isosceles spherical triangles, the angles at the base are equal.

Let ABC be an isosceles triangle, and AC, CB the equal sides; the angles BAC, ABC, at the base AB, are equal.

Let D be the centre of the sphere, and join DA, DB, DC; in DA take any point E, from which draw, in the plane ADC, the straight line EF at right angles to ED, meeting CD in F, and draw in the plane ADB, EG at right angles to the same ED; therefore the rectilineal angle FEG * is the inclination of the planes ADC, ADB, and therefore is the same with the

*6 Def. 11.

spherical angle BAC: from F draw FH perpendicular to DB, and from H draw, in the plane ADB, the straight line HG at right angles to HD, meeting EG in G, and join GF. Because DE is at right angles to EF and EG, it is perpendicular to the plane FEG *, and therefore the plane FEG is perpendicular to the plane ADB, in which DE * is: in the same manner, the plane FHG is perpendicular to the plane ADB; and therefore GF, the common section of the planes FEG, FHG, is perpendicular to the plane ADB *; and because the angle FHG is the inclination of the planes BDC, BDA, it is the same with the spherical angle ABC; and the sides AC, CB of the spherical triangle being equal, the angles EDF, HDF, which stand upon them at the centre of the sphere, are equal; and in the triangles EDF, HDF, the side DF is common, and the angles DEF, DHF are right angles; therefore EF, FH are equal: and in the triangles FEG, FHG the side GF is common, and the sides EG, GH * will be equal; and therefore the angle FEG is equal to FHG *; that is, the spherical angle BAC is equal to the spherical angle ABC.

Fig. 27.

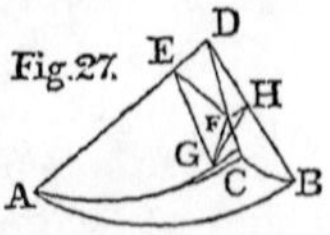

* 4. 11.
* 18. 11.
* 19. 11.
* 47. 1.
* 8. 1.

PROPOSITION V.

If, in a spherical triangle ABC, *two of the angles* BAC, ABC *be equal, the sides* BC, AC, *opposite to them are equal.* See fig. 27.

Read the construction and demonstration of the preceding proposition, unto the words, "and the sides AC, CB", &c. and the rest of the demonstration will be as follows, viz.

And the spherical angles BAC, ABC being equal, the rectilineal angles FEG, FHG, which are the same with them, are equal; and in the triangles FGE, FGH, the angles at G are right angles, and the side FG opposite to two of the equal angles is common; therefore * EF is equal to FH: and in the right angled triangles DEF, DHF, the side DF is common; wherefore * ED is equal to DH, and the angles EDF, HDF are therefore equal *, and consequently the sides AC, BC of the spherical triangle are equal.

* 26. 1.
* 47. 1.
* 4. 1.

PROPOSITION VI.

Any two sides of a spherical triangle are greater than the third.

Let ABC be a spherical triangle: any two sides AB, BC will be greater than the other side AC.

Let D be the centre of the sphere: join DA, DB, DC.

Fig. 28.

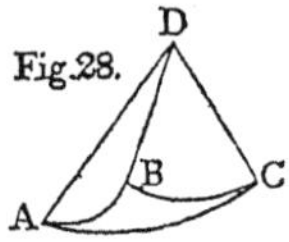

The solid angle at D is contained by three
* 20. 11. plane angles, ADB, ADC, BDC; and * any two of them ADB, BDC, are greater than the third ADC; that is, any two sides AB, BC of the spherical triangle ABC, are greater than the third AC.

PROPOSITION VII.

The three sides of a spherical triangle are less than a circle.

See fig. 28. Let ABC be a spherical triangle as before: the three sides AB, BC, AC are less than a circle.

Let D be the centre of the sphere: the solid angle at D is contained by three plane angles BDA, BDC, ADC, which to-
* 21. 11. gether * are less than four right angles; therefore the sides AB, BC, AC together, will be less than four quadrants, that is, less than a circle.

PROPOSITION VIII.

In a spherical triangle, the greater angle is opposite to the greater side; and conversely.

Let ABC be a spherical triangle: the greater angle A is opposite to the greater side BC.

Fig. 29.

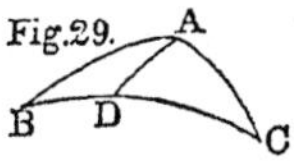

Let the angle BAD be made equal to the
* 5. of this. angle B, and then BD, DA will be equal *, and therefore AD, DC are equal to BC; but
* 6. of this. AD, DC are greater than AC *, therefore BC is greater than AC, that is, the greater angle A is opposite to the greater side BC. The converse is demonstrated as Prop. 19. 1. El. Q. E. D.

PROPOSITION IX.

In any spherical triangle ABC, if the sum of the sides AB, BC, be greater, equal, or less than a semicircle, the internal angle at the base AC, will be greater, equal, or less than the external and opposite BCD; and therefore the sum of the angles A and ACB will be greater, equal, or less than two right angles.

Let AC, AB produced meet in D.

1. If AB, BC be equal to a semicircle, that is, to AD; BC, BD will be equal, that is *, the angle D, or the angle A, will be equal to the angle BCD. * 4. of this.

2. If AB, BC together be greater than a semicircle, that is, greater than ABD, BC will be greater than BD; and therefore *, the angle D, that is, the angle A, is greater than the angle BCD. * 8. of this.

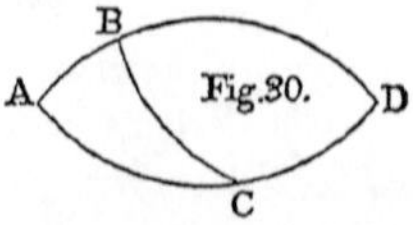

Fig. 30.

3. In the same manner it is shewn, that if AB, BC together be less than a semicircle, the angle A is less than the angle BCD. And since the angles BCD, BCA are equal to two right angles, if the angle A be greater than BCD, A and ACB together will be greater than two right angles. If A be equal to BCD, A and ACB together will be equal to two right angles; and if A be less than BCD, A and ACB will be less than two right angles. Q. E. D.

PROPOSITION X.

If the angular points A, B, C *of the spherical triangle* ABC, *be the poles of three great circles, these great circles by their intersections will form another triangle* FDE, *which is called supplemental to the former; that is, the sides* FD, DE, EF *are the supplements of the measures of the opposite angles* C, B, A *of the triangle* ABC, *and the measures of the angles* F, D, E *of the triangle* FDE, *will be the supplements of the sides* AC, BC, BA *in the triangle* ABC.

Let AB produced meet DE, EF, in G, M; and AC meet FD, FE, in K, L; and BC meet FD, DE, in N, H.

Since A is the pole of FE, and the circle AC passes through A, EF will pass through the pole of AC *; and since AC passes through C, the pole of FD, FD will pass through the pole of AC; therefore the pole of AC is in the point F, in which the arches DF, EF intersect each other. In the same manner D is the pole of BC, and E the pole of AB. * 2. of this.

Fig. 31.
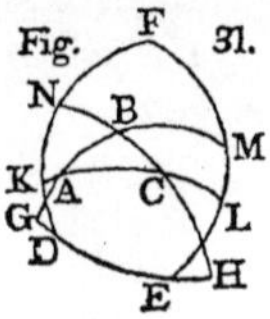

And since F, E are the poles of AL, AM, FL and EM are quadrants, and FL, EM together, that is, FE and ML together, are equal to a semicircle. But since A is the pole of ML, ML is the measure of the angle BAC, consequently FE is the supplement of the measure of the angle BAC. In the same manner, ED, DF are the supplements of the measures of the angles ABC, BCA.

Since likewise CN, BH are quadrants, CN, BH together, that is, NH, BC together, are equal to a semicircle; and since D is the pole of NH, NH is the measure of the angle FDE, therefore the measure of the angle FDE is the supplement of the side BC. In the same manner, it is shewn that the measures of the angles DEF, EFD are the supplements of the sides AB, AC, in the triangle ABC. *Q. E. D.*

PROPOSITION XI.

The three angles of a spherical triangle, are greater than two right angles, and less than six right angles.

See fig. 31. The measures of the angles A, B, C in the triangle ABC, together with the three sides of the supplemental triangle DEF, are * equal to three semicircles; but the three sides of the triangle FDE, are * less than two semicircles; therefore the measures of the angles A, B, C, are greater than a semicircle; and hence the angles A, B, C are greater than two right angles.

* 10. of this.
* 7. of this.

All the external and internal angles of any triangle are equal to six right angles; therefore all the internal angles are less than six right angles.

PROPOSITION XII.

If from any point C, *which is not the pole of the great circle* ABD, *there be drawn arches of great circles* CA, CD, CE, CF, *&c. the greatest of these is* CA, *which passes through* H, *the pole of* ABD, *and* CB, *the remainder of* ACB *is the least, and of any others* CD, CE, CF, *&c.* CD, *which is nearer to* CA, *is greater than* CE, *which is more remote.*

Let the common section of the planes of the great circles ACB, ADB, be AB; and from C, draw CG perpendicular to AB, which will also be perpendicular to the plane ADB *; join GD, GE, GF, CD, CE, CF, CA, CB.

* 4 Def. 11.

Of all the straight lines drawn from G to the circumference ADB, GA is the greatest, and GB the least*; and GD, which is nearer to GA, is greater than GE, which is more remote. The triangles CGA, CGD are right-angled at G, and they have the common side CG; therefore the squares of CG, GA together, that is, the square of CA is greater than the squares of CG, GD together, that is, the square of CD: and CA is greater than CD, and therefore the arch CA is greater than CD. In the same manner, since GD is greater than GE, and GE than GF, &c. it is shewn that CD is greater than CE, and CE than CF, &c., and consequently, the arch CD greater than the arch CE, and the arch CE greater than the arch CF, &c. And since GA is the greatest, and GB the least of all the straight lines drawn from G to the circumference ADB, it is manifest that CA is the greatest, and CB the least of all the straight lines drawn from C to the circumference: and therefore the arch CA is the greatest, and CB the least of all the circles drawn through C, meeting ADB. Q. E. D.

* 7. 3.

Fig. 32.

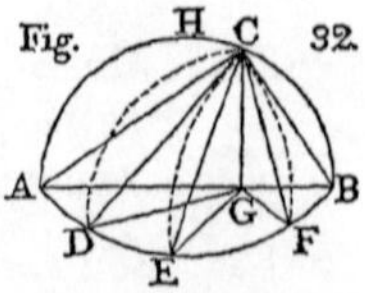

PROPOSITION XIII.

In a right-angled spherical triangle, the sides are of the same affection with the opposite angles; that is, if the sides be greater or less than quadrants, the opposite angles will be greater or less than right angles.

Let ABC be a spherical triangle, right-angled at A: any side AB, will be of the same affection with the opposite angle ACB.

Case 1. Let AB be less than a quadrant, let AE be a quadrant, and let EC be a great circle passing through E, C. Since A is a right angle, and AE a quadrant, E is the pole of the great circle AC, and ECA a right angle; but ECA is greater than BCA, therefore BCA is less than a right angle. Q. E. D.

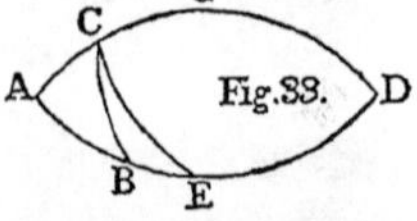

Fig. 33.

Case 2. Let AB be greater than a quadrant, make AE a quadrant, and let a great circle pass through C, E, ECA is a right angle as before, and BCA is greater than ECA, that is, greater than a right angle. Q. E. D.

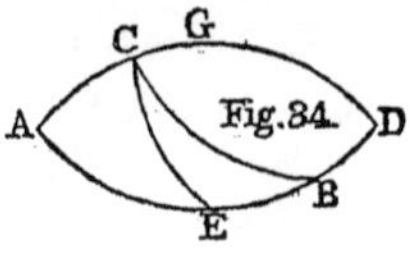

Fig. 34.

PROPOSITION XIV.

If the two sides of a right-angled spherical triangle be of the same affection, the hypothenuse will be less than a quadrant; and if they be of different affection, the hypothenuse will be greater than a quadrant.

Let ABC be a right-angled spherical triangle: if the two sides AB, AC be of the same or of different affection, the hypothenuse BC will be less or greater than a quadrant.

See fig. 33. Case 1. Let AB, AC be each less than a quadrant. Let AE, AG be quadrants; G will be the pole of AB, and E the pole of AC, and EC a quadrant; but by Prop. 12, CE is greater than CB, since CB is farther off from CGD than CE. In the same manner it is shewn, that CB in the triangle CBD, where the two sides CD, BD are each greater than a quadrant, is less than CE, that is, less than a quadrant. Q. E. D.

See fig. 34. Case 2. Let AC be less, and AB greater than a quadrant; then the hypothenuse BC will be greater than a quadrant: for let AE be a quadrant, then E is the pole of AC, and EC will be a quadrant. But CB is greater than CE by Prop. 12, since AC passes through the pole of ABD. Q. E. D.

PROPOSITION XV.

If the hypothenuse of a right-angled spherical triangle be greater or less than a quadrant, the sides will be of different or the same affection.

This is the converse of the preceding, and demonstrated in the same manner.

PROPOSITION XVI.

In any spherical triangle ABC, *if the perpendicular* AD *from* A, *on the base* BC, *fall within the triangle, the angles* B *and* C *at the base will be of the same affection; and if the perpendicular fall without the triangle, the angles* B *and* C *will be of different affection.*

* 13. of this. 1. Let AD fall within the triangle; then *, since ADB, ADC are right-angled spherical triangles, the angles B, C must each be of the same affection as AD.

Fig. 35.

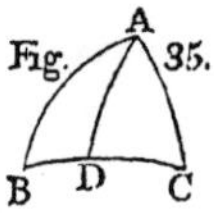

* 13. of this. 2. Let AD fall without the triangle, then * the angle B is

of the same affection as AD; and by the same the angle ACD is of the same affection as AD; therefore the angle ACB and AD are of different affection, and the angles B and ACB of different affection.

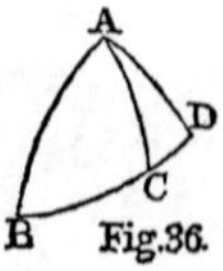

Fig.36.

COR. Hence if the angles B and C be of the same affection, the perpendicular will fall within the base; for, if it did not *, B and C would be of different affection. And if the angles B and C be of opposite affection, the perpendicular will fall without the triangle; for, if it did not *, the angles B and C would be of the same affection, contrary to the supposition.

* 16. of this.
* 16. of this.

PROPOSITION XVII.

In right-angled spherical triangles, the sine of either of the sides about the right angle, is to the radius of the sphere, as the tangent of the remaining side is to the tangent of the angle opposite to that side.

Let ABC be a triangle, having the right angle at A; and let AB be either of the sides: the sine of the side AB will be to the radius, as the tangent of the other side AC to the tangent of the angle ABC, opposite to AC.

Let D be the centre of the sphere; join AD, BD, CD, and let AE be drawn perpendicular to BD, which therefore will be the sine of the arch AB; and from the point E, let there be drawn in the plane BDC, the straight line EF at right angles to BD, meeting DC in F, and let AF be joined. Since therefore the straight line DE is at right angles to both EA and EF, it will also be at right angles to the plane AEF *; wherefore the plane ABD, which passes through DE, is perpendicular to the plane AEF *, and the plane AEF perpendicular to ABD: the plane ACD or AFD is also perpendicular to the same ABD: therefore the common section, viz. the straight line AF, is at right angles to the plane ABD *: and FAE, FAD are right angles *; therefore AF is the tangent of the arch AC, and in the rectilineal triangle AEF having a right angle at A, AE will be to the radius as AF to the tangent of the angle AEF *; but AE is the sine of the arch AB, and AF the tangent of the arch AC, and the

Fig.37.

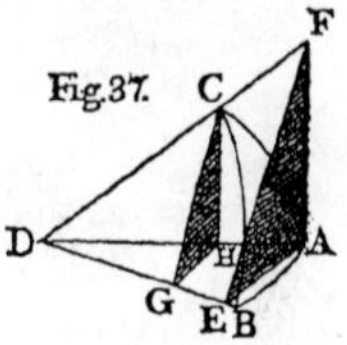

* 4. 11.
* 18. 11.
* 19. 11.
* 3 Def. 11.
* 1 Pl. Trig.

* 6 Def. 11. angle AEF is the inclination of the planes CBD, ABD *, or the spherical angle ABC: therefore the sine of the arch AB is to the radius, as the tangent of the arch AC to the tangent of the opposite angle ABC.

Cor. 1. If therefore of the two sides, and an angle opposite to one of them, any two be given, the third will also be given.

Cor. 2. And since by this proposition the sine of the side AB is to the radius, as the tangent of the other side AC to the tangent of the angle ABC opposite to that side; and as the radius is to the cotangent of the angle ABC, so is the tangent of the same angle ABC to the radius *, by equality, the sine of the side AB is to the cotangent of the angle ABC adjacent to it, as the tangent of the other side AC to the radius.

* Cor. 2 Def. Pl. Trig.

PROPOSITION XVIII.

In right-angled spherical triangles, the sine of the hypothenuse is to the radius, as the sine of either side is to the sine of the angle opposite to that side.

See fig. 37. Let the triangle ABC be right-angled at A, and let AC be either of the sides: the sine of the hypothenuse BC will be to the radius, as the sine of the arch AC is to the sine of the angle ABC.

Let D be the centre of the sphere, and let CG be drawn perpendicular to DB, which will therefore be the sine of the hypothenuse BC; and from the point G, let there be drawn in the plane ABD, the straight line GH perpendicular to DB, and let CH be joined; CH will be at right angles to the plane ABD, as was shewn in the preceding proposition of the straight line FA: wherefore CHD, CHG are right angles, and CH is the sine of the arch AC; and in the triangle CHG, having the right angle CHG, CG is to the radius as CH to the sine of the angle
* 1 Pl. Trig. CGH *: but since CG, HG are at right angles to DGB, which is the common section of the planes CBD, ABD, the angle
* 6 Def. 11. CGH will be equal to the inclination of these planes *, that is, to the spherical angle ABC. The sine therefore of the hypothenuse CB, is to the radius as the sine of the side AC is to the sine of the opposite angle ABC. Q. E. D.

Cor. Of these three, viz. the hypothenuse, a side, and the angle opposite to that side, any two being given, the third also is given by Prop. 2.

PROPOSITION XIX.

In right-angled spherical triangles, the cosine of the hypothenuse is to the radius, as the cotangent of either of the angles is to the tangent of the remaining angle.

Let ABC be a spherical triangle, having a right angle at A: the cosine of the hypothenuse BC will be to the radius, as the cotangent of the angle ABC to the tangent of the angle ACB.

Describe the circle DE, of which B is the pole, and let it meet AC in F, and the circle BC in E; and since the circle BD passes through the pole B of the circle DF, DF will also pass through the pole of BD. (13. 15. 1. Theod. Sph.) And since AC is perpendicular to BD, AC will also pass through the pole of BD; wherefore the pole of the circle BD will be found in the point where the circles AC, DE meet, that is, in the point F: the arches FA, FD are therefore quadrants, and likewise the arches BD, BE: in the triangle CEF, right-angled at the point E, CE is the complement of the hypothenuse BC of the triangle ABC, EF is the complement of the arch ED, which is the measure of the angle ABC, and FC the hypothenuse of the triangle CEF, is the complement of AC, and the arch AD, which is the measure of the angle CFE, is the complement of AB.

Fig.38.

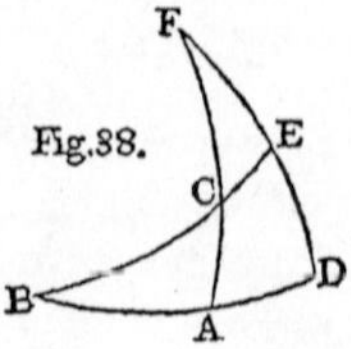

But * in the triangle CEF, the sine of the side CE is to the radius as the tangent of the other side is to the tangent of the angle ECF opposite to it, that is, in the triangle ABC, the cosine of the hypothenuse BC is to the radius, as the cotangent of the angle ABC is to the tangent of the angle ACB. Q. E. D. * 17. of this.

Cor. 1. Of these three, viz. the hypothenuse and the two angles, any two being given, the third will also be given.

Cor. 2. And since by this proposition the cosine of the hypothenuse BC is to the radius, as the cotangent of the angle ABC to the tangent of the angle ACB; but as the radius is to the cotangent of the angle ACB, so is the tangent of the same to the radius *; and, ex æquo, the cosine of the hypothenuse BC is to the cotangent of the angle ACB, as the cotangent of the angle ABC to the radius. * Cor. 2 Def. Pl. Trig.

PROPOSITION XX.

In right-angled spherical triangles, the cosine of an angle is to the radius, as the tangent of the side adjacent to that angle, is to the tangent of the hypothenuse.

See fig. 38. The same construction remaining; in the triangle CEF *,
* 17. of this. the sine of the side EF is to the radius, as the tangent of the other side CE is to the tangent of the angle CFE opposite to it; that is, in the triangle ABC, the cosine of the angle ABC is to the radius as (the cotangent of the hypothenuse BC to the cotangent of the side AB, adjacent to ABC, or as) the tangent of the side AB to the tangent of the hypothenuse, since the tangents of two arches are reciprocally proportional to their cotangents *.

* Cor. 1 Def. Pl. Trig.

COR. And since by this proposition, the cosine of the angle ABC is to the radius, as the tangent of the side AB is to the tangent of the hypothenuse BC; and as the radius is to the cotangent of BC, so is the tangent of BC to the radius; by equality, the cosine of the angle ABC will be to the cotangent of the hypothenuse BC, as the tangent of the side AB, adjacent to the angle ABC, to the radius.

PROPOSITION XXI.

In right-angled spherical triangles, the cosine of either of the sides is to the radius, as the cosine of the hypothenuse is to the cosine of the other side.

See fig. 38. The same construction remaining; in the triangle CEF, the sine of the hypothenuse CF is to the radius, as the sine of the
* 18. of this. side CE to the sine of the opposite angle CFE *, that is, in the triangle ABC, the cosine of the side CA is to the radius, as the cosine of the hypothenuse BC to the cosine of the other side BA. Q. E. D.

PROPOSITION XXII.

In right-angled spherical triangles, the cosine of either of the sides is to the radius, as the cosine of the angle opposite to that side is to the sine of the other angle.

See fig. 38. The same construction remaining; in the triangle CEF, the sine of the hypothenuse CF is to the radius, as the sine of the side EF is to the sine of the angle ECF opposite to it; that is, in the triangle ABC, the cosine of the side CA is to the radius, as the cosine of the angle ABC opposite to it, is to the sine of the other angle. Q. E. D.

OF THE CIRCULAR PARTS.

In any right-angled spherical triangle ABC, the complement of the hypothenuse, the complements of the angles, and the two sides, are called *the circular parts of the triangle,* as if it were following each other in a circular order, from whatever part we begin: thus, if we begin at the complement of the hypothenuse, and proceed towards the side BA, the parts following in order will be the complement of the hypothenuse, the complement of the angle B, the side BA, the side AC (for the right angle at A is not reckoned among the parts), and, lastly, the complement of the angle C. And thus, at whatever part we begin, if any three of these five be taken, they either will be all contiguous or adjacent, or one of them will not be contiguous to either of the other two: in the first case, the part which is between the other two is called the *Middle part,* and the other two are called *Adjacent extremes.* In the second case, the part which is not contiguous to either of the other two is called the *Middle part,* and the other two *Opposite extremes.* For example, if the three parts be the complement of the hypothenuse BC, the complement of the angle B, and the side BA; since these three are contiguous to each other, the complement of the angle B will be the middle part, and the complement of the hypothenuse BC and the side BA will be adjacent extremes: but if the complement of the hypothenuse BC and the sides BA, AC be taken, since the complement of the hypothenuse is not adjacent to either of the sides, viz. on account of the complements of the two angles B and C intervening between it and the sides, the complement of the hypothenuse BC will be the middle part, and the sides BA, AC opposite extremes. The most acute and ingenious Baron Naper, the inventor of Logarithms, contrived the two following rules concerning these parts, by means of which all the cases of right-angled spherical triangles are resolved with the greatest ease.

Fig. 39.

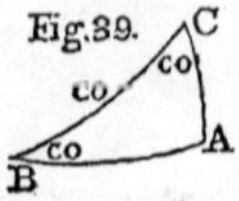

RULE I.

The rectangle contained by the radius and the sine of the middle part, is equal to the rectangle contained by the tangents of the adjacent part.

RULE II.

The rectangle contained by the radius and the sine of the middle part, is equal to the rectangle contained by the cosines of the opposite parts.

These rules are demonstrated in the following manner:

First, let either of the sides, as BA be the middle part, and therefore the complement of the angle B, and the side AC, will be adjacent extremes. And by Cor. 2. Prop. 17. of this, S, BA, is to the Co-T, B, as T, AC is to the radius, and therefore R$\times$S, BA$=$Co-T, B$\times$T, AC.

Fig.40.

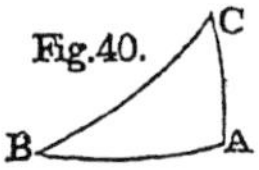

The same side BA being the middle part, the complement of the hypothenuse, and the complement of the angle C, are opposite extremes: and by Prop. 18. S, BC is to the radius, as S, BA to S, C; therefore R$\times$S, BA$=$S, BC$\times$S, C.

Secondly, let the complement of one of the angles, as B, be the middle part, and the complement of the hypothenuse, and the side BA will be adjacent extremes: and by Cor. Prop. 20. Co-S, B is to Co-T, BC as T, BA is to the radius, and therefore R$\times$Co-S, B$=$Co-T, BC$\times$T, BA.

Again, let the complement of the angle B be the middle part, and the complement of the angle C, and the side AC will be opposite extremes: and by Prop. 22. Co-S, AC is to the radius, as Co-S, B is to S, C; and therefore R$\times$Co-S, B$=$Co-S, AC$\times$S, C.

Thirdly, let the complement of the hypothenuse be the middle part, and the complements of the angles B, C, will be adjacent extremes: but by Cor. 2. Prop. 19. Co-S, BC is to Co-T, C as Co-T, B to the radius: therefore R$\times$Co-S, BC$=$Co-T, C$\times$Co-T, B.

Again, let the complement of the hypothenuse be the middle part, and the sides AB, AC will be opposite extremes: but by Prop. 21. Co-S, AC is to the radius, as Co-S, BC to Co-S, BA; therefore R$\times$Co-S, BC$=$Co-S, BA$\times$Co-S, AC. Q. E. D.

Solution of the Sixteen Cases of Right-angled Spherical Triangles.

GENERAL PROPOSITION.

In a right-angled Spherical triangle, of the three sides and three angles, any two being given, besides the right angle, the other three may be found.

In the following Tables the solutions are derived from the preceding propositions. It is obvious that the same solutions may be derived from Baron Naper's two rules above demonstrated, which, as they are easily remembered, are commonly used in practice.

See. fig. 40.

Case.	Given.	Sought.	
1	AC, C	B	R : CoS, AC : : S, C : CoS, B ; and B is of the same species with CA, by 22. and 13.
2	AC, B	C	CoS, AC : R : : CoS, B : S, C ; by 22.
3	B, C	AC	S, C : CoS, B : : R : CoS, AC; by 22. and AC is of the same species with B. 13.
4	BA, AC	BC	R : CoS, BA : : CoS, AC : CoS, BC. 21. And if both AB, AC be greater or less than a quadrant, BC will be less than a quadrant. But if they be of different affection, BC will be greater than a quadrant. 14.
5	BA, BC	AC	CoS, BA : R : : CoS, BC : CoS, AC. 21. And if BC be greater or less than a quadrant, BA, AC will be of different or the same affection. 15.
6	BA, AC	B	S, BA : R : : T, CA : T, B. 17. And B is of the same affection with AC. 13.

Case.	Given.	Sought.	
7	BA, B	AC	R : S, BA :: T, B : T, AC. 17. And AC is of the same affection with B. 13.
8	AC, B	BA	T, B : R :: T, CA : S, BA. 17.
9	BC, C	AC	R : CoS, C :: T, BC : T, CA. 20. If BC be less or greater than a quadrant, C and B will be of the same or different affection. 15. 13.
10	AC, C	BC	CoS, C : R :: T, AC : T, BC. 20. And BC is less or greater than a quadrant, according as C and A or C and B are of the same or different affections. 14. 15.
11	BC, CA	C	T, BC : R :: T, CA : CoS, C. 20. If BC be less or greater than a quadrant, CA and AB, and therefore CA and C are of the same or different affections. 15.
12	BC, B	AC	R : S, BC :: S, B : S, AC. 18. And AC is of the same affection with B.
13	AC, B	BC	S, B : S, AC :: R : S, BC. 18.
14	BC, AC	B	S, BC : R :: S, AC : S, B. 18. And B is of the same affection with AC.
15	B, C	BC	T, C : R :: CoT, B : CoS, BC. 19. And according as the angles B and C are of different or the same affection, BC will be greater or less than a quadrant. 14.
16	BC, C	B	R : CoS, BC :: T, C : CoT, B. 19. If BC be less or greater than a quadrant, C and B will be of the same or different affection. 15.

The second, eighth, and thirteenth cases, which are commonly called ambiguous, admit of two solutions: for in these it is not determined, whether the side or measure of the angle sought be greater or less than a quadrant.

PROPOSITION XXIII.

In spherical triangles, whether right-angled or oblique-angled, the sines of the sides are proportional to the sines of the angles opposite to them.

First, let ABC be a right-angled triangle, having a right angle at A; therefore, by Prop. 18. the sine of the hypothenuse BC is to the radius (or the sine of the right angle at A) as the sine of the side AC to the sine of the angle B. And in like manner, the sine of BC is to the sine of the angle A, as the sine of AB to the sine of the angle C; wherefore* the sine of the side AC is to the sine of the angle B, as the sine of AB to the sine of the angle C. (See fig. 40.) (* 11. 5.)

Secondly, let BCD be an oblique-angled triangle, the sine of either of the sides BC, will be to the sine of either of the other two CD, as the sine of the angle D opposite to BC is to the sine of the angle B opposite to the side CD. Through the point C, let there be drawn an arch of a great circle CA perpendicular upon BD; and in the right-angled triangle ABC* the sine of BC is to the radius as the sine of AC to the sine of the angle D, and in the triangle ADC (by 18. of this): and, by inversion, the radius is to the sine of DC as the sine of the angle D to the sine of AC: therefore, ex æquo perturbato, the sine of BC is to the sine of DC, as the sine of the angle D to the sine of the angle B. Q. E. D. (* 18. of this.)

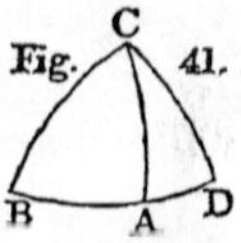

Fig. 41.

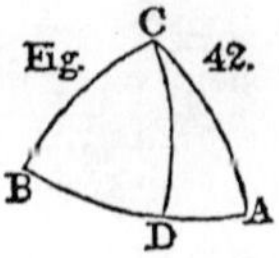

Fig. 42.

PROPOSITION XXIV.

In oblique-angled spherical triangles, having drawn a perpendicular arch from any of the angles upon the opposite side, the cosines of the angles at the base are proportional to the sines of the vertical angles.

Let BCD be a triangle, and the arch CA perpendicular to the base BD; the cosine of the angle B will be to the cosine of the angle D, as the sine of the angle BCA to the sine of the angle DCA. (See fig. 41. and 42.)

For by 22. the cosine of the angle B is to the sine of the angle BCA as (the cosine of the side AC is to the radius; that is, by Prop. 22. as) the cosine of the angle D to the sine of the angle DCA; and, by permutation, the cosine of the angle B is to the cosine of the angle D, as the sine of the angle BCA to the sine of the angle DCA. Q. E. D.

PROPOSITION XXV.

The same things remaining, the cosines of the sides BC, CD, *are proportional to the cosines of the bases* BA, AD.

See fig. 41. and 42.

For by 21. the cosine of BC is to the cosine of BA, as (the cosine of AC to the radius; that is, by 21. as) the cosine of CD is to the cosine of AD: wherefore, by permutation, the cosines of the sides BC, CD are proportional to the cosines of the bases BA, AD. Q. E. D.

PROPOSITION XXVI.

The same construction remaining, the sines of the bases BA, AD *are reciprocally proportional to the tangents of the angles* B *and* D *at the base.*

See fig. 41. and 42.

For by 17. the sine of BA is to the radius, as the tangent of AC to the tangent of the angle B; and by 17. and inversion, the radius is to the sine of AD, as the tangent of D to the tangent of AC: therefore, ex æquo perturbato, the sine of BA is to the sine of AD, as the tangent of D to the tangent of B. Q. E. D.

PROPOSITION XXVII.

The cosines of the vertical angles are reciprocally proportional to the tangents of the sides.

See fig. 41. and 42.

For by Prop. 20. the cosine of the angle BCA is to the radius as the tangent of CA is to the tangent of BC; and by the same Prop. 20. and by inversion, the radius is to the cosine of the angle DCA, as the tangent of DC to the tangent of CA: therefore, ex æquo perturbato, the cosine of the angle BCA is to the cosine of the angle DCA, as the tangent of DC is to the tangent of BC. Q. E. D.

LEMMA.

In right-angled plane triangles, the hypothenuse is to the radius, as the excess of the hypothenuse above either of the sides to the versed sine of the acute angle adjacent to that side, or as the sum of the hypothenuse and either of the sides to the versed sine of the exterior angle of the triangle.

Let the triangle ABC have a right angle at B: AC will be to the radius as the excess of AC above AB to the versed sine of the angle A adjacent to AB; or as the sum of AC, AB to the versed sine of the exterior angle CAK.

With any radius DE, let a circle be described; and from D the centre, let DF be drawn to the circumference, making the angle EDF equal to the angle BAC, and from the point F, let FG be drawn perpendicular to DE: let AH, AK be made equal to AC, and DL to DE: DG therefore is the cosine of the angle EDF or BAC, and GE its versed sine: and because of the equiangular triangles ACB, DFG, AC or AH is to DF or DE, as AB to DG: therefore* AC is to the radius DE, as BH to GE, the versed sine of the angle EDF or BAC: and since AH is to DE as AB to DG*, AH or AC will be to the radius DE as KB to LG, the versed sine of the angle LDF or KAC. Q. E. D.

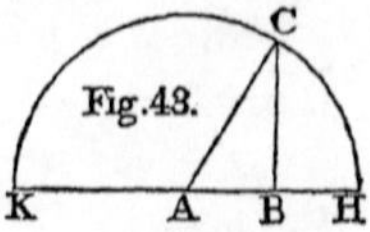

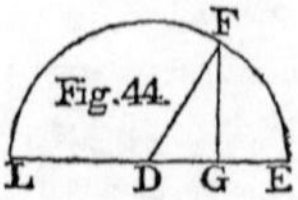

* 19. 5.

* 12. 5.

PROPOSITION XXVIII.

In any spherical triangle, the rectangle contained by the sines of two sides is to the square of the radius, as the excess of the versed sines of the third side or base, and the arch which is the excess of the sides, is to the versed sine of the angle opposite to the base.

Let ABC be a spherical triangle: the rectangle contained by the sines of AB, BC will be to the square of the radius as the excess of the versed sines of the base AC, and of the arch, which is the excess of AB, BC to the versed sine of the angle ABC opposite to the base.

Let D be the centre of the sphere, and let AD, BD, CB be joined, and let the sines AE, CF, CG of the arches AB, BC, AC be drawn; let the side BC be greater than BA, and let BH be made equal to BC; AH will therefore be the excess of the sides BC, BA; let HK be drawn perpendicular to AD, and since AG is the versed sine of the base AC, and AK the versed sine of the arch AH, KG is the excess of the versed sines of the base AC, and of the arch AH, which is the excess of the sides BC, BA: let GL likewise be drawn parallel to KH, and let it meet FH in L, let CL, DH be joined, and let AD, FH meet each other in M.

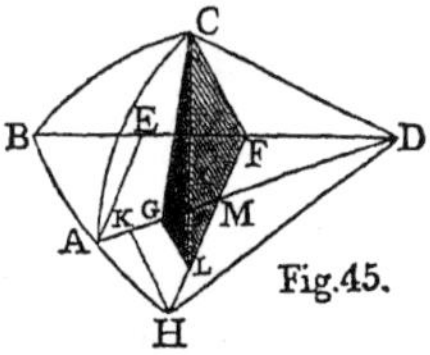

Fig.45.

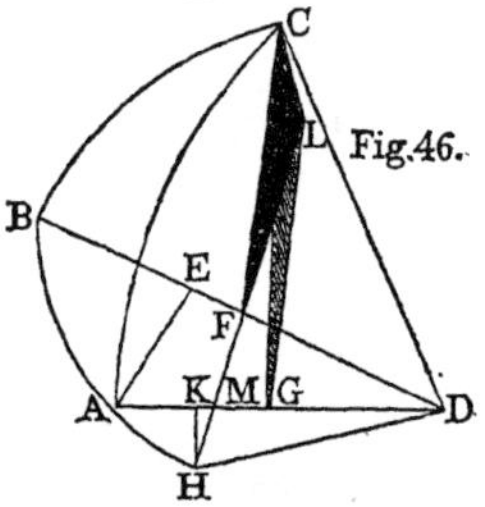

Fig.46.

Since therefore in the triangles CDF, HDF, the sides DC, DH are equal, DF is common, and the angle FDC equal to the angle FDH, because of the equal arches BC, BH, the base HF will be equal to the base FC, and the angle HFD equal to the right angle CFD: the
* 4. 11. straight line DF therefore* is at right angles to the plane CFH: wherefore the plane CFH is at right angles to the plane BDH,
* 18. 11. which passes through DF*. In like manner, since DG is at right angles to both GC and GL, DG will be perpendicular to the plane CGL: therefore the plane CGL is at right angles to the plane BDH, which passes through DG: and it was shewn that the plane CFH or CFL was perpendicular to the same plane BDH: therefore the common section of the planes CFL, CGL, viz. the straight line CL, is perpendicular to the plane
* 19. 11. BDA*, and therefore CLF is a right angle: in the triangle CFL having the right angle CLF, by the Lemma, CF is to the radius as LH, the excess, viz. of CF or FH above FL, is to the versed sine of the angle CFL; but the angle CFL is the inclination of the planes BCD, BAD, since FC, FL are drawn in them at right angles to the common section BF: the spherical angle ABC is therefore the same with the angle CFL; and therefore CF is to the radius as LH to the versed sine of the spherical angle ABC; and since the triangle AED is equiangular (to the triangle MFD, and therefore) to the triangle MGL, AE will be to the radius of the sphere AD (as MG to ML; that is, because of the parallels) as GK to LH: the ratio therefore which is compounded of the ratios of AE to the
* 23. 6. radius, and of CF to the same radius: that is*, the ratio of the rectangle contained by AE, CF, to the square of the radius, is the same with the ratio compounded of the ratio of

GK to LH, and the ratio of LH to the versed sine of the angle ABC; that is, the same with the ratio of GK to the versed sine of the angle ABC; therefore, the rectangle contained by AE, CF the sines of the sides AB, BC, is to the square of the radius as GK, the excess of the versed sines AG, AK, of the base AC, and the arch AH, which is the excess of the sides of the versed sine of the angle ABC opposite to the base AC. *Q. E. D.*

PROPOSITION XXIX.

The rectangle contained by half of the radius, and the excess of the versed sines of two arches, is equal to the rectangle contained by the sines of half the sum, and half the difference of the same arches.

Let AB, AC be any two arches, and let AD be made equal to AC the less; the arch DB therefore is the sum, and the arch CB the difference of AC, AB: through E the centre of the circle, let there be drawn a diameter DEF, and AE joined, and CD likewise perpendicular to it in G, and let BH be perpendicular to AE, and AH will be the versed sine of the arch AB, and AG the versed sine of AC, and HG the excess of these versed sines: let BD, BC, BF be joined, and FC also meeting BH in K.

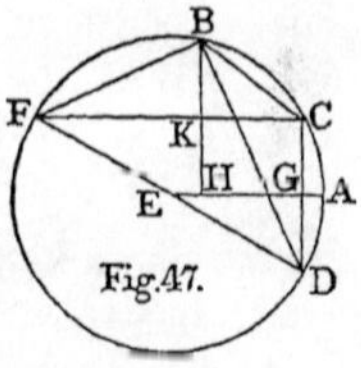

Fig.47.

Since, therefore, BH, CG are parallel, the alternate angles BKC, KCG will be equal; but KCG is in a semicircle, and therefore a right angle; therefore BKC is a right angle; and in the triangles DFB, CBK, the angles FDB, BCK in the same segment are equal, and FBD, BKC are right angles; the triangles DFB, CBK are therefore equiangular; wherefore DF is to DB, as BC to CK, or HG; and therefore the rectangle contained by the diameter DF and HG, is equal to that contained by DB, BC; wherefore the rectangle contained by a fourth part of the diameter and HG, is equal to that contained by the halves of DB, BC: but half the chord DB is the sine of half the arch DAB, that is, half the sum of the arches AB, AC; and half the chord of BC is the sine of half the arch BC, which is the difference of AB, AC. Whence the proposition is manifest.

PROPOSITION XXX.

The rectangle contained by half of the radius and the versed sine of any arch, is equal to the square of the sine of half the same arch.

Let AB be an arch of a circle, C its centre, and AC, CB, BA, being joined, let AB be bisected in D, and let CD be joined, which will be perpendicular to BA,

* 4. 1. and bisect it in E*, BE or AE therefore is the sine of the arch DB or AD, the half of AB: let BF be perpendicular to AC, and AF will be the versed sine of the arch BA; but, because of the similar triangles CAE, BAF, CA is to AE as AB, that is, twice AE to AF; and by halving the antecedents, half of the radius CA is to AE the sine of the arch AD, as the same AE to AF the versed

* 16. 6. sine of the arch AB. Wherefore*, the proposition is manifest.

Fig.48.

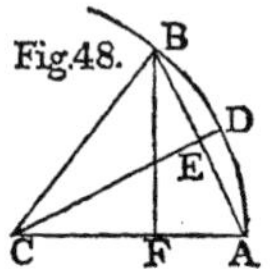

PROPOSITION XXXI.

In a spherical triangle, the rectangle contained by the sines of the two sides, is to the square of the radius, as the rectangle contained by the sine of the arch which is half the sum of the base and the excess of the sides, and the sine of the arch, which is half the difference of the same to the square of the sine of half the angle opposite to the base.

Let ABC be a spherical triangle, of which the two sides are AB, BC, and base AC, and let the less side BA be produced, so that BD shall be equal to BC: AD therefore is the excess of BC, BA; and it is to be shewn that the rectangle contained by the sines of BC, BA is to the square of the radius, as the rectangle contained by the sine of half the sum of AC, AD, and the sine of half the difference of the same AC, AD to the square of the sine of half the angle ABC opposite to the base AC.

Fig.49.

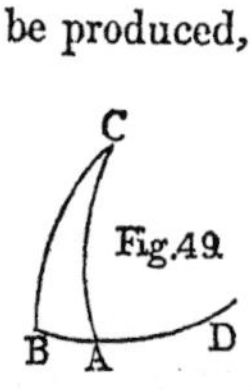

Since by Prop. 28, the rectangle contained by the sines of the sides BC, BA is to the square of the radius, as the excess of the versed sines of the base AC and AD, to the versed sine of the angle B; that is* as the rectangle contained by half the radius, and that excess, to the rectangle contained by half the radius, and the versed sine of B; therefore*, the rectangle contained by the sines of the sides BC, BA is to the square of the radius, as the rectangle contained by the sine of the arch which is half the sum of AC, AD, and the sine of the arch which is half the difference of the same AC, AD is to the square of the sine of half the angle ABC. Q. E. D.

* 1. 6.

* 29. and 30 of this.

Solution of the Twelve Cases of Oblique-angled Spherical Triangles.

GENERAL PROPOSITION.

In an oblique-angled spherical triangle, of the three sides and three angles, any three being given, the other three may be found.

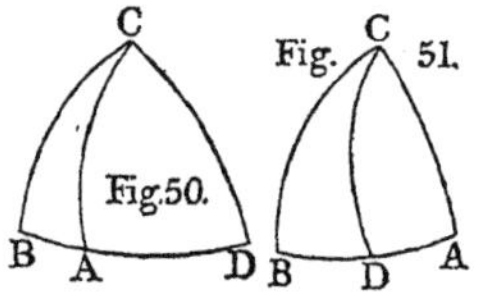

Fig. 51.

Cases.	Given.	Sought.	
1	B, D, and BC, two angles and a side opposite to one of them.	C.	CoS, BC : R : : CoT, B : T, BCA. 19. Likewise by 24. CoS, B : S, BCA : : CoS, D : S, DCA ; wherefore BCD is the sum or difference of the angles DCA, BCA according as the perpendicular CA falls within or without the triangle BCD ; that is *, according as the angles B, D are of the same or different affection. (* 16. of this.)
2	B, C, and BC, two angles, and the side between them.	D.	CoS, BC : R : : CoT, B : T, BCA. 19. and also by 24. S, BCA : S, DCA : : CoS, B : CoS, D; and according as the angle BCA is less or greater than BCD, the perpendicular CA falls within or without the triangle BCD ; and therefore*, the angles B, D will be of the same or different affection. (* 16. of this.)
3	BC, CD, and B.	BD.	R : CoS, B : : T, BC : T, BA. 20. and CoS, BC : CoS, BA : : CoS, DC : CoS, DA. 25. and BD is the sum or difference of BA, DA.
4	BC, DB, and B.	CD.	R : CoS, B : : T, BC : T, BA. 20. and CoS, BA : CoS, BC : : CoS, DA : CoS, DC. 25. and according as DA, AC are of the same or different affection, DC will be less or greater than a quadrant. 14.

Cases.	Given.	Sought.		
5	B, D, and BC.	DB.	R : CoS, B : : T, BC : T, BA. 20. and T, D : T, B : : S, BA : S, DA. 26. and BD is the sum or difference of BA, DA.	
6	BC, BD, and B.	D.	R : CoS, B : : T, BC : T, BA. 20. and S, DA : S, BA : : T, B : T, D; and according as BD is greater or less than BA, the angles B, D are of the same or different affection. 16.	
7	BC, DC, and B.	C.	CoS, BC : R : : CoT, B : T, BCA. 19. and T, DC : T, BC : : CoS, BCA : CoS, DCA. 27. the sum or difference of the angles BCA, DCA is equal to the angle BCD.	
8	B, C, and BC.	DC.	CoS, BC : R : : CoT, B : T, BCA. 19. also by 27. CoS, DCA : CoS, BCA : : T, BC : T, DC. 27. if DCA and B be of the same affection; that is (13), if AD and CA be similar, DC will be less than a quadrant; 14. and if AD, CA be not of the same affection, DC is greater than a quad rant. 14.	
9	BC, DC, and B.	D.	S, CD : S, B : : S, BC : S, D.	
10	B, D, and BC.	DC.	S, D : S, BC : : S, B : S, DC.	
11	BC, BA, AC.	B.	S, AB × S, BC : Rq : : S, $\frac{AC+AD}{2}$ × S, $\frac{AC-AD}{2}$: Sq $\frac{ABC}{2}$. AD being the difference of the sides BC, BA.	See fig. 49.
12	A, B, C.	The sides	In the triangle DEF, DE, EF, FD are respectively the supplements of the measures of the given angles B, A, C, in the triangle BAC; the sides of the triangle DEF are therefore given, and by the preceding case, the angles D, E, F, may be found; and the sides BC, BA, AC are the supplements of the measures of these angles.	See fig. 31.

The 3d, 5th, 7th, 9th, and 10th cases, which are commonly called ambiguous, admit of two solutions, either of which will answer the conditions required; for, in these cases, the measure of the angle or side sought, may be either greater or less than a quadrant, and the two solutions will be supplements to each other *.

* Cor. Def. 4. & 6 Pl. Trig.

If from any of the angles of an oblique-angled spherical triangle, a perpendicular arch be drawn upon the opposite side, most of the cases of oblique-angled triangles may be resolved by Napier's rules.

THE END.

G. Woodfall, Printer, Angel Court, Skinner Street, London.

EXERCISES upon the DIFFERENT PARTS of ITALIAN SPEECH, with References to Veneroni's Grammar, and an abridgement of the Roman History, intended at once to make the Learner acquainted with History, and the Idioms of the Italian Language. By MR. BOTTARELLI. New edition, much improved, by G. HEAVEN, 12mo. 3s. 6d. bound.

A KEY to BOTTARELLI'S EXERCISES, with a few Extracts in Prose and Verse, with English Translations, for the use of Beginners. By P. R. ROTA. New edition, by G. HEAVEN. 12mo. 2s. 6d. bound.

An ITALIAN and ENGLISH GRAMMAR, from Virgani's Italian and French Grammar, exemplified in Twenty Lessons. With Exercises, Dialogues, and Historical Anecdotes. A new edition, corrected and improved, in English and Italian, with Notes and Remarks, calculated to facilitate the Study of the Italian Language. By J. GUICHET, Professor of Languages. 12mo. 5s. A KEY to the above work. 12mo. 3s.

A NEW DICTIONARY of the Italian and English Languages, in Two Parts;—1. Italian and English;—2. English and Italian. By JOSEPH BARETTI. A new edition, corrected and greatly improved, in 2 large vols. 8vo. 1l. 4s. boards.

COMMEDIE SCELTE DI CARLO GOLDONI, Avvocato Veneto. Nuova edizione, 12mo. 6s. By A. GOMBERT.

GRAGLIA'S POCKET DICTIONARY, Italian and English, and English and Italian. A new edition, corrected, and very neatly printed, 7s.

LEGATO D'UN PADRE ALLE SUE FIGLIE, sopra la Religione; la Condotta; il Contegno; i Divertimenti; l'Amicizia; l'Amore, ed il Matrimonio, tradotto dall' Inglese del DOTTORE GREGORY. 12mo. 3s.

The same Work in Italian and French, 3s. 6d.

ECONOMIA DELLA VITA UMANA, tradotta dall' originale Inglese, 12mo. 3s.

LETTERE D'UNA PERUVIANA, tradótte dal' Francése in Italiano, di cúi si sóno accentáte tútte le vóci, per facilitar ágli straniéri il módo d'impará la prosodía di quésta língua. Dal SIGNOR DEODATI. Nuóva edizione, 12mo. 3s. 6d.

FABLES CHOISIES à l'Usage des Enfans, &c. avec un Index Alphabetique de tous les Mots traduits en Anglais, par LEWIS CHAMBAUD. 12mo. A new Edition, with very considerable Additions, Improvements, and Corrections; with a short Introduction, explaining and illustrating the Nature and Object of Fable. By G. HEAVEN. 2s.

A GRAMMAR of the FRENCH TONGUE, with a Preface, containing an Essay on the proper Method of Teaching and Learning that Language. By LEWIS CHAMBAUD. A new Edition, greatly improved. 5s.

SYLLABAIRE FRANCOISE; or a French Spelling Book, calculated to lead young Beginners with ease from the Knowledge of single Letters to the Reading of the longest and most difficult Polysyllables; to which is annexed, an Introduction to French Grammar, a Vocabulary, &c. The whole in French and English, 12mo. By Mr. PORNY. 21st edition, considerably improved, by G. HEAVEN. 2s.

MAGASIN DES ENFANS, ou Dialogues entre une sage Gouvernante et plusieurs de ses Elèves de la première Distinction. Par Madame de BEAUMONT. Nouvelle édit. soigneusement corrigée. Par G. HEAVEN. 12mo. 5s.

Printed in Dunstable, United Kingdom

91226075R00279